Cyber Security and Safety of Nuclear Power Plant Instrumentation and Control Systems

Michael A. Yastrebenetsky
State Scientific and Technical Centre for Nuclear and Radiation Safety, Ukraine

Vyacheslav S. Kharchenko
National Aerospace University KhAI, Ukraine

A volume in the Advances in Information Security, Privacy, and Ethics (AISPE) Book Series

Published in the United States of America by
IGI Global
Information Science Reference (an imprint of IGI Global)
701 E. Chocolate Avenue
Hershey PA, USA 17033
Tel: 717-533-8845
Fax: 717-533-8661
E-mail: cust@igi-global.com
Web site: http://www.igi-global.com

Library of Congress Cataloging-in-Publication Data

Names: Yastrebenetsky, Michael, 1934- editor. | Kharchenko, Vyacheslav, 1952- editor.
Title: Cyber security and safety of nuclear power plant instrumentation and control systems / Michael A. Yastrebenetsky and Vyacheslav S. Kharchenko, editors.
Description: Hershey, PA : Information Science Reference, an imprint of IGI Global, [2020] | Includes bibliographical references and index. | Summary: "This book examines methods of cyber security and safety in nuclear power plant instrumentation and control systems"-- Provided by publisher.
Identifiers: LCCN 2019051009 (print) | LCCN 2019051010 (ebook) | ISBN 9781799832775 (hardcover) | ISBN 9781799832782 (paperback) | ISBN 9781799832799 (ebook)
Subjects: LCSH: Nuclear power plants--Instruments--Security measures. | Nuclear power plants--Computer networks--Security measures.
Classification: LCC TK9152.1645 .C93 2020 (print) | LCC TK9152.1645 (ebook) | DDC 621.48/35--dc23
LC record available at https://lccn.loc.gov/2019051009
LC ebook record available at https://lccn.loc.gov/2019051010

This book is published in the IGI Global book series Advances in Information Security, Privacy, and Ethics (AISPE) (ISSN: 1948-9730; eISSN: 1948-9749)

British Cataloguing in Publication Data
A Cataloguing in Publication record for this book is available from the British Library.

For electronic access to this publication, please contact: eresources@igi-global.com.

Advances in Information Security, Privacy, and Ethics (AISPE) Book Series

Manish Gupta
State University of New York, USA

ISSN:1948-9730
EISSN:1948-9749

Mission

As digital technologies become more pervasive in everyday life and the Internet is utilized in ever increasing ways by both private and public entities, concern over digital threats becomes more prevalent.

The **Advances in Information Security, Privacy, & Ethics (AISPE) Book Series** provides cutting-edge research on the protection and misuse of information and technology across various industries and settings. Comprised of scholarly research on topics such as identity management, cryptography, system security, authentication, and data protection, this book series is ideal for reference by IT professionals, academicians, and upper-level students.

Coverage

- Computer ethics
- Telecommunications Regulations
- Cookies
- Electronic Mail Security
- Risk Management
- Device Fingerprinting
- Privacy-Enhancing Technologies
- IT Risk
- CIA Triad of Information Security
- Cyberethics

IGI Global is currently accepting manuscripts for publication within this series. To submit a proposal for a volume in this series, please contact our Acquisition Editors at Acquisitions@igi-global.com or visit: http://www.igi-global.com/publish/.

The Advances in Information Security, Privacy, and Ethics (AISPE) Book Series (ISSN 1948-9730) is published by IGI Global, 701 E. Chocolate Avenue, Hershey, PA 17033-1240, USA, www.igi-global.com. This series is composed of titles available for purchase individually; each title is edited to be contextually exclusive from any other title within the series. For pricing and ordering information please visit http://www.igi-global.com/book-series/advances-information-security-privacy-ethics/37157. Postmaster: Send all address changes to above address.

Titles in this Series

For a list of additional titles in this series, please visit: http://www.igi-global.com/book-series/advances-information-security-privacy-ethics/37157

Large-Scale Data Streaming, Processing, and Blockchain Security
Hemraj Saini (Jaypee University of Information Technology, India) Geetanjali Rathee (Jaypee University of Information Technology, India) and Dinesh Kumar Saini (Sohar University, Oman)
Information Science Reference • © 2020 • 300pp • H/C (ISBN: 9781799834441) • US $225.00

Advanced Security Strategies in Next Generation Computing Models
Shafi'i Muhammad Abdulhamid (Federal University of Technology Minna, Nigeria) and Muhammad Shafie Abd Latiff (Universiti Teknologi, Malaysia)
Information Science Reference • © 2020 • 300pp • H/C (ISBN: 9781799850809) • US $215.00

Advancements in Security and Privacy Initiatives for Multimedia Images
Ashwani Kumar (Vardhaman College of Engineering, India) and Seelam Sai Satyanarayana Reddy (Vardhaman College of Engineering, India)
Information Science Reference • © 2020 • 300pp • H/C (ISBN: 9781799827955) • US $215.00

Internet Censorship and Regulation Systems in Democracies Emerging Research and Opportunities
Nikolaos Koumartzis (Aristotle University of Thessaloniki, Greece) and Andreas Veglis (Aristotle University of Thessaloniki, Greece)
Information Science Reference • © 2020 • 272pp • H/C (ISBN: 9781522599739) • US $185.00

Safety and Security Issues in Technical Infrastructures
David Rehak (VSB – Technical University of Ostrava, Czech Republic) Ales Bernatik (VSB – Technical University of Ostrava, Czech Republic) Zdenek Dvorak (University of Zilina, Slovakia) and Martin Hromada (Tomas Bata University in Zlin, Czech Republic)
Information Science Reference • © 2020 • 499pp • H/C (ISBN: 9781799830597) • US $195.00

Cybersecurity Incident Planning and Preparation for Organizations
Akashdeep Bhardwaj (University of Petroleum and Energy Studies, Dehradun, India) and Varun Sapra (University of Petroleum and Energy Studies, India)
Information Science Reference • © 2020 • 300pp • H/C (ISBN: 9781799834915) • US $215.00

701 East Chocolate Avenue, Hershey, PA 17033, USA
Tel: 717-533-8845 x100 • Fax: 717-533-8661
E-Mail: cust@igi-global.com • www.igi-global.com

Editorial Advisory Board

List of Reviewers

Table of Contents

Detailed Table of Contents

the event of its failure, when it is required to be performed, and by the consequences in the event of a spurious actuation.

The requirements for safety important instrumentation and control (I&C) systems, their components (software and hardware complexes, equipment), and processes of their development are provided. The reliability of I&C systems and their components are ensured by the requirements for prevention and protection against common cause failures, compliance with the single failure and redundancy criteria, and prevention of human errors. The requirements for operational stability of the components cover resistance to environmental impacts, mechanical (including seismic) and other external impacts, and insensitivity to changes in power supply parameters and electromagnetic interference. Requirements for the performance quality include characteristics pertaining to the accuracy and performance time of functions. Requirements for independence ensure that the system remains operable if its components fail and that the effects of electromagnetic radiation, fires, breakdown of insulation, and leaks in cables are minimized.

Features of software as a component of instrumentation and control (I&C) systems are analyzed. Attention is paid to the importance of functions performed by software and hazards of such software. Requirements for characteristics of software as a component of I&C systems are analyzed. Different regulatory documents are considered in order to disclose common approaches to the use of dedicated software and off-the-shelf software components. Classification of software, as well as classification of requirements, is described. Criteria of selection and structuring of requirements, as well as criteria for software verification, are defined. As long as the characteristics of software components directly depend on the quality of the processes of software development and verification, requirements for software life cycle processes are considered.

The chapter discusses the importance of assessment of interference degree for various attributes of safety-critical instrumentation and control (I&C) systems and proposes applicable metrics. An approach to analysis of safety-critical I&C systems is presented. Such approach relies on performance of gap analysis and consideration of influence of human, technique, and tool. The approach is applicable to cyber security assessment for various safety-critical I&C systems, including complex instrumentation and control systems and field-programmable gate arrays (FPGA)-based systems.

Analysis of the STUXNET attacks on the Natanz gas centrifuge plant illustrated the hazards of a cyber physical attack. STUXNET demonstrated that a cyber-attack can introduce new and malicious function into I&C systems. Cyber robust nuclear power plant systems may be able to provide a truly independent level of defense in depth against cyber-attacks. The development of cyber robust systems involves identifying the plant's vulnerabilities and using non-digital means where such features can defeat malicious attacks. This is a relatively new idea, so a complete roadmap for this is not available. Nevertheless, some principles can be stated, and some methodologies can be discussed.

The accurate availability and safety assessment of a reactor trip system for nuclear power plants instrumentation and control systems (NPP I&C) application is an important task in the development and certification process. It can be conducted through probabilistic model-based evaluation with variety of tools and techniques (T&T). As each T&T is bounded by its application area, the careful selection of the appropriate one is highly important. This chapter presents the gap-analysis of well-known modeling approach—Markov modeling (MM), mainly for T&T selection and application procedures—and how one of the leading safety standards, IEC 61508, tracks those gaps. The authors discuss how main assessment risks can be eliminated or minimized using metric-based approach and present the safety assessment of typical NPP I&C system. The results analysis determines the feasibility of introducing new regulatory requirements for selection and application of T&T, which are used for MM-based assessment of availability and safety.

Safety assessment of nuclear power plant instrumentation and control systems (NPP I&Cs) is a complicated and resource-consuming process that is required to be done so as to ensure the required safety level and

The smart grid (SG) is a movement to bring the electrical power grid up to date so it can meet current and future requirements to fit customer needs. Disturbances in SG operation can originate from natural disasters, failures, human factors, terrorism, and so on. Outages and faults will cause serious problems and failures in the interconnected power systems, propagating into critical infrastructures such as nuclear industries, telecommunication systems, etc. Nuclear power plants (NPP) are an intrinsic part of the future smart grid. Therefore, it is of high priority to consider SG safety, mutual influence between NPP and SG, forecast possible accidents and failures of this interaction, and consider the strategies to avoid them.

This chapter contains information on the platforms (equipment families) that increasingly serve as basic tools in the development of instrumentation and control (I&C) systems for different applications, including nuclear power plants (NPPs). The advantages of the platforms are indicated, the stages of their design and development are shown, and major differences between the platforms belonging to different generations are addressed. The main characteristics and features of the modern platforms used in NPP automation are provided. The procedure for development and qualification of the platforms delivered to NPPs is discussed.

This chapter describes an element base of new generation for NPP I&C, namely field programmable gate array (FPGA), and peculiarities of the FPGA application for designing safety critical systems. FPGA chips are modern complex electronic components that have been applied in nuclear power plants (NPPs) instrumentation and control systems (I&CSs) during the last 15-17 years. The advantages and some risks caused by application of the FPGA technology are analyzed. Safety assessment techniques of FPGA-based I&CSs and experience of their creation are described. The FPGA-based platform RadICS and its application for development of NPP I&CS is described.

Foreword

Safeguarding nuclear power plants from accidents and attacks is critical for maintaining global security. The events at Chernobyl in 1986 and Fukushima in 2011 demonstrate all too clearly that any nuclear disaster has the potential to impact the entire world. Today, nuclear power plants and other industrial facilities face new threats from cyber criminals eager to exploit peaceful technologies for their own nefarious ends. Protecting these facilities from cyber security threats requires a new generation of instrumentation and control systems (I&C) that can withstand cyber attacks and ensure nuclear power does not fall into the wrong hands.

This book, *Cyber Security and Safety on Nuclear Power Plant Instrumentation and Control Systems*, covers a wide variety of topics related to the development and implementation of modern I&C; especially as it relates to cyber security in nuclear power plants. It is an invaluable resource for anyone looking to learn more about the assessment and assurance of I&C safety and cyber security, platforms (equipment families) for nuclear power plant I&C systems, diversity for nuclear power plant safety and cyber security, and even the use of drones to help monitor the safety and security of nuclear power plants.

Ukrainian specialists were the primary authors of this work. Because of Chernobyl, Ukraine has a great deal of experience responding to nuclear accidents and understands the importance of nuclear safety in a unique way. The close relationship between Ukrainian and American experts in nuclear power plant I&C remains a crucial asset for the global nuclear industry. Ukraine is a pioneer in applying Field Programmable Gate Array (FPGA) to nuclear power plants, and the FPGA platform RadICS—developed by the Ukrainian company Radiy—received a certificate from the U.S. Nuclear Regulatory Commission (U.S. NRC). Since 1994, a joint venture between the Westinghouse Electric Company (USA) and Kharton (Ukraine) has successfully operated in the Ukraine to create I&C for nuclear power plants around the world. This joint venture also took part in an initiative to update I&C at all Ukrainian power units, working with the U.S. Department of Energy (DOE) and Ukrainian State Scientific and Technical Center for Nuclear and Radiation Safety (SSTC NRS) participated in introduction of safety parameter display systems. Furthermore, the U.S. NRC has collaborated with the State Nuclear Regulatory Inspectorate of Ukraine (SNRIU) on issues such as safety I&C systems and cyber security.

Furthermore, as an active member of the American Nuclear Society (ANS), I am happy to report that Ukrainian specialists have actively participated in each of our "Nuclear Power Instrumentation, Control, and Human – Machine Interface" conferences since 2000. At these conferences, experts from the Ukraine provided authoritative reports on developments related to nuclear power plant safety, the reliability of I&C systems at nuclear facilities, the modernization of I&C at nuclear power plants in the Ukraine, and American and Ukrainian joint projects. Ukraine five times hosted the "Nuclear Power Plant Instrumentation and Control Systems: Safety Aspects" Scientific Technical conference for five years, providing an international audience of nuclear scientists and engineers with a forum for discussing pressing matters related to I&C.

I recommend *Cyber Security and Safety on Nuclear Power Plant Instrumentation and Systems* not only to I&C specialists working in the nuclear industry, but also to any professional working in an industry where I&C is essential. Aviators, railroad engineers, and production specialists from the chemical and gas industries can all benefit from the insights in this volume. This book will be a valuable tool for professionals looking to create innovative I&C solutions for protecting critical industries from accidents and cyberattacks.

I am honored to have worked with the author of this book and can endorse this volume with great pleasure.

H. M. Hashemian
AMS, USA

Preface

MOTIVATION

It is difficult to overestimate the importance of safety problems at Nuclear Power Plants (NPP) for all countries in which they are operated. This issue is extremely important in the context of energy safety and mankind safety in whole. Unfortunately, the accident at Fukushima-1 NPP (2011), which extended the list of major nuclear power accidents such as the Three Mile Island (1979) and Chernobyl NPP (1986) accidents, confirmed this conclusion.

Instrumentation and control (I&C) systems, which prevent incidents in NPP operation and mitigate the accident consequences, play a key role in NPP safety assurance.

In past decades, the main technological process of electric power production at operating NPPs has not undergone significant changes. However, in recent years, significant changes have been made in the design of I&C systems and technologies of development, verification and implementation. More than 40 years ago the digital technology in NPPs was applied for performing information functions in contrast to control functions. However, after the accumulation of essential experience in design of computer control systems for critical safety objects and experience in justification that such systems meet safety requirements have been accumulated, intensive implementation of computer systems for control and protection of nuclear reactors has begun.

Significant toughening of requirements for NPP safety has led to modification of international and national standard bases related to NPPs as a whole. This fact, together with the use of modern information and electronic technologies (in particular, such as Field Program Gate Array, FPGA), have also caused the necessity to revise international standards for I&C systems, determined, first of all, by documents of the International Atomic Energy Agency (IAEA) and International Electrotechnical Commission (IEC). For example, IAEA standards related to NPP I&C systems were issued in 1980 (IAEA, 1980) and 1984 (IAEA, 1984), updated in 2002 (IAEA, 2002) and later in 2016 (IAEA, 2016).

According to international and national regulatory documents, all NPP I&C systems (as well as other systems) are classified depending on their safety impacts. Methodology of their differentiation is distinct in different countries, but the concept of system differentiation by safety impacts is conventional.

This book mainly focuses on NPP I&C systems relevant to safety and named “safety important systems”. The examples of such systems are the reactor trip system, refueling machine control system, reactor power regulation and limitation system, etc. Safety important systems play a crucial role in control and monitoring of NPP operation; they also detect conditions in which power unit operation becomes unsafe and, if necessary, shut the reactor down.

Safety important NPP I&C systems have a set of different peculiarities: necessity to ensure high operating reliability and to meet a set of requirements essential in the safety context. For all NPP safety important systems, including I&C systems, the presence of a national regulatory body – government organization, independent from NPP, designers and manufactures of I&C systems and equipment – is very important. The names of these organizations are different in various countries, e.g. U.S. Nuclear Regulatory Commission in the USA, State Nuclear Regulatory Inspectorate in Ukraine. According of the Convention on Nuclear Safety (IAEA, 1994), which is approved by all countries operating NPPs, the main tasks of the regulatory body are to establish:

- Applicable national safety requirements and regulations (including requirements to I&C systems);
- A system of licensing with regard to nuclear installations and the prohibition of the operation of a nuclear installation without a license (including licensing related to installation of I&C systems);
- A system of regulatory inspection and assessment of nuclear installations to ascertain compliance with applicable regulations and the terms of licenses (including inspection and assessment of operating NPP I&C systems).

High requirements to NPP I&C systems take place not only for nuclear domain, but to I&C systems in many other sensitive applications, where the safety problem is essential one, such as railway, aviation and piloted space systems, automotive and health systems. These systems have been called critical safety systems or safety related systems. Concepts of safety assurance for critical I&C systems in different branches of activities have considerably fewer differences than concepts of equipment safety controlled by these systems. For example, the comparison of safety concepts for the NPP unit control system and a dangerous weapon such as a missile career with nuclear warheads, displayed the considerable community of such concepts with all the variety of controlled objects (Aizenberg, 2002).

The main particularities of modern I&C systems of critical objects common to various fields of their use are as follows:

- The widespread use of modern computer technology with developed software and a highly reliable electronic component base, a high degree of integration, which helps to reduce the number of elements and reduce the number of connections between them;
- Increase in productivity of computing, memory and storage devices, embedded IP cores which makes it possible to carry out complex control functions and complex calculations;
- Widespread implementation of both self-monitoring and self-testing components and platforms in I&C systems including diagnostic of peripheral equipment;
- The use of local area networks for the exchange of information between individual parts of the systems; a significant improvement in the interface with personnel, improving the quality of information display devices;
- Increase of data transmission speed through communication channels due to the use of optoelectronic and fiber optic lines; and
- Replacement of analog devices with digital ones on control panels, execution of peripheral products (for example, sensors, positioners) as hardware and software that implement all (or part of) the prescribed functions under the control of a program stored in their permanent memory and so on.

It is the generality of the properties of the I&C systems for various critical objects that allows us to offer this book to the attention of specialists involved in the development and operation of I&C systems not only for nuclear power plants, but also for various other critical domains: chemical and petrochemical industries, many types of transport, such as air, sea and railway transport (especially high-speed transport), some types of medical equipment which monitor patients and performs control function, storage facilities for hazardous substances, etc.

Safety is a predominant attribute of NPP I&C. Other attributes such as reliability, maintainability, availability and security are "slave" to safety. In a set of these attributes, it is necessary to mark out security and its attributes integrity, accessibility/availability and confidentiality as its most important component. Due to the fast evolution of methods and technologies of unauthorized information intrusions, the set of vulnerabilities of I&C, and their components, as well as attention to development of regulatory requirements on NPP I&C security, methods and means of its analysis and assurance, have increased significantly. This fact determines a necessity to consider answers for challenges in this field in the context of general problems of NPP I&C safety. Increased influence of security related factors causes development and implementation of a set of new methodologies to assess and improve safety such as security informed safety approach, assurance (safety and security) case techniques, complexing of safety and security analysis techniques and so on.

FEATURES AND CHANGES

The book predominately is written by authors from Ukraine: staff of State Scientific and Technical Centre for Nuclear and Radiation Safety (SSTC NRS, technical support organization of the Ukrainian Regulatory body), Research and Production Corporation Radiy (the biggest Ukrainian company developing, manufacturing and implementing NPP I&C systems) and related enterprise RPC Radics, Department of Computer Systems, Networks and Cybersecurity (CSNCS) Department of National Aerospace University "Kharkiv Aviation Institute" (KhAI) and Centre for Safety Infrastructure-Oriented Research and Analysis (division of the RPC Radiy at the CSNCS Department). The book summarizes experience, first of all, of Ukrainian specialists. This experience is of interest for USA and international community due to the following reasons.

Firstly, Ukraine has undergone a severe accident at the Chernobyl NPP. This had a significant impact on the evolution of nuclear power engineering in Ukraine, including progress of NPP I&C. The lessons of this accident have significantly contributed to the solution of numerous issues connected with NPP safety assurance in Ukraine: improvement of NPP operation culture, improvement of NPP platforms, application of more stringent requirements on safety (including requirements on I&C). Qualitative changes were made in governmental safety regulation. Therefore, specialists who directly participated in the mitigation of this accident consequences came to operators, to the Ukrainian regulatory body and to SSTC NRS, and companies developing and implemented I&C systems. The experience of these people was invaluable and was transferred from this generation to the next one.

Secondly, Ukraine is a pioneer in wide application of Field Programmable Gate Arrays (FPGA) in safety important NPP I&Cs. Nowadays, reactor protection systems, engineering safety feature actuation systems, and others were systems and equipment and software designed at the RPC Radiy and successfully operated in NPPs. These systems are applied not only in NPPs in Ukraine, but also in other countries (Bulgaria, Canada, Brasilia, Argentina etc.). The main advantages of these systems are high reliability, availability and safety, confirmed by 15-year experience of successful application, relative simplicity and "clarity" in verification of control safety functions, equipment compactness, and short terms of I&C systems replacement during modernization. Developed by Radiy&Radics designers FPGA platform RadICS for creation of I&C systems has been certified against requirements of the standard IEC 61508 (2015) and US NRC (2019).

Thirdly, Ukraine has experience in complete modernization of nearly all NPP I&C systems by new modern computer systems, performed at all 15 power units within last years.

This book "Cyber Security and Safety of Nuclear Power Plant Instrumentation and Control Systems" is updating and added version of the book "Nuclear power plant instrumentation and control systems for safety and security", which was issued by "IGI Global" in 2014. The main differences between new book and the book issued in 2014 are the follow:

- The title of the book was changed and was began now from the words "Cyber security". This change is connected with considerable increase of the number of the chapters (7 out of 17) devoted to cyber security assessment and assurance problems;
- The description of a few new systems (past-accident monitoring systems, etc,) and new platforms (RadICS and others) for safety important I&C systems were included;
- The changings of the standards and requirements to I&C systems were overviewed and analyzed;
- Safety and security management processes were described and analyzed; and
- A few techniques for safety and security assessment and assurance, different options of their joint application were described.

STRUCTURE OF THE BOOK

The book is organized into 17 chapters which divided on four sections: Requirements to NPP I&C systems safety and security (Chapters 1-4); Assessment of NPP I&C systems Safety and Cyber Security (Chapters 5-9); Assurance of NPP I&C systems Safety and Cyber Security (Chapters 10-14); NPP I&C Systems: Industrial Cases (Chapters 15-17). A brief description of each chapter follows.

Chapter 1 contains definitions of the main terms in this book - instrumentation and control (I&C) system, individual and overall I&C system, software-hardware complex (SHC), etc. Boundaries of I&C systems and their typical parts are described. General information about I&C systems, based on the use of up-to-date digital methods, is provided. The main peculiarities of such systems, which are described in more detail in further chapters of this book, are considered.

Chapter 2 describes the main standard bases for NPP I&C systems, documents of the International Atomic Energy Agency (IAEA) and International Electrotechnical Commission (IEC). Classifications of I&C systems and their components are given on the basis of their safety impact. All systems are divided into safety important and non-safety important. Thus safety important systems can be safety systems and safety related systems. According to IEC, functions to be performed by I&C systems shall be assigned to categories according to their importance to safety. Comparison of different types of classification of I&C systems is shown.

Chapter 3 provides the main properties of safety important NPP I&C. These properties are divided into groups: related to reliability (redundancy, single-failure criterion, protection against common cause failures, etc.); related to resistance (resistance to environmental impacts, mechanical impacts, seismic impacts, electromagnetic compatibility, change of power supply parameters); related to operation quality (accuracy, time characteristics, human-machine interface); independence of functions performed.

Chapter 4 contains classification and description of requirements on safety important NPP I&C software (SW) including requirement of standards IEC, IAEA, IEEE. SW peculiarities as an object of safety assessment are analyzed. The facts illustrating increasing of SW faults influence on reliability and safety as NPP I&C and computer-based systems for different critical applications are discussed. The criteria applied to assess SW are described. The methods and tools for evaluation of SW reliability and safety are analyzed.

Chapter 5 discusses importance of assessment of interference degree for various attributes of safety-critical instrumentation and control (I&C) systems, as well as proposes applicable metrics. An approach to analysis of safety-critical I&C systems is presented. The approach is applicable to cyber security assessment for various safety-critical I&C systems, including complex instrumentation and control systems and Field-Programmable Gate Arrays (FPGA)-based systems. Classification of life cycle models for safety-critical I&C systems including V-models is described. Importance of security environment establishment process is discussed, as well as its particular stages, including security-oriented analysis and assessment of safety-critical instrumentation and control systems. In order to establish secure development and operational environment for I&C systems, the requirements of the international standards to Nuclear Power Plants (NPPs) I&C systems security are analyzed and the proposed technique is illustrated using security case for RadICS FPGA-based platform.

Chapter 6 discuss the vulnerability of the current approach to cyber security assurance. Analysis of the STUXNET attacks on the Natanz gas centrifuge plant illustrated the hazards of a cyber physical attack. STUXNET demonstrated that a cyber attack can introduce new and malicious function into I&C systems. Cyber robust nuclear power plant systems may be able to provide truly independent level of defense in depth against cyber attack. The development of cyber robust systems involves identifying the plant's vulnerabilities and using non-digital means where such features can defeat malicious attacks. This is a relatively new idea, so a complete roadmap for this is not available. Nevertheless, some principles can be stated and some methodologies can be discussed.

Chapter 7 The chapter presents the gap-analysis of well-known modeling approach using Markov modeling (MM), mainly for technique and tool selection and application procedures, and how one of the leading safety standard IEC 61508 tracks those gaps. The task of accurate availability and safety assessment of a Reactor Trip Systems for Nuclear Power Plants Instrumentation and Control Systems (NPP I&C) applications is discussed in context of development and certification processes. Risks of inaccurate assessment can be eliminated or minimized using metric-based approach and present the safety assessment of typical NPP I&C system. The results analysis determines the feasibility of introducing new regulatory requirements for selection and application of T&T, which are used for MM-based assessment of availability and safety.

Chapter 8 presents developed framework that aggregates the most appropriate safety assessment methods typically used for NPP I&Cs. Key features that this framework provides are the formal descriptions of all required input information for every safety assessment method, possible data flows between methods, possible output information for every method. Such representation allows to obtain possible paths required to get necessary indicators, analyze the possibility to verify them by application of different methods that provide same indicators etc. Proposed approach can be used to specify and utilize required consequence of methods application so as to increase faithfulness of safety assessment results.

Chapter 9 gives a description of main aspects of cyber security assessment of NPP I&C systems. Cyber threats identification (at the development and operation of I&C systems) is considered as a part of cyber security assessment. Approaches to assessment of vulnerabilities of NPP I&C systems are given. Methods of assessment of completeness and sufficiency of realized measures for cyber security assurance of I&C systems are presented. Also the chapter contains the general principles of cyber security risk assessment which is necessary in case of using a risk-informed approach.

Chapter 10 analyzes the diversity as one of the main principles of NPP I&C safety assurance. A graphical model representing different variants of diversity during the development of safety-critical systems is suggested. The model addresses diversity types that are the most expedient in providing required reliability, safety and cyber security. The challenges addressed are related to factors of scale and dependencies among diversity types, since not all combinations of used diversity are feasible. This chapter also presents a set of models of multi-version systems and related techniques of diversity level and multi-version systems safety assessment. Approach to metric-probabilistic assessment is proposed and considered in the context of evaluating the software-based and FPGA-based MVS. A cost effective approach to selection of the most diverse NPP Reactor Trip System under uncertainty is presented. Case-study of the techniques is illustrated by example of assessment of multi-version FPGA-based NPP I&C developed by use of RadICS Platform including safety and security issues.

The aim of the Chapter 11 is to give details of safety and security management process based on the IEC 61508 requirements and applicable for Nuclear Power Plants Instrumentation and Control systems. The main contribution of this study comprises a set of detailed contents for safety and security management. The following aspects of safety and security management requirements for NPP I&C systems are considered: relation between safety and security management; safety and security management plan; human resource management; configuration management; computer tools selection and evaluation; documentation management; planning of safety and security assessment.

Chapter 12 contains a description of the existing notations for the Assurance Case (Claim, Argument and Evidence, CAE and Goal Structuring Notation, GSN). Supporting software tools for development of the Assurance Case are considered. Some ways for improvement and modification are proposed for both Assurance Case notations (CAE and GSN). For CAE we obtained annex with acceptance and coverage criteria as well as an algorithm of the Assurance Case update through life cycle stages. For GSN we improve structured argumentation with support of structured text using. Recommendations for using the Assurance Case notations and tools for I&C systems are formulated.

Chapter 13 describes principles of categorization of NPP I&C systems on cyber security levels based on consequences of possible malicious actions and cyber attacks. The general approaches to cyber security assurance (e.g., defense-in-depth, graded approach, cyber security culture, cyber security policy, etc.) are considered. The detailed set of cyber security measures according to cyber security level of NPP I&C system is presented.

Chapter 14 is devoted to analysis of interconnection and interaction of NPPs and smart grid (SG) in point of view security and safety. The complex nature of NPP and SG mutual interaction calls for the need of development of new approaches to NPP and smart grid safety/cyber assessment. The chapter considers an approach to influences formalization and series integration of safety assessment methods. A comprehensive description of SG safety and security modeling considering influences between SG and NPP is given. Challenges in smart grid safety and security in context of NPP risks (local and emergent) are discussed. This chapter study also gives the short overview of safety and security assessment and assurance's approaches, presents SG safety strategies. The role of IoT devices (such as sensors) in implementation of safety/security management is described.

Chapter 15 contains information about platforms (equipment families), which nowadays is the main tool in the development of instrumentation and control (I&C) systems in different applications, including NPP. The creation and use of the platform allows to use hardware and software developed as part of the corresponding platform instead of segmental products from different suppliers, often not compatible with together. The history of platform creation is described. There are information about I&C platforms that have been widely used at present in NPPs: TELEPERM XS (Siemens Energy, Germany / AREVA NP, France), ALS (Westinghouse Electric Company, USA), Spinline & Hardline (Rolls-Royce Civil Nuclear, UK), FirmSys & FitRel (China Techenergy Co., Ltd., China),POSAFE-Q (POSCO ICT Co., Republic of Korea), Vulcan/Vulcan M (Westron, Ukraine), MSKU-4/PS 5140 (Impuls, Ukraine). (Platform RadICS, elaborated by Radics (Ukraine), will be described in the Chapter 16.)

Chapter 16 describes element base of new generation for NPP I&C, namely Field Programmable Gate Array and peculiarities of FPGA application for designing of the safety critical systems. FPGA chips are modern complex electronic components which are applied in NPP I&C during last 15-17 years. The advantages and some risks caused by application of the FPGA technology are analyzed. Safety assessment techniques of FPGA-based I&C systems and experience of their creation are described. FPGA based platform RadICS and their application for development of NPP I&C is described.

Chapter 17 describes a structure of a multi-fleet of drones for monitoring of NPP accidents, consisting of main drone fleets and a reserve drone fleet, is considered. Various structures for systems of control stations for the multi-fleet of drones are suggested. Reliability models for the multi-fleet of drones with centralized (irredundant), centralized (redundant), decentralized, and partially decentralized systems of control stations are developed and analyzed. The dependency of the probability of failure-free operation of the multi-fleet of drones with 3 drones in the reserve drone fleet on the number of the main drone fleets is obtained for centralized (irredundant), centralized (redundant), and decentralized systems of control stations. Recommendations for choice of structures for the systems of control stations are formulated.

The book is intended for specialists involved in:

- Development and manufacture of components for safety important I&C;
- Design and operation of safety important NPP I&C systems;
- Licensing of NPP I&C systems and their components.

The book may also be useful for specialists who participate in the development and operation of safety control I&C systems, e.g. in airspace, railway, chemical industry, etc. Experience in safety important functions performed with FPGA may be interesting to developers, researchers, auditors from different branches what use these elements.

The book may also be of interest to students and lecturers at universities in specialties related to computer and software engineering and its critical applications and to nuclear engineering.

Michael A. Yastrebenetsky
State Scientific and Technical Centre for Nuclear and Radiation Safety, Ukraine

Vyacheslav S. Kharchenko
National Aerospace University KhAI, Ukraine

REFERENCES

Aizenberg, Y., & Yastrebenetsky, M. (2002). *Comparison of safety assurance principle of control systems for missile careers and nuclear power plants*. Kiev: Cosmic Science and Technics. (in Russian)

Babeshko, E., Kharchenko, V., Sklyar, V., Siora, A., & Tokarev, V. (2011). Combined Implementation of Dependability Analysis Techniques for NPP I&C Systems Assessment. *Journal of Energy and Power Engineering, 5*(42), 411-418.

IAEA. (1980). *Safety Guide No. 50-SG-D3. Protection system and related features in nuclear power plants*. Vienna, Austria: IAEA.

IAEA. (1984). *Safety Guide No. 50-SG-D8. Safety-related instrumentation and control systems for nuclear power plants*. Vienna, Austria: IAEA.

IAEA. (1994). *Convention on Nuclear Safety*. Vienna: IAEA.

IAEA. (2002). *Safety Guide No. NS-G-1.3. Instrumentation and control systems important to safety in nuclear power plants*. Vienna, Austria: IAEA.

IAEA. (2016). *Specific Safety Guide No. SSG-39. Specific safety guide. Design of instrumentation and control systems for nuclear power plants*. Vienna, Austria: IAEA.

Kharchenko, V. (Ed.). (2011). Safety of Critical Infrastructures. Mathematical and Engineering Methods. National Aerospace University KhAI. (in Russian)

Yastrebenetsky, M., Rozen, Y., Vinogradska, S., Johnson, G., Eliseev, V., Siora, A., & Kharchenko, V. (2011). *Nuclear Power Reactors Control and Protection Systems*. Kiev, Ukraine: Osnova-Print. (in Russian)

Acknowledgment

The authors thanks colleagues of the State Scientific and Technical Centre for Nuclear and Radiation Safety, RPC Radiy, RPC Radics, Centre for Safety Infrastructure-Oriented Research and Analysis, and Computer Systems, Networks and Cybersecurity (CSNCS) Department of National Aerospace University "KhAI" for their attention to work.

Colleagues of the State Scientific and Technical Centre for Nuclear and Radiation Safety (V. Filon, I. Chervonenko, Y. Malinovska, Y. Yesypenko) and Centre for Safety Infrastructure-Oriented Research and Analysis assisted in the preparation of the manuscript for publication.

Colleagues of the RPC Radiy, RPC Radics, CSNCS Department of National Aerospace University "KhAI" participated in discussions of a lot problems considered in the book. They concerns application of FPGA and FPGA based platforms in NPP I&C and other critical domains; software and hardware components safety assessment using a lot of techniques and their complexing; I&C security assessment including vulnerability and intrusion analysis and choice of countermeasures; research, development and implementation of multi-version systems considering challenges and solutions of the last years; smart grid safety and security analysis; application of drones for monitoring of NPP accidents and so on. Besides, authors thank to participants of the standing monthly seminar CriCTechS (Critical Computer Technologies and Systems) which works at the CSNCS Department during 19 years.

Contacts with the International Electrotechnical Commission Technical Committee TC 45 "Nuclear Instrumentation" Subcommittee SC 45A "Instrumentation, control and electric power systems of nuclear facilities" offered the possibility to attend a course on the modern state of work related to NPP I&C systems.

The authors and editors express a special thankfulness to reviewers of the chapters, which represents universities, research centers and industrial companies from Belarus, Bulgaria, Germany, Greece, Italy, Poland, Russia, Slovak Republic, Sweden, UK, Ukraine, USA.

The remarks and recommendations of the reviewers were very helpful and improved quality of the presented chapters.

The authors are thankful in advance for critical reviews and suggestions, which further will be taken into account during design, analysis, and assurance of safety and cybersecurity of NPP I&Cs.

Michael A. Yastrebenetsky
State Scientific Technical Centre on Nuclear and Radiation Safety, Ukraine

Vyacheslav S. Kharchenko
National Aerospace University KhAI, Ukraine

Chapter 1
NPP I&C Systems:
General Provisions

Michael A. Yastrebenetsky
State Scientific and Technical Center for Nuclear and Radiation Safety, Ukraine

Yuri Rozen
https://orcid.org/0000-0002-9366-5794
State Scientific and Technical Center for Nuclear and Radiation Safety, Ukraine

Oleksandr Klevtsov
https://orcid.org/0000-0001-5665-5039
State Scientific and Technical Center for Nuclear and Radiation Safety, Ukraine

ABSTRACT

The first chapter contains definitions of the main concepts in this book: instrumentation and control (I&C) system, individual and overall I&C system, central part of the I&C system and peripheral equipment, software-hardware complex (SHC), commercial of the shelf products (COTS), platforms (equipment family), functional safety. Differences between the SHC and I&C systems are explained. General information about I&C systems, based on the use facilities for obtaining, transfer, processing, and display of data, is provided. The main peculiarities of I&C systems, which are described in more detail in further chapters of this book, are considered. Information about functional safety of NPP I&C systems is described. A brief description of general aspects of cyber security of NPP I&C systems is given.

INTRODUCTION

I&C systems play significant role in assurance of safety and security of nuclear power plants (NPP). These systems perform automatic control of technological processes and equipment in operational modes; support operating personnel that monitor equipment condition and/or control processes; take part in performance of functions related to prevention of nuclear accidents, which could harm the population, personnel, environment and NPP itself; archive, display and record data required for analysis of causes,

DOI: 10.4018/978-1-7998-3277-5.ch001

progression of accident and recovery of safe controlled mode at NPP. In this connection, the role of I&C systems is steadily increasing, while improvement of digital technology, of application new complex electronic components, including Field Programmable Gate Arrays (FPGA), use of optoelectronic channels of communication, etc.

The object of the chapter is to explain the basic concepts required for better understanding of future chapters of the book to readers.

BACKGROUND

Formation of a new scientific and technical trend is nearly always followed with discussions, occurred due to initial uncertainty of used terminology. I&C systems intended for nuclear power plants are not an exception. There are several glossaries related to this trend, in particular: Rozenberg & Bobryakov, 2003; IAEA, 2007, developed by International Atomic Energy Agency (IAEA); IEC, 2019, developed by International Electrotechnical Commission (IEC). In the book IAEA, 2011 prepared by an international group of, specialists under the auspices of the IAEA, general concepts applicable to NPP I&C systems, and given description of I&C systems and their life cycle are considered. Moreover, definitions of used terms are provided in many international and national regulatory documents related to individual tasks of design and/or operation of NPP I&C systems.

Authors decided to describe their vision of development trends of NPP I&C systems, which are performed in practice in the process of modernization of current and designing of new systems.

Recent years are characterized by rapid development of electronics, digital computer technologies, information technologies, which are more widely used in the process of design of I&C systems. Advantages of these systems are the following:

- Increase of operating speed and storage capacity allowing on-line complex calculations in real time, required for implementation of control and security algorithms.
- High reliability of component parts (electronic components and computer equipment), having a crucial influence on the reliability of devices and systems based on them, which can be applied for solving more responsible tasks related to safety assurance.
- A high degree of integration of applied electronic components, which allows a significant decrease in the number of components and use simple connections between them, that also facilitate the increase of reliability while the requirements on functionality are complied with (or even amplified).
- One of the results is a possibility of wider use of redundancy, including structures with multiple redundancy, performing logic conditions "two-out-of-three", "two-out-of-four", etc., which were earlier used only in the most responsible cases.
- Simplicity of software modification determines the required flexibility of I&C systems, allowing adapt of one and the same hardware to solve programmatically a wide variety of control and management tasks, increase functionality, change of system characteristics in the process of operation, etc.
- Adoption of high quality means of manual input and display of information (wide-screen liquid-crystal monitors with high resolution, touch panels, keyboards, manipulators, etc.) and software developed for them allow implementing "friendly" human-machine interfaces, comparable by

functionality and ergonomics with those that have become familiar to modern computers, in I&C systems.

- Use of local computer networks and relevant network equipment for data exchange between devices included in individual I&C system and also between component parts of overall I&C system promotes significant saving of cable products, decreases labor expenditures of mounting, adjustment and maintenance, allows a high speed of transfer and required reliability of messages with their transfer of messages over fiber-optic lines, as well as required noise immunity of transfer in the electromagnetic environment typical for industrial facilities.
- Nearly unlimited possibilities of up-to-date computer technology for long-term storage of large scope of information are also topical for I&C systems, which in the process of operation allow performing continuous archiving of current data that can be requested and used to determine cause of emergencies, analysis of the sequence of events, assessment of actions of mechanical systems and personnel, accident management and restoration of the controlled condition, elaboration of measures for improvement of NPP safety.
- Use of built-in hardware and software, providing technical diagnostics, display and record of diagnostic messages, allows quickly detecting faults, automatically defining their locations and planning required actions for recovery of operation.

In the process of design of I&C systems for NPP specified capabilities of modern information technologies and available for use of element base allowed:

- Use of the more reliable and saving digital logic circuits instead of logic circuits based on contact-relay elements.
- Use of more accurate digital calculations, implemented by universal software controlled computer equipment (microprocessors, single chip microcomputers), by complex programmable electronic components, etc., instead of analog calculations.
- Replacing analog measuring and control devices by digital ones with improved algorithms able to take consider specific characteristics of controlled technological equipment.
- Operating personnel to manage manually elements of technological equipment directly with keyboard workstations.
- Implementing software-hardware peripheral equipment (sensors, actuators) with "microprocessor intellect", in which specific functions (adjustment of measurement ranges, automatic diagnostics, calibrations, etc.) are performed under control of programs stored in their read-only memory.
- Equipping every I&C system with inbuilt means of technical diagnostics, which provide continuous automatic monitoring of technical state system and their components (including software self-control) with a depth of fault search to one removable component part.
- Implementing distributed control of technological processes and NPP equipment, using local computer networks and fiber-optic communication lines for message exchange between geographically dispersed individual I&C systems and/or between the constituent parts of each of these systems.

INSTRUMENTATION AND CONTROL SYSTEMS

All systems and equipment of nuclear power plants can be divided into two categories:

- Technological systems and equipment, providing transportation, storage, generation and transformation of energy, protecting nuclear fuel, equipment, piping from damage, as well as preventing from spread of radioactive substances and ionizing radiation over the specified boundaries.
- Instrumentation and control (I&C) systems, implementing information technologies, related to obtaining of input signals from technological systems and equipment and teams of operating personnel, transfer, storage, processing of obtained information and output of control impacts on technological systems and equipment at NPP.

The general concept of instrumentation and control systems covers:

- Overall I&C systems, providing monitoring and control of all technological systems and equipment at the NPP or an individual NPP unit.
- Individual I&C systems, together performing all I&C functions provided for overall I&C system, interacting with each other, with operating personnel and technological systems and equipment. Such systems may differ in function, nature of the functions performed and impact on safety. If necessary, this can be clarified in the species concept, for example, reactor protection system, reactor power control system, safety important I&C system, etc.

Each individual I&C system participates in performing one or several I&C functions, being operational autonomous, reasonably separate, and can be considered in the process of design, assembling, adjustment, testing and operation individually and independently from other component parts of overall I&C system, in which it is contained.

Based on functional characteristics, individual I&C systems can be divided into:

- Information systems (Figure 1 a): Intended for obtaining, processing, storage, transfer, display and/or recording of information about a state and/or operation of technological systems and equipment or other I&C systems (performance of information functions).
- Control systems (Figure 1 b): Intended for influencing the state and/or operation of technological system, equipment or other I&C systems (performance of control functions).

Figure 1. Structural schemas of individual I&C systems

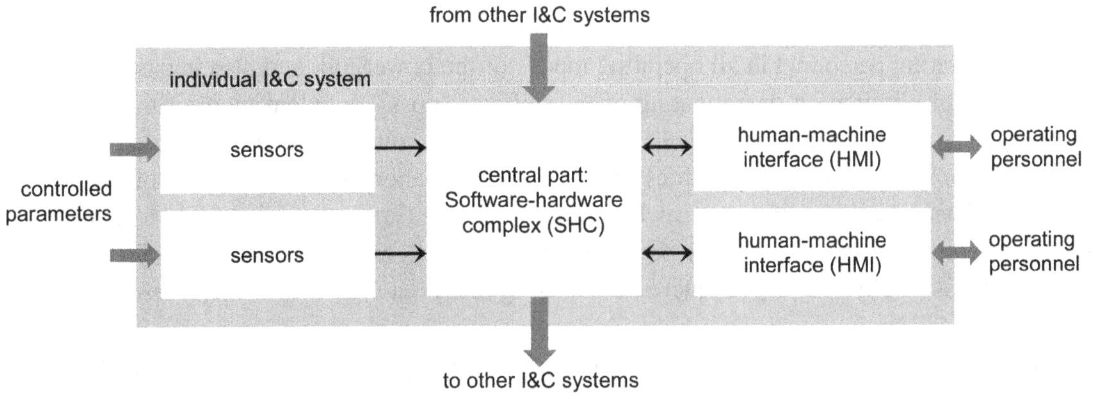

a – Information system, б – Control system

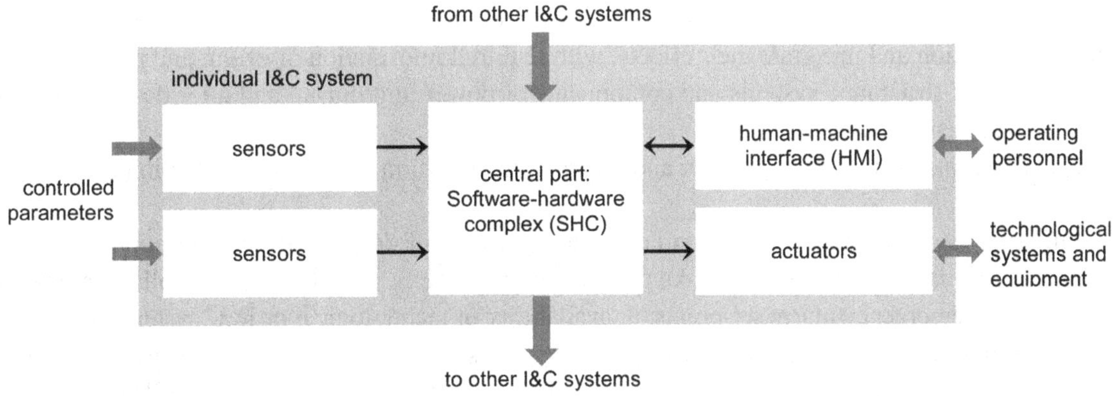

For example, in-core reactor monitoring system, neutron flux monitoring system, computer information system, safety parameters display system can be related to information systems. Examples of control systems are emergency protection system, reactor power control system, engineering safety features actuation system, standby diesel generator control system, refueling machine control system, etc.

However, it should be noted that this identification is not typical for modern I&C systems: in many cases information and control functions can be combined in one system. For example, reactor power control system, along with performance of control functions provides personnel with information of controlled parameters, state of technological equipment, provided control commands, i.e. also performs information functions. On the other hand, In-core reactor monitoring system performs information functions as well as individual control functions (for example, generating of commands, initiating actuation of preventive protection if a local energy in the reactor release is above acceptable rate).

Each individual I&C system is intended for performance of specific set of the main and auxiliary functions. The main functions of a system are determined by its designation:

Main *information functions* of a system are monitoring, display, alarm, recording and archiving.

Monitoring functions provide reception of current data of values of technological parameters, external and internal effects, state of structures, systems and components, initiating's events, commands of I&C systems and operating personnel in all operating modes of the power unit and also in case of accidents – in the process of elimination or mitigating of their effects (post – accident monitoring). These data can include, for example, neutron flux density, temperature, pressure and activity of the primary coolant, water level in steam generators and pressurizer, and other parameters that are determined by direct measurement. The values of parameters such as rate of neutron flux increase, stock before the boiling crisis, heat power, water vapor saturation steam temperature, etc., are determined by indirect measurements on the basis of physical dependencies, connecting that relate them with the measured values of other parameters.

Display functions provide visualization of current and archival data necessary for operating personnel for monitoring of processes operation, operation of I&C systems and results of their own actions in the process of power unit operation in operation modes, failure of normal operation and emergency. Display of post-accident monitoring data provides NPP personnel and involved safety experts, which control emergency situation and mitigrate their effects, with required information of origin and progression of accident, state of structures, systems and components of power unit during and after design basis and beyond design basis accidents.

Alarm functions provide visual and/or audio alarms to attraction of personnel attention to:

- Initiating event, which can cause an emergency, failures of design specific conditions for safety operation, external and internal risks (fire, earthquake, etc.), deviation of controlled parameters from their emergency alarm set-points, unavailability of technological or I&C systems for participation in performance of required safety functions (visual and audio emergency alarm).
- Failures of design specific operating limits and conditions, deviation of controlled parameters from their preventive alarm set-points, unavailability of technological or I&C systems for participation in performance of required safety functions (visual and, perhaps, audio preventive alarm).
- Actuation of automatic protection and interlocking, switching of operating modes, energizing or de-energizing, change of set-points, (visual indicator alarm).

Recording functions provide automatically printing documents of the specified format on paper carriers (on schedule or due to occurrence of predetermined events) and/or on call of personnel, continuous recording of graphs of variation of individual controlled parameters and/or groups of interconnected parameters with time.

Archiving functions provide monitoring data memorization in chronological order and storage of obtained data within a specified time period in the process of power unit operation, failures of normal operation, in emergencies, accidents and post-accident modes, with a possibility of its further display and/or recording. Data from an archive are used for assessment of power unit state, detection of short- and long-term changes (trends) of controlled parameters, preparation of reports, for further analysis of failure causes, accident progression and mitigratoin of their effects.

Main *control functions* of I&C systems are the following:

- Protcction, limitation, regulation, interlocking performed automatically.

- Discrete control, initiated by other control functions, commands of operating personnel or automatically.
- Remote control, initiated by operating personnel.

Performance of control functions provides in general:

- Assessment of state of controlled object and environmental conditions.
- Determination of impacts on the controlled object, which are required to achieve control purpose.
- Formation and output of commands for executive elements of technological equipment, which implement control impacts.
- Assessment of achieved control purpose.

Protection functions provide timely detection of failures of design specific operating limits and/or conditions of safety operation and output of commands for executive I&C systems and components of technological equipment initiated emergency reactor shutdown, emergency core cooling and residual heat removal, localization of radioactive materials and limitation of accidental release. Protection function has a higher priority compared to other functions of I&C systems.

Limitation functions provide timely detection of failures design specific operating modes and commands for executive I&C systems or technological equipment elements, on prohibition of modifications (increase or decrease) of specific parameters or initiated their forced return to the limits of the determined range, considered acceptable in current conditions. Limitation functions for example include: generation and output of commands, initiated forced reactor power decrease in the process of planned or unplanned disconnection of technological equipment; opening of pilot safety valve of pressure compensator if pressure in the first circuit reached design limit, etc.

Regulatory functions provide generation and output of command for executive I&C systems or executive elements of technological equipment to minimize the effect of external disturbances and/or transient processes in the controlled object to controlled parameters of this object. They are required in the process of power unit operating modes, in emergencies, design basic accidents (emergency control) and in the process of post-accident modes. Regulated parameters are, for example, neutron and heat power of reactor, level of coolant in the power compensator, levels of feed water in steam generators and regenerative heaters, pressure in the main steam header, turbine rotating velocity, etc. The regulation purpose can be maintenance of regulated parameters with a required accuracy on the level of the specified for them set-point or planned change of these parameters in time.

Interlocking functions detect failure of normal operation of technological equipment (overheating, overload, etc.) and/or conditions required for its operation (power supply, cooling, oiling, etc.) and output commands, which initiate load decrease or equipment disconnection, preventing its failure. Interlocking functions also include cancellation of fault command execution initiated by other control systems or operating personnel, which could cause damage of technological equipment.

Discrete control functions provide generation and output of commands initiated connection and disconnection of technological equipment or elements of pipe mounting, transfer and shutdown of mechanisms, reactivity control parts and other actions required for change power unit mode (operation in cold and hot standby, startup, power increase, planned shutdown and cooling) or for execution of commands obtained from other control I&C functions. Discrete control commands can be generated upon strict time schedule and/or depending on external events, state of other equipment and results of carrying out previous actions.

Remote control function provides execution of commands and directives of operating personnel by generation and output of control impacts directly to executive elements of technological equipment.

Each I&C usually performs not only main functions, but also auxiliary and service functions.

Auxiliary functions provide continuous automatic check of systems technical state and its components, related equipment and communication lines; diagnostics, display and archiving of messages of operation events; reconfiguration and restoration of operation after failures, etc.

Service functions support personnel actions in the process of specification and change of set-points, reconfiguration, periodic inspection of system component technical state, etc.

Complexity of I&C systems varies in a wide range – from single-circuit systems of automatic regulation of individual parameters to spatially distributed multiprocessor control computer systems. In addition, possibilities of up-to-date information technologies and available for use element base allow combining many functions, typical for safety systems as well as other (safety related or not important to safety) systems in one system. As an example can be a reactor power control, unloading, limitation and accelerated preventive protection system that combines monitoring, regulation of neutron or heat power of reactor, power limiting, discrete control of unloading reactor, archiving, display, warning and recording and also auxiliary and service functions (failure diagnostics, change of set-points, etc.).

Active in some countries (Ukraine, Russia, etc) regulatory documents (e.g. NP, 2008) divides all NPP systems including I&C systems into two categories: normal operation systems and safety systems.

I&C normal operation systems together with technological systems, equipment and operating personnel control and manage processes in operating mode of power unit. I&C systems, in particular, perform information and control functions required for automatic control of processes, state of technological systems and equipment, keeping technological parameters in the specified design boundaries, changing power unit operating modes, preventing from violations of operating limits.

For example, the in-core reactor monitoring system performs measurements of neutron physical, thermo-hydraulic and other parameters, defining the state of the core (distribution of neutron flux, energy release field, etc.); checks compliance between design and current characteristics of the core; alerts personnel about deviation of characteristics from design values and outputs signals into conjugate normal operation systems, that control reactivity; archives and displays the values of core parameters.

The radiation safety monitoring system provides continuous measurement of parameters, defining radiation environment in rooms and on NPP territory, in sanitary protection zone; archives and displays radiation environment data in each design specific control point; alerts personnel about exceeding of permissible radioactivity discharge into the environment and limits of radiation background.

The reactor power control, unloading, limitation and accelerated preventive protection system, providing automatic control of nuclear fuel fission processes, safety parameters display system are also related to normal operation systems.

Other examples of functions what I&C normal operation systems perform are:

- Automatic monitoring of construction and equipment state and control of technological processes (automatic level regulation in pressurizer; protection against unallowable increase of pressure in the first circuit; monitoring of coolant level in reactor vessel, identification of leak location and assessment of coolant flow due to leak; monitoring of activity and content of isotope-neutron absorbers in primary coolant, etc.).

- Automatic detection and limitation of effects of internal or external dangerous events considered in design (fire, earthquake, radioactivity release, etc.).
- Automated control of refueling, displaying data of position, transfer and direction of fuel assemblies; monitoring of neutron flux density and concentration of liquid neutron absorber solution; protection against damage, deformation, destruction, or fall of fuel assemblies in the process of their retrieval, relocation, and installation; mechanical interlocking on the margin of permissible relocation or detection of fuel assembly in non-design position, in case of power loss or occurrence of design specific initiating event.

I&C safety *systems* together with technological systems and equipment perform safety functions, in accordance with international standard IEC, 2011 (in Ukraine-according regulations NP, 2008; NP, 2015) as emergency reactor shutdown, emergency heat removal, residual heat removal from core and cooling pool, prevention or limitation of discharged radioactive substances over the specified boundaries. Operation of safety systems is required in cases, when normal operation systems are not able to keep controlled parameters in design specific operating limits, for example, due to failure, personnel fault or quick and reliable response to deviation of operating limits or safety operation conditions, to prevent escalation of emergency into an accident.

The I&C safety systems are involved when specific initiating events, violations of any safety operation conditions occurs, any of controlled parameters or specified combination of controlled parameters (emergency set-points) excess design specific limit, and/or on command obtained from other I&C safety system or from operating personnel.

The I&C safety system that has identified any of the listed causes:

- Alarms operating personnel and displays the cause that initiated the safety function.
- Generates and outputs command or a sequence of commands of design specific protective actions, to executive I&C systems or executive elements (starting devices) of technological safety systems.
- Prohibits executions of commands that could be outputted by other I&C safety systems, I&C normal operation systems and/or operating personnel, provided that they are not compatible with executed protective actions.
- Displays data required for operating personnel for monitoring of safety system operation, checks accuracy of its operation, and if necessary manually performs permitted safety assurance (duplicate, initiate or interlock commands of protective actions, initiate performance of new safety functions, etc.).

NPP I&C systems often combine performance of safety and normal operation functions. For example, neutron flux monitoring system on the basis of results of neutron flux density leak determines relative reactor power, calculates rate of its change, and when any of these parameters excesses the specified limit, neutron flux monitoring system generates signal initiated performance of preventive protection – normal operation functions, that provides reactor power reduction or prohibits its increase. The value of relative neutron density issued by this system is used in the process of performance of reactor power control function that is also related to normal operation functions. If relative neutron power or rate of its change excesses the specified emergency set-points, the neutron flux monitoring system generates signal to initiate performance of safety functions.

COMPONENTS OF I&C SYSTEMS

Individual I&C systems, being relatively isolated parts of the overall I&C system, together perform all I&C functions of the system, interacting with each other, with technological systems and equipment, operating personnel and related NPP I&C systems. Individual I&C system interact with its environment via interfaces separating them. The possibility and efficiency of such an interaction are provided by unification of relevant interfaces, which determine rules of interaction and requirements to devices, providing required interaction.

Each individual I&C system consists of hardware and software components required for implementation of specified main and auxiliary system functions. Two parts can be emphasized in the structure of modern I&C system:

- **Central Equipment:** Performing, in general, monitoring, archiving, display, recording and output of alarm signals (in control system – also generation and output of protection, limitation, regulation, interlocking, remote control commands). Central equipment of I&C system is a software-hardware complex – complex of independently operated devices, which interact with each other, with peripheral equipment and related I&C systems under control of application programs performing the specified operation algorithms.
- **Peripheral Equipment:** Providing conjugation of central equipment with operating personnel, technological systems and equipment. Peripheral equipment is a complex of devices, that are operated independently from each other and can be located on a distance from central equipment.

Moreover, I&C system includes: interface cables; service equipment used in the process of programming, debugging, checking, maintenance on the operating site; internal power supplies.

Peripheral Equipment

Peripheral equipment (peripheral or field devices) could be divided on three groups:

- Measurement field devices (sensors), intended for obtaining data of state and/or operation of technological structures, systems and components.
- Actuating field devices (actuators), directly influencing starting or regulating elements of technological structures, systems and components.
- Elements of manual input, display, alarm (HMI devices), used for visual and audible alarm of operating personnel, input of manual control commands, etc.

Sometimes the impulse line from technological equipment to valves of pressure sensors, differential pressure and level sensors (together with condensation and leveling vessels and other armature related to these lines) are also related to peripheral devices.

Measurement field devices include temperature, level, flow, pressure and differential pressure sensors, neutron flux measurements, transmitters, etc.

Actuating field devices include motor-, solenoid- and air-operated valves, valve control motors, switchgears, motor control centers, etc.

A peripheral device is usually designed as one independently operated device (rarely – as a group of independently operated parts, structurally and/or electrically interconnected). Operation conditions of peripheral device (operating and limiting parameters of environment, mechanical effects possible in the process of operation and abnormal natural phenomena, severity grade of the electromagnetic environment in the supposed location, etc.) are design specific that allows defining safety class, seismic category, group operating conditions and accommodation of each of peripheral devices and regulating requirements for it, resulting from this classification.

Peripheral devices of modern individual I&C can be divided into:

- *One-off items*, designed and supplied to NPP as components of specific I&C systems important to safety.
- *Replicated products*, specially designed as components of not previously determined set of I&C systems important to safety, and allowed for use at NPP.
- *Commercial of the shelf products* of general application, available on market, for design and manufacturing of which, there is no aim to use only in I&C systems important to safety, as components. In English literature, for examples, IAEA, 1999, such items are indicated by abbreviation COTS.

Application of COTS in the structure of I&C systems important to safety is one of the current trends caused by mass effect of their production and heavy competition of global leaders in information technology field, as a result there is a significant improvement of application properties (the accuracy, speed, noise immunity, etc.), relatively low cost, sufficient reliability, approved in different fields of application, independence form one manufacturer, high quality of company maintenance.

At the same time it should be considered that design and manufacturing of COTS are performed without taking into account international and national regulations, rules and standards of nuclear safety and control from state regulatory authorities. Information of scope, technique and results of production tests COTS are nearly inaccessible for nuclear users. The documents required to assessment of the possibility of safety use of products of general application in the designed I&C system are often absent. Relatively small consumption of NPP comparing to the overall scope of COTS production that does not contribute to establishment of required partnership between their designers and manufacturers, on the one hand, and with an end user – on the other hand. These limits use of COTS products in NPP I&C systems important to safety.

Central Equipment (Software-Hardware Complex)

Any I&C system is assembled directly on operating site of ready-made components, which in the recent past were individual instruments, automation devices and computer devices, communication cables, etc. Integration of hardware and software, debugging and testing of assembled system central part were performed directly on the place of future operation. However, a trend of increasing grade of manufacture components, composing central part of I&C systems, is clearly revealed currently.

Such components are designed, manufactured and supplied to users as plant-manufactured product – software-hardware complexes (SHC), which are more widely used in practice of building I&C systems in different industry branches, including nuclear power engineering (NP, 2000; NP, 2015, Yastrebenetsky, 2004; Yastrebenetsky, 2011). Testing of SHC in the factory environment guarantees that complex meets all specified requirements, stated in the documents. High grade of manufacture significantly simplifies and speeds up mounting and installation operations and integration of hardware and software components of central part of the I&C system before power unit start-up (Figure 2).

Figure 2. Central part and peripheral equipment of I&C system

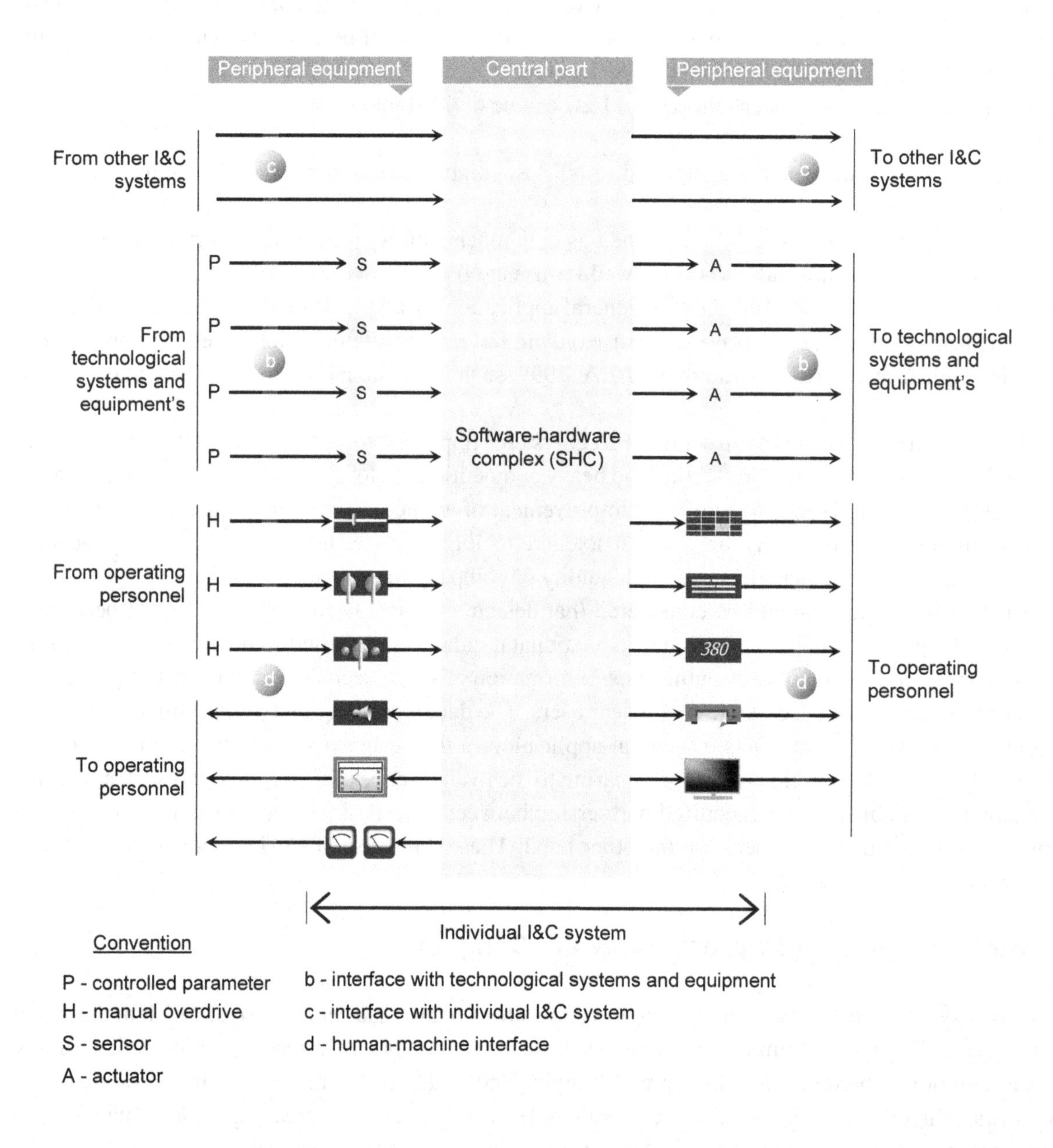

Each SHC is a functionally complete item in the form of one or several independently operated devices with built-in software, which are interconnected through electric and/or optic cables, with peripheral equipment of the same system and with other I&C systems. SHC is usually supplied to customer as a set with necessary service equipment, repair- and recovery reserve, operating and software documentation. To speed up mounting and decrease error probability, specially manufactured cable items, fully prepared for connecting all entering independently operated devices on the operating place, are often included in SHC structure.

One SHC, being a single component of the system central part that participates in implementation of all its main and auxiliary functions, is often contained in each I&C system. In any cases, these functions are distributed among several SHC, contained in one I&C system.

SHC performs:

- Receiving of signals from peripheral equipment (measurement field devices and/or elements of manual input).
- Normalization of discrete signals and transformation of analog signals into digital form.
- Exchange of digital messages with other SHC of the same I&C system and/or with SHC of other I&C systems via communication channels.
- Check adequacy of obtained information and its processing by the specified algorithms.
- Generation and output of control signals to the actuating field devices and signals to the remote HMI-devices, located in the main and emergency control rooms.
- Continuous monitoring (diagnostics) technical state of own parts, conjugate peripheral devices, and data transmission channels.
- Archiving of current and retrospective information.

SHC is composed of purchased electric and electronic products of general application and devices of own production, which are installed in manufacturing facility in supporting structures (shells) in accordance with SHC documentation. Before delivering to customer, manufacturer performs a full inspection of SHC operation with emulators of peripheral equipment and on the basis of results confirms compliance of supplied SHC with requirements of regulations, rules, standards and the specification, agreed with the customer. The manufacturer's warranties cover SHC as a whole, including purchased component parts applied in it. In the process of modernization of the I&C system the entire SHC is usually replaced.

In contrast, I&C system is composed of independently operated devices (products), obtained in accordance with customized specification; their mounting, connection and installation are performed in accordance with project documents of I&C system on NPP. Joint operation of devices, forming a central part of I&C system, can be inspected only on NPP, after mounting and integration. Supplier guarantees apply to each product individually, any but not for all I&C system. In the process of I&C system modernization only a part of equipment is usually replaced.

There are differences in both software SHC and I&C system. Software SHC is its integral part and designed by the same organization as for SHC. In the process of design of software its verification, including checking after integration with hardware SHC, is provided. Saving of program in read-only memory and further checking of SHC operation are performed on manufacturing facility before delivery to customer.

Software of I&C system are interpreted as a set of programs, saved in memory of SHC and all software-hardware peripheral devices (if applicable), and also service programs in machine-readable mediums, intended for checking of operation of I&C system and contained in it individual devices. These software components are designed independently by different organizations, component integration and further software checking can be performed only on operating place of the I&C system.

Independently operated devices, being component parts of SHC, can play different parts for safety and can be operated in different conditions. So the safety class, seismic category, operating conditions and resulting from this classification requirements for resistance to external influences. can be defined for each of component parts individually, but not for SHC as a whole.

SHC can be divided into:

- One-off items, each is designed, manufactured and supplied on NPP's by individual order as a component of individual I&C and cannot be directly (without any modifications) used in other systems.
- Replicated products, designed as components of I&C systems that were not preliminary defined, allowed for use at NPP and do not require any significant modifications of hardware and/or software in the process of manufacturing and supply for use in different industry branches.

Software of I&C Systems

Software of digital I&C systems participates in performing of main system functions. At Ukrainian NPPs such an approach was first implemented in information systems, developed on the basis of serial produced computer SM-2M, which was oriented for application in automated systems, manufactured in different industry branches. System software, provided operation and maintenance, was supplied in composition of SM-2M. Application software, directly connected with the tasks of process control and monitoring, was usually developed independently from supplier of SM-2M.

System software is programs directly executed in the process of operation. Examples of system software: input/output and communication drivers; interruptions management programs; job scheduling software; programs for the diagnosis and management of redundancy when failures. Integration of developed application software with system software and hardware was usually performed directly on operating place. Integration and further checking (verification) were executed in critical shortage of time that complicated detection of faults, which could be made during applications software design. Defects made during software design and not detected during verification, could reveal themselves under specific conditions during operation and cause failure of functions performed by the I&C system.

Typical features of the described approach to I&C systems software design: orientation for a single computer in the system ; storage of all executable programs in the memory of this computer; clear separation between system and application software; design of system and application software by different organizations. Insufficient reliability of the first universal computers, low operating speed, significant labor expenditures of design and debugging of application software, high probability of "hidden" defects, programming errors and impossibility of comprehensive testing of application software by developer did not allow using such computers in NPP safety systems.

Significant extension of a set of software performed or supported functions for NPP power units was provided, first, by element base development – occurrence of available for use microprocessors, single-chip microcomputer and microprocessor controllers, including one or several microprocessors together with a relevant memory, for which in IEC, 2008 a more general title programmable electronic was suggested. (In this standard control, protection and monitoring system, based on one or several programmable electronic devices, is called programmable electronic system).

There appeared a possibility to perform functional decomposition – division of complex functions in a set of significantly simpler. For implementation of each of them it was enough calculating possibilities and storage capacity of appeared at that time programmable electronic devices. For examples, functions of input and of input signals transformation; compare with set-points; functions of logic processing and generation of control commands; functions of continuous monitoring of technical state, etc. could be assigned. Along with functional decomposition, structural decomposition, providing participation in performance of one function of several sequentially or in parallel connected programmable electronic devices, was used. For example, input and transformation function in case of large number of input signals can be performed not by one, but by several simultaneously operating devices, among which all input signals are distributed; diagnostics function of technical state of system parts can be performed under control of software, directly contained in each of these parts, etc.

Such an approach for building programmable electronic system, called "distributed" or "decentralized" control, allowed to refuse from central computes, replacing them with a set of programmable electronic devices, distributed throughout the system. It is considered that such a device individually has low complexity in the sense of that all types of failures of each of its components are clearly determined, and device's behavior, in case of defects occurrence, is fully determined. Simplicity and relative independence of performed functions cardinally simplified design of software for programmable electronic devices of low complexity and for some functions it allowed to refuse form use of operation system, software interrupts, drivers, etc. Also an important factors are decreasing probability of errors during design and relatively simple determination of defects in the process of software testing (before supplying to customer). So, typical software features of modern I&C systems are: orientation for many programmable electronic devices, included in a system; storage of executable programs directly in the memory of these devices; combination of hardware and software design for these devices, simultaneously performed by one and the same organization; integration of hardware and software directly on manufacturing place and supply of programmable electronic devices with "built-in" software. This software can be changed only by replacing device, containing previous software, with a new one with modified software.

As a result of the specified peculiarities, software implementation of main functions of I&C systems important to NPP safety, became possible. In the second part of 80-s on French reactors programmable electronic systems, developed by Electricite de France, performed not only informational support functions, automated regulation, logic control, but also reactor protection. Similar tasks were solved by Westinghouse Electric Company at English NPP Sizewell B and also by Canadian corporation Atomic Energy of Canada Ltd. at NPP Darlington-A. In 2002-2003, Westinghouse Electric Company imbedded overall I&C systems in two power units of Czech NPP Temelín, in which software methods of performance of reactor protection, power regulation, reactor power limitation, in-core reactor monitoring, control of technological equipment, information display in main and emergency control rooms and radiation safety monitoring were used. Also programmable electronic devices were widely used in systems implemented by Siemens Power Corporation within last 12 years at a number of European NPPs and on two power units of Tianwan NPP in China.

Programmable electronic devices on USSR NPP were first used in automatic turbine control systems, developed in early 80-s of the previous century. Nowadays, on most Ukrainian power units all main functions of reactor control and protection are performed by software-hardware complexes based on programmable electronic devices.

Figure 3. Hardware and software of I&C system with distributed structure

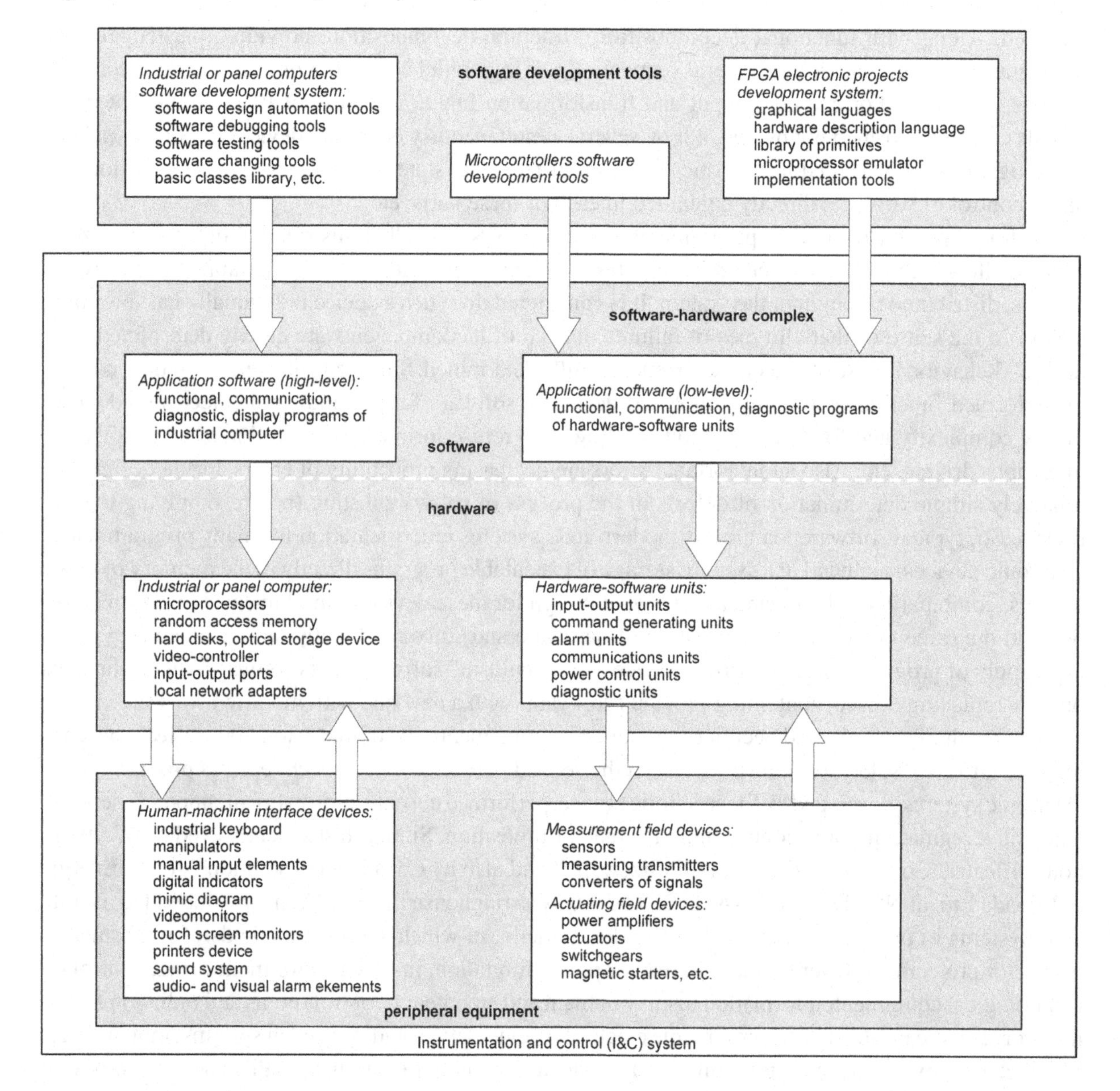

I&C system software is a set of programs, which:

- Control operation of all components of I&C system, provide their interaction with each other, conjugate peripheral devices and other I&C systems.
- Provide performance of human-machine interface functions.
- Diagnose of components of I&C system, conjugate peripheral devices and connecting lines during power-up and operation.
- Automate checking of proper operation of components of I&C system in the process of maintenance and after recovery.

- Support personnel actions during reconfiguration of I&C system (change of set-points, blocking of command generation, etc.).
- Prevent personnel errors during reconfiguration, maintenance and recovery of I&C system.

Application and system software are located in components of I&C system – on hard drives in read-only and random-access memory and/or in logic structures of complex programmable electronic components. I&C system software also includes copies of current program versions on external carriers and software documentation. Development of electronic projects of complex programmable electronic components is usually considered as one of programming types, though slightly different from point of view of requirements to design and verification.

In Figure 3 an example of interaction of hardware and software of I&C system is shown.

I&C PLATFORMS (EQUIPMENT FAMILIES)

I&C platform (equipment families) – set of hardware and software components that may work co-operatively in one or more defined configurations.

Platforms provide a possibility to use uniform hardware and software containing into platform during design and packaging of I&C systems instead from of separate items of different suppliers (often not compatible with each other) The platform is characterized by:

- Functional, structural and constructive completeness.
- Informational, power, constructive, operational and other types of compatibility.
- Advanced software, metrological, standard-methodological, informational and other types of support.

There are differences between platform and SHC. The platform is a conceptual object and exists only as an engineering, software or technological documentation, methods, instructions, standards, etc. The same platform generates a set of various SHCs, where each is a real product of industrial manufacturers The platform is not bound to any specific system: it is intended for typical functions, structures, environment, ways of use and operating conditions, common for a sufficiently wide, though restricted range of different systems. Each SHC that can be implemented on the basis of platform is usually devoted for one specific system. Separately every SHC uses only a part of possibilities of a relevant platform. (restricted nomenclature of hardware and software, supporting constructions, signals, interfaces, etc., required and sufficient for a specific application).

The platform is usually oriented for performance of a set of typical functions, which can be selected, gathered in various combinations, and customized according to tasks of the specific I&C system. Design of hardware and application software of SHC is supported by relevant tools and supports software, recommended or supplied by platform designer.

The examples of platforms and graphic images illustrating the structure of platforms would be described in the next chapters.

FUNCTIONAL SAFETY

Let's begin from common notion "safety". Classical definition of this notion according Webster dictionary "Safety – the condition of being safe; freedom from danger, injury of damage" (Webster, 1991). The later definition from Wikipedia is similar: "Safety is the state of being "safe", the condition of being protected from harm or other non-desirable outcomes" (Safety-Wikipedia).

The next step – definition of this notion for safety critical/safety related objects. The latter include technical objects (plants, equipment, machinery, equipment, enterprises), the malfunctioning of which in some adverse situations can lead to the threat (and in the worst case, loss) of human life, damage to the environment, destruction of structures and equipment, significant economic losses. Some examples of safety critical/safety related objects:

- Industrial – nuclear powers plants; chemical, petrochemical, metallurgy and the other plants.
- Transport – railway (especially high-speed), aviation, auto, sea and river transport, cosmic; equipment for transit of gas, oil, chemical substances to big distance.
- Telecommunication equipment.
- Military technique, etc.

Notion of safety of safety critical/safety related objects could contained from different interconnected components:

- Nuclear safety (the definition according to IAEA Safety Glossary "The achievement of proper operating conditions, prevention of accidents or mitigration of accident consequences, resulting in protection of workers, the public and the environment from undue radiation hazards", IAEA, 2007).
- Chemical safety.
- Fire safety.
- Radiation safety.
- Seismic safety.
- Functional safety.

Separate out the last component – functional safety/from this set. Functional safety is the property of the system (component) of safety critical objects, which consists in the ability to perform all the necessary functions important for safety, to maintain the required properties and to meet the specified characteristics in all operating modes and operating conditions stipulated by the project. This property can be applied to various systems (components) of different safety critical objects. However, it is most appropriate to use this property for information and control (I&C) systems, which has become more widespread in recent years (see, for example, IEC, 2008; Smith & Simpson, 2001, EXIDA (EXIDA). The notion "Functional Safety" for I&C systems appeared in 90-th years of the last century because wide using automatic systems and especially computers for safety assurance of safety critical objects.

We define the functional safety of the NPP I&C system as the property of this system to perform all the necessary functions important to safety, maintain the required properties and meet the specified characteristics in all modes and conditions stipulated by the project.

The definition of the notion "Functional Safety" in IEC, 2008 "Functional safety is a part of the overall safety relating to the EUC and the EUC control system that depends on the correct functioning of the E/E/PE safety-related systems and other risk reduction measures".

Let us explain this definition:

- EUC – equipment under control. In our case it is technological structures, systems and components of NPP. The EUC control system is separate and distinct from the EUC.
- E/E/PE based on electrical (E) and/or electronic (E) and/or programmable electronic (PE) technology. Electrical/electronic/programmable electronic devices include.
- Electro-mechanical devices (electrical – E).
- Solid-state non-programmable electronic devices (electronic – E).
- Electronic devices based on computer technology (programmable electronic – PE).

The following are programmable electronic devices:

- Microprocessors.
- Micro-controllers.
- Programmable controllers.
- Application specific integrated circuits (ASICs).
- Programmable logic controllers (PLCs).
- Other computer-based devices (for example smart sensors, transmitters, actuators).

Electrical/electronic/programmable electronic system (E/E/PES system) – is the system for control, protection or monitoring based on one or more electrical/electronic programmable electronic (E/E/PE) devices, including all elements of the system such as power supplies, sensors and other input devices, data highways and other communication paths, and actuators and other output devices.

Similarly to the situation in tasks of reliability when the determination and classification of failures are fulfilled before the selection of the reliability measures, the determination and classification of events - functional safety disturbances (violations, deviation) are realized before the introduction of the functional safety measures (with regard for specific features of the task formulated). International Scale on nuclear events (included events in NPP's) was published in IAEA, 2001. Events are classified on the scale at seven levels: the upper levels (4–7) ARE termed "accidents" and the lower levels (1–3) "incidents". Events which have no safety significance are classified below scale at level 0 and are termed "deviations". Events which have no safety relevance are termed "out of scale". The countries – NPP operators have own classification of these events. For example, in accordance with the Ukrainian normative document being in force since (NP, 2004) the events at NPP leading to deviations from limits and/or safe operation conditions, or to definite deviations from limits and/or conditions of normal operation (which list is specified in the above-mentioned document) are related to the disturbances in operation of NPP. Resulting from this, functional safety disturbance will mean the nuclear event or event which calculated by national regulation, caused by incorrect functioning of I&C (including failure of I&C systems). In such a way, functional safety disturbance is determined according the behavior I&C system in the closed loop "I&C system – technological equipment" (Yastrebenetsky, 2005)

The measures of functional safety of NPP I&C systems may be:

- Probability for absence of NPP disturbances caused by I&C system during the specified time period.
- Intensity of NPP disturbances, caused by I&C system.

For some types of consequences of the disturbances, connected with non-operability of the safety system channels, the availability coefficient could be selected as measures of functional safety. The example of change of NPP VVER disturbances intensity, caused by overall I&C systems, is shown on Figure 4. This information collected at 15 units of 4 Ukrainian NPP during 23 years.

Figure 4. Measure of functional safety – NPP disturbances intensity, caused by overall I&C systems

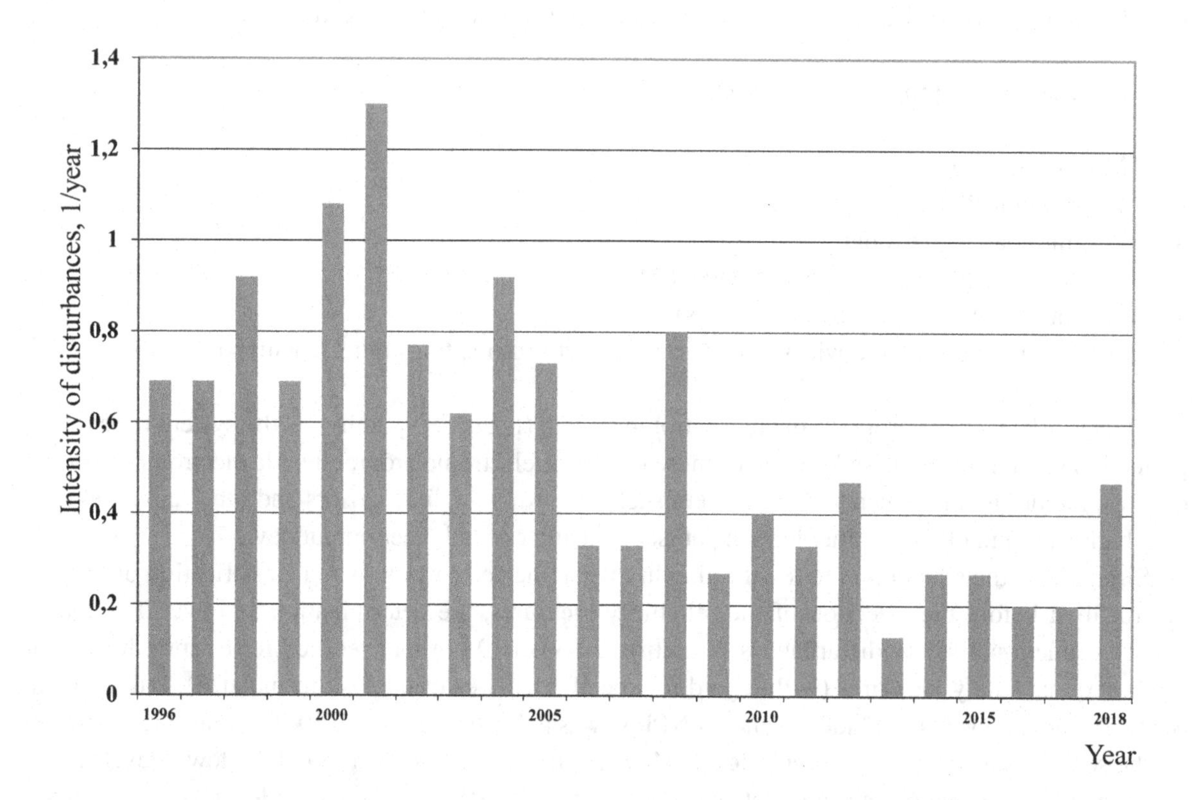

CYBER SECURITY

Modern NPP I&C systems are built with the wide use of digital technologies. Therefore, these systems are vulnerable for potential cyber threats. Cyber-attacks can lead to incorrect functioning of NPP I&C systems or to deviations of their technical characteristics from required. That can have negative impact on technological processes and consequently on NPP safety.

Many evidences of vulnerabilities in computer systems (in particular, at NPPs) has been obtained during last 10-15 years. Often cases of the malicious use of such vulnerabilities with serious consequences were registered. In a situation characterized by more and more complex threats, the possibility of cyber-attacks against critical infrastructure (including NPPs) is increased and became more dangerous than it was earlier. These circumstances require the release of cyber security regulations and the development, implementation and use the measures for protection against cyber threats at NPPs. Such activity is carrying out by competent international (IAEA, IEC, IEEE, ISO, etc.) and national organizations in many countries those use atomic energy (USA, Canada, Germany, UK, France, Japan, Ukraine, etc.).

The developed regulations establish cyber security requirements, affecting many levels and various life-cycle stages of NPP I&C systems. International regulatory framework covers such aspects as:

- General principles of cyber security assurance (defense-in-depth, graded approach, cyber security policy, cyber security culture, etc.).
- Categorization of NPP I&C systems on cyber security (defining of cyber security levels/degrees and caber security zones).
- Cyber security assessment (including cyber threats study, vulnerabilities assessment of computer system, assessment of adequacy of cyber security measures, cyber security risk assessment).
- Cyber security assurance (general protective measures and defensive measures graded according to cyber security levels of NPP I&C systems).
- Cyber security documentation (policy, program, plan, reports).

Cyber security should focus on prevention, detection and response measures against cyber threats that could directly or indirectly result in violations of safe operation of NPP and consequently lead to harmful effects on people, society or environment. Cyber security plays an increasingly important role in achieving these goals.

Cyber security assurance measures should be realized at all life-cycle stages of NPP I&C system: design, manufacturing of hardware, development of software, factory acceptance tests, implementation at NPP, site acceptance tests, operation, maintenance, recovery, decommissioning. The measures should be directed to protection of I&C systems and networks critical to safe and reliable operation of the nuclear facilities and prevention of theft, sabotage and other malicious acts against I&C systems.

SOLUTIONS AND RECOMMENDATIONS

There are common tendencies in development of I&C systems in all countries – NPP's operators. But a lot of these countries have own technical decisions in NPP I&C creation. Ukraine belongs to these countries: Ukraine imported I&C systems as far back as 15-20 years ago. After that full modernization of Ukrainian NPP I&C took place and Ukraine became to export NPP I&C systems and appropriated equipment to NPP in different parts of the world: not only to Europe, but to America and Africa.

Therefore authors can recommend this book to readers for acquaintance with Ukrainian experience in NPP I&C elaboration: this experience will be described in next chapters.

FUTURE RESEARCH DIRECTIONS

Let's mention some areas of activities from their range required for further development of NPP I&C systems.

1. New direction in nuclear energy in the world – creation of small modular reactors (SMR). SMR distribution states new problems for I&C specialists:
   a. High level of automatization.
   b. Decreasing of human factor influence.
   c. Level of defence-in-depth.
   d. Organization of control of mutualized MMR.
   e. Elaboration of international and national standards and rules, related to SMR.
2. Because of wide use of digital I&C systems at NPPs the issues of cyber security should be taken into account in the process of development, implementation, operation and safety assessment of these systems. Thus, the important directions for future research are:
   a. Development of requirements to cyber security and including them into international and national standards and regulations.
   b. Realization of cyber security measures in NPP I&C systems for prevention of malicious unauthorized access, modification or damage – of data and software, which could lead to accident, incorrect operation or loss of important information.
   c. Development of methods of assessment of safety, completeness and correctness of cyber security measures in the process of the licensing process.
3. Aging of specialist and loss of knowledge are the actual problems in field of NPP I&C. It is important to implement the knowledge management into activity of organizations which design, manufacture, operate or make safety assessment of NPP I&C systems. It should include the following:
   a. Development and implementation of knowledge management program in each organization which involved into actions in field of NPP I&C.
   b. Development of methods of capturing, documenting and saving the tacit knowledge of experts in field of I&C.
   c. Permanent training and tutoring for transfer of knowledge from experienced experts to young specialists.
   d. Development of national and international knowledge portals at NPP I&C for improving the cooperation between the different organizations.
4. One of research trends, dictated by effects of accident at NPP Fukushima, is the development of sensors able to sustain extreme environmental impacts, in case of accidents inside the reactor containment. These sensors are required to be qualified under conditions of high temperature, high radiation and aggressive environment.
5. Application of wireless sensors, decreasing costs for routing and maintenance of cables. Here are the following problems:
   a. Resistance of wireless devices to electromagnetic and radio frequency noise.
   b. Cyber security in case of unauthorized access to wireless channels, transmitting information from the outside, for interception, blocking, intentional distortion of signals or output of spurious signals to receiver.
   c. Integration of wireless equipment with current instrumentation and control systems and communication networks.

CONCLUSION

Chapter 1 contains common information about I&C systems and their functions, safety and safety related (normal operation) systems, components of I&C systems (central part, peripheral equipment), platforms, etc. However even from this chapter can be included that problems of NPP I&C safety and security were, are and will be actual. The confirmation of that is the experience of big accidents in NPPs, which changed requirements to NPP as whole and to NPP I&C particularly.

Chapter 1 with common information contained the main future directions. Note only three from them:

- Elaboration of I&C systems for small modular reactors – new direction in nuclear energy.
- Problem of NPP life extension are belonging more and more actual for the most countries – NPP-operators (USA, Canada, Russia, Ukraine, etc). Decision of this problem is impossible without wide I&C modernization.
- Computer technology is susceptible to a number of cyber threats (what are changed quickly - remember Stuxnet worm) and the issues of cyber security should be taken into account in the process of NPP I&C development, implementation and operation.

REFERENCES

EXIDA.com LLC. (n.d.). *Functional Safety - An IEC 61508 SIL 3 Compliant Development Process* (3rd ed.). Author.

IAEA. (1999). Modern instrumentation and control for nuclear power plants / Guidebook. Technical reports series, N°387. Vienna, Austria: IAEA.

IAEA. (2001). The international nuclear event scale (INES). User's manual. Vienna, Austria: IAEA.

IAEA. (2007). *Terminology used in Nuclear Safety and Radiation Protection: IAEA Safety Glossary*. Vienna, Austria: IAEA.

IAEA. (2011). *NP-T-3.12. Core knowledge on instrumentation and control systems in nuclear power plants*. Vienna, Austria: IAEA.

IEC. (2008). IEC 61508. Ed.2. Functional safety of electrical/electronic/programmable electronic safety related systems. IEC.

IEC. (2011). IEC 61513. Ed.2. Nuclear power plants – Instrumentation and control important to safety – General requirements for systems. IEC.

IEC. (2019). *IEC 60050–395. International electrotechnical vocabulary - Part 395: Nuclear instrumentation: Physical phenomena, basic concepts, instruments, systems, equipment and detectors*. IEC.

NP. (2000). *NP 306.5.02/3.035. Nuclear and radiation safety requirements to instrumentation and control systems important to nuclear power plants safety*. Kiev, Ukraine: State Nuclear Regulatory Committee.

NP. (2004). *NP 306.2.100. Provisions on procedure of investigation and calculation of the disturbances in nuclear plants operation*. Kiev, Ukraine: State Nuclear Regulatory Committee.

NP. (2008). *NP 306.2.141. General provisions on the safety of nuclear power plants*. Kiev, Ukraine: State Nuclear Regulatory Inspectorate.

NP. (2015). *NP 306.2.202. Nuclear and radiation safety requirements to information and control systems important to of nuclear power plants safety*. Kiev, Ukraine: State Nuclear Regulatory Inspectorate.

Rosenberg, M., & Bobryakov, S. (2003). *Elsevier's Dictionary on Nuclear Engineering*. London: Elsevier Science.

Safety. (n.d.). In *Wikipedia*. Retrieved from https://en.wikipedia.org/wiki/safety/

Smith, J., & Simpson, K. (2001). *Functional safety: A Straightforward Guide to IEC 61508 and related standards*. Oxford, UK: Butterworth Heinemann.

Webster's New World Dictionary. (1991). (3rd college ed.). Prentice Hall.

Yastrebenetsky, M., Butova, O., Inyishev, V., & Spector, L. (2005). Factors of functional safety of control systems for power units of NPP. *Nuclear Measurement and Information Technologies, 3*(15).

Yastrebenetsky, M., Rozen, Yu., Vinogradska, S., & Johnson, G. (2011). *Nuclear power plants safety: nuclear reactor control and protection systems*. Kiev: Osnova - Print.

Yastrebenetsky, M., Vasilchenko, V., & Vinogradska, S. (2004). *Nuclear power plants safety: instrumentation and control systems*. Kiev: Technika.

ADDITIONAL READING

Dunn, W. R. (2002). Practical Design of Safety-Critical Computer Systems, Reliability Press, Solvang. USA.

EPRI 1015313. (2010). *Computerized Procedure Systems: Guidance on the Design, Implementation, and Use of Computerized Procedure Systems, Associated Automation, and Soft Controls*, Electric Power Research Institute, Palo Alto, CA, USA.

Hashemian, H. M. (2006). *Maintenance of Process Instrumentation in Nuclear Power Plants*. Berlin, Heidelberg: Springer-Verlag.

Hashemian, H. M. (2011). *Measurement of Dynamic Temperature and Pressures in Nuclear Power Plants*. Canada: University of Western Ontario.

IAEA NP-T-2.11 (2018) *Approaches for Overall Instrumentation and Control Architectures of Nuclear Power Plants*. IAEA, Vienna.

IAEA NP-T-3.19 (2017) *Instrumentation and Control Systems for Advanced Modular Reactors*. IAEA, Vienna.

IAEA SSG-39. (2016) *Design of instrumentation and control systems for nuclear power plants.*, IAEA, Vienna.

O'Hara, J. (2009). *Applying Human Performance Models to Designing and Evaluating Nuclear Power Plants: Review Guidance and Technical Basis*. Upton, NY: Brookhaven National Laboratory. doi:10.2172/1013435

U.S. Nuclear Regulatory Commission. (2011). Regulatory Guide 1.152. Criteria for use of computers in safety systems of nuclear power plants (RG 1.152). Washington, DC.

U.S. Nuclear Regulatory Commission. (2018). NUREG-0800. Standard Review Plan for the Review of Safety Analysis Reports for Nuclear Power Plants. Chapter 7. Instrumentation and control., Washington, DC.

U.S. Nuclear Regulatory Commission Regulations. Title 10, Code of Federal Regulations, Part 50 (2015) – Domestic licensing of production and utilization facilities, Appendix A to Part 50 – General Design Criteria for Nuclear Power Plants. Washington, DC.

KEY TERMS AND DEFINITIONS

Component: A discrete element of a system. A component may be hardware or software and may be subdivided into other components. Examples are wires, transistors, integrate circuits, motors, relays, solenoids.

Functional Safety of the NPP I&C System: The property of this system to perform all the necessary functions important to safety, maintain the required properties and meet the specified characteristics in all modes and conditions stipulated by the project.

Human-Machine Interface (HMI): The interface between operating staff and I&C system and computer systems linked with plant. The interface includes displays, controls, and the Operator Support System interface.

Instrumentation and Control (I&C) System: System, based on electrical and/or programmable electronic technology, performing I&C functions as well as service and monitoring functions related to the operation of the system itself. The term is used as a general term which encompasses all elements of the system such as internal power supplies, sensors and other input devices, data highways and other communication paths, interfaces to actuators and other output devices.

Overall I&C System: A system providing monitoring and control of all technological systems and equipment at NPP or individual NPP power generating unit.

Platform (Equipment Family): Set of hardware and software components that may work co-operatively in one or more defined architectures (configurations). An equipment family usually provides a number of standard functionalities (e.g. application functions library) that may be combined to generate specific application software.

Software-Hardware Complex (SHC): Functionally complete item in the form of one or several independently operated devices with built-in software, which are interconnected through electric and/or optic cables, with peripheral equipment of the same system and with other I&C systems. SHC is usually supplied to customer as a set with necessary service equipment, repair and recovery reserve, operating and software documentation.

Chapter 2
International Standard Bases

Serhii Trubchaninov
State Scientific and Technical Center for Nuclear and Radiation Safety, Ukraine

ABSTRACT

The main standard bases for NPP instrumentation and control (I&C) systems are documents of the International Atomic Energy Agency (IAEA) and International Electrotechnical Commission (IEC). Standards are interconnected through the following: IAEA develops general safety principles for NPP I&C systems, and IEC develops technical requirements that use and specify safety principles. Structures of both bases are considered. IAEA and IEC safety classification principles are described. Classifications of I&C systems and their components are given on the basis of their safety impact. According to IEC, functions to be performed by I&C systems shall be assigned to categories according to their importance to safety. The importance to safety of a function shall be identified by means of the consequences in the event of its failure, when it is required to be performed, and by the consequences in the event of a spurious actuation.

INTRODUCTION

There is an expression that safety regulations are written with blood. It applies foremost to standards in nuclear power engineering, where accidents have a large-scale effect. Standardization in nuclear power engineering, including NPP I&C systems, has specific peculiarities in comparison with other branches of industry.

Firstly, in nuclear power engineering there is a strict prohibition on any actions that are not specified by the regulations: those actions are forbidden that are not allowed. In many other branches of industry there is a reverse statement: those actions are allowed that are not forbidden. Allowed actions are described in standards as NPP safety in general, and NPP I&C that provides the safety, in particular.

Secondly, in consideration with a global scale of accident international cooperation is used extensively in nuclear power engineering. It applies to development and use of NPP I&C safety standards. Standard bases of the International Atomic Energy Agency (IAEA) and International Electrotechnical Commission (IEC) are the most widespread in the world. These standards concentrate the best international experience

DOI: 10.4018/978-1-7998-3277-5.ch002

of development and operation of NPP I&C. This chapter is devoted to the description of these standard bases, especially classification of I&C systems and their components.

Besides international standards each country has a standard base concerning nuclear power engineering, including NPP I&C, e.g. American National Standards Institute (ANSI) in USA, Deutsches Institut fur Normung (DIN) in Germany, British Standards Institute (BSI) in UK. In addition, in USA standards of professional organizations are widespread – American Society of Mechanical Engineering (ASME) and, especially concerning NPP I&C, – Institute of Electrical and Electronic Engineers (IEEE).

Note, that international standards devoted to software will be described below in Chapter 4.

However, the main attention in this chapter is further given not to national, but to international standards, since harmonization of national standards with ones is currently a vital task.

BACKGROUND

Elaboration of international standards related to NPP I&C began immediately after commissioning of the first NPP. The IEC created a separate technical committee, devoted to NPP I&C, with name "Nuclear Instrumentation" in 1960. One of the first IEC publications – "General principles of nuclear reactor instrumentation" was issued in 1967. IEC understood importance of computer technique for NPP automatics: publication "Application of digital computer to nuclear reactor instrumentation and control" had data 1979. During more than 50 years IEC developed a lot of standards applicable to different types of I&C, to different aspects of NPP I&C design and operation. Special attention is paid to application of computer systems and the latest achievement in information technology.

The IAEA elaborated Codes on safety of NPPs, which related to all NPP systems, including I&C (e.g., "50-C-D. Code on Safety on Nuclear Power Plants. Design" (IAEA, 1988). Special safety guides related exclusively to protection systems and related features IAEA 50-SG-D3 and to safety related instrumentation and control systems IAEA 50-SG-D8 were issued in 1980 and 1984 (IAEA, 1980; IAEA, 1984). IEC and IAEA had close connections in the development of standard base for NPP I&C.

Many countries where NPPs are operated developed their national standard base related to I&C. One of the first American National Standards (based on IEEE Standard 323-1974) was ANSI/IEEE Std.323-1983 "IEEE Standard for Qualifying Class1 Equipment for Nuclear Power Generating Stations" which did not lose its importance for long time. The first USSR Regulation PBYa-04-74 "Rules of NPP Nuclear Safety" which contained requirements on reactor control and protection systems, was issued in 1974.

Ukraine after gaining independence (1991) used to 1999 USSR regulations and standards related to NPP I&C. Ukrainian regulation "NP 306.5.02/3.035-2000. Requirements for nuclear and radiation safety to information and control systems important to safety of nuclear power plants" was issued in 2000 (NP, 2000) and was used to 2015 when was replaced by (NP, 2015).

IAEA Standard Base

IAEA created in Vienna in 1957, is an international intergovernmental organization bound with the Agreement with United Nations Organization. The Agreement stipulates that IAEA acts as an autonomous international organization.

The objectives of the Agency are determined in its statute: "Article II: Objectives. The Agency shall seek to accelerate and enlarge the contribution of atomic energy to peace, health and prosperity throughout the world. It shall ensure, so far as it is able, that assistance provided by it or at its request or under its supervision or control is not used in such a way as to further any military purpose" (IAEA, 2013,b).

IAEA activity consist in providing emergency intervention in case of accidents, technical cooperation, information exchange, personnel training, and also in development of IAEA safety documents.

IAEA Safety Standards Series

According to the statute, IAEA determines safety standards and provides for their application. IAEA standards reflect the best experience and practices of countries, using nuclear power, and, as one of the main tasks, are intended to support formation of a proper national normative base. IAEA standards are not mandatory for IAEA member countries, but can be adopted by their own choice.

IAEA safety standards were always combined by a certain family.

Since 1996 the system of standards has been replaced by IAEA Safety Standards Series. Standards of the series "Nuclear Safety" included three levels, that were applied to NPP I&C:

- **"Safety Fundamentals"**: Setting main objectives, concepts and principles of safety assurance. There is one document in this level"- SF-1 "Fundamental Safety Principles" (IAEA, 2006) presents the fundamental safety objective and principles of protection and safety for nuclear facilities and activities.
- **"Safety Requirements"**: Setting requirements that must be met for safety assurance. They are expressed in imperative form and defined by objectives and principles represented in "Safety Fundamentals".
- **"Safety Guides"**: Containing recommendations based on international experience. These documents are less formal than "Safety Requirements" and specify actions, conditions and procedures to be complied with safety requirements.

Since 2009 a new structure of IAEA standards has started operating. These standards have three categories (see Figure 1).

"Safety Fundamentals": SF-1 "Fundamental Safety Principles" (IAEA, 2006) presents the fundamental safety objective and principles of protection and safety and provides the basis for the safety requirements.

"Safety Requirements": An integrated and consistent set of safety requirements establish the requirements that must be met to ensure the protection of people and the environment, both now and in the future. The requirements are governed by the objective and principles of the safety fundamentals. If the requirements are not met, measures must be taken to reach or restore the required level of safety. The format and style of the requirements facilitate their use for the establishment, in a harmonized manner, of a national regulatory framework.

The General Safety Requirements (GSR) are applicable to all facilities and activities. This document includes some parts. The examples are part 1 (IAEA, 2010), devoted to governmental, legal and regulatory framework for safety and part 4 (IAEA, 2009) devoted to safety assessment.

The Specific Safety Requirements (SSR) are applicable to specific installations and activities. Overarching requirements for NPP safety (including I&C systems) are contained in the documents from level 2 point 2 "Specific safety requirements, devoted to safety of nuclear power plants": "2.1 Design " (IAEA, 2012) and 2.2 "Commissioning and operation" (IAEA, 2011,b).

Figure 1. IAEA safety standards categories

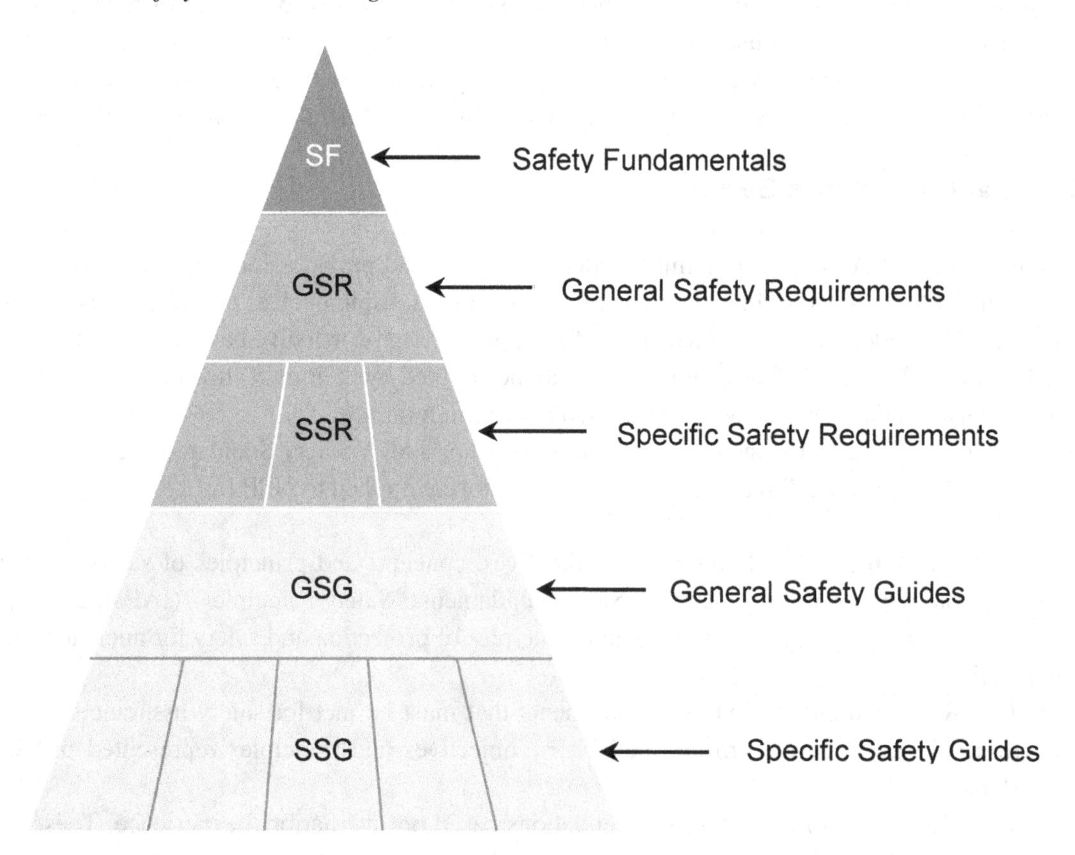

"Safety Guides": Safety Guides provide recommendations and guidance on how to comply with the safety requirements, indicating an international consensus that is necessary to take the measures recommended (or equivalent alternative measures).

General Safety Guides (GSG) are applicable to all facilities and activities. Document number 1 of this group – GSG-1 is the guide, devoted to classification of radioactive waste, the second GSG-2 contains criteria for use in preparedness and response for a nuclear and radiological emergency.

Specific Safety Guides (SSG) are applicable to specified facilities or activities. The recently developed new safety guide SSG-39 "Design of I&C Systems for Nuclear Power Plants" (IAEA, 2016). This document combines and supersedes two acting before safety guides: NS-G-1.3 (IAEA, 2002) which had recommendation regarding the implementation of IAEA requirements for NPP I&C systems and NS-G-1.1 (IAEA, 2000) which provided detailed guidance on the development of software for NPP I&C systems important to safety. SSG-39 (IAEA, 2016) was compiled by leading experts of I&C of different countries (USA, Germany, Great Britain, France, Ukraine, Japan, Canada and others) be supervision of G.Johnson and A. Duchac. Contents of SSG-39 is given on Table 1

Table 1. Contents of IAEA standard SSG-39 "design of I&C systems for nuclear power plants"

1	Introduction
2	The management system for instrumentation and control design
3	Design basis for instrumentation and control systems
4	Instrumentation and control architecture
5	Safety classification of instrumentation and control functions, systems and equipment
6	General recommendations for all instrumentation and control systems important to safety
7	Design guidelines for specific instrumentation and control systems, and equipment
8	Considerations relating to the human–machine interface
9	Software
ANNEX I	Bibliography of international instrumentation and control standards
ANNEX II	Correlation between this safety guide and IAEA safety standards series
ANNEX III	Areas where practices of member states differ

G.Johnson and A. Duchac in their paper devoted to main new idea of SSG-39 (Johnson, 2017) shown the main differences SSG-39 and previous guides NS-G-1.3 and NS-G-1.1:

- SSG-39 takes into account the continuing development of computer applications and the evolution of the methods necessary for their safe, secure and practical use.
- The document identifies two important interfaces with I&C design: Human Factors Engineering and Cyber Security. SSG-39 accounts for developments in human factors engineering and provides considerations on the interactions of I&C design with human factor engineering programs. The document gives recommendations on I&C design features that affect these topics, e.g. the human-machine interface in the design of the main and supplementary control rooms. Guidance is also given for user displays and controls.
- Criteria that existed in NS-G-1.1 and other criteria developed after the publication of that standard were included in SSG-39 to cover topics that mainly involved system performance requirements, communications systems, and cyber security considerations.
- When NS-G-1.1 was written most digital systems were being developed using general purpose microprocessors or programmable logic controllers. Since that time the industry has witnessed the use of other kinds of digital platforms such as systems programed using hardware description languages (e.g., FPGA) and industrial digital devices having limited functionality, The selection and use of such devices raise issues that are different from the older technologies.

Other IAEA Documents

International Nuclear Safety Advisory Group: INSAG, in which the most authoritative experts from different countries are working, was established in 1985. The group develops general safety concepts on the basis of analysis of activity results both within IAEA framework and on other information. INSAG documents formally have information status. However, in fact they should be applied in international practice, because they reflect international trends of nuclear safety. The first INSAG paper was devoted to causes and effects of the Chernobyl NPP disaster. For NPP I&C 75-INSAG-3 "Basic safety principles for nuclear power plants" is very important. It was issued in 1988, then revised in 1999 and reissued under the same title but with another number INSAG-12 (IAEA, 1999,a). The document contains objectives and fundamental principles of safety assurance.

Documents in IAEA Safety Report series – SRS contain papers made by of a group of experts. These publications describe practical experience, give some practical examples and detail methods, used to achieve safety requirements compliance.

In IAEA Technical Report Series – TRS let us note a NPP I&C reference book (IAEA, 1999, b) prepared by a group of authors from different countries. A significant part of the book is occupied by a description of NPP I&C in Finland (Loviisa), France (series N°4), Great Britain (Sizewell B), Canada (CANDU, series 6), Russia (WWER-1000), USA and others.

A significant number of documents, related to NPP I&C, was issued in *"TECDOC"* series. These documents are generally devoted to specific tasks and summarize experience of IAEA member countries. The following issues are considered:

- Modernization and new technologies, including: software issues (quality assurance, verification and validation, life cycle management and others); digital hardware (reliability, safety analysis, I&C computerized hardware and others).
- Ageing and operating experience.
- Human factor, including operator support systems, man-machine interface, simulators and others.

An example of document in TECDOC series is IAEA-TECDOC-1016 "Modernization of instrumentation and control in nuclear power plants" (IAEA, 1998). It contains information about management of modernization, design criteria, requirements and restrictions, operating and licensing aspects, description of I&C modernization in different countries – USA, France, German, Korea, Russia, Ukraine, etc.

In the development of TECDOC documents a significant role was played by the Technical Working Group on NPP Instrumentation and Control, uniting experts from a number of countries. This group was created in 1970. The main objective of the Working Group is to promote the exchange of information on NPP I&C and to stimulate and, if possible, coordinate research in field of NPP I&C in interested Member States and international organizations. The scope of work covers all aspects of the life cycle of I&C systems and equipment from feasibility study and design through installation, commissioning and licensing to operation, maintenance and decommissioning. In 2011 a report (IAEA, 2011, a) was prepared by the experts of this group, which is an introductory description of I&C systems and their life cycle. It compiles the necessary basic information to understand I&C systems in NPPs, an explanation of the significant role of I&C systems in maintaining and improving safety, plant performance, and economic (chairpersons of this report preparation meeting were R.Wood-USA and J.Eiler - Hungary). This report received wide circulation among specialists.

IAEA Safety Classification Principles

Classification on NPP components according to its importance to NPP safety began since the earliest days of NPP creation (First IAEA Safety Guide on safety functions and component classification was issued in 1979 (IAEA, 1979). Safety class is an attribute that is taken into account by equipment developers, system designers, operating organization, other participants in NPP creation and modernization, nuclear regulatory authorities. One follows safety class in setting requirements on structure, systems and components (SSC), developing, producing, product testing, designing, checking and providing maintenance, and also during assessment of their compliance with regulatory requirements. It should be taken into account that a degree of "rigidity" of requirements depends considerably on a type of class, to which an object safety assessment is referred. A general NPP safety concept is described in INSAG report (IAEA, 1999,a), where it is indicated that "all safety related components, structures and systems are classified on the basis of their functions and significance with regard to safety, and they are so designed, manufactured and installed that their quality is commensurate with that classification".

IAEA standards which contained information about NPP I&C safety classification are situated on 3 levels (Figure 2).

The Specific Safety Requirements SSR 2/1 (IAEA, 2012) located on high level. This document applicable to all NPP structures, systems and components (SSC's) and covering not only the I&C, but also various all items important to safety. These items shall be identified and shall be classified on the basis of their function and various NPP their safety significance. The method for classifying the safety significance shall be based primarily on deterministic methodologies complemented where appropriate by probabilistic methods, with account taken of factors such as:

- The safety function to be performed.
- The consequences of failure to perform the safety function.
- The frequency with which the item called to perform the safety function.
- The time following a postulated initiating event at which, or the period for which, it will be perform the safety function.

Document of the next level – Specific Safety Guides SSG-30 (IAEA, 2014) devoted to safety classification of NPP SSC's important to safety. Big part of this document contains categorization of the functions. The functions (including, of course, I&C functions) should be categorized into a limited number of categories on the basis of their safety significance, using an approach that takes account of the following factors:

- The consequences of failure to perform the function.
- The frequency of occurrence of the postulated initiating event for which the function will be called upon.
- The significance of the contribution of the function in achieving either a controlled state or a safe state.

Figure 2. IAEA standards contained information about NPP I&C safety classification

IAEA Safety standard SSG-2/1. Safety of nuclear power plants: design. Specific safety requirements. 2012

↓

IAEA Safety standards SSG-30. Safety classification of structures, systems and components. 2014

↓

IAEA Safety standards SSG-39. Design of instrumentation and control systems for nuclear power plants. Specific safety guide. 2014

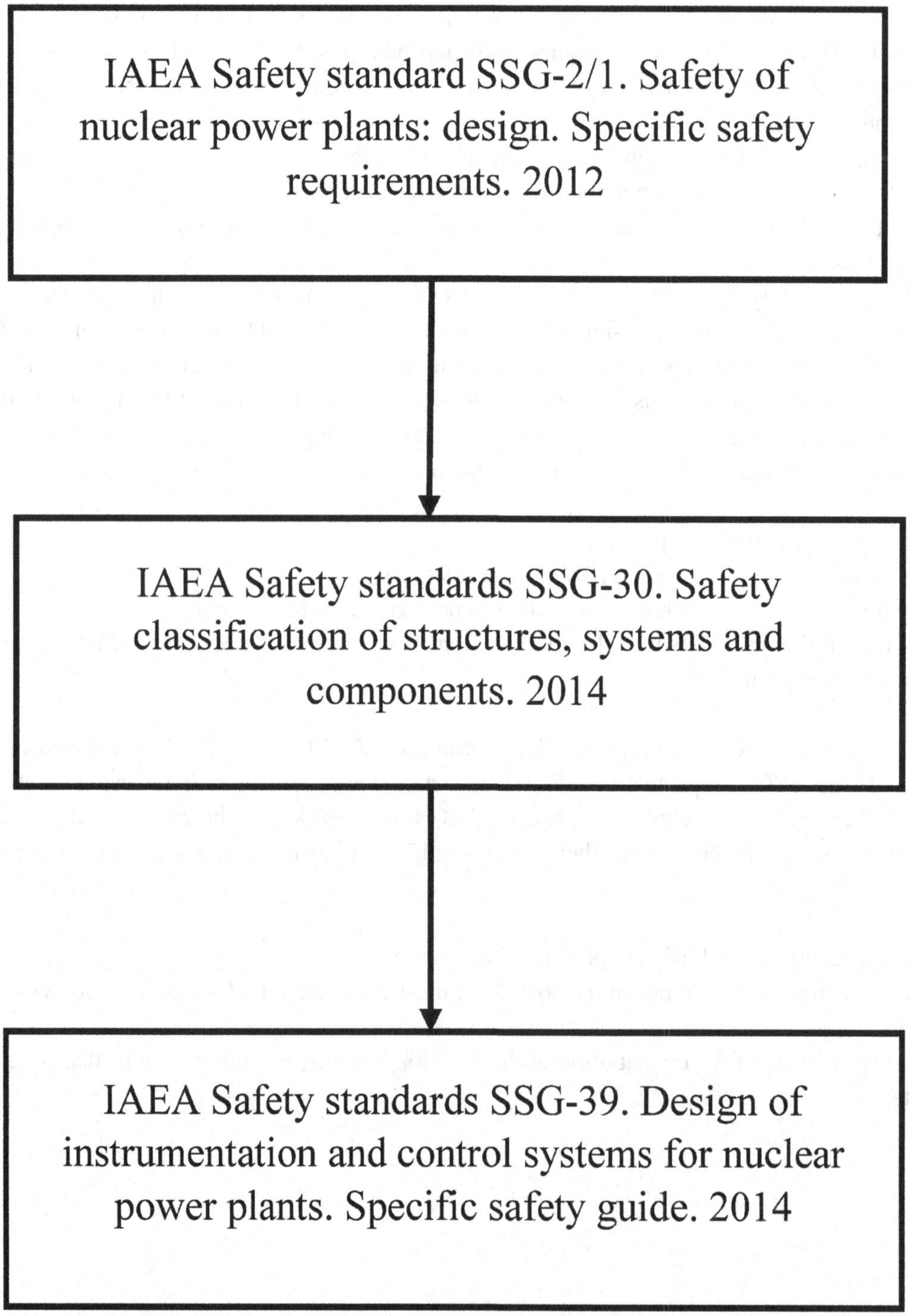

Three safety categories for functions and three safety classes for SSCs important to safety are recommended in SSG-30:

- **Safety Category 1:** Any function that is required to reach the controlled state after an anticipated operational occurrence or a design basis accident and whose failure, when challenged, would result in consequences of 'high' severity.
- **Safety Category 2:** There are three possibilities in this category:
    - Any function that is required to reach a controlled state after an anticipated operational occurrence or a design basis accident and whose failure, when challenged, would result in consequences of 'medium' severity; or
    - Any function that is required to reach and maintain for a long time a safe state and whose failure, when challenged, would result in consequences of 'high' severity; or
    - Any function that is designed to provide a backup of a function categorized in safety category 1 and that is required to control design extension condition without core melt.
- **Safety Category 3:** There are five possibilities in this category:
    - Any function that is actuated in the event of an anticipated operational occurrence or design basis accident and whose failure, when challenged, would result in consequences of 'low' severity; or
    - Any function that is required to reach and maintain for a long time a safe state and whose failure, when challenged, would result in consequences of 'medium' severity; or
    - Any function that is required to mitigate the consequences of design extension conditions, unless already required to be categorized in safety category 2, and whose failure, when challenged, would result in consequences of 'high' severity; or
    - Any function that is designed to reduce the actuation frequency of the reactor trip or engineered safety features in the event of a deviation from normal operation, including those designed to maintain the main plant parameters within the normal range of operation of the plant; or
    - Any function relating to the monitoring needed to provide plant staff and off-site emergency services with a sufficient set of reliable information in the event of an accident (design basis accident or design extension conditions), including monitoring and communication means as part of the emergency response plan, unless already assigned to a higher category.

SSG-39 (IAEA, 2016) devoted to I&C systems located on the low level on Table 2 and includes Chapter 5 "Safety Classification on instrumentation and control functions, systems and equipment. This Chapter based on SSR 2/1 and SSG-30. Some differences between SSG-39 and SSG-30:

- The third factor of categorization I&C functions in SSG-39: "The time following a postulated initiating event at which, or the period of time during which, the function will be required to be performed".
- In SSG-30 three safety categories for functions and three safety classes for classes for SSC's are recommended. According SSG-39, a larger or smaller number of categories and classes may be used.

IEC STANDARD BASE

IEC Standards Related Directly to Safety of NPP I&C Systems

IEC was founded in 1906 in order to facilitate international cooperation in the field of electrical and ccommunication technology. Afterwards IEC sphere of activity was extended by electronics, instrument engineering, computers and other branches of modern technology, connected with information technologies. IEC is the oldest international organization of standardization. IEC is the biggest and the most authoritative organization of standardization in the world.

IEC standardization covers two main aspects:

- The interchangeability of products;
- Standard methods of measuring and assessing quality and performance.

More information about general IEC activity can be found at http:\www.iec.ch. IEC work is organized among technical committees (TC), and each TC is responsible for a specific course. Many of technical committees contains of subcommittees (SC), which in turn are divided in a range of Working Groups (WG).

The subject of the book corresponds foremost to activities of TC 45 (Nuclear Instrumentation) and, in particular, to its subcommittee SC 45A. Name of SC-45 was "Instrumentation and control of nuclear facilities". This subcommittee was formed in 1963. Note, that area of interests and name of SC-45A were expended in 2013- electrical power systems were added. Structure of TC-45 is shown on Figure 3.

SC-45A standards cover the entire lifecycle of these I&C systems, from conception, through design, manufacture, test, installation, commissioning, operation, maintenance, aging management, modernization and decommissioning. A major aspect of SC-45A is the application of emerging electronic techniques in order to meet nuclear instrumentation and control requirements, particularly computer systems and advances in information processing and control. The Figure 4 shows the structure of standards, developed SC-45A and related to NPP I&C systems safety and their components.

These standards are arranged in 4 levels. The top-level document is IEC 61513 (IEC, 2011). It provides general requirements for I&C systems and equipment that are used to perform functions important to safety in NPPs. IEC 61513 covers:

- Unit overall control systems and individual information and control systems.
- Systems, using computers and software, and systems that do not use them.

In IEC 61513 two safety lifecycles are considered:

- Common, covering overall unit control systems.
- Individual information and control systems.

Figure 3. Structure of IEC technical committee TC- 45 "nuclear instrumentation"

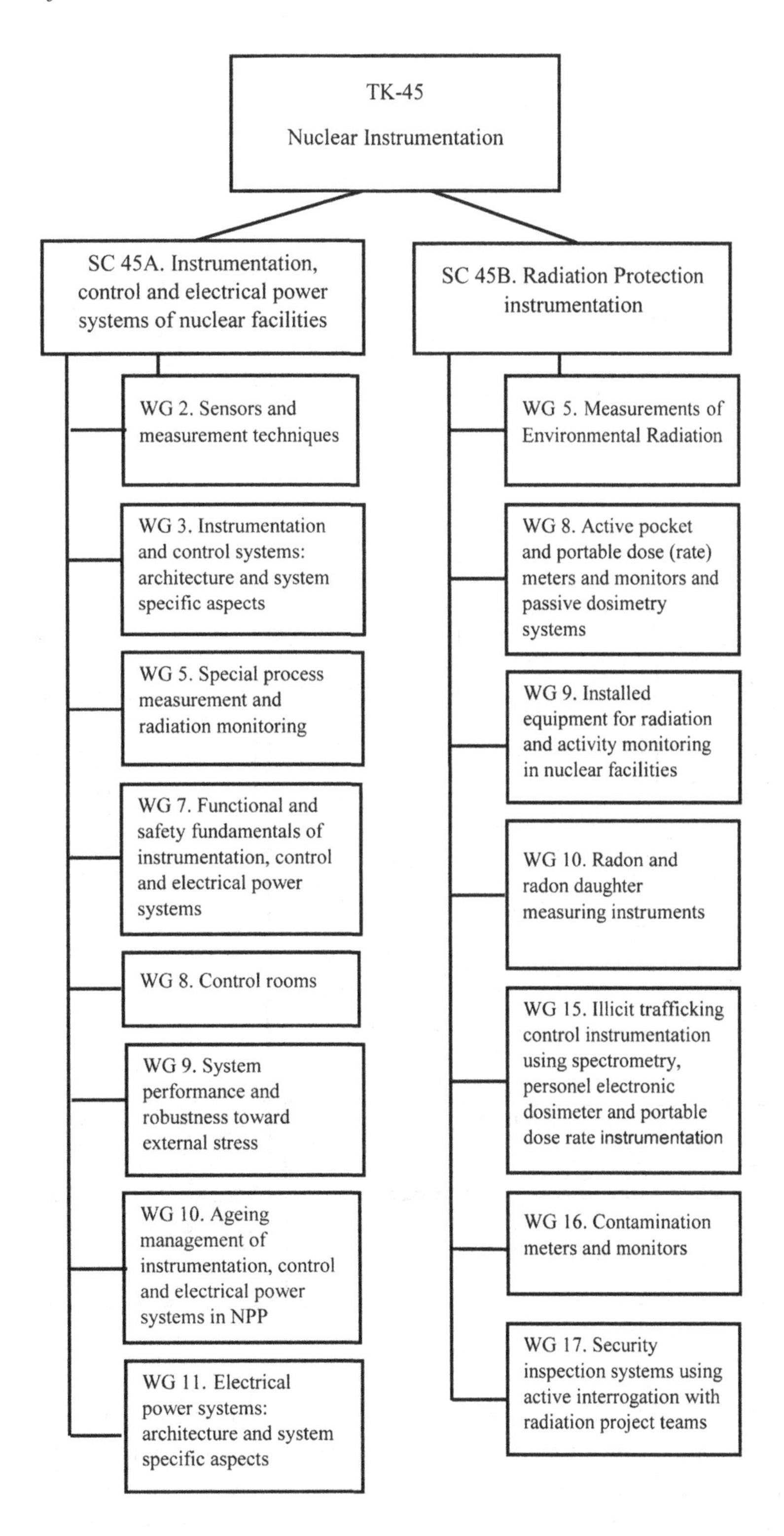

Figure 4. Structure of IEC standards devoted to safety important NPP I&C systems and their components

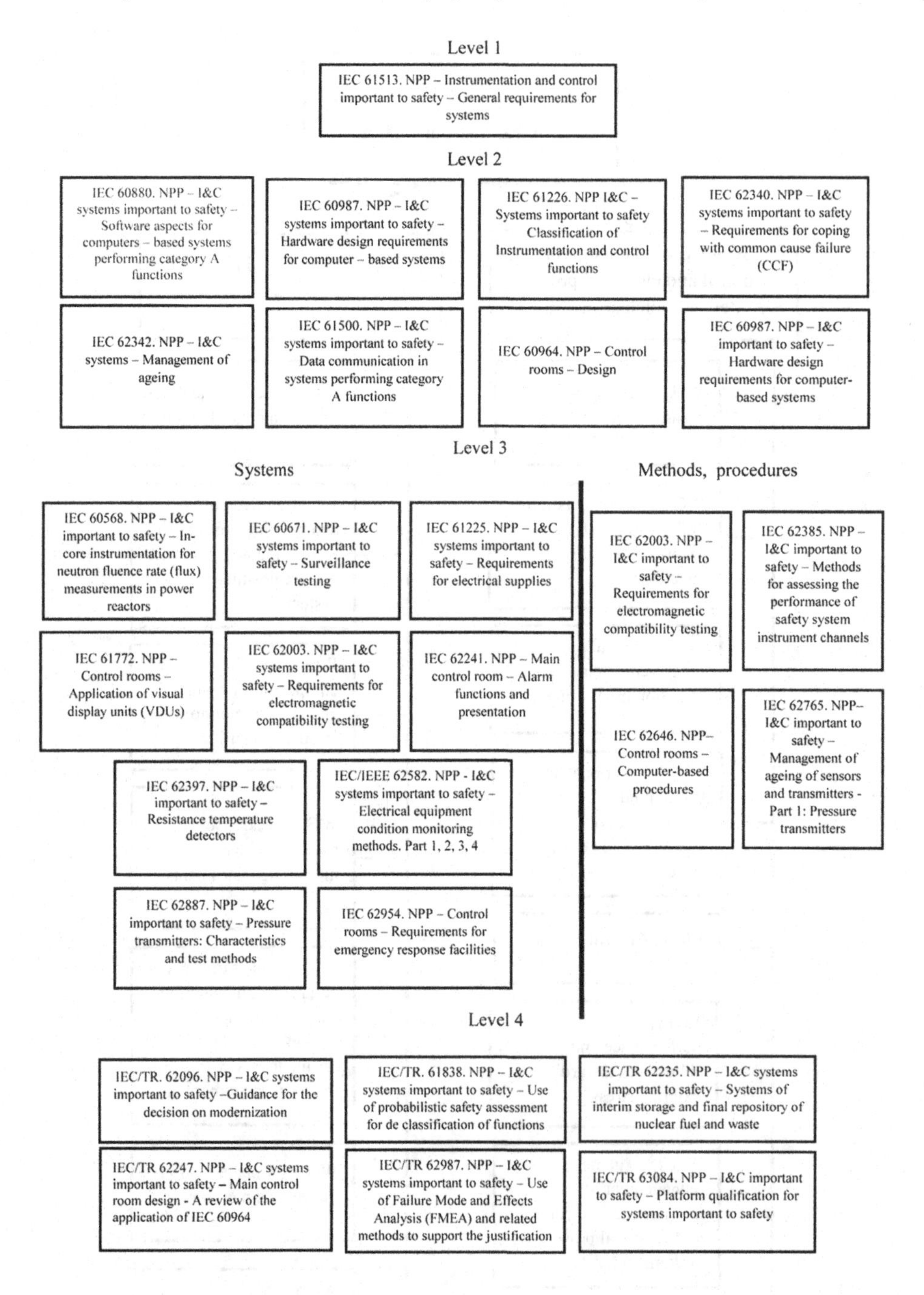

System requirements in IEC 61513 are supplemented with requirements for system design, integration, validation, assembling, adjustment and operation, including requirements for their compliance assessment. Detailed documentation requirements at all stages of system lifecycle should be mentioned.

IEC 61513 contains description of a top level of requirements, regardless of reactor type and used design solutions: specification of a set of requirements contains in other IEC standards are at the 2-nd and

3-rd levels, as shown in Figure3. IEC 61513 refers directly to other IEC SC-45A standards for general topics related to categorization of functions and classification of systems, qualification, separation of systems, defense against common cause failure, software aspects of computer-based systems, hardware aspects of computer-based systems, and control room design. The standards referenced directly at this second level should be considered together with IEC 61513.

At the third level, IEC SC-45A standards not directly referenced by IEC 61513 are standards related to specific equipment, technical methods, or specific activities.

At the fourth level technical reports, not considering normative documents, are found. There are document IEC TR 61838, devoted to the use of probabilistic safety assessment for the classification, document IEC TR 62096, contained Guidance for the decision on modernization applies to all NPP I&C systems, regardless to safety importance, etc.

IEC Standards on Critical Systems

Standard IEC 61508 "Functional safety of Electrical/Electronic/Programmable Electronic safety-related systems" (IEC, 2008), developed by subcommittee SC- 65A "System aspects" of technical committee IEC TC 65 "Industrial process measurement, control and automation", applies to critical systems in different branches of industry (see chapter 1). This standard refers to a wide class of systems, including the following types of components: electrical (E) (e.g., electromechanical devices), electronic (E) (e.g., nonprogrammable transistor devices), programmable (PE) (e.g., microprocessors, microcontrollers, logic controllers). In IEC 61508 these components are indicated as E/E/PE and respective systems as E/E/PE systems or E/E/PES.

In the standard, the notion of safety lifecycle is introduced, it is an activity connected with implementation of safety related systems starting from the development of design concept till E/E/PE systems are not usable.

IEC 61508 requires that functional safety assessment is made for all parts of E/E/PE system at all lifecycle stages. This standard is a basic: it is not only used as an independent one in some branches of industry, but also forms a ground for development of branch standards. In IEC 61508 the main attention is paid to computer systems. In Application D of IEC 61513 the interconnection between IEC 61508 and nuclear application standards is considered. Standard IEC 61508 is an umbrella for standards of critical systems safety for different applications (Figure 5).

Figure 5. IEC Standards for different applications of functional safety

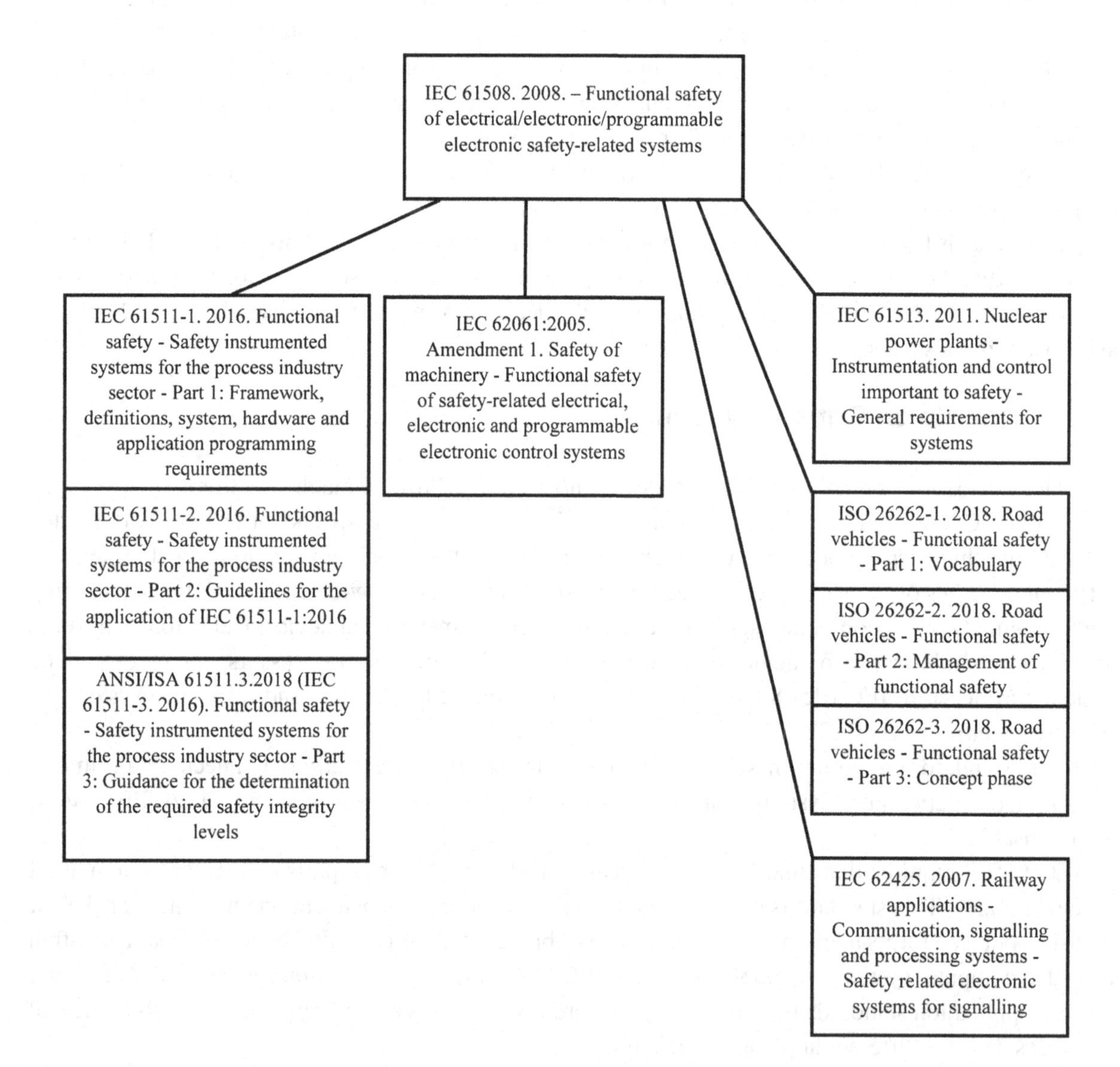

IEC Common Technical Standards

In IEC membership there is also a range of technical committees that developed common technical standards for different branches of technique, not only for NPPs. TC 77 "Electromagnetic compatibility" elaborated a set of standards with common name "Electromagnetic compatibility (EMC) – Testing and measurement techniques –" 61000-4- and additional name and number, which takes into account type of test (e. g. Electromagnetic compatibility (EMC) – Testing and measurement techniques – Electrostatic discharge immunity test IEC 61000-4-2).

Standards of technical committee IEC TC 56 "Dependability" have a great importance for NPP I&C. Dependability covers the availability performance and its influencing factors: reliability performance, maintainability performance and maintenance support performance (including management of obsolescence). The standards cover generic aspects on reliability and maintainability management, testing and analytical techniques, software and system dependability, life cycle costing, technical risk analysis and project risk management.

Interconnection Between IEC Standards and Other International Standards

IEC and IAEA closely interact with each other in the context of NPP I&C systems, though, their functions differentiate. According to a formal cooperation agreement between IAEA and TC 45 IEC, concluded in 1981, IAEA is responsible for development of general concepts for NPP I&C safety, TC-45 IEC is responsible for development of technical requirements, using and detailing the safety concepts mentioned above. Terms and definitions, used in IEC standards, correspond to IAEA standards.

A considerable part of European standards, developed by European Committee of Electrotechnical Standardization – CENELEC, is identical to IEC standards or slightly differs. So, all 7 parts of IEC 61508 are ratified by CENELEC, and this standard is published under EN 61508.

IEC is closely connected with the International Organization for Standardization – ISO. Taking into account rapid growth of computerization, IEC and ISO in 1986 created a joint technical committee (ISO/IEC Joint Technical Committee for Information Technology – JTC1). This committee consists of subcommittees. Standards applied in the field of NPP I&C are developed by subcommittee 7 on software engineering (JTC1/SC7 Software Engineering). A necessity of analysis of this group of standards is caused by a wide use of the latest achievements of information technologies in the I&C systems.

The main part of standards on software engineering is devoted to the description of software lifecycle processes (supply and purchase, requirement analysis, design, coding and testing, integration, operation and maintenance, documentation, configuration management, quality assurance, verification and validation, project management).

IEEE standards form their own system inadequate to IEC standards. In particular IEC focuses on important to safety systems, IEEE focuses on safety systems. It notes (Johnson, 2002), that the collection of IEEE and IEC standards have some overlap, but in many cases cover significantly different topics. Collaboration between IEC with IEEE was realized as follows: two high level agreements signed in 2007 and 2008 between the IEC and the IEEE; technical collaboration started in 2009 between IEC/SC45A and IEEE/NPEC; publication of IEC/IEEE standards on condition monitoring (IEC/IEEE, 2011), basic requirements to qualifying electrical equipment important to safety and interfaces (IEC/IEEE, 2016). post-accident monitoring (IEC/IEEE, 2017).

IEC SAFETY CLASSIFICATION PRINCIPLES

Categorization of Functions

IEC 61226 (IEC, 2009,b) defines a function determined as a specific purpose or objective to be accomplished, that can be specified or described without reference to the physical means of achieving it. This standard extends the classification strategy presented in IAEA and establishes the criteria and methods to be used to assign the I&C functions of an NPP to one of the three categories A, B and C, depending on their importance to safety, or to an unclassified category for functions.

Category A denotes the functions that play a principal role in the achievement or maintenance of NPP safety to prevent design basis event (DBE) from leading to unacceptable consequences. This role is essential at the beginning of the transient when no alternative actions can be taken, even if hidden faults can be detected.

Category B denotes functions that play a complementary role to the category A functions in the achievement or maintenance of NPP safety, especially the functions required to operate after the non-hazardous stable state has been achieved, to prevent DBE from leading to unacceptable consequences, or mitigate the consequences of DBE. The operation of a category B function may avoid the need to initiate a category A function. Note that Draft of IEC 61226 Edition 4 proposed prevent not only DBE, but anticipate operation occurrence (AOO) also for categories A and B.

Category C denotes functions that play an auxiliary or indirect role in the achievement or maintenance of NPP safety. Category C includes functions that have some safety significance, but are not category A or B. If a function does not meet any of the criteria given below, then it shall be "non-classified".

IEC 61226 uses the approach based on qualitative criteria and not on probabilistic estimates (though, it indicates that probabilistic estimates can complete qualitative criteria). Most of the classification criteria are given in such a way that no additional analysis is required. For all categories examples of functions and systems, performing the functions, are given.

An I&C function shall be assigned to category A if it meets any of the following criteria:

- Functions required to reach the non-hazardous stable state, to prevent a DBE from leading to unacceptable consequences, or to mitigate its consequence.
- Functions, whose failure or spurious actuation would lead to unacceptable consequences, and for which no other category A function exists that prevents the unacceptable consequences.
- Functions required to provide information and control capabilities that allow specified manual actions necessary to reach the non-hazardous stable state.

The I&C typical functions assigned to category A are necessary for:

- Reactor shutdown and maintenance of sub-criticality.
- Isolation of containment.
- Provision of information for essential operator actions.
- Decay heat transport the ultimate heat sink.
- Typical I&C systems are as follows:
- Reactor protection system.
- Safety actuation system.
- Key instrumentation and displays to permit pre-planned operator actions that are defined in the NPP operating instructions, and that are required to ensure NPP safety in the short term.

An I&C function shall be assigned to category B if it meets any of the following criteria and is not otherwise assigned to category A:

- Functions required after the non-hazardous stable state of a DBE has been reached, to prevent it from leading to unacceptable consequences.

- Functions required to provide information or control capabilities that allow specified manual actions necessary after the non-hazardous stable state has been reached to prevent a DBE from leading to unacceptable consequences.
- Functions, the failure of which during normal operation, would require the operation of a category A function to prevent an accident which study is required.
- Functions to reduce considerably the frequency of a DBE as claimed in the safety analysis.
- Functions that provide continuous or intermittent tests or monitoring of functions in category A to indicate their continued availability for operation and alert control room staff to their failures, when no alternative means (e.g. periodic tests) are provided to verify their availability.

The I&C typical functions assigned to category B are as follows:

- Spent fuel pool cooling.
- Main cooling system isolation.
- Post-accident monitoring system.
- Functions credited in the safety analysis to prevent from escalating to accidents.
- Monitoring/controlling the handling of fuel where failure could cause radiation release or fuel degradation outside the limits and conditions of normal operation.

Typical I&C systems are as follows:

- NPP automatic control system or preventative protection system.
- Instrumentation needed to apply operating procedures for DBE.
- Safety circuits and interlocks of fuel handling systems used when the reactor shut down.

An I&C function shall be assigned to category C if it meets any of the following criteria and is not otherwise assigned to category A or category B:

- Plant process control functions so that the main process variables are maintained within the limits assumed in the safety analysis.
- Functions that provide continuous or intermittent tests or monitoring of functions in category A and B to indicate their continued availability for operation and alert control room staff to their failures, and are not classified category B.
- Functions to monitor and take mitigating action following internal hazards within the NPP design basis (e.g. fire, flood).
- Functions to warn personnel or to ensure personnel safety during or following events that involve or result in release of radioactivity in the NPP, or risk of radiation exposure.
- Functions to monitor and take mitigating action following natural events (e.g. seismic disturbance, extreme wind), etc.

The typical I&C functions assigned to category C are as follow:

- Monitoring and controlling performance of individual systems and items of equipment during the post-accident phase to gain early warning of the onset of problems, and to keep radioactive releases ALARA.
- Limiting the consequences of internal hazards.
- Those for which operating mistakes could cause minor radioactive releases, or lead to radioactive hazard to the NPP operating staff.
- Those necessary to warn of internal or external hazard.
- Access control.
- Communication to warn of significant on- or off-site releases for the purposes of implementing the NPP's emergency plan.

Typical I&C systems are as follows:

- Alarm system.
- Access control system.
- Emergency communication systems.
- Control room data processing system.
- Fire suppression systems.

Classification of Systems and Their Components

According IEC 61513 (IEC, 2011), functions, systems and equipment of NPPs may be considered from two points of view: functional or system. Categorization of functions is shown hereinbefore.

From the second point of view, a system is classified (i.e. a total set of interconnection components, for which a composition, limits and a set of functions are specified). Each I&C important to safety should be classified according to categories of its functions, which are often related to different categories. IEC 61513 introduces a concept of system class and sets a relation between a category of function and minimum required I&C class that can perform this function (Table 2).

Table 2. Correlation between classes of I&C systems and categories of I&C functions

Categories of I&C Functions Important to Safety			Classes of I&C Systems
A	(B)	(C)	1
	B	(C)	2
		C	3

As it seen from the table, the standard provides division of system into three classes, unambiguously connecting them with three function categories important to safety. Each class is characterized by a specific set of requirements for system features and capabilities, and also for design, production and quality of components. Satisfaction of these requirements allows considering a system as a proper one for performance of functions of a specific category

The requirements for the function with the highest safety category determine the class of the system.

Before classification of components notes that functions of I&C components (software-hardware complex SHC, hardware-HW, and software-SW) are offered to be considered as components of those I&C functions, in performance of which they are participating. The classification of I&C systems component is based on the principle that each system function can be assigned to unambiguously specific set of components required and sufficient for realization of this function.

If one component, for example, a sensor is included in several systems, it is referred to the highest of safety classes that it could have as a component of each of the systems separately (Figure 6). Hardware that is directly connected with components of a specific safety class is referred to the same class. An example could be signal galvanic isolation devices (also called "isolators"), providing a possibility to use the same sensors in the systems of different safety class, as shown in Figure 6. Communication lines, connecting components of one safety class, are referred to the same class; connecting components of different safety class — to the highest of the classes. The same rule is applied to equipment of data communication channels and local networks.

Figure 6. Safety classification of components, common for different I&C systems

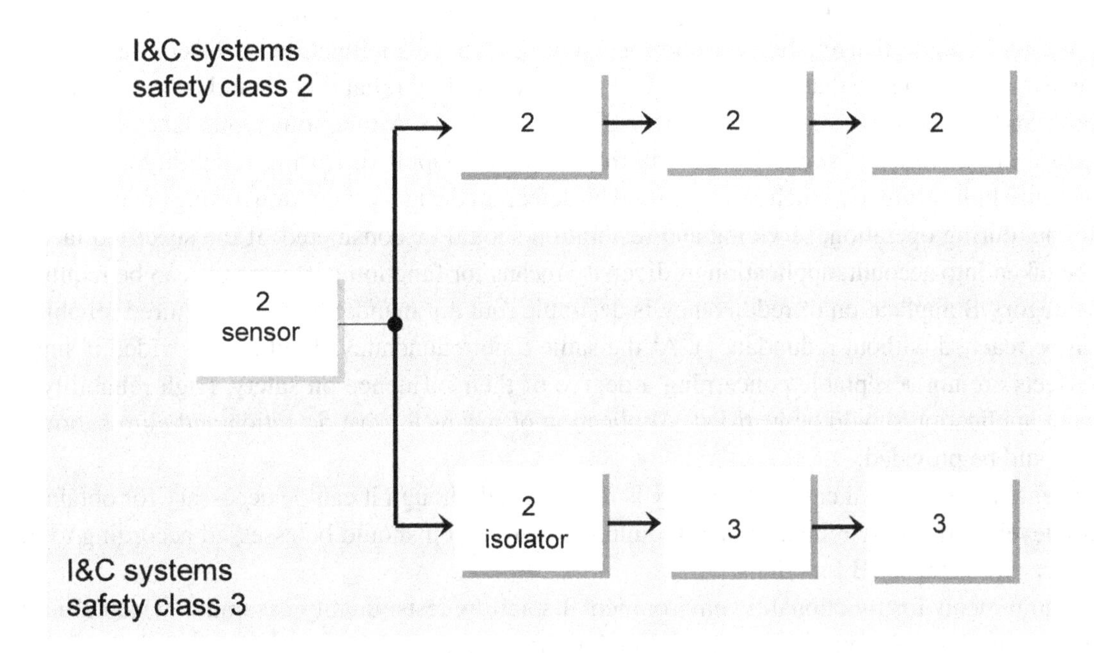

1, 2 – safety classes of components

Safety class of service items (including tools and service software) is proposed to install, taking into account:

- Safety class of I&C, SHC and HW, in which the service items is included or applied;

- Nature of performed service functions;
- Operation condition of I&C system, SHC or HW while using service item as intended;
- Type of service item connection (permanent or only within a time period of its use as intended).

Safety class of service item, if permanently connected and allowing a direct influence on parameters and characteristics of I&C, SHC, HW or SW (change of modes, set-points, adjustments, installations or program modifications etc.), are identical to the safety class of system or component. In other cases the use of service items, related to a lower safety class, can be justified.

SAFETY CLASSIFICATION AND QUALITY REQUIREMENTS

To provide proper guarantee of operability and quality of I&C system operation, corresponding to their significance with regard to safety of systems and components of different classes, considerably distinct requirements should be applied.

IEC 61226 differentiates requirements for systems and equipment, depending on a category of the function they are performing. For each category general and specific requirements are determined. It is determined, for example, that reliability requirements can be identical for functions of different categories, though a level of confidence that the function will have the required reliability should be the highest for category A and can decrease for categories B and C. For category A functions redundancy, providing performance of, at least, the single failure criterion, should be applied. During reliability assessment common cause failure effects, which can be caused by defects of design, production, fitting or errors made by personnel during operation, servicing and restoration should be considered. If the specified factors cannot be taken into account, application of diversity means for function performance can be required.

For category B application of redundancy is desirable (but not mandatory if the required reliability level can be reached without redundancy). At the same time redundancy should be provided if single failure effects are not acceptable concerning a degree of their influence on safety. High reliability of component application should be justified. Application of means for fast detection and elimination of failures should be provided.

For category C in general case redundancy is not required, though it can be necessary for obtaining reliability level, in this case reliability and redundancy estimation should be assessed according to the same rules as for category B functions.

The requirements for functionality, environmental stability, tests, quality assurance which are differentiated for each safety category are determined in the same way.

An unreasonable setting a safety class of system and its components too high will cause a considerable increase of costs at all lifecycle stages: at development, validation, and verification of new hardware and software; production, equipment tests and acceptance; design, integration, checking, installation and commissioning; servicing and maintenance during operation. On the other hand, if systems (components) are related to a lower safety class, than is actually required, insufficiently strict requirements would be specified for them; in this case the quality of such systems (components) may not be complied with their actual significance with regard to safety and (or) a sufficient confidence in such compliance would not be obtained. Unreasonable understating of class should be considered as a hidden lack of safety, i.e. violation of one of the fundamental safety concepts.

Figure 7. Ukrainian standard base, related to safety and cyber security of NPP I&C systems

		NPP functional safety	NPP life extension	Cyber security
Level 1	Ukrainian Laws	About use of nuclear energy and radiation safety		About the basic principles of cyber security assurance of Ukraine
Level 2	Regulations	NP 306.2.202. Requirements for safety of I&C systems [1]	NP 306.5.02/2.068. Requirements to safety of I&C systems life extension	*NP. Requirements to cyber security of NPP I&C systems*
Level 3	Recommended documents	GND 306.7.02/2.041 Methodic of evaluation of conformity I&C systems to safety requirements		
Level 4	Standards of NPP operating organization	SOU NAEK100. I&C. Technical requirements [8]	MT-T.0.08.168. Methodic of statistical analysis for I&C life extension *Requirements to online monitoring for I&C maintenance and life extension*	
Out of levels	SSTC NRS documents	*SSTC NRS. Methodics for I&C safety assessment in the licensing process*		

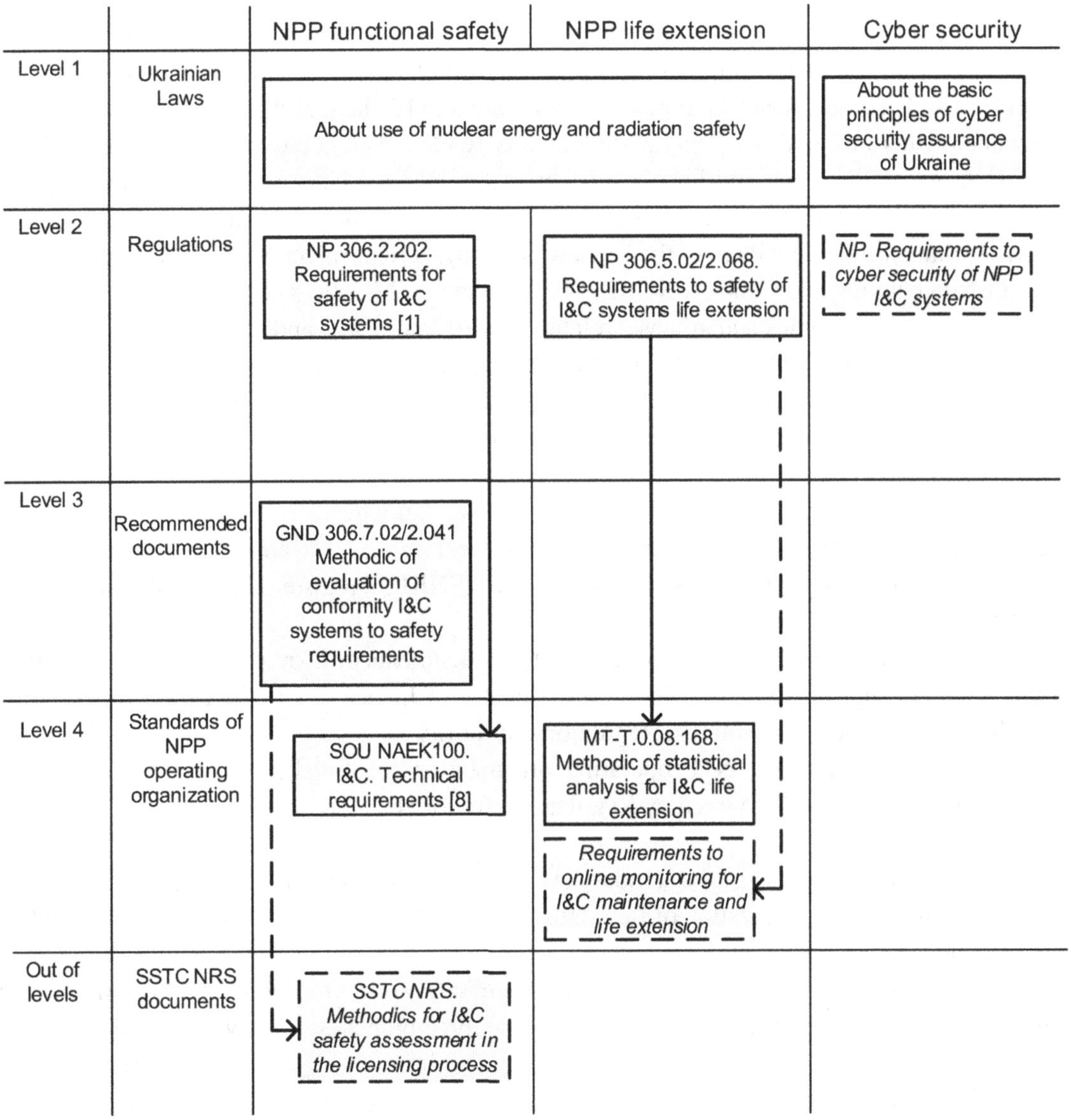

SHORT REMARKS ON THE NATIONAL STANDARD BASES

Every country, where there are NPP's, has its own regulatory authority and own standard base related as to NPP safety as whole, as to NPP I&C systems.

A number of IEC standards devoted to software (IEC, 2006, IEC, 2004,b), to hardware (IEC, 2007,c), to classification of functions (IEC, 2009,b), to control rooms (IEC 2009,a), to separation (IEC, 2004,a), to coping with common cause failures (IEC,2007,b), to data communication (IEC, 2009,c), etc., have been published as EN standards and implemented as nationally in 31 European countries.

Though, it should be noted that the IEC I&C classification is not used, for example, in USA, where only one safety class – 1E is determined. It includes electrical equipment and systems required for emergency reactor shutdown, isolation of containment, reactor core cooling, heat removal from containment and reactor or other actions important for prevention of radioactive materials into environment. (ANSI-IEEE). Other systems and equipment are not related to IE class. NPP I&C standard base in USA includes documents of Codes of Federal Regulations (CFR), American National Standard Institute (ANSI), Institute of Electrical and Electronics Engineers (IEEE), United States Nuclear Regulatory Commission (U.S.NRC), that included Regulatory Guides, Publications prepared by NRC staff, where we emphasis Standard Review Plan for the Review of Analysis Reports for Nuclear Power Plants: LWR Edition- Instrumentation and Controls NUREG-0800, Chapter 7 (NUREG, 2016), etc.

As an example of national standard base, related to NPP I&C safety and security, we show structure of Ukrainian standard base, what was created after Ukrainian independence in 1991. These documents are located on different levels of hierarchical pyramid of documents, which are founded Ukrainian regulatory framework on nuclear and radiation safety (Figure 7):

- Level 1. Ukrainian laws, international conventions and agreements of Ukraine.
- Level 2. The regulations- obligatory normative-legal acts of Ukrainian Regulatory Authority (State Nuclear Regulatory Inspectorate of Ukraine – SNRIU), registered in the Ministry of Justice of Ukraine.
- Level 3. Recommended, explanatory, information documents of Ukrainian Regulatory Authority, which are not subjects to registration in the Ministry of Justice of Ukraine; Ukrainian state standards, recommended by Ukrainian Regulatory Authority.
- Level 4. The standards of NPP operating organization (National Nuclear Energy Generating Company "Energoatom"), agreed by Regulatory Authority.

Furthermore, there are group of the standards which are situated out of level, e.g. the documents issued by technical support organization of Regulatory Authority – State Scientific and Technical Center for Nuclear and Radiation Safety (SSTC NRS).

The documents which have a immediate attitude toward safety important NPP I&C systems are located at level 2, 3, 4 and out of level. These documents define three branches of activity:

- NPP I&C systems safety.
- Life extension of NPP I&C systems.
- Computer security of NPP I&C systems.

The documents are shown at Figure 7 on intercrossing of the corresponding level and the branch.

I&C systems safety. The first Ukrainian regulations with requirements to NPP I&C safety NP 306.5.02/3.035 (NP, 2000) "Requirements to nuclear and radiation safety of instrumentation and control systems important to NPP safety" was issued in 2000. This document was used for design, manufacturing and safety assessment of I&C systems for new NPPs units: Khmelnitsky-2, Rivne-4 and for I&C systems modernization on the other Ukrainian NPPs. There were many changes in computer technique, as well as in international standards since 2000. These changes stimulated the necessity to update this document. New regulations NP 306.2.202 (NP, 2015) was issued in 2015, is located in level 2 at Figure 7.

The base of elaboration of standard (NP, 2015) were:

- Experience of using NP 306.5.02/3.035-2000 for 14 years at 15 NPP units.
- IAEA safety standard SSG-39 (IAEA, 2016) (this standard was elaborated with participation of SSTC NRS representative).
- The standards of International Electrotechnical Commission, Technical Committee TC45 "Nuclear Instrumentation" (in the first case- IEC 61513, ed.2 (IEC, 2011).
- Publications, devoted to NPP I&C systems (e.g. Yastrebenetsky (2004), Yastrebenetsky (2011).

The main differences of regulations NP 306.2.202 in comparison with NP 306.5.02/3.035 are as follow:

- NP 306.2.202 contains the requirements not only to I&C systems and their components, but also to all stages of their life cycle –the processes of system design, manufacturing of new components, qualification, implementation, operation, etc.
- The requirements to systems and their components were divided on general regulatory requirements, which related to level 2 (document of Regulatory Authority), and document of operating organization with more detailed quantity requirements, which related to level 4.
- The requirements took in account the lessons due Fukushima accident.
- Safety classification of systems functions became similar to last edition of standard IEC 61226 "Nuclear power plants. Instrumentation and control systems important to safety. Classification of instrumentation and control functions" and to corresponding European standard EN IEC 61226.
- NP 306.2.202 establishes the requirements to post-accident monitoring systems, including the requirements for archiving and storage of data needed to analyze of the causes and the process of passing of the accidents.
- The list of electromagnetic interferences possible in normal operation and limiting conditions is expanded.
- The classification criteria's and testing methods (seismic resistance, electromagnetic disturbances, etc.) are more tougher.
- Use of FPGA-technology for the implementation in safety important I&C systems, including the features of the development and implementation of FPGA-technology are taken into account.

Document GND 306.7.02/2.041 "Methodic of evaluation of information and control systems conformity to safety requirements", was issued simultaneously with NP 306.5.02/3.035. GND 306.7.02/2.041 was located at level 3. Now SSTC NRS is currently reviewing this document taking in account international and national practice.

Life extension of I&C systems. As rule, the life time of units more then life time of NPP I&C systems and their components. Furthermore, unit life extension is an urgent problem for majority of countries, including Ukraine. The requirements to safety important I&C life extension are contained in the documents:

- Regulation NP.306.5.02/2.068 "The Requirements to order and content of work for life extension of safety important instrumentation and control systems" (level 2).
- Operator's methodic MT-T.08.168-05 "Methodic of statistical analysis of data about equipment of information and control systems, electrical systems and armature electrics drives for failure point process trend detection for definition of life extension possibility".

The main actions which need for the definition of I&C life extension possibility are the inspection of technical state of I&C system equipment and analysis of operating reliability, including statistical analysis of trend of failures

Cyber security of NPP I&C systems. NPP and other nuclear facilities are the objects of safety critical infrastructure, which should be protected from cyber-attacks. I&C systems which take part in the control and monitoring of technological processes at NPP require special attention. Information about regulatory requirements to cyber security of NPP I&C systems will be described in the next chapters.

SOLUTIONS AND RECOMMENDATIONS

International normative bases of IAEA and IEC are sufficiently coordinated and continue developing actively. New IEC standards and regular revisions of active ones, periodic redevelopment of IAEA documents prove these statements.

Intensive development of information technologies, Industrial 4.0 appearance set new tasks for nuclear applications. The most important directions of work which than have to be reflected for standardization applicable to NPP I&C systems can be recommended

- More attention is given to harmonization of national documents of different countries with international standards.
- Use big data methods during creation new I&C systems, as for improvement of operating I&C systems maintenance.
- Use drones fleet in the first case in post-accident monitoring systems.

One of the interesting subjects for research directions – comparison of standards devoted to safety important/safety related systems for different applications (e.g. Biscoglio & Fusani, 2010).

FUTURE RESEARCH DIRECTIONS

- Research and development, elaboration of international and national standards devoted to I&C systems for small modular reactors (SMR's). SMR's one of the perspective ways in nuclear industry in many countries.
- Elaboration of the set of standards devoted to cyber security (requirement to cyber security, cyber security control, coordination safety and cyber security, etc.).
- Research and development, elaboration of standards about wireless devices.
- Elaboration of standards about qualification of platforms for NPP I&C systems important to safety.
- Elaboration of standards devoted to spent fuel pools instrumentation, to accident monitoring instrumentation.

CONCLUSION

The goals of standards, that pertain to NPP I&C systems, are:

- Establish requirements for I&C systems that are needed to assure NPP safety.
- Make available to designers sufficiently effective methods for elaboration of I&C systems and their components in accordance with the acting requirements.
- Establish methods for checking conformity of I&C systems with the requirements.
- Assure certain frameworks for interface between NPP plant, different participants of I&C systems development, and the Regulatory Authority.
- Establish requirements for I&C systems operation.

The International standard bases of IAEA and IEC related to NPP I&C systems, are described in this chapter. The IEC standard base is very advanced and includes more than 70 standards. These bases systematically are improved and supplemented. A great part of new additions concerns to new safety requirements after accident on Fukushima-1. The main position in IAEA standard base, related to NPP I&C, will rank IAEA safety standards SSG-39 (IAEA, 2016) "Design of Instrumentation and Control Systems for Nuclear Power Plants".

Besides international standards, all countries used national ones. The harmonization of national and international standards is an urgent task.

Note that some of the standards in force in the United States (developed by the American Institute of Electrical and Electronics Engineers – IEEE, and the American Society of Mechanical Engineers – ASME and others) de facto have become international standards and have been widely adopted not only in the United States but in other countries, successfully supplementing the IEC standards. Many documents of the U.S. Nuclear Regulatory Commission (U.S. NRC) have received international use. An example of a U.S.NRC document which has been widely disseminated is NUREG-0800, which contains standard plans for safety analysis of different structures, components, equipment and systems. Section 7.0 of this document is devoted to I&C systems.

ACKNOWLEDGMENT

The author of this chapter would like to recognize Prof. Michael Yastrebenetsky, State Scientific and Technical Center for Nuclear and Radiation Safety, Ukraine, who was instrumental in the researching and writing of this chapter.

REFERENCES

Biscoglio, I., & Fusani, M. (2010). Analyzing quality aspects in safety-related standards. In *Seventh American Nuclear Society International Topical Meeting on Nuclear Plant Instrumentation, Control and Human-Machine Interface Technologies NPIC&HMIT 2010*. Las Vegas, NV: American Nuclear Society.

IAEA. (1980). *50-SG-D3. Protection systems and related features in nuclear power plants*. Vienna, Austria: IAEA.

IAEA. (1984). *50-SG-D8. Safety related instrumentation and control systems for nuclear power plants*. Vienna, Austria: IAEA.

IAEA. (1988). *50-C-D. Code on the safety on nuclear power plants: Design*. Vienna, Austria: IAEA.

IAEA. (1998). *TECDOC-1016. Modernization of Instrumentation and control in nuclear power plants*. Vienna, Austria: IAEA.

IAEA. (1999a). INSAG-12. Basic safety principles for nuclear power plants. 75-INSAG-3, Rev. 1, Vienna, Austria: IAEA.

IAEA. (1999b). *Modern instrumentation and control for nuclear power plants: A Guidebook*. Technical report series N°387. Vienna, Austria: IAEA.

IAEA. (2000). *NS-G-1.1. Software for computer based systems important to safety in nuclear power plants: Safety guide*. Vienna, Austria: IAEA.

IAEA. (2002). *NS-G-1.3. Instrumentation and control systems important to safety in nuclear power plants: Safety guide*. Vienna, Austria: IAEA.

IAEA. (2006). *SF-1. Fundamental safety principles*. Vienna, Austria: IAEA.

IAEA. (2009). Safety assessment for facilities and activities: General safety requirements. IAEA safety standards series No. GSR Part 4. Vienna, Austria: IAEA.

IAEA. (2010). Governmental, legal and regulatory framework for safety: General safety requirements. IAEA safety standards series No. GSR Part 1, Vienna, Austria; IAEA.

IAEA. (2011, a). SSR-2/2. Safety of nuclear power plants: commissioning and operation. Vienna, Austria: IAEA.

IAEA. (2011, b). *NP-T-3.12. Core knowledge on instrumentation and control systems in nuclear power plants*. Vienna, Austria: IAEA.

IAEA. (2012). *SSR-2/1. Safety of nuclear power plants: design*. Vienna, Austria: IAEA.

IAEA. (2013b). *The Statute of the IAEA*. www.iaea.org/About/statute.html

IAEA. (2014). *SSG-30. Safety classification of structures, systems and components in nuclear power plants*. Vienna, Austria: IAEA.

IAEA. (2014). IEC (1980). IEC 60780. Nuclear power plants – Electrical equipment of the safety system – Qualification. Ed. 2. IAEA.

IAEA. (2016). *SSG-39. Design of instrumentation and control systems for nuclear power plants*. Vienna, Austria: IAEA.

IEC. (2004a). IEC 60709. Nuclear power plants – Instrumentation and control systems important to safety – Separation. Ed. 2. IEC.

IEC. (2004b). *IEC 62138. Nuclear power plants – Instrumentation and control important for safety – Software aspects for computer-based systems performing category B or C functions*. IEC.

IEC. (2006). *IEC 60880. Nuclear power plants – Instrumentation and control systems important to safety – Software aspects for computer-based systems performing category A functions*. IEC.

IEC. (2007a). *IEC 62342. Nuclear power plants – Instrumentation and control systems important to safety – Management of aging*. IEC.

IEC. (2007b). *IEC 62340. Nuclear power plants – Instrumentation and control systems important to safety – Requirements for coping with common cause failure (CCF)*. IEC.

IEC. (2007c). *IEC 60987. Nuclear power plants – Instrumentation and control important to safety – Hardware design requirements for computer-based systems*. IEC.

IEC. (2008). *IEC 61508. Functional safety of Electrical/Electronic/Programmable Electronic safety-related systems*. IEC.

IEC. (2009a). *IEC 60964. Nuclear power plants – Control rooms – Design*. IEC.

IEC. (2009b). IEC 61226 Nuclear Power Plants – Instrumentation and control important to safety–Classification of instrumentation and control functions. Ed. 3. IEC.

IEC. (2009c). IEC 61500. Nuclear power plants – Instrumentation and control important to safety – Data communication in systems performing category A functions. Ed. 2. IEC.

IEC. (2011a). IEC 61513. Nuclear power plants – instrumentation and control important to safety – General requirements for systems. Ed. 2. IEC.

IEC/IEEE. (2011). *IEC/IEEE 62582 Nuclear power plants – Instrumentation and control important to safety – Electrical equipment condition monitoring methods – Part 1: General. Part 2: Indenter Modulus. Part 4*. IEC/IEEE.

IEC/IEEE. (2016). *IEC/IEEE 60780-323 - Nuclear facilities - Electrical equipment important to safety-Qualification*. IEC/IEEE.

IEC/IEEE. (2017). IEC 63147 Edition 1.0 2017-12 IEEE Std 497: IEEE/IEC International Standard - Criteria for accident instrumentation for nuclear power generation stations. IEC/IEEE.

Johnson, G. (2002). Comparison of IEC and IEEE standards for computer-based control systems important to safety. In *CNRA/CSNI workshop on licensing and operating experience of computer-based I&C systems. Workshop Proceedings*. AEN/NEA.

Johnson, G., & Duchac, A. (2017). The Development of the New Idea Safety Guide for Design of Instrumentation and Control Systems for Nuclear Power Plants. *Reliability, Theory & Applications*, *1*, 2017.

NP. (2000). NP 306.5.02/3.035. Requirements for nuclear and radiation safety to information and control systems important to safety of nuclear power plants. Kyiv, Ukraine: NP.

NP. (2015). NP 306.2.202 Requirements for nuclear and radiation safety to information and control systems important to safety of nuclear power plants. Kyiv, Ukraine: NP.

NUREG. (2016). *NUREG-0800. U.S. Nuclear Regulatory Commission Regulations: Standard Review Plan for the Review of Safety. Analysis Reports for Nuclear Power Plants: LWR Edition*. NUREG.

Yastrebenetsky, M. (Ed.). (2004). *Nuclear power plants safety. Instrumentation and control systems*. Kyiv, Ukraine: Technica.

Yastrebenetsky, M. (Ed.). (2011). *Nuclear power plants safety. Nuclear reactors control and protection systems*. Kyiv, Ukraine: Osnova-Print.

ADDITIONAL READING

IEEE Std. 1023 (2004). *IEEE Recommended Practice for the Application of Human Factors Engineering to Systems, Equipment, and Facilities of Nuclear Power Generating Stations and Other Nuclear Facilities.*

IEEE Std. 308 (2001). *IEEE standard criteria for class 1E power systems for nuclear power generating stations.*

IEEE Std. 344 (2004). *Recommended Practice for Seismic Qualification of Class 1E Equipment for Nuclear Power Generating Stations.*

IEEE Std. 352 (2001). *Guide for General Principles of Reliability Analysis of Nuclear Power Generating Station Safety Systems.*

IEEE Std. 379 (2000). *Standard Application of the Single-Failure Criterion to Nuclear Power Generating Station Safety Systems.*

IEEE Std. 384 (2008). *Standard Criteria for Independence of Class 1E Equipment and Circuits.*

IEEE Std. 577 (2004). *IEEE Standard Requirements for Reliability Analysis in the Design and Operation of Safety Systems for Nuclear Facilities.*

IEEE Std. 603 (2009). *IEEE Standard Criteria for Safety Systems for Nuclear Power Generating Stations.*

IEEE Std. 627 (2010). *IEEE Standard for Qualification of Equipment Used in Nuclear Facilities.*

IEEE Std. 7-4.3.2 (2016). *Standard Criteria for Digital Computers in Safety Systems of Nuclear Power Generating Stations.*

KEY TERMS AND DEFINITIONS

Category of I&C Functions: Group of I&C functions, defined according to their importance to safety.

Function: Specific purpose or objective to be accomplished, that can be specified or described without reference to the physical means of achieving it.

I&C Systems Important for Safety: I&C systems whose malfunction or failure could lead to undue radiation exposure of the site personnel or member of public, and those I&C systems that prevent anticipated operation occurrences from leading to f unacceptable consequences.

National Regulatory Authority: Authority designated by government for regulatory purposes for safety assurance.

Regulatory Requirement (Regulations): A requirement which is established by the National Regulatory Authority.

Safety Classification: Differentiation of systems or their components into classes, depending on their impact on NPP safety.

Safety Fundamentals: A document which contains fundamental principles of safety assurance and safety objectives.

Standard Base: A set of standards/regulations which is established by government or international organizations for specific area of activity.

Chapter 3
Description of Requirements to Safety Important I&C Systems

Yuri Rozen
https://orcid.org/0000-0002-9366-5794
State Scientific and Technical Center for Nuclear and Radiation Safety, Ukraine

Serhii Trubchaninov
State Scientific and Technical Center for Nuclear and Radiation Safety, Ukraine

ABSTRACT

The requirements for safety important instrumentation and control (I&C) systems, their components (software and hardware complexes, equipment), and processes of their development are provided. The reliability of I&C systems and their components are ensured by the requirements for prevention and protection against common cause failures, compliance with the single failure and redundancy criteria, and prevention of human errors. The requirements for operational stability of the components cover resistance to environmental impacts, mechanical (including seismic) and other external impacts, and insensitivity to changes in power supply parameters and electromagnetic interference. Requirements for the performance quality include characteristics pertaining to the accuracy and performance time of functions. Requirements for independence ensure that the system remains operable if its components fail and that the effects of electromagnetic radiation, fires, breakdown of insulation, and leaks in cables are minimized.

INTRODUCTION

The steadily growing role of modern instrumentation and control systems in ensuring the safety and security of nuclear power plants requires that:

- the properties of each system and its components comply with requirements of national and international regulations, rules, and standards on nuclear safety at all life cycle stages and

DOI: 10.4018/978-1-7998-3277-5.ch003

- the procedure for the development, design, manufacture, testing, acceptance, and operation of the system itself and its components, established by regulations, rules, and standards on nuclear safety, be observed.

The purpose of this chapter is to show the requirements for the properties (parameters and characteristics) of I&C systems and their components that ensured safety when used for automation of nuclear power plants (NPPs).

BACKGROUND

The experience of the world community in standardizing, ensuring, and assessing the safety of I&C systems has been accumulated in the requirements of international and national standards and regulatory documents and taken into account in the large-scale upgrading process carried out as part of the programs to improve the safety and extend the life of existing power units. The upgrade process was based on the use of modern information technologies, new electronic components and computer facilities, optical transmission networks, and diagnostic, display, and archive computer tools.

The requirements set forth below for I&C systems important to safety and their components are determined by the applicable requirements presented in:

- Safety Standards of International Atomic Energy Agency (IAEA): IAEA, 2016a; IAEA, 2016b; IAEA, 2016c; IAEA, 2010, etc.
- Standards of the International Electrotechnical Commission (IEC): IEC, 2011; IEC, 2009; IEC, 2007; IEC, 2006; IEC, 2005, etc., and identical European standards related to of I&C systems important for NPP safety: EN, 2013; EN, 2010a; EN, 2010b; EN, 2015, etc.
- Reports of the Western European Nuclear Regulators Association (WENRA): WENRA, 2014a; WENRA, 2014b, WENRA, 2012.
- Regulations, rules, and standards on nuclear and radiation safety in force in Ukraine: NP, 2008; NP, 2015; NP, 2016a; NP, 2016b, etc.
- IEC industrial standards, European standards, and state standards of Ukraine identical to them.

Requirement Selection Criteria

General criteria for the selection of safety requirements for the I&C systems of NPP have been developed upon analysis and summary of national and international standards, regulatory documents, safety recommendations of international organizations, as well as other sources. Although these criteria were identified for I&C systems, they generally show how the choice of requirements for each safety important system and piece of equipment can be justified. The following criteria were used in the development of regulatory documents:

- **Necessity:** criterion according to which the use of an object that does not comply with a given requirement can cause violation of limits and/or conditions of NPP safe operation.
- **Completeness:** criterion according to which the use of an object that complies with all requirements, in all probability, does not cause violation of limits and/or conditions of NPP safe operation.

- **Sufficiency:** completeness criterion that belongs to an individual requirement.
- **Ambitionless:** criterion according to which the least possible rigid requirements satisfying the sufficiency criterion ("reasonable sufficiency") are accepted.
- **Correctness:** consistency of a requirement with identical requirements established in current regulatory documents of the same or higher hierarchical level.
- **Progressivity:** compliance of a requirement with the achieved level of science and technology.
- **Pithiness:** distinctness and objectivity of an object's properties for which the requirement is established.
- **Verifiability:** potential for evaluating compliance of an object with each requirement based on facts, tests, and/or analysis.
- **Traceability:** verifiability criterion that belongs to different stages of an object's lifetime.
- **Clarity:** quality criterion for the wording of requirements to be understood by specialists without further explanations.
- **Unambiguity:** quality criterion for the wording to prevent different interpretations of the same requirement.
- **Openness:** criterion according to which a requirement is to be known to each participant of the work performed at any stage of an object's life cycle.
- **Categorical:** criterion according to which each object should meet all requirements established.

The necessity, completeness, and sufficiency criteria were taken into account in determining the composition of requirements; the ambitionless, correctness, progressivity, pithiness, verifiability, and traceability criteria in establishing the semantic content of each requirement; and the clarity and unambiguity criteria in drafting requirements in the documents being developed. The pithiness, verifiability, traceability, clarity, unambiguity, openness, and categorical criteria make it possible to assess the compliance of systems and equipment with established requirements in the licensing process.

Separation of Requirements

In development of the regulatory framework, a principle was proposed and implemented for the separation of requirements for I&C systems important to safety and their components into:

- *Regulatory*, which are based on well-established, universally recognized, and proven fundamental and organizational and technical principles and do not need frequent revision and are recognized as necessary to ensure safety.
- *Technical*, which establish standards, test methods, evaluation rules, and criteria for conformity of products and documents subjected to more frequent changes and not excluding alternative recommendations that allow the operating organization and/or the end user to make independent decisions, if necessary, aimed at ensuring safety.

Regulatory requirements are established in regulatory documents that are developed, approved, and updated by a state organization dealing with the development and implementation of state policy for the safety of nuclear energy (regulatory body). Technical requirements ensuring compliance of I&C systems and their components with regulatory requirements are established in standards of the operating organization. For I&C systems and their components, such regulatory documents in Ukraine are, respectively, NP-2015 and SOU-2016 (Figure 1).

Figure 1. Norms, rules and standards governing the requirements for I&C systems safety

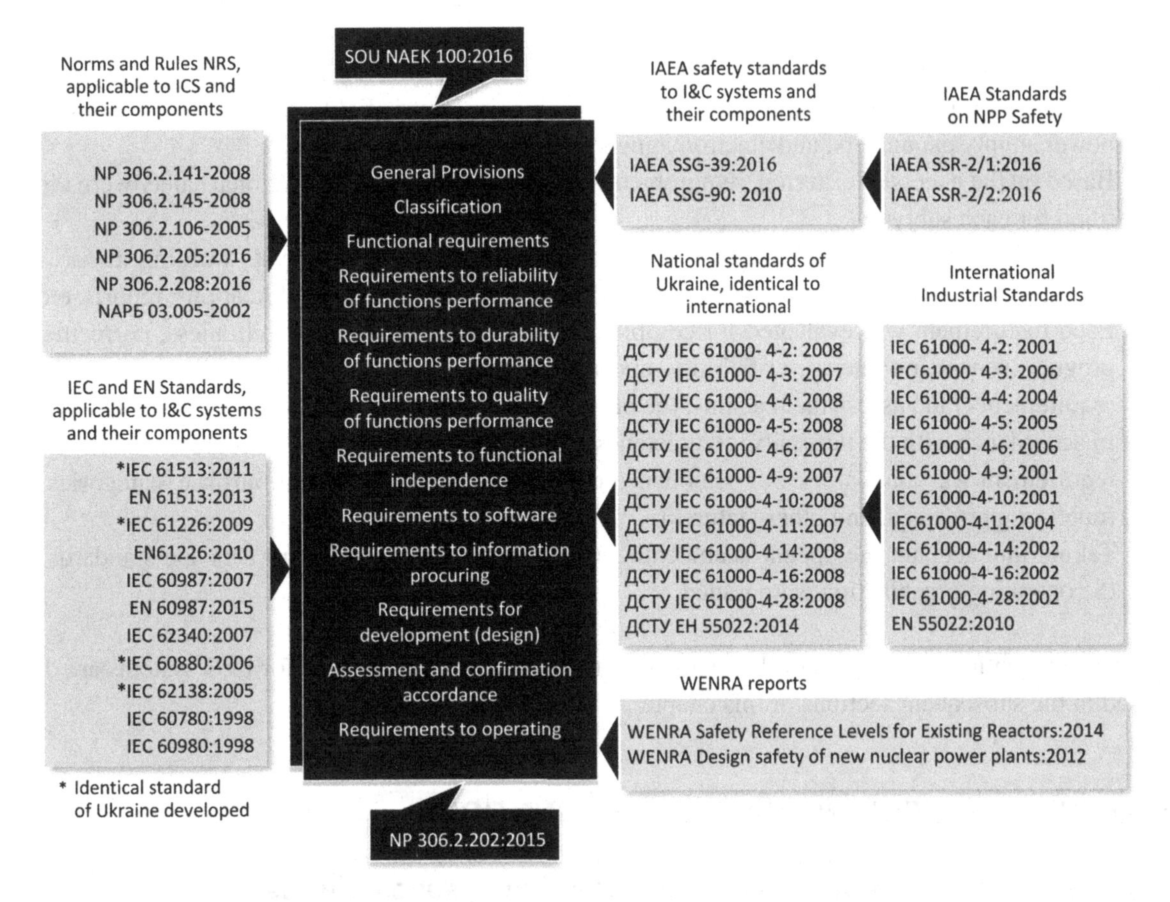

Regulatory and technical requirements are differentiated in the regulatory documents taking into account the classification adopted according to the categories of functions performed. The classification in accordance with IEC-2009 takes into account the role of each function in ensuring and maintaining safety and consequences caused by failure to perform or erroneous fulfillment of the function.

Procedure for the Development of Regulatory and Technical Requirements

The experience of international society in ensuring the safety of NPPs, including the safety of I&C systems and their components and software, accumulated in requirements of international and national regulations, rules, and standards on nuclear safety, was considered in the large-scale upgrade process carried out at all Ukrainian NPPs within the programs for safety improvement and lifetime extension for operating power units. These activities contributed to the accumulation of national experience in standardization, assurance, and assessment I&C safety.

The following procedure was used in the development of regulatory and technical requirements:

- A list of regulatory documents (international and national regulations, rules, and standards) with requirements for I&C systems and their components and/or software was compiled, and full lists of such requirements were prepared.
- The requirements were structured by type (requirements for reliability, life, quality, independence of functions, software and information support, development, evaluation, and confirmation of compliance, operation).
- Each type of requirement was divided into subtypes (for example, resistance requirements covered resistance to environmental impacts, mechanical stresses, electric fields, special media, changes in power supply parameters, and electromagnetic interference).
- Based on the necessity criterion, requirements important for ensuring functional safety were identified for each subtype.
- A thorough analysis of each requirement was carried out by comparing its wording in various regulatory documents with involvement, if necessary, of other sources (publications, reports, etc.).
- Each requirement was evaluated for compliance with the sufficiency, ambitionless, correctness, progressiveness, pithiness, verifiability, and traceability criteria to propose the wording of the requirements (and assess them against the clarity and unambiguity criteria).
- In accordance with the necessity, completeness, and sufficiency criteria, the degree to which each requirement was mandatory was established depending on the functional purpose, categories of functions performed, and other features).
- Taking into account the openness and categorical criteria, draft regulatory acts and standards of the operating organization were issued.

The composition and content of the regulatory and technical requirements for I&C systems are discussed in the subsequent sections of this chapter.

REQUIREMENTS FOR RELIABILITY OF FUNCTIONS

Prevention of and Protection Against Common Cause Failures

Measures must be provided and implemented for protection of I&C systems and SHCs against loss of their ability to perform the necessary category A and B functions in case of a simultaneous failure of two or more elements caused by the same event or cause. For I&C systems and SHCs participating in the performance of category C functions, requirements for the prevention of and protection against common cause failure are recommended.

The following cases are considered common cause failures:

- Appearance of unrevealed (latent) errors that may occur in the design and development of hardware and software, production, delivery, assembly, integration, adjustments, maintenance and/or recovery of the I&C system (SHC).
- Interference of I&C system or SHC components through common parts of input, output, power supply, and ground circuits or over the space of rooms that can be manifested during operation, connection, and disconnection or duc to a failure.
- Influence of closely located electrical equipment and/or electrical power cables.

- Deviation from normal operating conditions that can be caused by process equipment failures, abnormal acts of nature (earthquake, lightning stroke), dangerous external or internal events (fire, flood), failures of power supply systems, ventilation systems, etc.

Prevention of and protection against common cause failures are ensured by compliance with the requirements of regulatory documents for the development (design), assessment and confirmation of compliance, and inspections and maintenance of I&C systems and SHCs and compliance with the requirements for diversity, independence, prevention of human errors, and resistance to external factors.

Compliance With the Single Failure Criterion

I&C systems and SHCs should perform all specified functions of categories A and B in any postulated initiating event (PIE) combined with a failure of one (any) element or one human error independent of this PIE, which is caused by a wrong action or failure to perform the prescribed action. Additionally, potential failures that may be caused by an initiating event and latent failures not detected by the built-in diagnostic tools are taken into account.

Single failures of passive elements that are properly designed, manufactured, and controlled can be not considered if the probability of their failures (considering loads and environmental conditions, including impact of the PIE itself) does not exceed the minimum allowed value for the entire period of time after PIE within which operation of these elements is required. The criterion is applied independently of a single failure type (failure to operate, spurious operation) and should consider cases when a failure of one element directly or indirectly causes dependent failures of other elements.

Compliance with the single failure criterion means that an I&C system or SHC can perform all required category A and B functions in case of the worst possible configuration: for example, if an initiating event occurred when individual redundant parts of the I&C system or SHC were taken out of service for inspection, maintenance, or recovery. As an exception, incompliance with the single failure criterion is allowed for a limited time necessary for inspection, maintenance, or recovery upon agreement with the State Nuclear Regulatory Inspectorate. This time is defined through engineering evaluation of reliability so that the potential for a single failure within this period does not exceed the minimum allowed value established for the appropriate function.

The outputs of the failed or affected redundancy part should be automatically established and maintained in the sates that are defined as the most acceptable in terms of safety.

The single failure criterion is also used with respect to category B functions, but the potential for latent impairment of performance is not commonly taken into account.

Compliance With the Redundancy Principle

Compliance with the reliability indicators and single failure criterion for category A and B functions is ensured by redundancy (use of additional tools and/or capabilities that are redundant to those that are minimally required for function performance). Redundancy involves several identical or different components that form redundant channels of an I&C system (or SHC), where each may perform the required function regardless of the technical condition of other channels. Redundant power supply, sources and receivers of data, and connecting lines used for transmission of signals and messages between I&C systems, SHCs, or channels participating in the performance of category A functions, especially those access to which in power unit operation is impossible (for example, inside the containment), are also provided.

The redundancy method is selected so that increase in the reliability of the required functions is not accompanied by increase in the likelihood of erroneous actions, meaning that the acceptable ratio should be ensured between the probability of failures to actuate and spurious actuation. The efficiency of redundancy should be ensured by:

- Observance of the independence principle for redundant channels, data sources and receivers, and power supply connecting lines.
- Continuous automatic monitoring of technical condition of redundant channels and diagnostics of operability failures at the level of removable parts of each channel.
- Operability recovery by rapid replacement of the failed removable part of the redundant channel without taking other channels out of service.

Redundancy of the emergency reactor protection (ERP) function is provided in the following way:

- ERP function realized by at least has two independent SHCs, each SHC having not less than three independent redundant channels.
- Power supply of each channel is provided through two inputs from different reliable sources.
- Each channel has a complete set of input signals and generates an output signal upon any condition for initiating protective actions.
- In case of disconnection or failure of one channel (without taking the entire SHC out of service), a protection trip signal should be set automatically at the output of this channel.
- Each SHC should initiate protective actions upon signals for actuation of the channels in accordance with the logical condition selected by reliability analysis (at least "two-out-of-three").
- A command initiating protective actions should be transferred from each SHC to the executive system through at least two lines.
- The executive system should perform protective actions upon a command obtained from any SHC.
- SHC taken out of service should not, in any conditions, issue commands to initiate protective actions and should not prevent the executive system from performing commands received from another SHC.

The redundancy of neutron flux monitoring function provides that:

- The neutron flux density and rate of its change should be monitored by two independent sets of equipment (SHC), each having at least three independent channels.
- As a data source for backup control panel, an individual set of equipment with three independent monitoring channels is provided.
- For monitoring of the neutron flux density in refueling of nuclear fuel, an additional system having at least three independent channels can be provided.

For category B functions, redundancy can be justified, for example, by requirements for reliability indicators, lack of time for alternative actions in case of failure to perform the functions, and severity of potential consequences of failures.

The redundancy of category C functions is not commonly provided but can be required in some cases to ensure the reliability (non-failure operation) of these functions.

Compliance With the Diversity Principle

Diversity is the property that refers to a group of two or more I&C systems or SHCs that simultaneously and independently perform functions that are identical in relation to the safety objectives. The diversity principle envisages that parts of this group differ from each other by the performance, structure, components, software, and/or other features or achieve their goals in different ways. Actual differences between the parts of this group determine the type of diversity:

- Design: different methods (approaches) are applied to the design of hardware and/or software of each element in the group.
- Functional: different algorithms are implemented by group elements.
- Signal: each element of the group initiates one and the same protective action with different sets of input information.
- Hardware: component parts in different elements of the group differ by the performance principle, are manufactured by different techniques, and/or supplied from different manufactures.
- Software: different software modules, programming languages, and instrumental tools for software development in each element of the group are used.
- Subjective: elements that form the group and/or their component parts are developed by different teams.

The most efficient combination includes several types of diversity. Diversity is used to:

- Minimize the effect of latent errors that can occur in the design and packaging of I&C systems and/or development and manufacture of SHCs and manifest themselves as a common cause of simultaneous failure of several redundant group elements.
- Overcome difficulties related to ensuring and/or demonstrating necessary confidence in the absence of latent errors.
- Compensate for cases when complex I&C systems or SHCs are not sufficiently proven by actual operating experience.

Compliance with the diversity principle is mandatory for the group of I&C systems or SHCs taking part in the performance of the emergency reactor protection function. For other functions of category A, the need for or suitability of adequate diversity is determined upon probability analysis of “latent” errors made in the development (design) and manufacture that may cause simultaneous failures of several group elements, severity of probable failure effects, degree of operating approval, etc.

The diversity principle may be not observed if the risk of possible common cause failures is recognized as more acceptable compared to the significant increase in the cost of design, development, and operation of diversified I&C systems and SHCs performing the same functions. For a group of independent elements participating in the performance of category B and C function(s), compliance with the diversity principle is not mandatory.

Prevention of Human Errors

Prevention of human errors is one of the important factors of safety assurance in power unit control in operational states, accident management, as well as maintenance of I&C systems and SHCs and changes in their configuration.

Operational personnel should receive full, timely, and accurate data of set and current values of monitored process, neutronic, and other parameters, state of power unit structures, systems, and equipment, initiating events, actions of control systems sufficient for power unit control, timely detection and elimination of operational occurrences, prevention of emergencies, accident management, and performance evaluation.

The redundancy of I&C components that take part in data display functions related to category A (including facilities placed in the main control room) is provided. To eliminate the potential for false interpretation of data when one of the redundant channels fails or is taken out of service, the values of each parameter obtained from all channels are displayed simultaneously or only values that are considered reliable upon automatic check or are obtained from operable channels are shown.

Data on human actions that can affect safety should be immediately transferred to the main control room and emergency control room. If process systems and equipment important to safety can be controlled not only form the main or emergency control room but also from other places, automatic visual identification of the place from where control is currently performed (the potential for simultaneous control from different places should be excluded) is provided.

Personnel should be immediately notified about malfunctions of components in I&C systems (SHCs, peripheral equipment) or software that impede the performance of their category A and B function(s).

Errors that can lead to a nuclear accident and cause risk of personnel exposure to radiation in the refueling process are prevented by the refueling machine control system. Operational personnel and safety experts that manage accidents and mitigate their effects obtain required data on radiation environment and states of systems, equipment, and physical barriers on the path of ionizing radiation and radioactive releases from the post-accident monitoring system.

Personnel who monitor the condition of I&C systems and deal with their maintenance and recovery should receive timely and full diagnostic messages containing data on inoperable components of I&C systems, parts of SHC, and power sources and/or those intentionally taken out of service. The diagnostic messages should be displayed to facilitate and accelerate the process of making decisions to recover the operability of the failed I&C component or SHC part.

In the development and operation of an I&C system, measures preventing errors when its configuration changes (adjustment of setpoints and control law parameters, conditions for initiation of protections, interlocks, and alarms, removal from service and introduction into operation of individual components for check) should be envisaged and implemented. The configuration can be changed only in compliance with the rules established in operational documents and only by trained personnel that use dedicated hardware and software. Attempts to make any changes beyond the allowed limits should be automatically blocked and followed by an error signal. Operational personnel are warned in advance about the supposed change in configuration and informed on its start and completion. The I&C systems performing category A functions should envisage local warning alarm and notification of operational personnel about an attempt to remove a redundant channel from service, which is not authorized and/or cannot be detected from the main control room, and should exclude the potential for simultaneous removal of two redundant channels (two SHCs) from service.

In the recovery of an I&C system (SHC), its component parts should be commonly replaced without power dump. In this case, any adjustments in conjugate parts of the system and adjacent products are not required. Specific design solutions prevent the potential for failure in the replacement of component parts and connection of external cables. Products and their removable component parts related to 2(A) safety class are labeled so that they can be distinguished from those related to lower safety classes. Diversity component parts are labeled to identify to what I&C system or SHC they belong.

Protection Against Unauthorized Access

For prevention of intentional or unintentional deactivation, reconfiguration, input of interferences, damage, or theft that can pose a threat to safety, protection against unauthorized access is provided and focuses on the following:

- Operating stand-alone devices (hardware).
- Removable parts of devices and resident programs.
- Switching elements to connect devices to external circuits.
- Elements intended for reconfiguration of I&C systems (SHCs).
- Power switches, elements for mode selection and manual control.
- Embedded tools for technical diagnostics.
- Tools for data input, elements of access to software, database, and archive.
- Operation recovery tools contained in storages.

For protection against unauthorized access, the following is provided: administrative measures (restricted access to rooms); physical protection (seals on door locks, safety cabinets, etc.); software methods (use of passwords, access restriction through external interfaces); location of programs and information in write-protected memory spaces; and alarm for access to devices and/or attempt of unauthorized modification of programs and information. Access inside the devices for maintenance, restoration, reconfiguration, program adjustment, etc., should be provided without compromising the effectiveness of protection.

Dependability Measures

Dependability measures define the reliability, maintainability, and durability of I&C systems and their components.

Reliability is standardized and estimated for the main functions performed by I&C systems and for their components and removable parts. Failure criteria and reliability measures are determined considering the nature of function (continuous or discrete) and type of potential failures.

Continuous functions are intended for monitoring, archive, display, register (analogue), and regulation, while discrete functions are intended for warning, digital register, protection, limitation, blocking, and discrete and remote control.

The failure criteria for continuous functions may include incompliance, wrong implementation, or discrepancy of requirements for characteristics of the performed function; the mean time between failures (MTBF) is taken as reliability measures. The failure criteria for discrete functions are failure to operate (lack of the output signal in the conditions specified for its generation) and spurious operation (presence of the output signal in the absence of conditions for its generation). The failure flow parameter is taken as a reliability measure for "spurious actuation" and the availability factor for "failure to operate".

In the standardization and assessment of each function for reliability, the dependability (reliability) of all devices involved in its performance (including peripheral equipment, component parts of SHC, data transmission facilities and connecting lines, power sources, etc.) is taken into account. The criterion for device failure is incompliance, wrong implementation, or discrepancy with the established characteristics of at least one of the required functions of the device, which necessitates its restoration or replacement. Reliability measures for devices restored directly at the operating site include MTBF or failure rate and for unrecoverable storable devices include the mean time to first failure. The required values of reliability measures are determined for normal operating conditions. Aging, wear, common cause failures, and human errors are considered when there are no approved methods and initial source data that allow the numerical estimate of their probability and influence the reliability.

Maintainability is standardized and estimated for devices recoverable at the operating site. Maintainability measures include the mean time to repair (MTTR) necessary for detecting the inoperative part of operating stand-alone devices, preparatory operations, replacement of the inoperative part with a serviceable one (from the operation recover kit), and final operations, including check of the device after recovery for correct performance. The MTTR does not include delays required for call and arrival of maintenance personnel, delivery of a serviceable part to the operating site, execution of documents before and after restoration, and repair of the failed part.

The MTTR required values are determined upon agreement between the designer of the I&C system (developer of SHC or peripheral device) and operating organization (customer) for all devices involved in the implementation of functions of categories A and B. They are usually supplemented with requirements for the operation recovery kit, depth and methods for displaying the results of diagnostics, automation of testing, service equipment, etc.

Durability is standardized for I&C systems, SHC parts, and peripheral devices. The durability measure includes the service life indicated by the developer, which determines the calendar duration of operation to be followed by upgrade of the I&C system, replacement of the SHC and/or peripheral devices, or acceptance or agreement of the decision to continue operation over a new specified lifetime.

The established lifetime of I&C systems and SHC is not less than 30 years. Over this time, parts of SHCs and peripheral devices whose service life (at least 15 years) is limited by their suppliers may be replaced.

Technical Diagnostics

Embedded technical diagnostic resources automatically detect failures to operate for parts of SHCs, redundant channels, peripheral devices, and command and signal transmission lines. The technical state is monitored after power is turned on, continuously during operation, and in a periodic manner. After turning on the power of SHC, the following is checked automatically:

- Compliance of the composition and configuration with design characteristics.
- Compliance of the loaded software version with the SHC configuration.
- Zero distortions of programs and data in read-only memory.
- Attachment of all connectors provided by the design.
- Operability of SHC parts.

- Proper condition of signal and command transmission circuits, when appropriate.
- Correct exchange of messages between SHC parts and with adjacent SHCs within the same and/or another I&C system.

In the process of SHC operation, the following is automatically monitored in a continuous manner:

- Voltage of main and redundant power supply.
- Proper condition of transmission circuits for signals and commands.
- Authenticity of continuous and discrete input signals.
- Correct performance of all parts with use of embedded tests provided by developers.
- Zero errors during data exchange.
- Zero failures in the execution of programs that terminated the performance.
- Temperature increase and/or smoke generation in operating stand-alone parts of SHCs, condition of remote control circuits, operability of diagnostic hardware, etc.).

If during monitoring of technical state after turning on the power supply or during operation, malfunction is found in at least one part of the SHC or peripheral device, an appropriate notification is automatically issued to personnel who monitor the condition and perform maintenance and restoration of the I&C system. Data on the time, place of occurrence, and type of each violation detected are achieved and displayed (automatically or on call) in the form that allows this violation to be estimated as soon as possible. Operating personnel are automatically given visual and audible warning (generalized malfunction signal) if the detected malfunction prevents the performance of category A or B function(s). Continuous automatic monitoring during operation, search for malfunctions and generation of alerts, archive and display of diagnostic information, and failures of embedded technical diagnostic resources should not affect the performance of functions important to safety.

Periodic monitoring is intended to verify the components of I&C systems, SHCs, and peripheral devices whose continuous automatic monitoring during operation is impossible, unfeasible, or not provided. Periodic monitoring is carried within the entire period of operation: in routine maintenance (in operating modes of the power unit) and at each in scheduled refueling outage (at the power unit in shutdown).

Routine maintenance is intended to verify the correct implementation of each category A discrete function, sequentially simulating the conditions that require its performance. Removal of a redundant channel participating in the performance of emergency protection function from for maintenance (periodic check) is accompanied by automatic transfer to and holding of each output in the state that corresponds to the initiation of commands for protective actions. In the SHC participating in the implementation of other safety functions, each output of the redundant channel removed from service is automatically transferred to and held in the state that corresponds to zero control command and/or signal.

Inspection is carried without action on the executive elements of process equipment, should not have any negative impact on the operation and safety of the power unit, and should not prevent the operational personnel from performing their functions. The inspection is completed by restoring the initial configuration of the I&C system. During the scheduled outage of the power unit, inspections cover all components involved in the implementation of the required (basic) I&C functions; for example, for the regulation function, from the sensor to the final actuating element of process equipment, including connecting cables. In general, the following actions are envisaged in periodic monitoring during scheduled outage of a power unit:

- Resistance measurement of electrical isolation and grounding.
- Check of the correct function performance by each redundant channel and system as a whole, including that with real actuators.
- Calibration of measuring channels, check of accuracy characteristics of control and alarm channels.
- Determination of delays in output signal (command) generation and/or duration.
- Executions of other inspections according to operational documents.

Inspections allow detecting hidden operability failures of components that are not detected during continuous automatic control and tendencies of changes (degradation) in quantitative characteristics, which in future can lead to a function failure.

Embedded technical diagnostic tools, service equipment, and supporting software supplied with SHCs provide automation of test actions (input signals), registration of associated reactions (output signals), determination of quantitative characteristics and archiving of periodic monitoring results to minimize the complexity and duration of inspections.

REQUIREMENTS FOR RESISTANCE OF FUNCTIONS PERFORMED

General Characteristics

Peripheral equipment, SHCs, and their operating stand-alone parts should be resistant (immune) to external factors (EFs) that can occur in places of their location and last as long as required (normal operating conditions) and within a limited time period envisaged by the design (extreme conditions). Extreme conditions can be caused, for example, by accidents in process equipment, connection, disconnection, reloading or short circuit of powerful electrical technical assemblies, failures of ventilation system, conditioning, and power supply used for normal operating conditions, abnormal natural phenomena (earthquake, strokes of lightning), or internal events (fire, flood).

In general, resistance (insensibility) should be provided in relation to:

- Temperature, humidity, barometric pressure, ionizing radiation, corrosive active agents, dust (environment EFs).
- Vibrations, strokes, seismic effects (mechanical EFs).
- Double-current electric fields (electrical EFs).
- Water and solutions that can affect devices in accidents and decontamination fluids (specific environment EFs).
- Long-term deviations from nominal values and short-term variation in power supply (power supply EFs).
- Impacts of electrical or electronic equipment and other sources of electromagnetic disturbances (electromagnetic EFs).

EF of each type is generally characterized by a set of qualitative features and quantitative parameters related to normal and extreme conditions (working and limiting EF values). For peripheral equipment, SHCs, and their operating stand-alone parts, the following is indicated:

- Low and/or high working values of all EFs that guarantee the operability of devices and preservation of their characteristics within a regulated service life.
- Allowed low and/or high limiting values of each EF which a device should have and can sustain within a certain time period, remaining operable and without irreversible degradation of specified properties.

Instead of working and limiting EF values, parameters of testing impacts that simulate the effect of these EFs during factory tests to check correct operation of peripheral equipment and SHCs is usually indicated.

Correct operation is usually understood as performance of all required functions, absence of spurious output signals, and absence of spurious signals and maintenance of properties within the specified limits. Violation of correct operation is considered errors in performing at least one function, deviation of any product property over the specified (allowed) limits, generation of a false output signal, loss or misrepresentation of information in the memory, and need for personnel intervention for reload and/or restart of software.

Test results are estimated by the following criteria:

- A – correct operation during and after impact.
- B – temporary disturbance of correct operation during impact and automatic (without personnel interference) recovery of correct operation after termination of the impact.
- C – temporary disturbance of correct operation during impact and recovery of correct operation by personnel interference after termination of the impact.
- D – disturbance of correct operation due to damage caused by the impact (requires recovery of the failed device).

Resistance (insensitivity) of peripheral equipment and SHCs is determined by A criterion. Test results obtained by B, C, and D criteria indicate incompliance with requirements for resistance or insensitivity to EFs.

In planning and performance of EF resistance (insensitivity) tests, the following rules are applied:

- For power supply, generation of input signals or commands, and monitoring of outputs (output signals), other operating stand-alone parts of the same SHC and/or simulators and measurement technologies (further – auxiliary equipment) are used.
- Auxiliary equipment and connecting cables should be EF resistant and immune or protected against EFs.
- Tests are performed under nominal parameters of input signals and values of all EFs (except for the tested one), which meet normal test conditions.
- Resistance (insensitivity) to EF of each type is sequentially checked (except for resistance to environment humidity and to irrigation of water and solutions in simulation of emergency conditions).
- For SHCs and peripheral equipment consisting of several operating stand-alone parts, each of them is individually subjected to tests.
- Checks are performed in each of the regulated operation modes (or in the mode when the product is most sensitive to EFs of particular type, provided that it was preliminary specified).

- Test impact is repeated the required number of times within a sufficient time period to estimate operation correctness with required reliability.

Resistance to Environmental Impacts

SHCs and peripheral equipment should be resistant to the expected working values of environment EFs (without time restrictions) and to limiting values of these EFs (within the expected maximal duration of their existence). Working and limiting values of environment EFs are determined by the group of operating conditions for peripheral equipment and SHCs (Table 1).

Table 1. Groups of operating conditions

Location area	Room category	Group of operating conditions
Strict access area	Rooms inside containment	E.1.1
	Rooms of process equipment	E.1.2
	Rooms of primary transducers	E.1.3
	Rooms of electrical equipment and SHC	E.1.4
Normal access area	Rooms of process equipment	E.2.1
	Rooms of electrical equipment and SHC	E.2.2
	Rooms with air condition (MCR, ECR, etc.)	E.2.3

In the regulation and check of resistance to environment EFs, one is guided by experimental data on the actual working values of these EFs in places of devices, estimates of their limiting values, and expected duration and frequency of their occurrence. If such data are missing, generic working (Table 2) and limiting (Table 3 and 4) values are used for relevant groups of operating conditions (sign "–" means that EF values are not regulated).

Resistance to environment EFs is confirmed by tests for impacts of temperature, humidity, barometric pressure, and radiation.

Resistance to Mechanical Impacts

SHCs and peripheral devices should be resistant to expected working values of mechanical EFs (without time restriction) and to limiting values that occur during earthquakes. Working values of mechanical EFs are determined by the group of location conditions to which operating stand-alone parts of SHCs (peripheral devices) are related (Table 5).

In the regulation and check of resistance to working values of mechanical EFs, experimental data on the actual values of these EFs at the locations of devices are used. If such data are missing, generic operating values of EFs for groups of placement conditions to which the devices belong are applied (Table 6).

For devices related to groups of placement conditions P1.1, P1.2, and P1.3, requirements for resistance to working values of mechanical EFs are not standardized.

Table 2. Generic working values of environment EFs

Type of EF and measurements unit	Working values of EFs for the group of operating conditions						
	E.1.1	**E.1.2**	**E.1.3**	**E.1.4**	**E.2.1**	**E.2.2**	**E.2.3**
Temperature:							
low value, °C	15	15	15	15	15	15	18
upper value, °C	60	60	30	30	60	30	27
Humidity:							
low value, %	5	5	10	10	5	10	20
upper value:							
% (at 15-30 °C)	100	90	75	75	90	75	80
g/m³ (at 30-60 °C)	36	32	–	–	32	–	–
Ionizing γ-radiation (upper value):							
absorbed dose rate, mGy/h	**	0.9	0.15	0.03	0.12	0.004	–
absorbed rate within 10 years, Gy	10^5	80	13	2.7	10.5	0.35	–

* For groups of operating conditions E.1.2 and E.2.1, upper working values of concentrations of other corrosive agents can be additionally defined.
** In case of location outside of process boxes 30 mGy/h and in boxes 3 Gy/h

Table 3. Generic limiting values of environment EFs for the group of operating conditions E.1.1

Type of EF and measurements unit	Limiting values of EFs for mode				
	A	**B**	**C**	**D**	**E**
Temperature (upper value), °C	75	90	150	60	60
Rate of change in temperature (upper value),°C/h	5	10	20	–	–
Humidity (upper value at an upper temperature value), %	100	steam-gas mixture		–	–
Barometric pressure:					
low value, κPa	50	86	86	50	–
upper value, κPa	130	180	560	130	560
Ionizing γ-radiation:					
absorbed dose rate, Gy/h	1.0	1.0	10^3	1.0	–
absorbed rate, Gy	15.0	5.0	10^4	$0.7 \cdot 10^3$	–
Duration (upper value), h	15	5	10	720	24

Note. Conventional symbols of modes:
A – violation of heat removal from containment; B – small leak; C – maximum leak; D – post-accident mode (effects of small and maximum leaks); E – check of impermeability

Table 4. Generic limiting values of environment EFs for groups of operating conditions E.1.2, E.1.3, E.1.4, E.2.1, E.2.2, E.2.3

Type of EF and measurements unit	Working values of EFs for the group of operating conditions					
	E.1.2	E.1.3	E.1.4	E.2.1	E.2.2	E.2.3
Temperature (upper value), °C	75	50	50	75	50	40
Rate of change in temperature (upper value),°C/h	10	5	5	10	5	5
Humidity (upper value at upper temperature value), %	steam-gas mixture		98*	100	98*	90
Barometric pressure (upper value), кPa	130	130	–	–	–	–
Mode duration, h	5	3	2	3	2	2

* Without moisture condensation
Note. Limiting values of EFs are defined: for groups E.1.2 and E.2.1 by leak of process equipment;
for group E.1.3 by line break from process equipment to sensors; for groups E.1.4 and E.2.2 by disconnection of ventilation; for group E.2.3 by conditioning system malfunction

Table 5. Groups of location conditions

Sources of mechanical EFs	Group of operating conditions	Mounting technique	Group of location conditions
Absent	Any	On civil structures	P.1.1
		On support structures	P.1.2
		On process equipment	P.1.3
Available	E.1.3, E.1.4, E.2.2	On civil structures	P.2.1
		On support structures	P.2.2
	E.1.1, E.1.2, E.2.1	On civil structures	P.3.1
		On support structures	P.3.2
		On process equipment	P.3.3

Resistance to mechanical EFs is confirmed by results of tests for impact of sinusoidal vibrations and mechanical shocks.

For each operating stand-alone device contained in peripheral equipment and SHC, seismic resistance category (I, II, or III) is established depending on the role of performed functions for safety and operability during and/or after an earthquake.

Devices participating in the performance of functions that should be initiated and/or performed during an earthquake (emergency reactor shutdown, blocking of moving mechanisms, etc.) or directly after an earthquake (emergency reactor cooling, residual heat removal, automatic control of critical parameters, radiation environment monitoring, etc.) relate to seismic resistance category I. Devices related to seismic resistance category I should be capable of performing all specified functions when exposed to EFs caused by an earthquake that can occur on the NPP site with a repeatability period of every 10,000 years (further - maximum calculation earthquake or MCE; in international standards, it corresponds to the concepts of "safe shutdown earthquake" (SSE) or "maximum potential earthquake" (S2)).

Table 6. Generic working values of mechanical EFs

Type of EF and measurements unit	Values of mechanical EFs for the group of placement conditions				
	P.2.1	**P.2.2**	**P.3.1**	**P.3.2**	**P.3.3**
Sinusoidal vibrations:					
upper value of displacement amplitude*, mm	0.75	1.5	3.5	3.5	7.5
upper value of acceleration amplitude**, m/sec^2	2	5	10	10	20
upper value of frequency, Hz	150	150	150	150	150
relative duration, %	100	100	100	100	100
impact direction	Z	Z	Z	Z	X, Y, Z
Mechanical shock:					
upper value of peak acceleration, m/sec^2	40	40	70	70	70
shock impulse duration, msec	100	100	50	50	50
impact direction	Z	Z	Z	Z	X, Y, Z

* At frequencies lower than transition frequency (9-10 Hz)
** At frequencies higher than transition frequency
Note. In the table the following conventional symbols are used: X is the direction along the horizontal plane; Y is the direction along the horizontal plane perpendicular to X; Z is the direction along the vertical plane

Seismic resistance category II includes devices that are not related to category I if disturbance of their performance or deterioration of their characteristics caused by an earthquake can lead to interruption in power generation by the power unit (control of reactor power, maintenance of process parameters within established limits, etc.). Devices related to seismic resistance category II should be resistant to seismic EFs caused by an earthquake that can occur on the NPP site with a repeatability period of every 500 years (further – design basis earthquake or DBE); in international standards, it corresponds to the concepts of "operating basis earthquake" (OBE) or "potential earthquake during operation" (S1). Devices that cannot be related to categories I or II by the above specified criteria are included in category III. Seismic resistance requirements are not regulated for them.

Seismic EFs at the location of devices define the response spectra that consider a potential response of civil and intermediate structures to horizontal and vertical earth seismic vibrations, which can be filtered or intensified using vibration rates and damping typical of these structures.

The response spectra are determined by calculation and/or modeling taking into account:

- Seismicity of the NPP site and parameters of seismic impacts during an earthquake, which are determined on the basis of seismological studies taking into account specific geodynamic, seismotectonic, seismological, soil, and hydrological conditions of the site.
- Margin relative to the established level of site seismicity, which is determined by the operating organization and agreed with the state nuclear regulatory body of Ukraine.
- Peak acceleration of the horizontal component of the ground vibrations during MCE, which is taken at least 1 m/sec^2 regardless of the site seismicity.

- Height of device location and mounting on a building support structure (ceiling, wall, column), an intermediate structure (in a control panel, cabinet, console), or process equipment (piping, piping valve, etc.).

If there are no results from calculations or modeling, the response spectra are determined using generic values of seismic EFs (Table 7).

A response spectrum is characterized by two horizontal and one vertical components acting simultaneously. Acceleration amplitudes in the vertical direction are accepted with a factor of 0.7 from the values indicated in Table 7.

Table 7. Generic values of seismic EFs (required response spectra)

Mounting technique	Height, m	Acceleration amplitude, m/sec^2, at frequency, Hz									
		0,5	1	2	3	4	5	6	10	15	30
Project earthquake (average repeatability 500 years)											
On civil structures	From 0 to 5 inclusive	–	0.2	0.3	0.3	0.3	0.3	0.3	0.3	0.2	0.1
	Over 5 to 10 inclusive	–	0.5	0.6	0.6	0.6	0.6	0.6	0.6	0.5	0.3
	Over 10 to 25 inclusive	–	1.0	1.2	1.2	1.2	1.2	1.2	1.2	1.0	0.5
	Over 25 to 35 inclusive	–	1.2	1.5	1.5	1.5	1.5	1.5	1.5	1.2	0.6
	Over 35 to 70 inclusive	–	1.6	2.0	2.0	2.0	2.0	2.0	2.0	1.6	0.8
On intermediate structures or process equipment	From 0 to 5 inclusive	–	0.4	0.6	0.6	0.6	0.6	0.6	0.6	0.4	0.2
	Over 5 to 10 inclusive	–	1.0	1.2	1.2	1.2	1.2	1.2	1.2	1.0	0.6
	Over 10 to 25 inclusive	–	2.0	2.4	2.4	2.4	2.4	2.4	2.4	2.0	1.0
	Over 25 to 35 inclusive	0.1	2.4	3.0	3.0	3.0	3.0	3.0	3.0	2.4	1.2
	Over 35 to 70 inclusive	0.2	3.2	4.0	4.0	4.0	4.0	4.0	4.0	3.2	1.6
Maximal calculation earthquake (average repeatability 10,000 years)											
On civil structures	From 0 to 5 inclusive	0.1	1.0	1.2	1.2	1.2	1.2	1.2	1.2	1.0	0.5
	Over 5 to 10 inclusive	0.2	2.0	2.4	2.4	2.4	2.4	2.4	2.4	2.0	1.0
	Over 10 to 25 inclusive	0.3	4.0	4.8	4.8	4.8	4.8	4.8	4.8	4.0	2.0
	Over 25 to 35 inclusive	0.4	4.6	5.7	5.7	5.7	5.7	5.7	5.7	4.6	2.3
	Over 35 to 70 inclusive	0.5	6.0	7.6	7.6	7.6	7.6	7.6	7.6	6.0	3.0
On intermediate structures or process equipment	From 0 to 5 inclusive	0.2	2.0	2.4	2.4	2.4	2.4	2.4	2.4	2.0	1.0
	Over 5 to 10 inclusive	0.4	4.0	4.8	4.8	4.8	4.8	4.8	4.8	4.0	2.0
	Over 10 to 25 inclusive	0.6	8.0	9.6	9.6	9.6	9.6	9.6	9.6	8.0	4.0
	Over 25 to 35 inclusive	0.8	9.2	11.4	11.4	11.4	11.4	11.4	11.4	9.2	4.6
	Over 35 to 70 inclusive	1.0	12.0	15.2	15.2	15.2	15.2	15.2	15.2	12.0	6.0

Seismic resistance of peripheral equipment and SHCs is determined upon tests, during which each of their operating stand-alone parts is subjected to sinusoidal vibrations simulating seismic EFs. Vibration parameters (impact spectrum) are determined taking into account the placement height and installation method, based on the floor response spectrum of civil structures approved for nuclear power plants. In the absence of the customer's data, the generic values of seismic EFs indicated in Table 7 are taken for the impact spectrum.

The tests simulate the method of fixing the device and use parts, materials, and technique specified in the documents.

The test impact is applied simultaneously along three mutually orthogonal (two horizontal and one vertical) directions. In the absence of necessary test equipment, a simultaneous impact along two (one horizontal and one vertical) directions is applied, while the acceleration amplitudes in the horizontal direction are increased by 1.4 times compared to the ones indicated in Table 7.

During and/after test impacts corresponding to MCE, EFs that can occur in emergencies or accidents and that adversely affect the seismic resistance are also simulated. To take into account possible factors of mechanical aging, devices of seismic resistance category I are subjected to test impacts that simulate DBE at least five times before testing their resistance to MCE.

Devices of seismic resistance category I and II should not have main frequencies of proper oscillations (resonant frequencies) in the range from 0.5 Hz to 30 Hz. The design, supports, and method of fixing the devices provided for in the documents should ensure their resistance to overturning under the static force applied to the center of mass and equal to the product of maximum seismic acceleration and the mass of the device.

Immunity to Electrical Impacts

Peripheral equipment, SHCs, and their operating stand-alone parts should be immune to electric EFs (low-frequency electric fields) in normal operating conditions.

Actual or expected working intensity values of low-frequency electric fields at locations of these devices are determined using experimental and/or calculated data. If such data are absent, the generic working parameters of electric EFs are used (Table 8).

Table 8. Generic working values of electric EFs

Type of EF and measurements unit	Working values of electric EFs for the group of operating conditions						
	E1.1	E1.2	E1.3	E1.4	E2.1	E2.2	E2.3
Electric field intensity, kV/m	5	5	5	–	5	–	–
Electric field frequency, Hz	50	50	50	–	50	–	–
Note. 1. Phase of electric field in relation to the voltage of supply alternating current and direction of the intensity vector can be any 2. Sign "–" – not regulated							

The immunity to electric EFs is defined upon tests in which operating stand-alone parts are sequentially exposed to test impacts simulating a low-frequency electric field that may occur in operating conditions. The phase of the electric field in relation to the voltage of supply alternating current is 0°, 90°, 270° (alternately) and the direction of the electric field intensity vector alternates along each of the three orthogonal planes.

Resistance to Special Environments

Peripheral equipment, SHCs, and their operating stand-alone parts related to the group of operating conditions E1.1, E1.2, E1.3, E1.4, or E2.1 should be resistant to:

- Irrigation of water and solutions, whose composition, temperature, direction, and duration of impact are determined by analysis of potential accident consequences in places where the devices are located.
- Deactivating solutions, the chemical compositions and mass fracture of each of their component are agreed with the operating organization or customer.

If there is automatic gas firefighting equipment in the rooms where the devices that perform the functions of category A are located, it should be resistant to the reagent that fills the room in case of firefighting system actuation (type of reagent used is agreed with the operating organization or customer).

Immunity to Changes in Power Supply Parameters

I&C systems, SHCs, and peripheral equipment that receive power from the power unit in-house network should be immune to long-term deviations of voltage, short-term changes and fluctuations of voltage, power interruption, and long- and short-term changes of supply frequency that may occur in normal and extreme operating conditions. Operating stand-alone parts that receive power from secondary power sources that are part of the I&C system and SHC should be immune to changes in the parameters of primary power supply of these sources. Immunity is estimated upon tests, during which the following rules are applied in addition to the general provisions:

- All operating stand-alone parts of SHC and peripheral equipment receiving power from one source are tested simultaneously.
- Test impacts are applied to power ports such as terminal screw, clamps, connectors, and other design elements defining the physical boundaries between the device and the power unit's in-house network (device can have one or several power ports).
- If the device has several mutually redundant power ports, test impacts are applied alternately to each port, the other ports should be disconnected from the in-house network in the process.

Immunity to Electromagnetic Interferences

I&C systems, SHCs, and peripheral equipment should be immune to electromagnetic interferences that affect or can affect them in normal and extreme conditions:

- Interference from electrostatic discharges.
- Interference from radio frequency electromagnetic radiation.
- Interference from rapid transient processes/packages of impulses.
- Interference from power and current spikes.
- Conductive interference brought by radiofrequency fields.
- Interference from magnetic fields of circuit frequency.

- Interference from impulse magnetic fields.
- Interference from damped oscillatory magnetic fields.
- Oscillatory damped interference.
- Conductive asymmetric interference in a range from 0 Hz to 150 кHz.
- Interference in ground lines.

Required immunity to interference should be provided during the design of I&C systems and the development, manufacture, and installation of SHCs and peripheral equipment and should be retained during operation through the creation and maintenance of proper electromagnetic environment at the location of their stand-alone parts.

Immunity to interference is regulated by establishing the parameters of test impacts simulating interference of each type and the places and methods of their application in which the I&C system (SHC, peripheral equipment) should function correctly. The places of applying test impacts include screw clamps, cleat, connectors, shells, and other structural elements that determine the physical boundaries between the device with its environment (further – ports, see Figure 2).

Figure 2. Places of application of test actions simulating interference

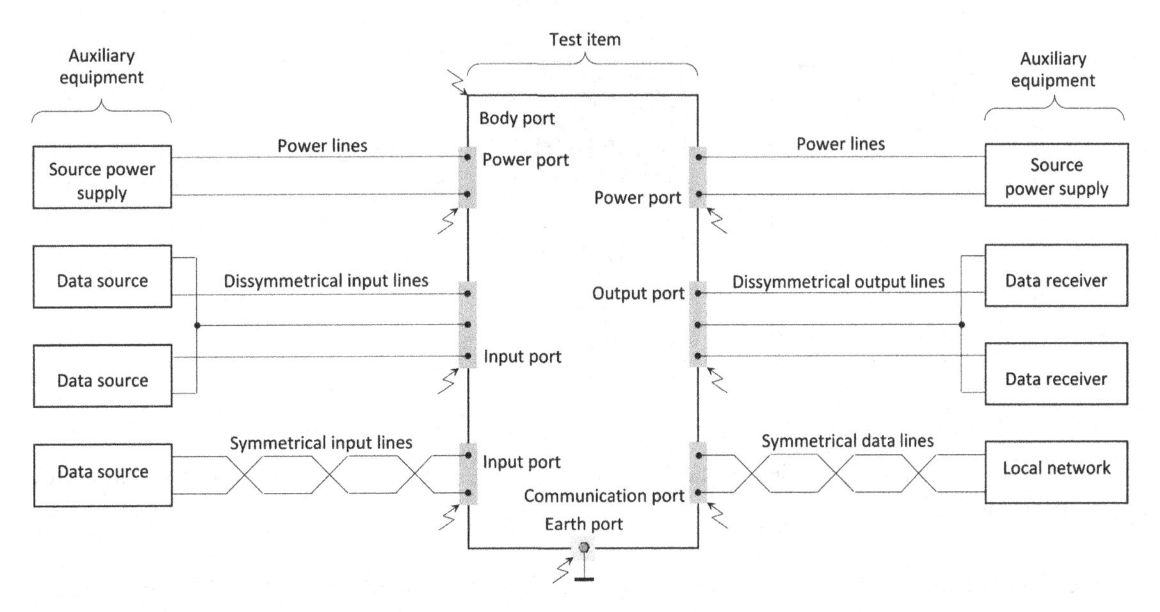

- Power port – in places of connection of two or more wires from one power source.
- Input port – in places of connection of two or more wires from one or several interconnected sources of input signals.
- Output port – in places of connection of two or more wires, connecting the device with one receiver of the output signal.
- Communication port – in places of connection to one communication channel or one local network.
- Ground port – in places of connection to the protective and, if available, signal grounding.

- Housing port – on the outer surface of the shell, metal noncurrent-carrying elements, and cable shields electrically connected to the shell.

To assess immunity to interference, tests are performed during which operating stand-alone parts of SHCs and peripheral equipment (further – devices) are subjected to test impacts that simulate the effect of interference of each type that may occur in normal and extreme conditions and can have a negative impact on the efficiency or characteristics of the device. In general, test impacts are applied to circuits connected to direct and alternating current power port(s), to unsymmetrical and/or symmetrical lines connected to input, output, and communication ports, and to ports of protective and (if available) signal ground and to the housing port (see Figure 3).

Figure 3. Interference simulating circuit

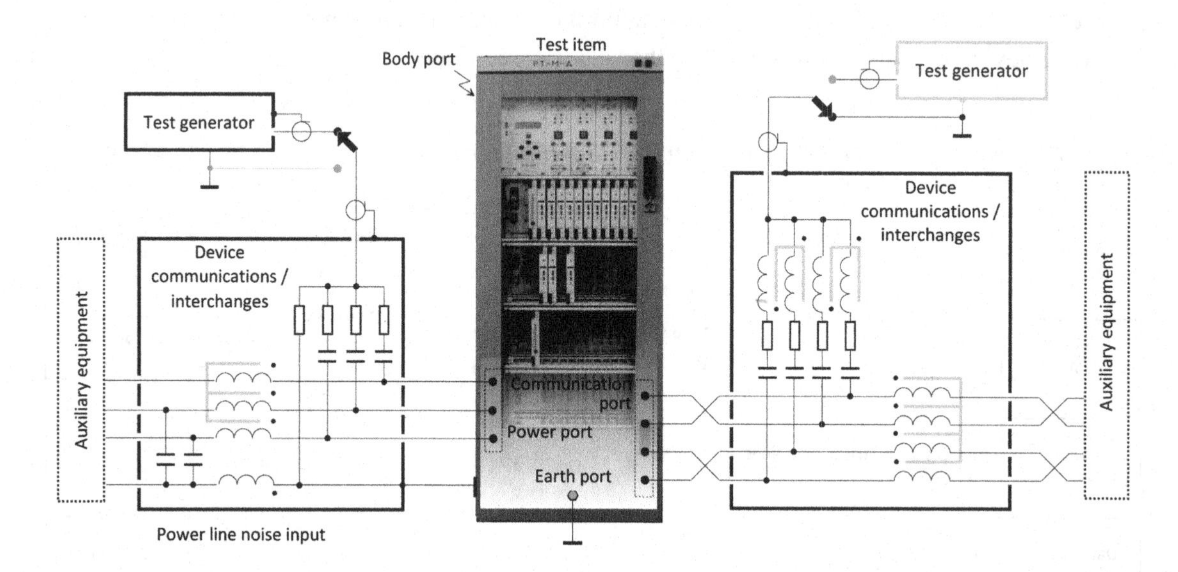

During tests of immunity to electromagnetic interference, the following rules are used in addition to general provisions:

- Electromagnetic environment in rooms where tests are performed should not affect their results.
- Immunity to each type of interference specified for the device is tested individually (alternately).
- Device location, types and length of connecting cables should, when applicable, meet actual operating conditions (cables included in the device or indicated in its documents are to be used).
- Test impacts are allowed to be applied simultaneously to several operating stand-alone devices that are part of the HSC, connected to each other by electric communication lines (no more than 1 m long), and/or nonelectric communication lines of any length.
- If a device has several power, ground, input, output, and/or communication ports, test impacts are applied to each of them separately.

- For devices that have a significant (more than five) number of identical ports or ports with a significant number of identical connecting lines, a limited number of ports (lines) that are most sensitive to this type of interference can be selected upon agreement with the customer.
- Degree of test severity is chosen taking into account the electromagnetic environment in places of actual or expected location of devices (qualitative features of the electromagnetic environment are given in SOU, 2016), and only those features are taken into account that can significantly affect the intensity of interference of respective type.

The interference immunity requirements and test methods and criteria for evaluating the results given in Rozen, 2007, and Rozen, 2008, correspond to SOU, 2016, international standards for electromagnetic compatibility, and identical state standards of Ukraine.

Resistance to Failures of Elements

Each function of the I&C system and SHC should be performed and meet specified requirements in case of failure of any elements of the same I&C system (SHC) participating in the performance of functions of lower category. Single failures of elements in primary and secondary power supply systems as well as short-term interruption of the supply voltage should not affect the functions of categories A and B (functions of category C should be performed after the power supply is restored). Resistant of elements to failures should be provided by:

- Adequate calculating and other resources (inputs, outputs, memory, power supply, etc.);
- Redundancy of components of I&C systems (parts of SHC) performing functions of categories A and B, as well as equipment and connecting lines, which are used for transmission of signals and messages between them and receive data from other systems.
- Availability of tools that can detect deviations the determined behavior and automatically recover normal operation.
- Availability of embedded technical diagnostic tools that provide continuous automatic testing of the technical condition and allow quick detection of the malfunction of parts that can lead to a failure in the initiation of a discrete function.
- The potential to rebuild the algorithm, which allows, in case of element failures, to restore operation in normal mode or switch to the previous standby mode, for example, if the received data are unreliable or loss of input data cannot be eliminated.
- Testing of input continuous signals for authenticity and automatic exclusion of the potential for using data represented by doubtful signals to ensure resistance of information sources or connecting lines to failures.
- Use of two-bit discrete signals for the generation, transmission, and input of data on the condition of a controlled object, logical levels 01 and 10 being interpreted as two different states of the object (for example, connected and disconnected) and logical levels 00 and 11 indicating an error of a signal source or damage of a connecting line.
- Use of keys with three positions to form a command by closing the control circuit in one of the last positions in opening the check circuit closed at the middle position of the key.
- Use of communication protocols capable of detecting and correcting errors in received messages.
- Automatic restoration of data authenticity in each of the redundant channels.

REQUIREMENTS FOR THE QUALITY OF FUNCTIONS PERFORMED

Accuracy

Accuracy requirements are regulated for measurement channels of the I&C system and SHC, which display, register, archive, and/or transmit numerical values of physical quantities in accepted units; for control channels (protection, blocking, control) and alarm; and (upon agreement between the developer and operating organization or customer) for components of the I&C system and component parts of SHC, forming measuring, control, and alarm channels.

In the quality standardization and assessment, the following rules are followed:

- Main safety important process parameters are preferably monitored by direct measuring than by calculation on the basis of measured values of other parameters (indirect measurement method).
- When the measurement range for each monitored parameter is chosen, the potential that the operating range of a parameter may be exceeded in emergencies and accidents is considered.
- If for a satisfactory coverage of all range of change in a monitored parameter, more than one sensor is needed, proper overlapping of adjacent measurement ranges and automatic switching of these sensors are provided so that saturation or the influence of distortion at the edges of the ranges does not interfere with obtaining the results of required accuracy.
- In case of inoperability or removal of one redundant measurement (control, signaling) channel from service, the remaining channels should comply with the established accuracy requirements.
- Measuring, control, and signaling channels in which redundancy of parts is provided should comply with the established accuracy requirements in the event any of these parts fails.

Accuracy of measuring channels is specified as metrological characteristics of the channels to which the measurement range, conversion function, and error characteristics are related. Error characteristics of a channel are standardized using:

- Allowed absolute or relative error in normal (or, upon agreement between the developer and the operating organization or customer, extreme) operating conditions or.
- Allowed main (absolute or relative) error in normal test conditions and allowed additional errors that can be caused by change of each EF in the range of working or limiting values established for it.

The conventional error of a measuring channel is calculated as the ratio of the absolute error to the range of change in the monitored parameter. The absolute error is determined as the difference between the measured and actual value of a monitored parameter, where the measured one is considered the parameter value: directly obtained for a database; displayed or registered in a digital or analog form; calculated by an output signal; or transmitted in an output digital signal (message). The actual value of a monitored parameter for the I&C system is its estimation with use of measurement tools and for SHC is the measured value of a proper analog input signal, considering the nominal static characteristic of the signal source (sensor, normalizing transducer, etc.).

Accuracy requirements of measuring channels are established upon agreement between the developer and the operating organization or customer, where metrological characteristics of updated I&C systems and their components are usually significantly higher than those operating previously. Before commissioning all measuring channels of the I&C system, they pass metrological certification according to the requirements of state standards.

In well-grounded cases, instead of metrological characteristics of measuring channels, metrological characteristics of component parts that are determined to ensure the metrological compatibility of these component parts and define the metrological characteristics of measuring channels by calculation can be regulated. The metrological characteristics of component parts of measuring channels are experimentally assessed during their development, after manufacturing, and during operation.

Accuracy of control and signaling channels is specified as accuracy characteristics of these channels in normal and extreme operating conditions. Accuracy characteristics include limits of the allowed absolute error of issue and withdrawal of a command (output signal) and limits of the allowed relative error in generation of time delays if they are provided in algorithms for the performance of control functions.

For the I&C system, the absolute error of issue and withdrawal of a command (output signal) is determined as the difference between the actual value of a monitored parameter that caused the indicated action and the specified value (setpoint) of this parameter at which such an action should be performed. For SHCs, in determination of the absolute error instead of a monitored parameter, values of the informative parameter of a proper analog input signal are considered. The error in generation of the time delay is calculated as the difference between the actual and predetermined values.

Limits of allowed errors in issue (withdrawal) of commands and output signals and generation of the time delay in change of any EF in the range of regulated values are determined upon agreement between the developer and the operating organization or customer. To confirm compliance of the control and signaling channels with specified requirements, their accuracy characteristic are checked in commissioning and periodic monitoring within scheduled outages of power units.

Function Performance Time

The function performance time is determined for each main function of the I&C system and SHC in the form of nominal or maximum allowed (upper and/or lower) values:

- Cycle time of data input from sensors and other I&C systems (SHC).
- Time resolution in data input and archiving.
- Delay of performance of discrete function.
- Speed of function performance.
- Rate of exchange through communication lines and local networks.
- Time of information storage in a database and an archive.
- Time of actuation after power restoration.

The nominal values of time resolution in data input and archiving are determined for accurate differentiation and registration of the order of initiating events, changes in the condition of process equipment and monitored parameters, and actions of control systems and personnel in archive (for further analysis and estimation):

- For information on initiating events, violations of design limits of safe operation, conditions for initiation of safety control systems, and commands of protective actions – not worse than 0.01 sec.

- For information on the condition of process equipment, values of monitored parameters and setpoints, violations of operational limits and conditions, commands for limitation, regulation, process protection and interlocking, and discrete and remote control – not worse than 0.1 sec.

Delays in the performance of discrete functions (maximum permissible upper values):

- Issue of protection commands – not more than 0.1 sec, commands of limitation and interlocking – not more than 0.1 sec from the occurrence of a specific design condition until a control signal at the input of the actuating system or an element of process equipment (except for the commands that should be generated with a specified delay).
- Execution of remote control commands – not more than 0.1 sec from the signal generation at the output of a manual control key before a control signal at the input of a proper actuating element of process equipment.
- Warning of personnel and display of data on dangerous initiating events, violations of design limits and conditions, changes of monitored parameters and condition of process equipment, and actuation of safety control systems and normal operation systems – not more than 1,0 sec from the occurrence (change) until actuation of preventive or emergency alarm at personnel workplaces and appearance of relevant information in the specified format of a data display device.
- Notification of personnel on operability failures that affect safety - not more than 10 sec (not more than 1 min for other failures) from the occurrence of the failures until actuation of preventive alarm at workplaces.
- Archiving of data on initiating events, violations of design limits of safety operation, conditions for generation of protective action commands – 0.01 sec, data on the condition of process equipment, values of monitored parameters and setpoints, violations of operational limits and conditions, commands of restriction, regulation, process protection and interlocking, and discrete and remote control – 0.1 sec between the time of event occurrence and the time at which it would be registered in a database and an archive.
- Sampling on the operator's call and display of information from a database or an archive – not more than 2 sec from completion of directive input before appearance of appropriate information in the specified format on a display device.

The function performance speed (maximum permissible number of functions performed within a time unit):

- Calculation of design parameters for protection functions – not less than 100 times per second for each parameter.
- Calculation of design parameters for functions of restriction, regulation, and interlocking – not less than 10 times per second for each parameter.
- Comparison of the monitored parameters with setpoints of preventive and emergency alarm – not less than 10 times per second for each parameter.
- Archiving of data on initiating events, failures of design limits of safe operation, conditions for generation of protective action commands – not less than 100 times per second, data on the condition of process equipment, values of monitored parameters and setpoints, violations of operational

limits and conditions, commands for restriction, regulation, process protection, and interlocking, discrete and remote control – not less than 10 values per second for each type of achieved data.
- Updating of video frame variable data displayed on the monitor screen – not less than once per second.

Time of information storage in a database (in the SHC operational memory) – not less than 24 hours, in an archive is usually within one fuel campaign of a reactor facility.

Time of actuation for peripheral equipment and SHC after short-term (not more than 10 min) intentional or unintentional disconnection and subsequent power restoration is determined as the maximum allowed upper value and, after its expiration, automatic recovery of specified functions interrupted due to power cut is ensured. The performance of category A functions in full scope with standardized properties is automatically resumed not later than in 1 min after power restoration and the performance of B and C functions not later than in 5 min (or a greater time interval, upon agreement between the designer of the I&C system, the developer of SHC, and the operating organization or customer).

Human-Machine Interface

The I&C system and SHCs directly interacting with NPP personnel should support the human-machine interface whose properties minimize workloads on personnel and decrease the probability of human errors. From the I&C and SHCs, human-machine interface is supported by:

- Alarm facilities (visual and audible alarm), multi-access information display (video monitors, informational screens and mimic panels), command input elements and operator's guidelines placed in MCR, ECR and local control panels.
- Automated workplaces of operational personnel and/or workstations placed in MCR and ECR.
- Workstations of operational personnel placed in rooms of the duty engineer and/or technical support center.
- Automated workplaces of personnel managing accidents and safety experts situated in emergency response centers.
- Elements of alarm, indication, display (video monitors, panel computers), and manual control elements are embedded into devices.

Human-machine interface facilities placed in MCR and ECR are labeled and placed in such a way that operational personnel could easily and accurately assess the state of the power unit and its systems, promptly detect changes, and take actions required for power unit control.

Workplaces of operational personnel in MCR and ECR should meet ergonomics requirements and take into account stereotypes of operators behavior and factors of engineering psychology. In the design of workplaces, the behavior of operational personnel in emergency situations when required actions should be simple, clear for understanding and execution, and performed within a short time period and should not last very long is considered.

Data display facilities and input elements of commands and guidelines must be structured and identified taking into account their functions and priorities and their location corresponds to the logical sequence of actions of operational personnel during power unit control. It is provided that data display facilities placed in MCR and ECR and contained in post-accident monitoring systems should visually differ from other display facilities, which are placed in the same rooms.

Information displayed on video screens should be in the form convenient for perception and analysis, proven in practice, and favorably evaluated by personnel. Each operator may choose to be provided with generic and/or detailed information in the form of mimic planes, histograms, graphs, tables, logic diagrams, text messages, etc. Data on the current values of monitored parameters, state of structures, systems, and elements, output signals and commands, and operability failures should be automatically updated on the screen. Symbolic representation of sensors and actuators should allow personnel to easily identify and accurately define their state (position) and operability.

The information displayed on video screens should be organized as a system of independent fragments (still images) with a hierarchical structure that provides a general view of a controlled object and its sequential specification at several disaggregation levels ("general-to-specific"). Still frames for display should be selected in simple and visual ways with the minimum number of actions required for it. Protection against the loss of important information due to overlaying of still images in their actuation and/or in changes of sizes and location of screens by the operator, in which still images are displayed, should be provided.

Alarm messages generated to workplaces of the operational personnel in detection of failures of design limits and/or normal and safe operating conditions, protection actuation, operability failure of a controlled object and in other specified cases are displayed on video monitors in a dedicated screen area, not overlapped by other images. The text of the alarm message should allow personnel to promptly and definitely detect the place, time, nature, and, if possible, the degree of failure hazard.

The issue of alarm messages should be followed by visual and/or audio signals, which have differences that allow personnel to qualitatively assess a degree of failure hazard. Tools for termination of audio signals to avoid unnecessary acoustic load and attract attention to new alarm messages must be provided. Visual alarm lasts until the causes that led to the alarm signal are eliminated and then the alarm terminates automatically. The time of alarm actuation and termination and the causes that led to the alarm message are registered in a database.

The operator must be able to confirm reception of each identified alarm message and authorized to prohibit the issue of individual alarm messages and cancel prohibition from his workplace. Confirmation that the alarm signal is received should be displayed on the screen (for example, change of color or conversion from blinking to a smooth glow) and should lead to termination of the audio signal. The alarm message whose reception has been confirmed is automatically deleted from the screen when the associated failure is eliminated. Prohibition to issue the alarm message is followed by its removal from the screen and termination of the audio signal. Upon the operator's call, a chronological list of all eliminated and remaining failures and time periods of their occurrence and elimination must be shown on the screen.

Facilities of human-machine interface at workplaces of personnel who control the condition and provides maintenance and recovery of the I&C system must be manufactured, labeled, and placed so that it would be possible to accurately assess the condition of the system and its components, detect places of occurrence and the nature of operability failures, and make a decision on their elimination.

REQUIREMENTS FOR THE INDEPENDENCE OF FUNCTIONS PERFORMED

Property of Independence

A group of mutually redundant I&C systems or SHCs should remain operational and maintain established characteristics in the performance of category A functions independently of potential (incorporated in the design) external factors and/or in the case of removal from service, malfunctioning of one of these systems or SHC, and as a result of erroneous actions of personnel during their maintenance or recovery. Each I&C system (SHC) should remain operational and maintain the established characteristics in the performance of category A functions independently of the state of any element, group of elements, or a channel of the same or another system intended for performance of lower category functions.

Independence from external factors (such as fire, flood, extreme temperature and humidity, electromagnetic disturbances, earthling break, etc.) and from erroneous actions of personnel who provide maintenance or recovery is ensured by physical separation of components that belong to different I&C systems, SHCs and their redundant channels, and related cables (for examples, location of SHC in different rooms, using redundant channels of one SHC in separate cases or support structures inside one case). The preferable way of physical separation of cables is the use of individual cables of channels and penetrations.

Independence from removal from service or malfunctioning of individual components of the I&C system, SHC, or redundant channels is provided by functional and/or electric separation. Functional separation ("functional isolation") ensures that each I&C system (SHC, redundant channel) has a full set of input data required for the performance of specified functions and electric separation provides galvanic isolation and circuit shielding (electric separation is provided in cases when a common source of input data, a common signal receiver, and/or one and the same power source is used for different I&C systems, SHCs, and redundant channels).

For galvanic isolation, electric, optic, and other separation devices ("isolators"), fiber optic lines, and local networks, transmitting information in the form of optic signals, are used. The quality of galvanic separation is defined by electric strength of isolation between galvanic isolated or isolating during the operation by electric circuits, as well as between the enclosure and all isolated from the enclosure electric circuits of device, and by resistance of electric isolation between the same circuits at normal and extreme values of external environmental factors.

Electric strength of isolation is defined by the test direct current voltage or the amplitude value of test alternating current voltage that does not cause breakdown or isolation breaking when induced for one minute between the tested circuit and interconnected clips of other circuits, including safety-ground clips. The test voltage depending on a device safety class, circuit nominal voltage, and conditions for the performance of tests should be not lower than the ones specified in Table 9.

Table 9. Electric strength of isolation (test voltage)

Category	Test conditions	Test voltage, V, for the circuit with rated voltage						
		DC					AC	
		12 V	24 V	48 V	110 V	220 V	220 V	380 V
A	Working *	1500	1500	1500	1500	1800	2100	2800
	Boundary **	500	500	500	800	1080	1270	1700
B, C	Working *	500	500	500	1300	1800	2100	2800
	Boundary **	300	300	300	800	1080	1270	1700
* At upper operating temperatures and humidity (see Table 2) **At upper temperature and humidity limits (see Tables 3, 4)								

Electric isolation resistance between the tested circuit and interconnected clips of other circuits, including ground clips is:

- Not less than 40 MOhm in normal test conditions (naturally established in the room).
- Not less than 10 MOhm at the upper temperature working value.
- Not less than 2 MOhm at the upper humidity working value.

Independence from adjacent I&C systems, SHCs, and redundant channels is ensured by:

- Selection of structure for connections, interfaces, and communication protocols through communication lines or the local network, which allow checking accuracy of obtained data and, in case of failures in any of the devices connected to the local network, maintain communication between other devices.
- Use of hardware and software for control of data flows, protocol processing, detection and correction of errors in order that any failures during transmission and reception of messages do not affect the performance of specified functions of the I&C system (SHC).
- Use of different data transfer paths between channels.

Permissible Disturbance Emission

Operating stand-alone parts of SHCs and peripheral equipment during operation, connection, and disconnection should not cause switching or other interference, which could cause operation failures of other components of the I&C system connected to the same primary power network or to the same power source.

Level of radiated disturbances during operation, connection and disconnection of operating stand-alone component parts of SHCs and peripheral equipment should not exceed the values specified in EN, 2010b, that regulates requirements for informational process equipment intended for operation in manufacturing facilities.

For devices with consumption current not more than 16 A (in a single phase) connected to the common primary power network, disturbance emission standards are specified (harmonic components of consumed current and/or voltage oscillations caused by this current) in the primary power network.

Fire Safety

Operating stand-alone component parts of SHCs and peripheral equipment should meet requirements of fire-prevention regulations of NAPB, 2002. Fire safety is provided under maximal permissible long- and short-term increase of power supply voltage, high voltage on inputs and outputs, short circuits inside a devices and output circuits. Fire prevention should be provided by:

- Use of fire-proof materials, coatings, and cables (noncombustible, hardly inflammable, flame-retardant and nonsmoking and without toxic discharges) that passed specific tests and were certified according to the established procedure.
- Use of component parts in which ignition sources are not generated during reloading, short circuits, or failures.
- Limitation of voltage that can occur in input or output circuits in case of adjacent equipment failures or as a result of human errors.
- Use of active facilities for control or protective termination of ignition sources, or automatic device de-energizing in case of fire hazards.

Probability of fire in any operating stand-alone device is not higher than10^{-6} a year. For prompt detection of ignition inside a device, continuous automatic monitoring and preventive alarm in case of detection of hazards (temperature increase, smoke in a case) should be provided. The requirements for monitoring and alarm when a fire is detected inside a device are mandatory for devices that perform category A functions and are recommended for devices that perform category B and C functions.

In order to quickly detect a fire that has begun and take measures to eliminate it, the fire alarm information systems and/or automatic fire extinguishing control systems should be provided in the power unit rooms. Fire alarm systems and automatic fire extinguishing systems must be designed, developed and placed so as to ensure that their spurious actuation does not adversely affect personnel and other systems.

SOFTWARE REQUIREMENTS

For software of the I&C system, SHC, and intellectual (software controlled) peripheral equipment, requirements are regulated for functions, structure and elements; for diagnostics and self-control; for protection against failures, corruptions, unforeseen actions; for protection against interference with the work and/or unauthorized changes to the software.

Similar requirements are provided for electronic projects of complex programmable components (FPGA).

Functions, Structure, and Elements

The software manages the performance of all the functions to be implemented by software or using software with given reliability and quality. The software of I&C systems, SHC, and peripheral equipment has a modular structure. The text of one module must contain a limited number of operators, have a clear structure, and be easy to modify and test.

In the software of I&C systems, SHC, and peripheral equipment involved in the performance of category A functions, the use of the operating system should be limited to only the simplest functions, and the use of interrupts is prohibited. For I&C systems, SHC and peripheral equipment involved in the performance of category B functions, the operating system and interrupts can only be used in justified cases.

Diagnostics and Self-Monitoring

The software should implement continuous automatic monitoring of the technical condition of the I&C systems, SHC and peripheral equipment and provide technical diagnostics with regulated completeness, depth, veracity, efficiency, and frequency.

The software should diagnose in-house tools ("self-monitoring"), for example, using methods of re-counting and comparing results, identifying prohibited situations, assessing the duration of programs, subprograms, procedures, etc., and providing automatic registration, storage and displaying data on the results of diagnosis and self-monitoring. The implementation of diagnostic and self-monitoring programs should not affect the main functions of the software or lead to degradation in their characteristics. Failures (errors in the performance) of the diagnostic and self-monitoring programs should not affect the implementation of the basic functions of the I&C systems, SHC, and peripheral equipment.

Service software automates the monitoring of the technical condition of I&C systems, SHC, and peripheral equipment during their maintenance and periodic inspections (tests).

Protection Against Failures, Distortions, and Unforeseen Actions

It is based on technical diagnostics and envisages reconfiguration of the structure of the I&C system (SHC) and restoration of the computing process. The software should verify incoming information, notify personnel when inaccuracy is detected, and protect against dangerous consequences that could be caused by data distortion.

To protect against computer viruses, the software should include:

- Integrity monitoring of system areas, application programs, and data used.
- Monitoring events critical to system safety.
- Prevention of negative consequences when unforeseen actions are started.
- Creation of a safe and isolated operating environment.
- Detection of infected files in the purchased software, in the constant memory of the purchased products, and in complex electronic components (FPGA).
- Software safety recovery.

Figure 4. Development of I&C system components

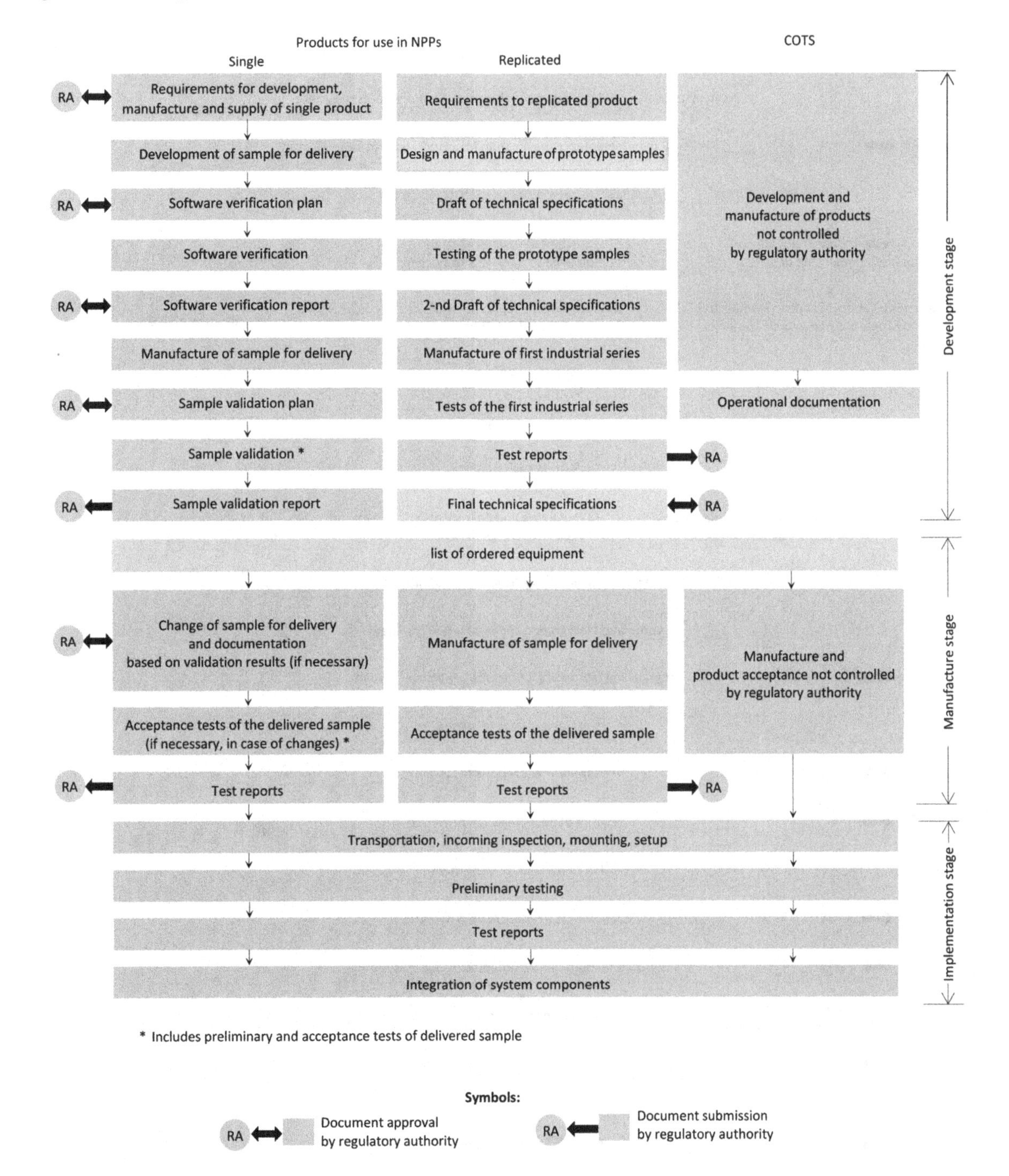

Figure 5. Stages of creating of important to safety I&C system

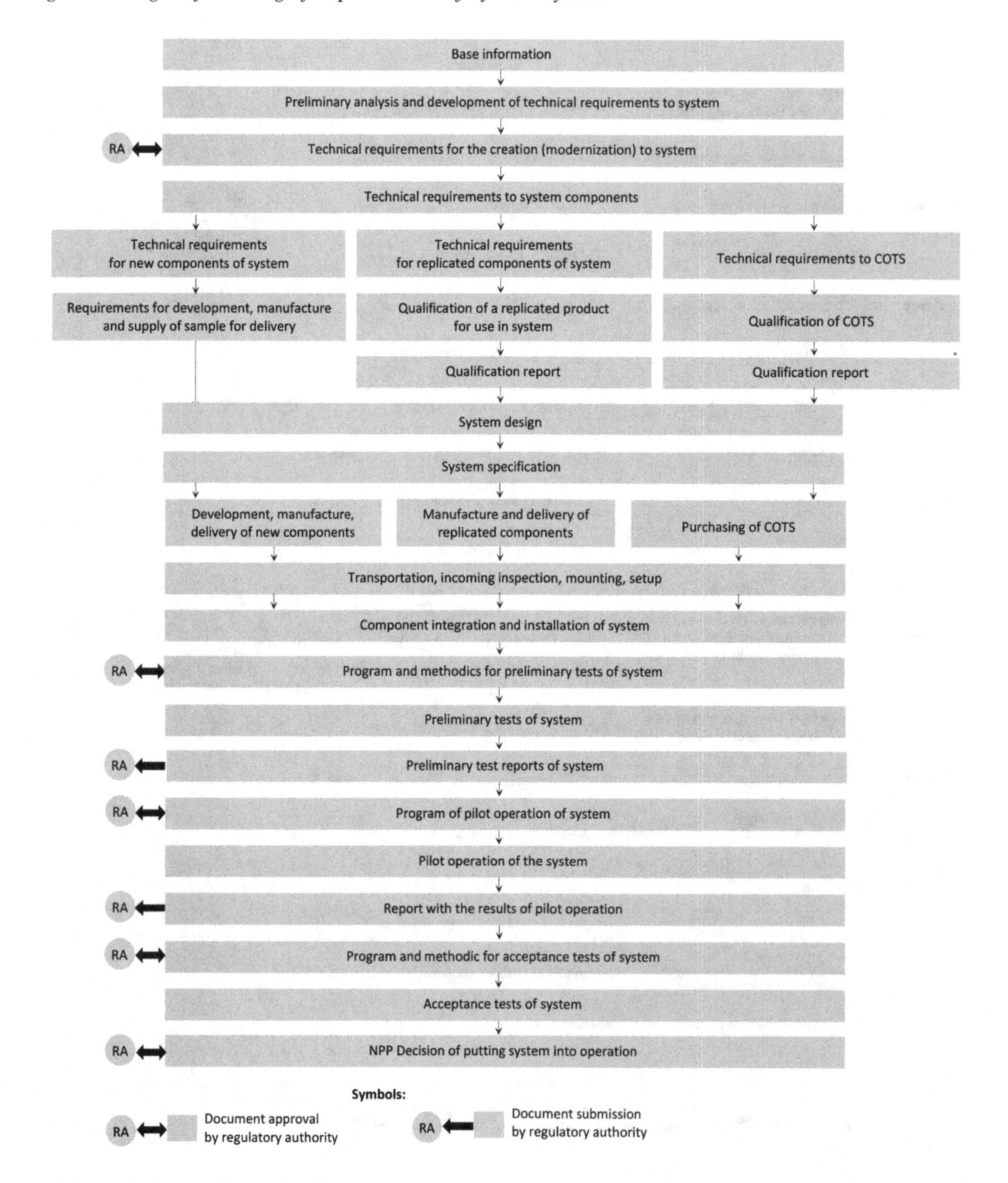

Protection Against Software Intervention

It is necessary to establish the requirements for protection against unwanted and dangerous intervention with the software, as well as from changing it through external computer networks and/or when non-resident storage media are used:

- Software that participates in the performance of functions related to category A is completely isolated from interaction with external computer networks.
- Software that participates in the performance of functions related to categories B and C is isolated from access to the Internet.
- Any changes to the software can be made only after appropriate authorization.
- Potential unauthorized changes to the software manually or using external storage media should be excluded.

The negative impact of measures taken for protection against interference with the software on the execution of programs and/or characteristics of functions performed, which are implemented by software or using software should be excluded.

In more detail, the software requirements (including requirements for elaboration and verification, etc.) are given in Chapter 4.

REQUIREMENTS FOR DEVELOPMENT OF COMPONENTS AND DESIGN OF SYSTEMS

Single, replicated, and commercial off-the-shelf (COTS) products that can be used as components of I&C systems should be distinguished. Based on the concept of safety, requirements for the I&C systems themselves and their components, as well as for the processes of their development, introduction, and operation, which cover all life cycle stages are regulated (Figures 4 and 5).

Single SHC and peripheral products are developed, manufactured, and delivered to NPP's as components of a specific system being developed (new or upgraded).

The initial data for the development of components for a new I&C system are based on the system design. The components intended for the upgraded I&C system are developed based on the initial data of the operating organization (user).

The preassigned initial data are supplemented and specified in the draft Technical Requirements for the development, manufacture, and supply of a single product, which is agreed with the user and the operating organization and pass the state review on safety. Compliance with applicable regulations, rules, standards, and draft specifications is confirmed by the results of product validation.

The replicated SHCs and peripheral products are developed as components for I&C systems whose number is not determined in advance, i.e., as a platform. The requirements for them cannot be "tied" to any one system: they are established by the developers of these products, based on their own representation of their purpose, applicability, and operating conditions. The manufacture (replicated) of such products and their delivery to NPP's is carried out according to the results of acceptance tests of a standard sample implemented on the basis of the developed platform, and reception is based on hand-off tests, confirming the safety of each delivered sample during operation as part of the I&C system for which it is intended.

Commercial off-the-shelf products are increasingly used in I&C systems and SHC. This is due to significant improvement in consumer properties, relatively low cost, sufficient reliability, testability, quality of branded after-sales services resulting from mass production and the competition of leading world manufacturers.

However, such products are not pre-oriented for use in safety-important I&C systems, their development, manufacture and delivery to NPP's is carried out without taking into account regulations, rules, and standards for nuclear safety and is not controlled by the regulatory agency, but information on the volume, methodology and results of their tests is practically inaccessible to system designers. For these reasons, the use of general industrial products as part of I&C systems and SHCs is limited to performing, although more complex, but less critical functions (transfer, archiving, processing and display of data not related to safety, etc.). The potential use of COTS (industrial computers, network equipment, means of the human-machine interface, etc.) as components of each safety-important I&C system and SHC should be justified and confirmed by their qualification.

In addition to factory tests and qualification, NP-2015 and SOU-2016 Ukrainian documents, for example, provide for:

- Check of the delivered components of the I&C system after mounting and adjustment at the place of operation.
- Tests of I&C system after integration of its components, in trial operation, and in introduction of the system into permanent operation.
- Maintenance, inspections, failure recovery, and changes during operation and system upgrades.

SOLUTIONS AND RECOMMENDATIONS

The compliance of the requirements for I&C systems and their components discussed in this chapter with national and international regulations, rules, and standards is a necessary but not sufficient condition for the safety of these systems: it have to be complemented with another condition, which is the fulfillment of requirements for the design, manufacturing, testing, inspection, and maintenance of I&C systems, hardware and software complexes, and peripheral equipment for these systems. This condition is mentioned but was not sufficiently detailed in this chapter and requires separate consideration.

In particular, it is recommend that the requirements for these processes at all life cycle stages of I&C systems and their components be included into new regulations and rules on I&C safety.

FUTURE RESEARCH DIRECTIONS

Further improvement of the I&C systems important for safety and their components should be accompanied by research and development aimed at:

1. Extension of requirements for the properties of NPP I&C systems and their components to be set forth in regulations, rules, and standards on nuclear and radiation safety and observed in the design, development, manufacturing, commissioning, and/or operation in relation to:
    a. Cyber security of I&C systems.

   b. Resistance and immunity of peripheral equipment to limiting values of EFs that may occur in design-basis and beyond design-basis accidents.
2. Presentation in regulatory documents (taking into account national and international experience):
   a. Rules for design of I&C systems important to safety, development of software and hardware complexes, peripheral equipment, and software for these systems, qualification of commercial-off-the-shelf products and programs applied in them.
   b. Criteria and rules for confirming the suitability of general industrial (commercial) equipment and purchased software for use in NPP I&C systems and SHCs.
   c. Methods to determine the properties of NPP I&C systems and their components at all life cycle stages, assessment criteria and rules to confirm compliance with specified requirements.
3. Regulatory and guidance support of lifetime extension activities for I&C systems and their components after expiration of their operational life established by their developers:
   a. Presentation of rules for establishing of qualification requirements for I&C systems, SHCs, and peripheral equipment operating at NPP in regulatory documents.
   b. Establishment of rules and criteria for assessing the results of qualification and making decisions on lifetime extension of I&C systems and their components whose properties correspond to established qualification requirements or on the replacement of products that have not passed qualification and are operated with a significant excess of the established lifetime.
4. Regulatory and guidance support that will be required for the wider expected application of new information technologies and equipment at nuclear power plants:
   a. Establishment of general requirements for relatively separate systems for monitoring the technical condition of structures, systems, process equipment, and automation means, as well as systems for collecting, archiving, displaying, and recording monitoring results.
   b. Development of general rules and instructions for in-service monitoring of the accuracy characteristics of sensors and criteria for their suitability for further intended use.

CONCLUSION

The requirements for I&C systems, their components, software, processes for their development, verification, and assessment, and confirmation of compliance have to be:

- Harmonized with international IAEA and IEC standards for NPP safety.
- Take into account the features of modern information technologies, national and foreign experience in the development, upgrade, and operation of I&C systems and their components at NPP.
- Specify the classification criteria for seismic resistance and the rules for modeling seismic effects during seismic resistance tests.
- Establish the complex of requirements for the development, implementation, and operation of I&C systems and their components, which cover all stages of their life cycle.

Further improvement of the I&C systems involves the establishment, implementation, and monitoring of the cyber security requirements for the I&C systems and the resistance and immunity of peripheral equipment to the boundary values of external factors that may occur in design-basis and beyond design-basis accidents.

REFERENCES

EN. (2010a). *EN 61226. Nuclear power plants. Instrumentation and control important to safety. Classification of instrumentation and control functions*. EN.

EN. (2010b). *EN 55022. Information Technology Equipment - Radio Disturbance Characteristics - Limits and Methods of measurement*. EN.

EN. (2013). *EN 61513. Nuclear power plants - Instrumentation and control important to safety - General requirements for systems*. EN.

EN. (2015). *EN 60987. Nuclear power plants. Instrumentation and control important to safety. Hardware design requirements for computer-based systems*. EN.

IAEA. (2010). *SSG-90. Seismic Hazards in Site Evaluation for Nuclear Installations*. IAEA.

IAEA. (2016a). *IAEA SSG-39. Design of Instrumentation and Control Systems for Nuclear Power Plants*. IAEA.

IAEA. (2016b). *IAEA SSR-2/1. Safety of Nuclear Power Plants: Design*. IAEA.

IAEA. (2016c). *IAEA SSR-2/2. Safety of Nuclear Power Plants: Commissioning and Operation*. IAEA.

IEC. (2005). *IEC 62138. Nuclear power plants- Instrumentation and control important for safety. Software aspects for computer-based systems performing category B or C function*. IEC.

IEC. (2006). *IEC 60880. Nuclear power plants - Instrumentation and control systems important to safety - Software aspects for computer-based systems performing category A functions*. IEC.

IEC. (2007). *IEC 60987. Nuclear power plants - Instrumentation and control important for safety - Programmed digital computers important to safety for nuclear power stations*. IEC.

IEC. (2009). *IEC 61226. Nuclear power plants - Instrumentation and control systems important to safety – Classification*. IEC.

IEC. (2011). *IEC 61513. Nuclear power plants - Instrumentation and control important to safety - General requirements for systems*. IEC.

NAPB. (2002). *NAPB 03.005. Fire protection. Development of firefighting standards for nuclear power plants with pressurized water reactors*. Kyiv, Ukraine: Ministry of Fuel and Energy of Ukraine.

NP. (2008). *NP 306.2.141. General principle safety of nuclear power plants*. Kiev, Ukraine: State Nuclear Regulatory Inspectorate of Ukraine.

NP. (2015). *NP 306.2.202. Requirements for nuclear and radiation safety instrumentation and control systems important to safety of nuclear power plants*. Kiev, Ukraine: State Nuclear Regulatory Inspectorate of Ukraine.

NP. (2016a). *NP 306.2.205. Requirements for power systems important for the safety of nuclear power plants*. Kiev, Ukraine: State Nuclear Regulatory Inspectorate of Ukraine.

NP. (2016b). *NP 306.2.208. Requirements for seismic design and assessment of seismic safety of nuclear power plant units*. Kiev, Ukraine: State Nuclear Regulatory Inspectorate of Ukraine.

Rozen, Y. (2007). Electromagnetic compatibility of instrumentation and control systems components (1). Rules for regulations and estimation. *Nuclear and Radiation Safety, 2*.

Rozen, Y. (2008). Electromagnetic compatibility of instrumentation and control systems components (2). Electromagnetic interference resistance. *Nuclear and Radiation Safety, 4*.

SOU. (2016). SOU NAEK 100. Engineering, scientific and technical support. Instrumentation and control systems important for the safety of nuclear power plants. General technical requirements. Kiev, Ukraine: National Nuclear Power Company "Energoatom".

WENRA. (2012). *Research of WENRA Reactor Harmonization Working Group. Design safety of new nuclear power plants*. WENRA.

WENRA. (2014a). *WENRA Safety Reference Levels for Existing Reactors. Update in relation to lessons earned from Tepco Fukushima Dai-Ichi Accident*. WENRA.

WENRA. (2014b). *WENRA Safety Reference Levels for Existing Reactors*. WENRA.

ADDITIONAL READING

IAEA. (2011). NP-T-3.12. *Core Knowledge on Instrumentation and Control Systems in Nuclear Power Plants*.

IEC. (1996). IEC 60980. *Recommended Practice for Seismic Qualification of Electrical Equipment for Nuclear Power Generating Stations*.

IEC. (1998). IEC 60780. *Nuclear power plants - Electrical equipment of the safety system – Qualification*.

IEC. (2009). IEC 61226. *Nuclear power plants -Instrumentation and control systems important to safety – Classification*.

KEY TERMS AND DEFINITIONS

Common Cause Failure: Simultaneous failure of two or more elements in different redundant channels resulting from the same cause, which may lead to a failure of the I&C (HSC) function.

Diagnosis: Determination of the technical condition of a system (component) and detection and identification of its inoperative components at the appropriate level of disaggregation.

Diversity: Is a way to reduce the likelihood of common cause failure of two or more redundant systems, components, or channels that independently perform the same function though specially provided differences between these systems (components, channels) implemented at the stage of their design, development, and/or manufacture.

External Factors (EFs): Factors that can occur in places of hardware location and include: environment EFs (temperature, humidity, barometric pressure, ionizing radiation, corrosive agents, dust), mechanical EFs (vibrations, strokes, seismic effects), power supply EFs, specific environment EFs (water and solutions that can affect devices in accidents and decontamination fluids), and electromagnetic EFs.

Life Cycle: A set of stages for development, commissioning, and use of a system (component) within a time period that that begins from the concept development and specification of technical requirements and ends when the system (component) is removed from service when its further use is impossible or inexpedient.

Redundancy: Application of additional tools and/or capabilities that are redundant to those that are minimally required for function performance.

Single Failure Criterion: Criterion that requires an I&C system (SHC) to perform all specified functions in any postulated initiating event (PIE) combined with a failure of one (any) element independent of this PIE.

Chapter 4
Requirements to Products and Processes for Software of Safety Important NPP I&C Systems

Vladimir Sklyar
National Aerospace University KhAI, Ukraine

Andriy Volkoviy
Mellanox Technologies Ltd., Kyiv R&D Center, Ukraine

Oleksandr Gordieiev
Banking University, Ukraine

Vyacheslav Duzhyi
National Aerospace University KhAI, Ukraine

ABSTRACT

Features of software as a component of instrumentation and control (I&C) systems are analyzed. Attention is paid to the importance of functions performed by software and hazards of such software. Requirements for characteristics of software as a component of I&C systems are analyzed. Different regulatory documents are considered in order to disclose common approaches to the use of dedicated software and off-the-shelf software components. Classification of software, as well as classification of requirements, is described. Criteria of selection and structuring of requirements, as well as criteria for software verification, are defined. As long as the characteristics of software components directly depend on the quality of the processes of software development and verification, requirements for software life cycle processes are considered.

DOI: 10.4018/978-1-7998-3277-5.ch004

INTRODUCTION

Regardless of the purpose and application area any modern digital systems has software as integral part of the system. Instrumentation and control systems are not exceptions and may include software in many various forms: firmware and embedded software (written for particular hardware and usually executed without an operating system), system software (e.g. operating systems and platforms), middleware and device drivers, application software (typically written to be run under operating systems and usually interact with users), configuration for FPGA devices, etc. Software of different forms and types has specific properties. Moreover functions that are performed by software impose constraints on both software as a product and software lifecycle as a processes. For example, use of operating systems and application software has a very limited scope in safety important systems.

In the context of safety important I&C systems, increase in portion of software-produced or software-supported functions requires more attention to software. In this chapter software (SW) for nuclear power plant's (NPP) instrumentation and control (I&C) systems is concerned. That means that references to specific regulations for nuclear power engineering are given, particular terminology and classifications are used.

BACKGROUND

The increase of the number of nuclear power plant I&C software executed functions causes an increase of the "weight" of software device defects and its possible sources of failures. Based on different estimates such defects cause up to 70% of the failures of computer systems of critical application complexes, of the total number of those attributed to nuclear power plant I&C systems (Everett, 1998) (Lyu, 1996). Given this, the present trend is having an increasing dynamic role over time.

In the 1960s software defects caused up to 15% of the failures, and in the 1970s it was 15-30%, and by the year 2000 they were the cause of up to 70% of computer system failures. This trend shows up even more in space rocket technology (Aizenberg, 2002). Analysis of the cause of accidents and catastrophes of space rocket systems, where on board and ground computer systems have already been in use for several decades, allows one to determine that in the past 40 years each fifth accident is related to failure of a digital control system. Six of seven failures of these systems were caused by the occurrence of software defects. One such defect of computer software of the Ariane-5 navigational system in 1997 led to an accident which cost nearly one half billion dollars (Adziev, 1998). In nuclear power generation programmable I&C systems have had a shorter history, however, here also there have been accidents due to software defects.

The reliability of software, as for the I&C system as a whole, depends on the design quality at stages that directly precede development of the software:

- Development of requirements for I&C system.
- Mathematical modeling.
- Software implemented functioning algorithms.

Errors committed at these stages become sources of complex defects in software. In this sense, software, on the one hand, accumulates the deficiencies of the preceding stages, and on the other hand, is the "field," in which they can show up and be eliminated. However, the efforts that must be made to do this, increase by an order of magnitude.

Consequently, software is becoming an even more important factor determining the safety of nuclear power plant I&C system. This explains the fact that software of nuclear power plant I&C system, in accordance with national and international normative documents, is a separate and very important object of safety standardization.

SOFTWARE OF NUCLEAR POWER PLANT I&C AS AN OBJECT OF SAFETY REQUIREMENT ESTABLISHMENT

Aspects of Software in Establishing Safety Requirements

Software has a number of important features that should be taken into account in establishing requirements for it. The main of these features are listed below.

1. On the one hand, software is a component of I&C system and shall comply with general requirements for the system, and on the other hand it is an independent and specific object for establishment of requirements, which is confirmed by a large number of international and national standards and methodological normative documents completely devoted to software.
2. Defects that are committed during the development and are not revealed during software verification, can be actuated under certain conditions in the I&C system operating process and lead to their failure. This failure cannot be compensated even if redundant channels are available. If that channels use identical software versions, software defects are in all channels and reveal themselves simultaneously leading to the same kind of distortion of information at the outputs. Therefore, software defects are potential and quite likely source of common cause failure. For this reason, on the one hand software requirements include both requirements for its characteristics (structure, functions and properties) and software lifecycle processes; on the other hand there is a requirement for whole I&C system related to adherence of diversity principle, that is addressed primarily to software, because the use of several program copies increases the likelihood of failures and faults, caused by their hidden defects.
3. At different stages of the software lifecycle (primarily design, coding, integration and testing) different tools are widely used. These tools are also software products, which are intended to reduce the number of defects and increase the reliability of I&C software. However, defects can also be introduced into the I&C software through the software tools. It is the common approach when control systems are based on programmable logic controllers (PLC) for which specialized computer-aided design (CAD) tools are used, and in view of the complexity of such CAD tools both intrinsic defects of a tool and improper use of a tool can be the source of I&C software defects. Therefore, requirements for software must include requirements for software tools used in development and verification.
4. Because documentation is an integral part of software, the requirements for I&C software also include requirements for documentation that is used at all stages of the lifecycle.

5. Software must be examined not only as an independent object of safety standardization, but as a necessary means that will ensure conformity of the I&C system to requirements established for it with regard to redundancy, maintainability, technical diagnostics and so forth.
6. Software requirements are not permanent. The experience with creation and use of I&C system as well as improvement of the information technologies lead to the necessity to improve the requirements. Therefore, requirements must reflect basic and most stable situations considering this experience and prospects of software development technologies.
7. Nuclear power plant I&C systems are complex systems which can consist of several subsystems, each produced with the use of one or several platforms. Consequently, software of I&C system is a set of various software components (computer programs), which differ in functional purpose, developer companies, programming languages and technologies used, etc. This causes asymmetry of requirements for different software components.
8. Quantitative requirements for reliability are difficult to establish for software, in contrast to I&C system hardware items. There are several factors causing the absence of common and standard methods of quantitative evaluation of software reliability. These factors include: uniqueness of software as an object of evaluation, in spite of actively continuing industrialization of development processes and introduction of numerous standards for techniques of developing software; insufficient development of theoretical aspects of this evaluation and lack of a mutual opinion about its expediency; complexity of representing objective and complete information on defects that are discovered at different stages of the software lifecycle, and others.

Classification of I&C Software

Specification of requirements for different kinds of software depends on and usually based on I&C software classification. The following classification features are recommended to use (see Figure 1):

- Affiliation of the software with various I&C system and subsystems.
- Functional purpose.
- Level of approval.
- Effect on safety.

The selection of these classification features is made on the basis of analyzing modern international standards for I&C software, which are important for nuclear power plant safety, in particular (IAEA, 2000), (IEC, 2006, a) and (IEC, 2008).

Based on these features software has a multidimensional (parallel) classification, in which individual groups of its types are relatively independent. The arrows between components of individual facets indicated the most preferred combinations of software types, which are classified according to different features. It should be noted that some facets can be more detailed and presented in the form of hierarchical classifications.

Figure 1. Classification of I&C software

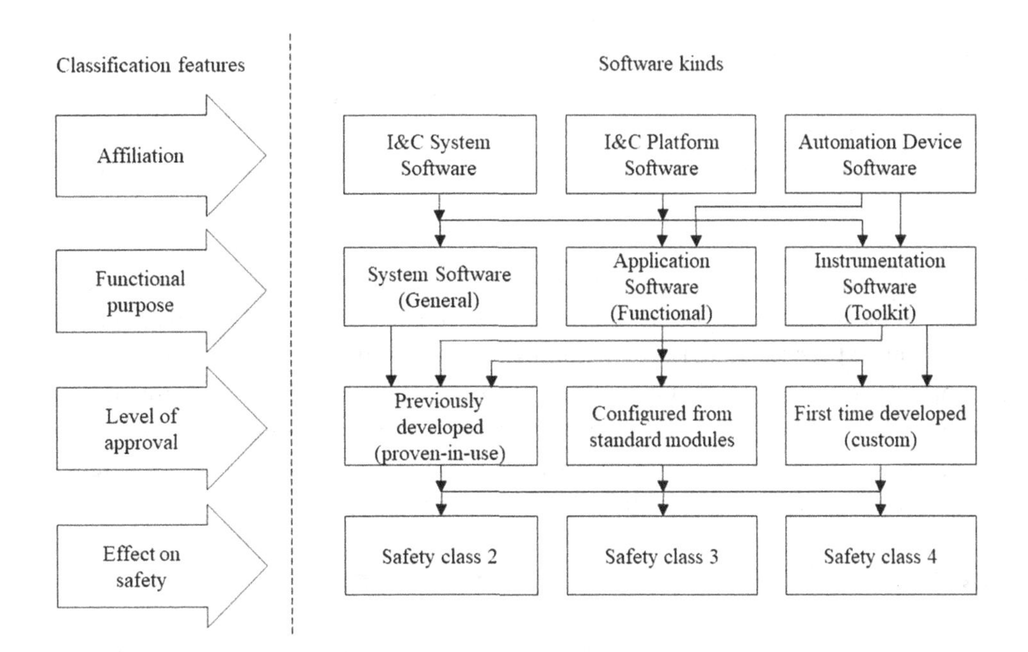

By affiliation software can be a part of: I&C system, I&C platform, some automation devices or equipment.

Based on purpose software is classified into: general (or system) software; application (or functional) software; instrumentation (or toolkit) software, which is used in development, testing and verification. The examples of instrumentation software include different tools, which are intended for processes of design, translation, configuration control, debugging, and verification.

Level of approval is an important classification feature according to which there are:

- Previously developed (proven-in-use) software, also known as off-the-shelf (OTS) software. This kind of software can include commercially accessible software, developed and supplied by other companies, and also standard application software, which is created and approved in similar or different projects.
- Software configured from standard (previously developed) software modules (library blocks). The configuration tools for such software usually is proven-in-use software.
- First time developed (custom) software. Such software is created especially for the given system and has no operational experience in other applications.

Previously developed (OTS) software further be classified by other features, such as source code availability (openness), possibility of changes, amount of operating experience, etc.

Influence on safety is determined by I&C system safety class in which this software is used. According to the Ukrainian legislation any I&C system must be assigned to one of three safety classes, denoted by numbers 2, 3 and 4. Moreover for functions performed by I&C system are assigned to the category denoted by letters A, B or C. Therefore I&C system can be:

- Safety class 2(A), if at least one function of that system has category A.
- Safety class 3(B), if system does not perform category A functions and at least one function of that system has category B.
- Safety class 3(C), if system does not perform category A and B functions and at least one function of that system has category C.
- Safety class 4, if none of its functions are classified by category (such systems are consider as non-safety).

It is important, that affiliation of software does not affect I&C software requirement directly. Purpose of software affect on the requirement, because special set of requirements is established for tools that are used for development and verification. Level of approval strongly influences the software requirements, e.g. required methods and scope of verification can be very different for proven-in-use and for custom software. But, of course, the greatest dependence is between software requirements and influence on safety, expressed by Safety Class. Moreover, safety class defined for I&C system imposes requirements for software of all components, platforms and even related automation devices.

The Criteria of Selection and Structuring of Requirements

Selection and any activities aimed at meeting requirements are impossible without establishing a clear classification features, determining factors and selection criteria. The main factors and criteria are considered below.

1. General criteria for selection of requirements or, in other words, "requirements for requirements." Among such criteria for nuclear power plant I&C software the most important are the criteria of necessity, completeness, adequacy, correctness, verifiability, and openness. These criteria are related to the criteria that were developed and are used for evaluating the execution of requirements for software during expert analyses (Vilkomir, 1999), (Vilkomir, 2000). For example, in accordance with the criterion of completeness during generation of many requirements for software elements must be separated and taken into account that reflect "covering" by requirements of these components such as: completeness of conformity to specifications; completeness of consideration of software lifecycle stages; completeness of the diagnostics, and so forth.
2. Classification and content of I&C system requirements as a whole. The full set of these requirements includes:
    a. Requirements for the composition of the functions.
    b. Requirements for quality of the execution of these functions.
    c. Requirements for reliability of function execution.
    d. Requirements for stability of function execution against external influences.
    e. Requirements for lack of influence on other systems.
    f. Requirements for procedures and processes that support meeting requirements for functions, quality, reliability and stability.

This set should be designed for the full set of software requirements and should be correspondingly supplemented and specified. In particular, the subsets of requirements for processes of software development and verification, which play a priority role from the standpoint of assurance of reliability and safety, should be expanded and worked out in maximum degree.

3. Particular features of software as an object of safety standardization. The following set of the software features have a direct effect on the selection of classification features and generation of subsets of requirements:
    a. Software is both a component of the system for which regulatory requirements have been established and a means that assures fulfillment of the regulatory requirements for I&C system. Consideration of this feature is most important in defining requirements for monitoring and diagnosis, reliability and stability. In doing so different external disturbing influences for software should be examined.
    b. Software is a possible source of common cause failure. Nature of software makes it necessary to have requirements for protection from common cause failures due to improvement of software development and verification processes and use of the diversity principle, which in turn determines the necessity of classification features for methods and means of diversity implementation.
    c. Software is a multi-component system. During the statement and classification of requirements the purpose, level of approval and safety class of different software components have to be considered.
    d. Software is a product and a process. This feature of software is one of the critical ones in selecting classification features of requirements and generation of their complete sets, which considers the certain influence of development and verification processes on software characteristics.
4. The existing regulations, which include standards determining software requirements. On the basis of these standards requirements for I&C software can be selected as the socalled normative profile for software. In the general case normative profile is a subset and/or combinations of the positions of basic standards for a specific subject area, which are required for implementation of the required functions in the system. In this case, we mean the normative profile of requirements for I&C software that is important for nuclear power plant safety. The said standards form the profile-forming base for producing the normative profile of software requirements (for example, software lifecycle models, structure of requirements for software, set of metrics and methods of evaluation, requirements for tools, etc.).
5. Possible variants of requirement structuring. This factor is conceptual in nature, because it determines the general approaches, priority and interconnection between different requirements for I&C software. Several variants of software requirements structuring are possible:
    a. Product-oriented: requirements that determine characteristics for software as a component of I&C system. It does not take into account the fact that software characteristics are built in and implemented at different stages of the software lifecycle.
    b. Process-oriented: requirements correspond to software lifecycle processes and define features of process and intermediate product of each stage in the form of "stage-tasks-requirements" statements. This approach is widely used and allows clear process management and quality assurance, but complicates the definition of software product features and for complex software can lead to difficulties with integration.

c. Mixed process-product-oriented: requirement are divided in two groups and describe both features of development processes and features of final product. In this case the advantages of the first two approaches are used.

General and Functional Requirements

The classification of software requirements can be performed in two stages: in the first stage, which corresponds to the upper level of the hierarchy, we determine the place of normative requirements among the full set of requirements for software (classification of kinds of requirements for software); in the second stage, which corresponds to the lower level of the hierarchy, we carry out the classification of general requirements for software, based on the process-product approach.

The set of requirements for software corresponds to the set of requirements for I&C system, because it contains both requirements and functions, and for their quality (properties), and to reliability, stability, and processes (both development and verification). The particular features of this full set for software consist of:

Figure 2. Classification of software requirements

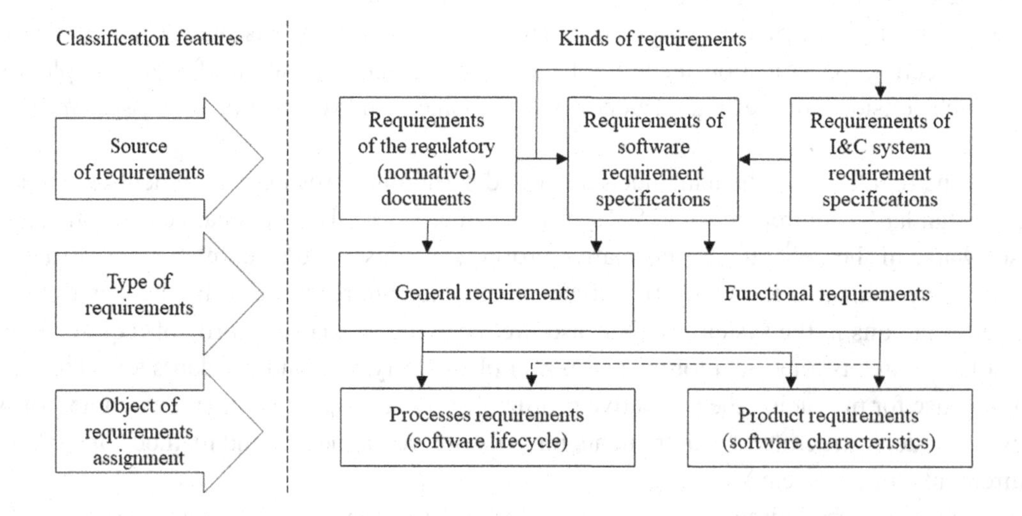

1. Requirements for software as an element of I&C system are determined based on requirements for the I&C system as a whole.
2. Requirements for the structure in software elements precede requirements for functions. In this case we are speaking of general requirements for software functions that are important for safety, and not about functions that are determined by its purpose.

3. Full set of quality characteristics we have separated out one, which is most important from the standpoint of safety standardization, which determines the requirements for monitoring and diagnostics.
4. Requirements for processes are determining to a great extent, because they are expanded and worked out in detail with consideration of safety assurance.

In order to conduct classification for requirements we shall distinguish three features: source of requirements; type of requirements; object of requirements assignment (Figure 2).

For the first of them we can distinguish requirements of the regulatory (normative) documents and requirements that are contained in the requirement specifications for development of I&C system and development of software.

In the development of specifications for software or I&C system (as a whole) requirements of the regulatory documents must be taken into account. Requirements of the specifications of the software are developed with consideration of the specifications for I&C system.

According to the type of requirement software requirements are divided into general and functional. General requirements do not depend directly on what the functions are implemented in I&C software, but are determined only by the safety class, level of approval and its purpose. Functional requirements depend completely on the purpose of the I&C and tasks which are solved by the software. The functional category normally includes requirements for productivity, synchronization, information protection, required service lives, portability and so forth.

Depending on the object of assignment one can distinguish requirements for software lifecycle processes (development and verification) and product requirements (software characteristics). In the regulatory documents general requirements are normally given as those pertain to processes and products. Functional requirements as a rule pertain to software and to the product, although they can determine some of the requirements for the process of software creation with consideration of specific features of the design, the tools used and so forth.

The classification of the general requirements is considered below.

Results of the classification of general requirements for software characteristics (software as a product) and processes of its creation are given in Figures 3 and 4 respectively. In the classification of requirements for software characteristics two groups of requirements are delineated: for structure and for properties. The first group includes requirements for features of the construction and functioning of software. The second of these groups brings together requirements for software properties such as requirements for its sufficiency and adequacy for functions execution, monitoring and diagnosis, reliability and stability, protection against cyber threats.

Figure 3. Classification of requirements for products (software characteristics)

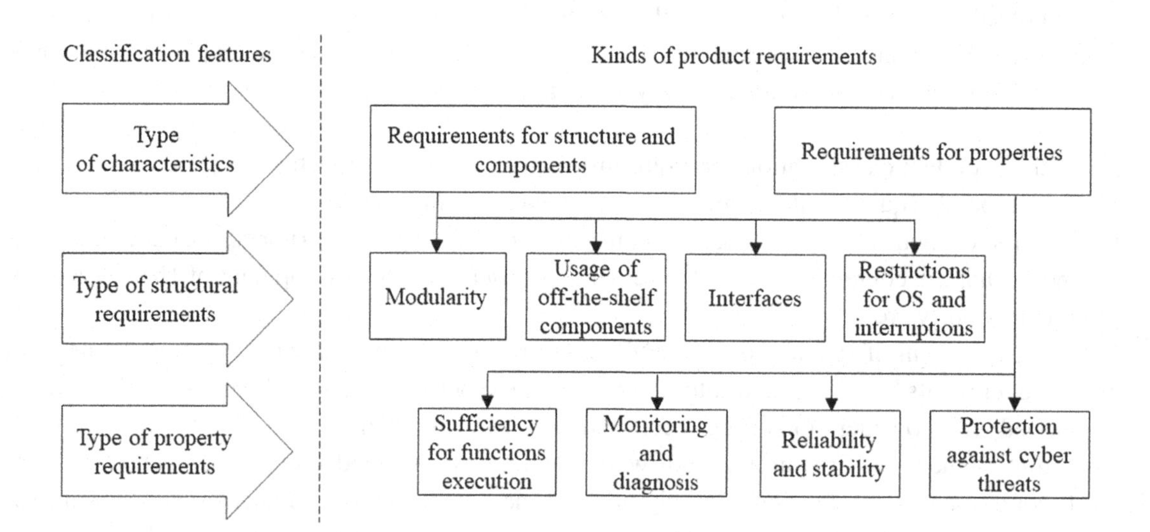

Requirements for Software Characteristics

Requirements for structure and components includes the following requirements:

1. Requirements to modularity.
2. Requirements to use of off-the-shelf components (pre-developed software).
3. Requirements to interfaces.
4. Restrictions for use of the operating system and interrupts.

The first subgroup of the requirements is due to the need to present software in the form of a modular structure. In doing so the source code of one module must contain a limited number of operators, and the modules must have a clear structure, be easily modifiable and tested.

The second subgroup determines the preference of the use of previously developed software. Using OTS software components one must: evaluate its conformity to the functions and characteristics of I&C system, where the use of OTS components is preferred for which one should determine the functions and characteristics of the OTS components and correlate them with specifications for I&C software; analyze the results of OTS components operation from the standpoint of its conformity to the adopted criteria, norms and rules of safety; develop, if necessary, a list of the required modifications for adaptation of the OTS components to conditions of its use in I&C system; execute such adaptation and perform testing; develop and implement the plan for verification of the changes made. The amount and extent of evaluation of conformity of OTS software components to these criteria are determined by the safety class (I&C safety class). The importance of requirements for the use of OTS components (both developed as special purpose and COTS-components) for safety of the I&C systems as a whole should be emphasized. According to existing estimates (Kersken, 2001), the amount of OST components in software of mature systems can reach 80-85% of the total amount of software.

Figure 4. Classification for requirements for processes (software lifecycle)

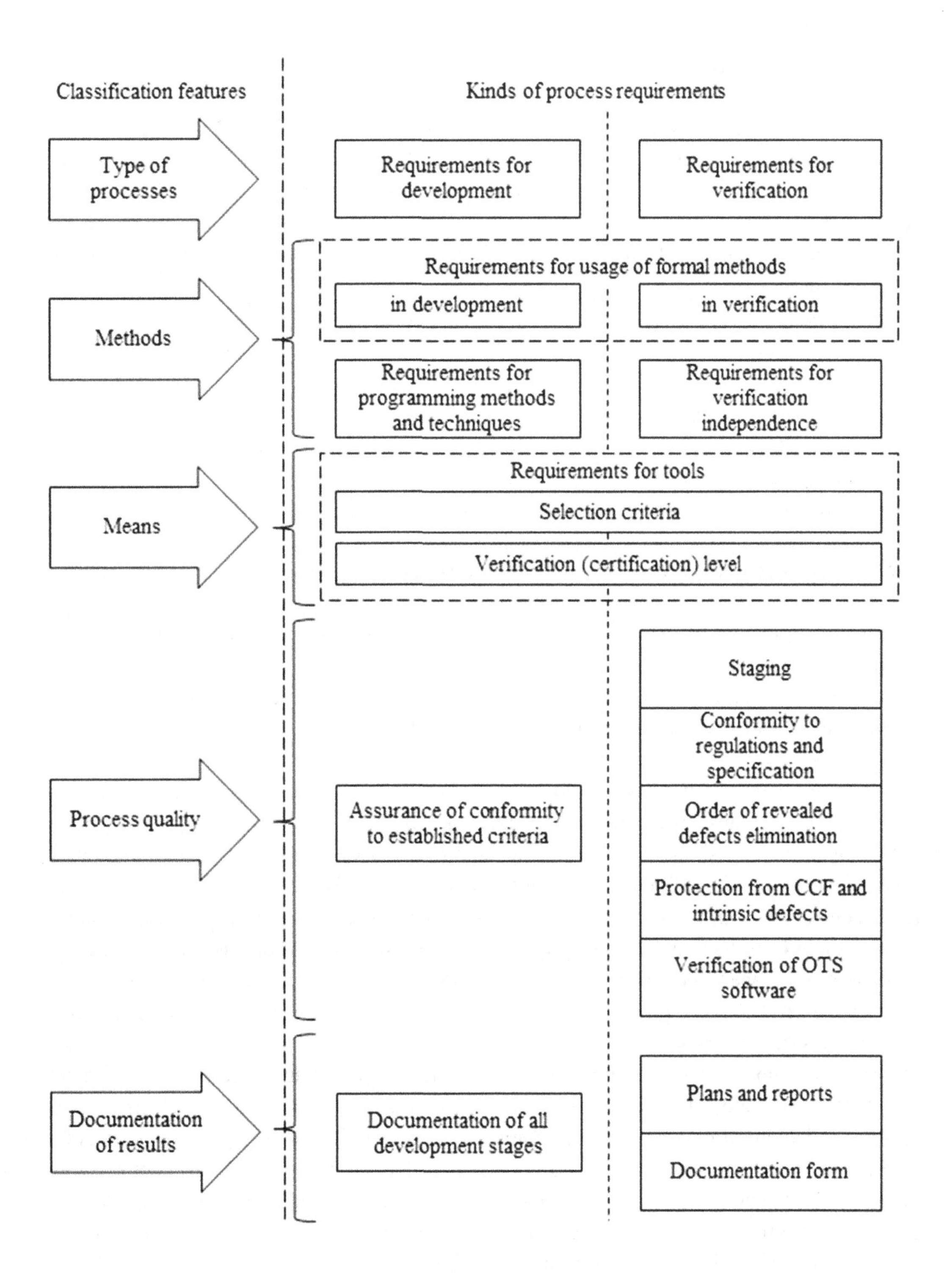

The third of the listed subgroups of requirements determines the need for complete and clear description of the interfaces between the software being examined and the operator (also known as human-machine interface), hardware platform and peripheral hardware (sensors, drives and so forth) of a given I&C system, and also other I&C devices, systems and subsystems. This description determines the limits of the software being analyzed.

Restrictions for use of the operating system and interrupts are included in requirements for I&C software of safety class 2. If the use of the operating system is deemed necessary, it should execute only the simplest functions. The use of interrupts in the course of executing the most critical functions should be prohibited. One should note that in order to fulfill this requirement more precise criteria should be developed, after providing a detailed explanation of the functions for which the use of the operating system must be limited.

Requirements for monitoring and diagnostics can be divided into four groups with consideration of the kinds of processes and objects, for evaluation of the state of which software is used:

1. Requirements to monitoring I&C system by programming means.
2. Requirements to diagnosis (search for malfunctions) of I&C system by software means.
3. Requirements to self-monitoring of software.
4. Requirements to self-diagnostics of software.

In other words the main requirements for monitoring and diagnosis are:

1. Software should perform (a) continuous automatic monitoring of operating condition and (b) periodic function checks of the I&C system.
2. Software should provide diagnostics of I&C system at the level required by specification.
3. Software should provide self-monitoring and self-diagnosis.

For this purpose, the following should be used: monitoring of intermediate and the final results of the execution of programs and their allowable duration; repeated counting and comparison of the results; discovery of prohibited situations; monitoring data in memory and so forth. For monitoring of I&C software of safety class 2 different types of diversity can be used.

It is necessary that in the process of monitoring and diagnosis: all functions are checked that are important for I&C system safety; during periodic testing it is mandatory to check devices which are not built-in or permanently connected monitoring devices; all degradations of characteristics of safety functions are discovered on a timely basis; if any failure is discovered timely automatic actions that correspond to the situation are generated.

Moreover, an important part of requirements for monitoring and diagnosis are the requirements related to execution of mandatory limitations and procedures during their implementation: implementation of monitoring and diagnostic programs (self-monitoring and self-diagnostics) should not affect fulfillment of programs of the main information and control functions and/or lead to unacceptable degradation of a characteristic; one should make an analysis of the situations and procedure, which allow to avoid false errors; the software should provide automatic recording, storage and display of data on results of monitoring and diagnostics (self-monitoring and self-diagnostics).

Requirements for reliability and stability: By reliability of software we mean its property of preserving serviceability and converting raw data to the result being sought under the given conditions in the assigned time. By stability of software we mean its ability to execute its functions in anomalous situations (during breakdowns and failures of hardware devices, operator errors and errors in the raw data).

Requirements for software related to assurance of reliability and stability can be classified according to a scheme, whose basic elements are: sources of failures and influences on software and I&C system; kind of failures and influences; methods of protection from them.

Sources of failures can be: internal sources with respect to I&C system (both software and hardware); external sources with respect to the I&C system (other I&C systems; operating personnel; repair personnel).

By kind of failures, which should be compensated by means of programming devices, we can distinguish: failures (breakdowns) of hardware devices; failures (breakdowns) caused by the appearance of software defects, which are introduced at the design stage and are not detected during testing and verification.

In turn, software anomalies that can be the cause of I&C system failure are classified into:

1. Defects that appear under certain conditions of the system, its individual components and sets of input signals.
2. Defects that appear during non-standard functioning of hardware of the I&C system.
3. Defects caused by incorrect or incomplete specifications of the software.
4. Defects introduced in development of the software (at all stages of the lifecycle).
5. Defects related to the use of tools and that depend on other software and interfaces between parts of the software or other systems.

The main kinds of influences, resistance to which should be assured by software are the following: unintentional or intentional errors of personnel; unauthorized actions or unauthorized access to programs, data, operating systems; malicious software, including viruses, spyware and trojans, which are sets of instructions that execute actions not stipulated by the specifications and that represent a threat to safety; distortions of incoming information that arise from measurement devices (sensors) and along communication channels from other systems.

Thus, requirements for software related to reliability and stability consist in that the software must implement protection from all of the listed kinds of failures and actions. In this case protection should be assured from failures by general factors, which are due to the appearance of intrinsic defects of the software, by failures and breakdowns of hardware devices of the I&C system.

The following methods are used to protect from the listed kinds of failures and influences on software.

1. Technical diagnostics (monitoring and determination of the cause of a failure or breakdown), reconfiguration of the structure and restoration of the computational process or control process. This method is universal and by appropriate loading of its constituents can assure protection from a broad class of failures. In the I&C system it should be used for protection from hardware failures.
2. Software, functional or other kinds of diversity. The use of diversity is a systems requirement, which is aimed at protection from common cause failures and is related to the use of different kinds of redundancy in the process of creation (development and verification) of software and in the final product, i.e. the software itself. Software diversity (usage different software versions) is achieved by using different algorithms, languages, libraries, programming approaches, operating systems and so forth. Functional diversity is assured by using more than one criterion for identification of each situation that requires the initiation of control actions.

It should be emphasized that for software of safety control systems, which execute emergency protection functions, the emergency situations must be discovered by several methods based on different physically interconnected production parameters, while the analysis of data on the values of these parameters should be performed by different software modules.

For I&C software of safety class 2, in addition, when using software, functional or other kinds of diversity, one must: evaluate the degree of correlation of different versions (analyze the actual level of diversity), their capability for joint compensation of software defects; analyze the substantiation and influence on safety for additionally introduced components – different software or hardware versions. Software diversity should not create a danger of non-fulfillment of functional requirements.

3. Establishment of access categories, application of different password systems, digital signature procedures, use of special encoding algorithms and others. These methods assure protection from errors of personnel and unauthorized actions.
4. Monitoring reliability and protection of incoming information from distortions. In this case one should check: incoming signals being present in zones of access, established in accordance with the specification; logical non-contradictory nature of values of input variables and so forth.

Note that during the use of all protection methods software is an object of protection and a means of assuring reliability and safety. In this case, there must be the introduction of additional software components, which in turn can be sources of failures and therefore they must be carefully analyzed. Functional characteristics of the software and I&C system as a whole should not be degraded to an unacceptable value (just as during monitoring and diagnostics).

Protection against cyber threats: The software of I&C system shall be protected from undesirable and unsafe interference to work and unauthorized changes via external computer networks and the use of non-resident storage media.

To achieve such protection connection with Internet shall be excluded and any changed can be possible only after appropriate authorization. Also special methods of protection from viruses and other malware should be used.

At the same time, measures against cyber threats should not affect the execution of applications software and deteriorate performance of the functions that are implemented by software.

Requirements for Development of Software

Requirements for methods of software development are divided into two main groups:

1. Requirements to use of formal methods.
2. Requirements to programming methods and approaches.

The first group of requirements is to a certain extent recommendatory in nature and indicates the need (expediency) of using at all stages of development and verification formal methods that are based on rigorous mathematical description of formulations of problems related to different stages of software improvement and verification with use of a theoretical apparatus of algorithms, mathematical logic, graph theory and so forth, and also on proof of the correctness of solving these problems by means of standard procedures. Such methods are used in particular for:

1. Transition from verbal to formal description of general and functional requirements for software and development of its formal specifications.
2. Mathematical proof of the conformity of software to specifications or requirements of previous stages of development.
3. Development of application programs using formal procedures of synthesis.
4. Analysis of syntactic and semantic correctness and carrying out test verifications of execution of functional requirements for software.
5. Improvement of the verifiability of software and formalization of the evaluation of results during independent verification and validation.

The best developed and best known methods of formalized checks of software are methods based on formal procedures of logical output, proof of the correctness of algorithms and programs (Anderson, 1979), and also FTA- and FME(C)Aanalysis methods that are widely used to analyze hardware (IEC, 2006, c) and (IEC, 2006, b). The first of them is based on constructing fault tree and events analyses. The second is based on analysis of the fault modes and effects criticality analyses.

The second group of requirements is determined by the preference of using standardized designations of variables in software, files of constant and predefined length, subroutines with minimal number of parameters (e.g. with one output and one input), etc. Moreover, this group of requirements is related to the need to exclude methods in programming development that complicate the software, e.g. complex branches and cycles in the programs, complex indexes in the files and so forth. Note that in the methodological normative documents, which are used in some countries, requirements are contained regarding the need of use of systems in software development that are important for nuclear power plant safety, special methods that improve its reliability, in particular the so-called method of defensive programming (Lawrence, 2002), (Ben-Ari, 2000).

Requirements for tools used to develop software reflect two aspects that are related to their usage:

1. Determination of the criteria for selecting automated development and verification tools.
2. Degree of verification of these tools.

It should be noted, that in the existing normative documents the selection criteria of tools are not given, but the need for the software developer to provide substantiation of such criteria and demonstrate proof that the devices used conform to them is postulated.

The main principle applicable for the tools is that tools used to generate code, must pass through verification with the same requirements as the I&C software itself.

Requirements for Software Verification

Software verification is an important part of I&C software lifecycle. Verification of software is defined as the process of proving the conformity of results obtained at a certain stage of software development with the requirements established in the preceding stage. As noted earlier, the majority of requirements for methods and means of development and verification are uniform. An important distinguishing feature of this group of requirements for verification is the necessity of assuring its independence, that is, carrying it out by persons who are not direct developers of the software.

Requirements for Verification Independence: The integral requirements of independence are level of independence of the experts (organizations) that conduct software verification, and the agreement of these levels with the software safety classes.

The following levels of independence are possible:

1. **Maximum Independence**: Verification is conducted by experts or organizations that administratively and/or financially are independent of the software developers. This level of independence can be broken down into two sublevels:
    a. Administrative independence.
    b. Administrative and financial independence. In this case we are speaking of conducting verification by representatives of a different organization, which specializes in solving such tasks.
2. **Partial Independence**: Verification is carried out by other experts of the same organization, and their administrative and/or financial independence from the software developers is not required. In this case there can be partial administrative and/or financial independence, if the verification is conducted by experts of a different subdivision of the organization, for example by representatives of quality control service, are subordinate directly to the director.
3. **Minimal Independence**: Verification is conducted by the developers themselves, and the review of its results is performed by other experts.

By means of the technology of independent verification and validation (IV&V) one can implement the principle of diversity with respect to the software creation process. In order to assure the highest degree of verification it is necessary that one use tools (utilities), that are different from those which the developer used.

It should be noted that conducting an independent verification can be accomplished according to different systems and with different depth, which depends on the software safety class, worthwhile tasks and existing resources.

Actually three basic scenarios of verification implementation are possible:

1. Full verification and validation of the entire project is carried out, which repeats practically all stages of verification within limits of the project, using intrinsic (diverse) tools and methodologies.
2. Independent consecutive evaluation (rechecking) of all results of the verification performed by the developer organization is carried out. In this case all checks are conducted that are stipulated by the verification and validation plans, and also checks proposed by specialists of the expert analysis organization, and tools of both the inspected and inspecting organizations are used.
3. Independent sampling evaluation (recheck) of results of the verification of the most important functions from the safety standpoint is carried out, which is made by the developer organization.

Quality of Verification: The use of independent verification and validation techniques allows one to improve the quality of this process. By software verification quality we can mean the degree of conformity of software to regulatory requirements after it is carried out and elimination of any discovered defects.

The verification quality is evaluated by analyzing fulfillment of the following requirements:

1. Requirements to staging of the process. The essence of the requirement consists in that the verification must be carried out after each software development stage (specification, design, and coding and others).
2. Requirements to verification of software conformity to requirements of normative documents (general requirements for characteristics and software development, described earlier) and specifications (functional requirements).
3. Requirements to order of elimination of any discovered defects and malfunctions. Components of this requirement are constituents of the process of elimination of defects, time periods for defect elimination, conformity of the time periods of elimination of defects to the software safety class. The process of eliminating defects, independent of the software safety class, includes that a mandatory stabilizing when discovered in the process of development, testing and verification; analysis of the causes, degree of influence on safety; introduction of the necessary changes to the software; repeated check of software with documentation of the results.
4. Requirements to protection from intrinsic defects and common cause failures. Elements of this check are discovery of potential sources of CCF, caused by defects of the software or other components; analysis of their influence on safety of the software and I&C systems as a whole; evaluation of the effectiveness of using devices to protect against these failures.
5. Verification of different kinds of software, including previously developed (OTS) software.

Requirements for Documentation: Documenting is an important part of the verification process and implies the development of two basic documents (groups of documents):

1. The software verification plan, which can consist of a general (coordination) and several particular verification plans and test methods;
2. The software verification report (reports and test protocols) for software verification.

Requirements for documenting software verification results include requirements for the presence, structure and content of a plan (plans), produced before the beginning of verification, and report (reports), which is produced based on results of verification and requirements for the form of material presentation.

All documentation related to development and verification should be set forth in an accessible form, understood by experts, who did not participate in creating the software. The given requirements imply, in particular, traceability of all actions executed in the verification process, which allows one to establish a comparison between the input and output elements at each of the software creation stages and to make a transparent check of the completeness of execution of all requirements, beginning from requirements for the I&C system, then general and functional requirements for the software and ending in reports on verification (tests) of different subsystems or software functions.

The software verification plan should determine: choice of verification strategy and sequence for conducting it; methods and devices used in the verification; sequence of documenting actions and evaluation of verification results.

EVALUATION OF SOFTWARE FOR NUCLEAR POWER PLANT I&C SYSTEM

Criteria and Principles of Evaluation

The goal of the software evaluation is to check conformity to established requirements. This evaluation is conducted by analyzing the documentation submitted by the software developers, and also by verification of software using special tools. The project documentation (for example, the design description) and documents issued by the developer particularly for the licensing purposes (for example, safety analysis report) can be examined. During the expert evaluation some additional information can be requested from developer to clarify issues of the main documentation.

The purpose of the expert work is to improve the level of quality and reliability of the software. Therefore, all comments and recommendations of the experts should be transferred to the developers for timely elimination of any discovered defects. As a result of the joint activity of developers and experts corrections can be made to the design and, thereby, reduce the number of software defects.

The basis of the software expert evaluation methodology is assessment of the meeting the requirements for software at different stages of the lifecycle. In this case it is necessary to evaluate functional and general requirements for software, and also requirements for development and verification. The indicated requirements are to be combined in the criteria, which the software must satisfy, as well as processes of development and verification.

It is suggested that the following five criteria be used (Vilkomir, 1999):

- Completeness
- Documentation
- Accessibility
- Independence
- Successfulness

Software meets the criterion of completeness if its specifications completely correspond to the specifications of the I&C system and the software meets general and functional requirements of the specification, including requirements for development and verification.

Software meets the criterion of documentation, if the composition and structure of the documents developed for all stages of design, verification and operation, correspond to requirements of standards, norms and rules. The documentation criteria and completeness are interconnected: in accordance with the completeness criterion the content aspect of software development is analyzed; in accordance with the documentation criterion the formal aspect of evaluation is evaluated.

Software meets the criterion of accessibility if the documentation for development and verification of software is presented in a form that is clear and understandable to experts, who do not participate directly in their development. Moreover, in accordance with this criterion traceability (transparency, verifiability, checkability) of step by step execution of requirements for software at different stages of the lifecycle must be assured.

Software meets the criterion of independence if the degree of independence of software checking corresponds to the safety class of the system. For systems of safety class 2 the evaluation must be performed by a group of experts (organization), which is administratively and/or materially independent of the experts (organizations) which developed the software. For systems of safety class 3 the development and verification must be carried out by different specialists, however the administrative and financial independence is not required.

Software satisfies the criterion of successfulness if the inspection was successfully completed before beginning of system usage and if by that time all discovered defects and deficiencies have been analyzed and eliminated.

The criteria are an important part of the overall system of software evaluation. Conformity of the criteria and evaluated requirements can be given in the form of a matrix, which contains particular evaluations of the meeting individual requirements and summary evaluations based on the criteria. At the outset the evaluation in accordance with each of the five previously described criteria is formed on the basis of analyzing individual requirements, and then a concluding evaluation is produced.

Along with the general principles of systems approach and expert knowledge additional principles shall be implemented in the expert evaluation of software:

1. The principle of diversity of methods, hardware, actions of experts, methods of generating expert evaluations of software. This requirement determines the internal diversity of the evaluation process, thereby supplementing external diversity, which results in the fact that the expert evaluation and independent verification assure increasing reliability of software evaluation.
2. The principle of asymmetry of efforts distribution. A particular feature of software evaluation is the fact that due to its complexity it is impossible to assure complete testing of the behavior of software for all theoretically possible sets of input data. Therefore, while carrying out expert evaluations under conditions of limited time and resources the main efforts must be concentrated in critical steps and results of software development, analysis of the completeness and reliability of tests.

Next we will propose the content of operations for software evaluation at all basic stages of the life-cycle: development of requirements for software, design and coding of software, verification (development of the verification plan, preparation of the verification report).

In the stage of software requirements development the evaluation contains three steps: evaluation of the conformity of requirements for software to the requirements for the system; evaluation of the representation and specifications of requirements for software and general requirements; transfer of findings to the developer and obtaining back the corrective and additional requirements.

At the software design stage the evaluation contains four steps: evaluation of implementation of software requirements in the design; analysis of the structure for a subject of assured protection from common cause failures due to software errors; listing of requirements and functions of software for use in subsequent stages of the evaluation; transfer of findings to the developer and obtaining from him information and corrective actions for software design and evaluation of their adequacy.

At the stage of software verification plan development the evaluation contains four steps: evaluation of the existence in the verification plan of programs and methods for software testing; evaluation of accessibility of the verification plan; evaluation of reflection in the verification plan the requirements from the detailed list, which is compiled in the preceding stage of the evaluation; evaluation of the completeness and adequacy of the number of tests included in the verification plan. If necessary these stages can be supplemented by defining recommendations for additional testing of the functions more important for safety, transmission of comments and additions to the developer and obtaining a corrected and supplemented verification plan from him.

In the stage of software verification report preparation the evaluation contains six steps: evaluation of the existence in the verification report of protocols and official statements for each program and method of testing; evaluation of the completeness of the tests carried out; evaluation of independence of the verification conducted; evaluation of the software tools used in the development and verification; evaluation of the rate of success of completion of all tests; statement of recommendations for the regulatory body on the possibility of using the software.

The stages of evaluating the plan and the report on software verification are the most important. At these stages the regulatory body has the opportunity to receive evidence of achievement of the required level of quality and reliability of the software. For this purpose the plans, programs and methods of software testing are evaluated before the beginning of the tests, and the additions and comments made are transmitted to the software developer for their consideration.

Evaluation of Software Characteristic

Tasks and approaches to evaluation: Evaluation of the software characteristics includes the following tasks.

The first task is analysis of software conformity to general requirements defined in national and international standards. These requirements do not depend directly on the functional purpose of the software, but are determined by the designation and safety classification of systems.

The second task is evaluation of the completeness and quality of implementation of functional requirements in the software, which are defined in the software requirement specification. For reusable software an evaluation of the conformity of functions implemented in the software to the requirements is determined based on the context of the intended usage.

The first of the listed tasks is universal in nature and therefore can be partially or completely formalized. To this end, general methodology should be defined. Such methodology can include: obtaining the normative profiles (requirements) of the software; development (systematization, profiling, selection) of parameters for evaluation of characteristics (properties) of software and the established requirements; analysis of the results of evaluation and determination of the level of satisfaction of the requirements established for software; procedures (algorithms) of software evaluation using different parameters.

The evaluation of software requires determination of the composition of the corresponding characteristics and parameters, and also methods of their evaluation. Software quality can be evaluated by several characteristics, among which there are functionality, reliability, usability and so forth. The significance of each of these characteristics depends on the area of software application. For software systems that are important to safety a determining characteristic is reliability.

Given the controversies regarding the understanding and use of the term "software reliability" two existing approaches to its evaluation, which conventionally are called "qualitative" and "quantitative," can be considered.

The qualitative approach is used everywhere and is oriented to a system (hierarchy) of requirements, i.e. profiles determined by standards, industry regulations and normative documents of companies, the fulfillment of which is checked during software reliability evaluation. Results of reliability evaluation in this case are formulated in the form of the conclusions "corresponds" or "does not correspond" for individual components, which directly or indirectly affect reliability.

The quantitative approach to evaluation is oriented to the development of models, that receive as input parameters characteristics of both processes (development and verification) and software itself, and gives as output the indicators that characterize reliability (Lyu, 1996).

These indicators most frequently are analogues of reliability indicators of equipment with the difference that such events as "component failures," are formulated as "manifestation of software defects." Moreover, special indicators (metrics) are also used that determine the level of residual defects, rate of their discovery during testing and so forth.

Nevertheless, the instability of the manifestation of defects in sophisticated software systems and their uniqueness do not allow one with high degree of accuracy to determine quantitative values of the characteristics of quality and reliability. To solve this problem, the special methods of analysis such as FTA (Fault Tree Analysis), RBD (Reliability Block Diagram), FME(C)A (Failure Modes, Effects (and Criticality) Analysis) and others, can be used.

Metrics, indicators and raw data for evaluation: There are different approaches to defining of metrics and their relationship to the concepts of software quality and reliability indicators (Pressman, 1997).

The first approach (in accordance with the IEEE standards (IEEE, 1990), (IEEE, 1988, a) and (Pressman, 1997)) views this relationship on the basis of the categories "absolute-relative" and is based on the following definitions.

Absolute indicators (measures) are quantitative indicators that characterize absolute values of different attributes of software and the development process (for example, the number of defects discovered in each software module, the number of lines of initial software text and so forth). In this approach the metrics, in contrast to absolute indicators, are intended especially for comparison of different software designs. For example, comparison of two software applications based on an absolute indicator such as total number of defects discovered is not informative because lack of possibility for judging the size (measured in the lines of source code or number of operators) of comparable programs, their complexity, conditions of development, testing and other characteristics. It is obvious that it is more expedient in this example to use metrics that determine the relationship of the total number of defects to the size of software, quality of programming modules, test time and so forth.

The special feature of the second approach is the fact that metrics are interpreted as dedicated indicators (supplemental with respect to known indicators), which can be given as absolute or relative evaluations of software.

It should be emphasized that the boundary between metrics and reliability indicators of software is quite difficult to draw. Reliability indicators are primarily quantitative characteristics similar to indicators used in classic reliability theory (probability of no-failure, mean time before failure and so on), while metrics are specific indicators for software, which can evaluate reliability indirectly or with respect to other products (reference standards).

Next we will discuss metrics with consideration of the more common second approach. It should be noted that metrics can also give a quantitative evaluation of any given property as well as requirements for software. In this case, by raw data, or parameters (primitives) of metrics we mean the initial quantitative values that are needed for their calculation. The raw data can be other indicators or metrics as well as different constants, coefficients and so forth.

Software developer organizations should be encouraged to use various metrics, because they allow one to evaluate the level of quality and reliability of software being developed and their design processes, and also to discover existing problems (for example, inadequate testing of software, failure to follow the standards, ineffective work of individual groups of developers and so forth) and to take the necessary measures to solve them. Moreover, the need for calculation and analysis of various metrics arises during verification and validation of software, because these processes must rely to a greater degree on accurate quantitative evaluations, and not on subjective opinion of developers or customers.

Basic standards that define metrics and sequence of their computing are:

- IEEE standard 982.1-1988 (IEEE, 1988, a), which defines the list and order of reliability metric calculations.
- IEEE standard 982.2-1988 (IEEE, 1988, b), which clarifies the sequence of using the standard IEEE 982.1-1988.
- ISO/IEC standard 25010 (ISO, 2010), which defines the software quality model.
- ISO/TEC standard 25023 (ISO, 2011), which establishes the basic nomenclature of external software quality metrics, including metrics of reliability, and defines basic principles of their selection and evaluation.
- Ukrainian standard DSTU ISO/IEC 25010:2016 (DSTU, 2016), which is harmonized national edition of international standard ISO/IEC 25010 (ISO, 2010), that began to operate in 2018.

A quality model is presented in standard (ISO, 1999), according to which software is evaluated with a set of internal, external and quality in use metrics. In this case software quality is defined as the total set of properties that determine software capability to satisfy assigned requirements in accordance with its purpose.

The application area of external quality metrics is validation and expert evaluation of the software. The group of external quality characteristics and metrics corresponding to them describe the programming product that is completed and ready for use. In order to evaluate software quality the standard (ISO, 1999) defines six groups of external and internal characteristics.

1. Functionality is a set of software properties that determines its ability to execute the established functions.
2. Reliability is the set of properties that enable the software to retain its serviceability and to convert raw data into the desired result under predetermined conditions in an established period of time.
3. Usability is the set of properties that characterizes the necessary conditions of software use by users.
4. Efficiency is the set of properties that characterizes conformity of the software resources used to quality of execution of its functions.
5. Maintainability is the set of properties that characterizes the level of efforts needed to execute the required software modifications.
6. Portability is the set of properties that characterizes the adaptability of software to work in different functional environments.

The set of metrics that pertain to each group of higher level characteristics is again divided into several sub-characteristics. For example, the software reliability, which is defined as the capability of software to maintain its level of performance under stated conditions for a stated period of time, includes the following sub-characteristics:

Figure 5. Interconnection and scopes of standards IEEE 982.1and ISO/IEC 25010

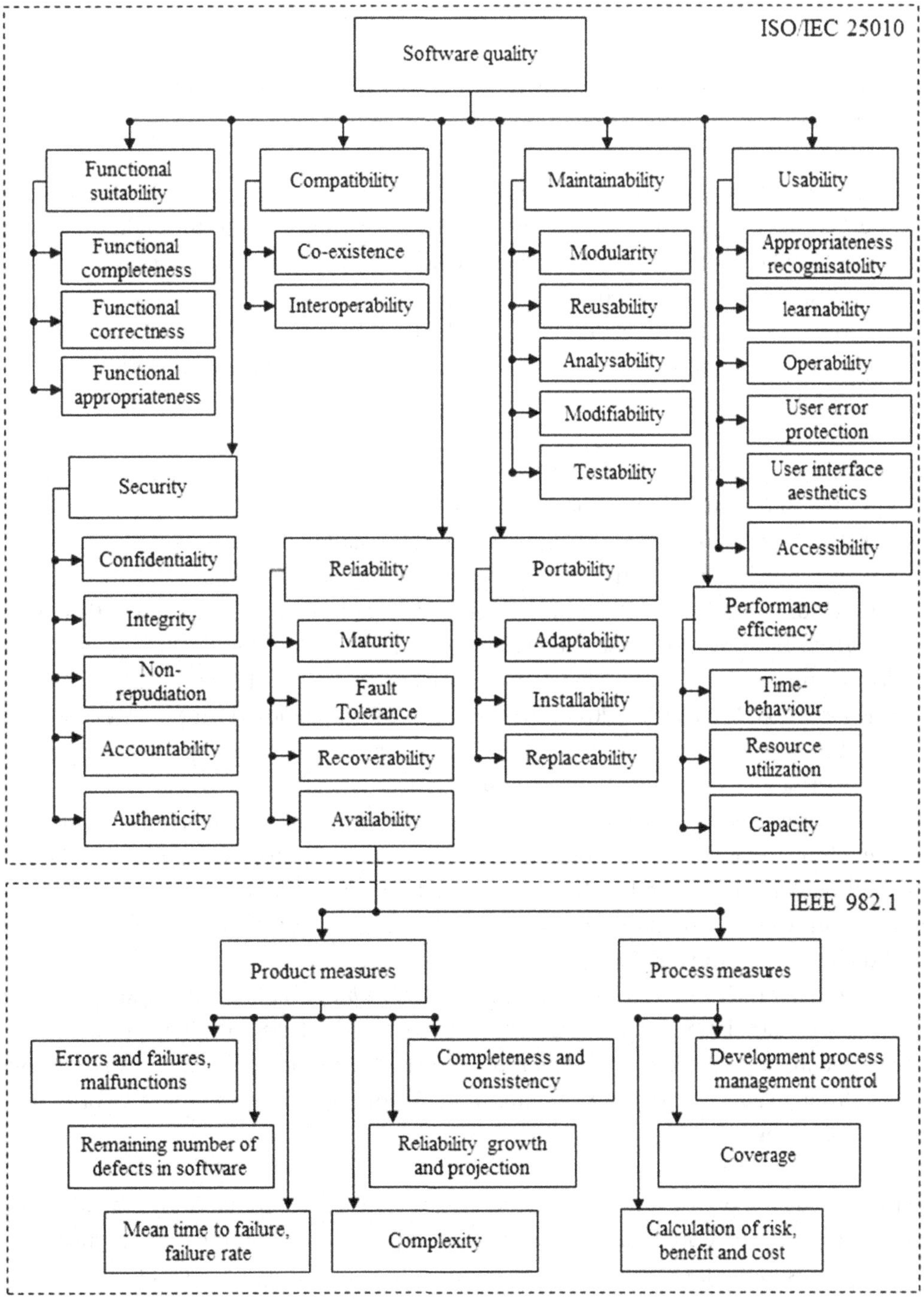

ISO/IEC 25010
Software quality
Functional suitability
Functional completeness
Functional correctness
Functional appropriateness
Compatibility
Co-existence
Interoperability
Maintainability
Modularity
Reusability
Analysability
Modifiability
Testability
Usability
Appropriateness recognisatolity
learnability
Operability
User error protection
User interface aesthetics
Accessibility
Security
Confidentiality
Integrity
Non-repudiation
Accountability
Authenticity
Reliability
Maturity
Fault Tolerance
Recoverability
Availability
Portability
Adaptability
Installability
Replaceability
Performance efficiency
Time-behaviour
Resource utilization
Capacity

IEEE 982.1
Product measures
Errors and failures, malfunctions
Remaining number of defects in software
Mean time to failure, failure rate
Complexity
Reliability growth and projection
Completeness and consistency
Process measures
Development process management control
Coverage
Calculation of risk, benefit and cost

1. Maturity is the set of indicators that describe frequency of occurrence remaining in the software.
2. Fault tolerance is the ability of software to retain a certain functioning level during the onset of software malfunctions.
3. Recoverability is the property of software to restore its ability to work (assigned level of functioning), and also program data.
4. Reliability compliance is the degree of software conformity to normative requirements for reliability (standards), and also to customer requirements.

The basic nomenclature, calculation sequence and scale of possible values for metrics of software quality that pertain to each group of quality products with reference to the lifecycle process in which the metric is used, and composition of the necessary documentation for determining input parameters for calculation are defined in (ISO, 2000).

In addition to the listed categories the standard (IEEE, 1988, a) determines a number of functional groups that characterize different properties of reliability of the software itself (indicators, or product measures), as well as the design process (process metrics), based on which the reliability metrics are classified.

Examined standards can be used to create profiles of the evaluation and quality assurance of software. Figure 5 shows the interconnection of the standards IEEE 982.1-1988 (IEEE, 1988, a) and ISO/IEC 25010 (ISO, 2010). The standard ISO/IEC 25010 is fundamental and assures comprehensive inclusion of software quality. Evolution of software quality models in context of the standard ISO/IEC 25010 (Gordieiev Oleksandr, Kharchenko Vyacheslav, Fominykh Nataliia & Sklyar Vladimir, 2014) and features of standard ISO/IEC 25010 were represented in (Gordieiev Oleksandr, Kharchenko Vyacheslav & Fusani Mario, 2015), (Gordieiev Oleksandr, Kharchenko Vyacheslav & Vereshchak Kate, 2017), (Gordieiev Oleksandr & Kharchenko Vyacheslav, 2018). At the same time the standard IEEE 982.1-1988 allows one to assure more thorough analysis of software reliability as one of the top priority quality characteristics of software of information and control critical systems.

Based on an analysis of the classifications presented above for the systematic description of quality metrics and reliability a unified system of classification features is proposed (Figure 6). Development of the systemic classification of reliability metrics and software quality is a necessary condition of successful harmonization of normative documents and creation of effective methods of evaluation and assurance of quality of the software being developed.

Production and analysis of initial information for determining metrics: Analysis of the standards IEEE 981.1-1988 (IEEE, 1988, a) and IEEE 981.2-1988 (IEEE, 1988, b) allows to conclude that in order to determine input parameters for calculation of different metrics various information sources are required at the different stages of the software lifecycle. Among the main sources of information we can distinguish the following:

- General information on the software project. This includes dates of the beginning and completion of each step of the software lifecycle, description of processes of verification, number of releases (that is versions, outputs) of software and so forth. The reporting of general information on the course of execution of the software design process is executed by the manager of the software project.

Figure 6. Classification scheme of reliability metrics and software quality

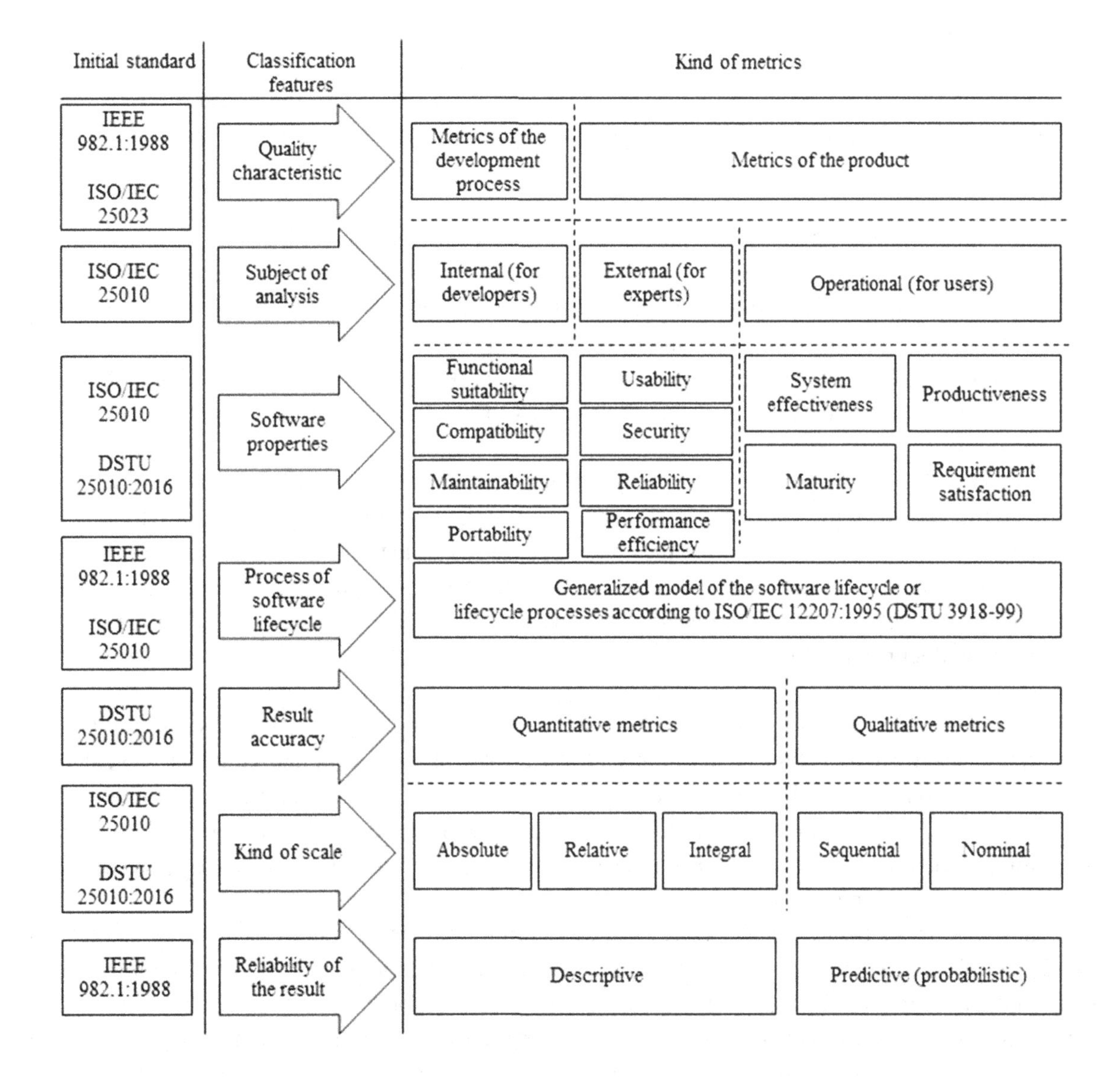

- The report on results of performing software verification after completion of each step of the software lifecycle. The report should include information on the nature of defects discovered, reason for their introduction and later discovery, time lost for preparation and conducting verification, and also information. The report is compiled directly by analysts-experts, as a rule manually.
- The report on software testing results. This includes information on time of discovery, level of seriousness and nature of software defects and information on reasons and time (date or stage of the software lifecycle) of the introduction of the corresponding defects, time of their correction,

and also the test kits used, number of successful runs of the program and so forth. The report is compiled directly by test engineers manually or using automated recording devices.

- Technical description of the software release that is provided for testing or verification. This should include information on the overall number of procedures and functions (modules), number of modified, supplemented and deleted modules in comparison with the preceding release, and also the modular structure of the software for the possibility of converting the software to a state graph. The technical description is compiled directly by programming engineers, manually as a rule.
- The source code and executable code, which allow one to determine objective characteristics of the software; number of lines in the source code, size of the program, number of operators and operands used, and also the total number of their occurrence on the program and so forth. The software source code can be used along with corresponding utilities to produce a state graph of software as a whole and of individual modules. The software presentation in the form of a graph is necessary for calculating individual metrics and for developing test benches and executing the testing itself.
- Technical documentation for the software, which includes requirements for the software (specification), technical description and so forth. In calculating certain metrics the production of a comprehensive result requires that the experts propose evaluations or assignments of weighting factors (significance) for different software characteristics, input parameters or intermediate results. These factors can be determined based on analysis of corresponding sections of the software technical documentation.

Thus, the executed analysis allows one to insert additional classification features for quality metrics and their parameters: source of information for the determination of input parameters to calculate quality metrics (reports on fulfillment of verification and testing processes, technical description of the software release, source code and so forth); information compiler (project manager, system analysts, programmer-managers and test engineers and so forth); degree of objectivity (reliability and completeness) of the presented information.

It should be emphasized that the process of determining numerical values of input parameters for calculation of reliability metrics manually is unfeasible because of complexity and considerable amount of initial data. Therefore, the task of developing (or selection) support tools for gathering the initial information and automatic determination of input parameters for the calculation to metrics is needed. For this reason software source code analyzers are used as objective sources of information for determining input parameters for calculation of software quality metrics.

Based on the examined standards and analysis of publications (IEEE, 1988, a), (ISO, 2000) and (Pressman, 1997) a database of reliability indicators (metrics) and tools has been developed, which allows one to make a choice of indicators with consideration of the previously listed classification features, and also the stage of the lifecycle at which the reliability evaluation is made, and to produce their quantitative value. The calculation of measures is carried out by using deterministic methods of evaluation described in a standard (IEEE, 1988, b), while for calculation of predictive measures various probability reliability models can be used.

Evaluation of Software Development and Verification Processes

An evaluation of software development processes is accomplished by examination and analysis of documentation in accordance with the requirements for methods, devices and documentation of development described above (see Figure 4). In this case, one can use special metrics, and the evaluation overall can be carried out and presented by means of radial metric diagrams.

The ratio of the number of software modules (or subsystems), which have been developed using such methods, to the total number of modules (subsystems) can be used as metrics for evaluating the fulfillment of requirements for the use of formal methods.

The quality of software development, fulfillment of requirements for the number of modules, complexity of relations among them can also be evaluated by using special metrics (for example, the Halstead metric, McCabe metric and others (Pressman,1997)).

It should be emphasized that evaluation of the fulfillment of requirements for software by nature is subordinate with respect to evaluation of verification, because these requirements are overall requirements for development and verification, or are checked directly together with evaluation of software verification. This pertains, in particular, to evaluation of the development results themselves, execution of requirements for methods and tools.

Tasks and criteria of software verification evaluation: The quality of conducting verification of software is of great significance for reliability and safety of the I&C system. The tasks of software verification evaluation are: analysis of software requirements based on requirements for the system and general requirements, which are determined by normative documents; check of the conformity of task formulations for software development to these requirements; quality check of the verification plan, test methods and their completeness in accordance with the tasks assigned to the software; the quality check of verification reports and their conformity to plans and methods.

One should note that these tasks, and also the tasks of evaluating the quality of their solution from the standpoint of fulfillment of requirements for safety, are not easy to formalize. Usually the verification analysis is accomplished by traditional methods of documentation analysis, and individual results can be checked by using specially developed tools. At the same time, considering the high criticality and importance of a maximally objective and complete evaluation of software verification, one must find approaches to the development of models that describe this process and allow one to improve its quality.

The process of software verification evaluation of I&C systems that are important for nuclear power plant safety can be constructed by means of (Kharchenko, 2000): formation and structuring of the full set of requirements for software, which must be checked in the verification process at different stages of the lifecycle; development of a system of criteria for evaluating software verification; compilation of a system of verification evaluation criteria and set of requirements established for the software; formalization of the verification analysis processes and its evaluation for basic criteria; creation and use of tools for support of software safety analysis during verification, licensing and expert analysis.

Criteria for verification is similar to the criteria of software evaluation represented above and includes the criteria of completeness, independence, successfulness, documentation and accessibility.

Software verification corresponds to the completeness criteria if during the verification the conformity of software to all requirements of specifications, standards and other normative documents was tested.

Software verification corresponds to the independence criteria in accordance with the software safety class. Thus, for class 2 verification is conducted by a group of specialists (organization), which are administratively and/or financially independent of the specialists (organization), which developed the software.

Software verification corresponds to the criterion of successfulness, if the verification was finished completely prior to placing the system in use, that is by this time all defects found were analyzed and eliminated (or a well-founded decision for their subsequent elimination was made).

Software verification corresponds to the criterion of documentation if a plan and report were issued which describe in detail the course and results of verification. In this case individual parts of a verification plan and report, which have individual significance (programs, test methods, protocols and so forth) can be issued in the form of individual documents.

Software verification corresponds to the accessibility criterion if all of the documentation on software verification was set forth in a form understandable to specialists who do not participate in conducting the development and verification.

Software verification corresponds to the accessibility criterion if all of the documentation on software verification was set forth in a form understandable to specialists who do not participate in conducting the development and verification.

Evaluation of Documents Related to Software Development: During the expert evaluation of documents on software development the requirement specification and design documentation are evaluated. In the expert analysis of the requirement specification the following are established and evaluated: the extent to which software requirements are correct and not contradictory; to what extent the functional requirements for software correspond to requirements for the I&C system; how fully are general requirements for software reflected in the feasibility study (requirements, which are established independently of specific functional purpose of the I&C system).

The evaluation of software requirements is conducted with consideration of: safety class of the I&C system, which includes software is a component; level of software approval; software purpose.

During expert examination of the design documentation one will analyze: description of the composition, structure and functions of component parts of the software; information on methods and means of testing and running experiments with software.

The evaluation of verification documents: The basic verification documents for evaluation are verification plan and verification report.

The software verification plan (SVP) evaluation is conducted based on criteria of documentability, accessibility, completeness, independence.

1. In the evaluation based on documentability criterion one establishes that the SVP was issued prior to the beginning of software verification and defines: choice of verification strategy; sequence of conducting verification; methods and devices used in the verification process; sequence of documentation of verification actions; sequence of verification results evaluation.
2. In the evaluation based on accessibility criterion one establishes that the SVP is set forth in a form understandable for specialists who did not participate in the software development process.
3. Based on the completeness criterion one carries out the evaluation of the following items stipulated in the SVP:
    a. Sequence of conducting verification. In evaluating the sequence of conducting verification one must establish that the verification stipulates after each step of the software development: generation of requirements for software; design; coding.

   b. Completeness of tests. In evaluating the completeness of tests it should be established that the sets of tests that are selected for verification will assure the possibility of checking all stipulated functions and interfaces of the I&C system, and also the check of fulfillment of requirements for software.
   c. Software verification and development tools. In evaluating the tools of software verification and development it should be established that the SVP stipulates the use of automated design and testing tools and indicates the selection criteria for them. One should evaluate the conformity of the proposed criteria to the requirements of norms, rules, standards and recommendations; in particular, when using automated tools to generate code one must check that the given tools have gone through verification with the same requirements as the software itself.
   d. Particular features of the verification of different kinds of software.
4. For an evaluation based on the independence criterion it should be established that in the SVP: for software of I&C system of safety class 2 it is stipulated that verification will be carried out by a group of specialists (organization), who are administratively and/or financially independent of the specialists (organization) that developed the software; for software of I&C system of safety class 3 it is stipulated that the verification will be conducted by specialists who have not participated directly in development of the software (administrative and financial independence is not required), or by software developers on the condition that the review and evaluation of the verification results will be done by independent specialists.

Comments and recommendations made during the expert examination must be considered in the final version of the SVP, which is used in performing the software verification.

Evaluation of the software verification report (SVR) is conducted based on criteria of documentability, accessibility, completeness, independence and successfulness.

1. In the evaluation based on the criterion of accountability one should establish the fulfillment of requirements for the SVR structure, which should contain: lists of input and output signals during software tests; results of tests and their evaluations; deficiencies discovered during tests; conclusions based on results of analysis of the discovered deficiencies and measures to eliminate them, and also to evaluate the degree of detail of the documentation of all stages of the software verification process.
2. In the evaluation based on the criterion of accessibility one should establish that the SVR is set forth in a form that can be understandable for specialists who did not participate in the software verification process.
3. According to the completeness criterion one will evaluate the conformity sequence, strategy and order of conducting verification, methods, tests and software verification tools used that are stipulated in the SVP and actually used (reflected in the SVR).
4. In the evaluation based on the independence criterion one should compare the independence of the specialist (organization) that conducted the verification as stipulated in the SVP and the actual dependence from the specialist (organization) that developed the software.
5. Based on the successfulness criterion one will check correctness of the evaluation of results of each test and establish that all deficiencies discovered in the course of the software verification are recorded, analyzed, eliminated and results of subsequent evaluations are presented.

Tools to Support Evaluation

There are the following classification features for tools: functional purpose of the tool; degree of process automation; number and nomenclature of lifecycle stages and processes, supported by the tool; project components (its components or stages of development, verification or expert examination), supported by the tool; degree of intelligence; possibility of integrating a given tool with other tools.

These features allow classifying of tools as follow.

1. By functional purpose one can distinguish tools for informational, analytical, and organizational support.
2. By degree of process automation tools can be subdivided into manual (partially formalized), automated, and automatic. In determining the type of tool based on this feature one must consider the degree of automation of preparation, input, analysis, documentation and display of information.
3. By the number of supported stages tools are divided into local, compositional, and end-to-end. This feature determines the boundaries and scope of the operation of a tool.
4. By kind of project components one can distinguish tools that support evaluation of: products (requirements, specifications, design components, codes, methods, reports and so forth); processes (specification, design, coding, testing, verification and so forth).
5. By level of intelligence one can distinguish: non-intelligent tools, or traditional kind of tools (without using knowledge-based methods); intelligent tools; and combined tools.
6. According to possibility of integration tools are divided into integratable, which allow one to use a given tool together with other ones, and non-integratable.

By grouping the different (by purpose, level of intelligence and so forth) tools it is possible for carrying out various scenarios of the expert examination, which require use of analytical, information and organizational type tools, intelligent or combined tools and so forth.

Informational tools are intended for the generation, preliminary processing and analysis of information required for carrying out independent verification and expert examination of software and, as a rule, are automated, local or composite tools of non-intelligent or combined type, which support, above all, verification and expert examination of products.

Tools of this kind provide:

1. Generation of a profile-like base of national and international normative documents, which determine the software requirements and order of evaluating their execution during verification and software expert examination.
2. The generation of general and particular normative profiles of software based on an analysis of profile-like documents:
    a. Software requirements (structure and properties; inspection and diagnostics; reliability and tolerance; development; verification).
    b. Methods of evaluating the fulfillment of requirements.
    c. Evaluation of the quality of the process and expert examination results.
3. The formalized analysis of the general and functional requirements for the software that has been examined by experts based on the submitted documents.

4. Formalized preparation of data on the expert analyzed software and processes of its development and verification based on templates (questionnaires).
5. Databases on software expert examination that have been carried out and are being carried out, which include full systematized information on tasks, expert analysis object, course and results of the expert examination.
6. A transition from verbal to formal description of software requirements (partially formalized verbal matrices, semantic trees, product rules, Z-notations).
7. Databases of quality metrics and software reliability.
8. Database of software reliability models.

Results of execution of the informational tools are databases of profile-like documents, metrics, models, expert examinations; normative profiles; standardized plans of the verification and its evaluation; completed templates (questionnaires).

The questionnaires and templates are used to run individual procedures of the expert examination and verification in accordance with chosen methods.

Analytical tools are intended for carrying out direct analysis of software and to evaluate conformity of the verified (expert analyzed) software to established requirements, reliability of verification and expert examination and are automatic local tools that are both intelligent (combined) and nonintelligent, which support primarily verification and expert examination of products.

Analytical tools support:

1. Verification of normative profiles (completeness, correctness and consistency of the general requirements) of the software designs being developed and analyzed by experts.
2. Statistical analysis of software.
3. Dynamic testing of software.
4. Selection and rating of metrics and reliability measures (quality) of software.
5. Selection and verification of software reliability models.
6. Analysis of the fulfillment of general and functional requirements for expert analyzed software.
7. Analysis of the reliability and safety of software based on standardized methods.
8. Evaluation of the completeness, reliability and other characteristics of independent verification and expert examination.

Results of running the analytical tools are technical reports on verification of normative profiles; static and dynamic testing of software; selection and rating of metrics, measures and models of software reliability; analysis of the fulfillment of requirements for expert analyzed software; evaluation of characteristics of reliability, completeness and resources for carrying out independent verification and expert examination.

Organizational support tools are intended for planning, organizing and controlling the process of independent verification and expert examination and are automated or manual, local or composite tools of the non-intelligent type, which support verification and expert examination of processes and products.

Tools of this kind use as initial information normative documents; design documentation; data and results of running information and analytical tools.

Organizational support tools provide:

1. Planning of the expert examination (tasks, schedules, resources, personnel).
2. Timely analysis of the course and results of the expert examination.
3. Management of expert analysis process.
4. Documenting the results of the expert examination (partial and summary reports).
5. Analysis and evaluation of the quality of the process for conducting independent verification and expert examination of software.

Results of running the organizational support tools are general planning documents; diagrams of work execution while conducting independent verification and expert examination of software; technical reports on the course, results of independent verification and expert analysis, evaluation of the process of carrying them out; summary reports.

It should be emphasized that at the present time tools of the analytical type have been the most popular for evaluation of software, which support the solution of statistical analysis and software dynamic testing tasks.

SOLUTIONS AND RECOMMENDATIONS

Specific features of software development and usage require proper regulation requirements towards the program components.

At the same time requirement to software should be agreed with requirements to I&C system. Categories of functions performed by I&C system have influence with software requirements, including requirements for composition of the functions, quality, reliability, stability, interaction with other components, procedures and processes.

Therefore developing of requirements for software components shall be done taking into account features of target I&C system, international and national regulatory requirement. For the solution of this issue systematic approach and methods, supported with appropriate tools, are required.

Modern model-based methods and techniques should be applied to assess NPP I&C software, in particular, model-checking (Lahtinen et al., 2010) and invariant-oriented evaluation (Kharchenko (Ed), 2012), software safety analysis techniques (Hui-Wen Huang et al., 2011) etc.

FUTURE RESEARCH DIRECTIONS

To match the latest trends and industry requests software components of I&C systems become more complex. Development of software engineering technologies also opens up new aspects and generates new issues for designing and implementation of software. Therefore possible implications of new programming technologies must be analyzed to ensure timely and adequate adaptation and clarification of regulatory frameworks.

Also attention should be paid to the fact that in large projects of I&C systems several organizations with different background and possibly from different countries can participate. Thus harmonization of requirements and ensure their adequate interpretation may be beyond the common regulatory aspect. In this scope, issues of personnel training, establishment of effective communications between the development teams and the utilities and other become important.

All these issues require systematic study and comprehensive scientific researches.

CONCLUSION

1. Software is a specific object for safety regulation. It is a component of I&C system to which requirements are applicable, and also it is a means of ensuring the satisfaction of regulatory requirements. At the same time software is the most likely sources of common cause failures. Therefore the need to minimize risks of common cause failures is reflected in the requirements to processes of software development and verification, as well as application of diversity.
2. Standardization, evaluation and assurance of software safety should be based on processand-product-approach. I.e. harmonized requirements for the program as a product and the processes related to the creation, evaluation and use of programs at various stages of the lifecycle should be used.
3. Degree of completeness, adequacy and correctness of requirements to software is the determining factor in assessing their compliance, and thus ensure the quality, reliability and safety of both software and I&C system of NPP.
4. Methods that are used for software evaluation should be standardized and cover all aspects of software development and application. If it is necessary, correct application of such methods can be evaluated by experts. From this point of view special significance is acquired by criterion of documentation.

REFERENCES

Adziev, A. V. (1998). Myths about software safety: Lessons of famous disasters. *Open Systems, 6*. Retrieved December 16, 2012, from http://www.osp.ru/os/1998/06/179592/

Aizenberg, A., & Yastrebenetsky, M. (2002). Comparison of safety management principles for control systems of carrier rockets and nuclear power plants. *Space Science and Technology*, *1*, 55–60.

Anderson, R. B. (1979). *Proving Programs Correct*. New York: Wiley.

Ben-Ari, M. (2000). *Understanding programming languages*. Wiley.

DSTU-25010. (2016). *Systems and software engineering – Systems and software Quality Requirements and Evaluation - System and software quality models*. Author.

Everett, W., Keene, S., & Nikora, A. (1998). Applying Software Reliability Engineering in the 1990s. *IEEE Transactions on Reliability, 47*(3-SP), 372-378.

Gordieiev Oleksandr & Kharchenko Vyacheslav. (2018). IT-oriented software quality models and evolution of the prevailing characteristics. In *Proceedings of the 9 International Conference Dependable Systems, Services and Technologies - DESSERT'2018*, (pp. 390-395). Kyiv, Ukraine: Academic Press.

Huang, H.-W., Wang, L.-H., Liao, B.-C., Chung, H.-H., & Jiin-Ming, L. (2011). Software safety analysis application of safety-related I&C systems in installation phase. *Progress in Nuclear Energy, 6*(53), 736–741. doi:.2011.04.002 doi:10.1016/j.pnucene

IAEA (2000). IAEA NS-G-1.1. Software for computer based systems important to safety in nuclear power plants.

IEC (2006,a). IEC 60812. *Analysis technique for system reliability – Procedure for Failure Mode and Effects Analysis (FMEA).*

IEC (2006,b). IEC 60880. *Nuclear power plants – Instrumentation and control systems important to safety – Software aspects for computer-based systems performing category A functions.*

IEC (2006,c). IEC 61025. *Fault tree analysis.*

IEC (2008). IEC 61508. *Functional Safety of Electrical/Electronic/Programmable Electronic Safetyrelated Systems.*

IEEE (1988,a). IEEE 982.1. *Standard Dictionary of Measures to Produce Reliable Software.*

IEEE (1988,b). IEEE. 982.1. *Standard Guide of Measures to Produce Reliable Software.*

IEEE (1990). IEEE 610.12. *Standard Glossary of Software Engineering Terminology.*

ISO/IEC (2010). ISO/IEC 25010. *Systems and software engineering – Systems and software Quality Requirements and Evaluation (SQuaRE) – System and software quality models.*

ISO/IEC (2011). ISO/IEC 25023. *Systems and software engineering – Systems and software Quality Requirements and Evaluation (SQuaRE) – Measurement of system and software product quality.*

Kersken, M. (2001). *Qualification of pre-developed software for safety-critical I&C application in NPP's.* Paper presented at CNRA/CSNI Workshop on Licensing and Operating Experience of Computer-Based I&C Systems, Hluboka-nad-Vltavou, Czech Republic.

Kharchenko, V. S. (Ed.). (2012). CASE-assessment of critical software systems. Quality. Reliability. Safety. Kharkiv, Ukraine: National Aerospace University KhAI.

Kharchenko, V. S., & Vilkomir, S. A. (2000). *The Formalized Models of Software Verification Assessment.* Paper presented at 5th International Conference Probabilistic Safety Assessment and Management, Osaka, Japan.

Lahtinen, J., Valkonen, J., Bjorkman, K., Frits, J., & Niemela, I. (2010). *Model checking methodology for supporting safety critical software development and verification.* Paper presented at ESREL 2010 Annual Conference, Rhodes, Greece.

Lawrence, S., Hatton, L., & Howell, C. (2002). *Solid Software.* Prentice Hall.

Lyu, M. R. (1996). *Handbook of software reliability engineering.* McGraw-Hill Company.

Oleksandr, G., Vyacheslav, K., & Kate, V. (2017). Usable Security Versus Secure Usability: an Assessment of Attributes Interaction. In *Proceedings of the 13th International Conference on ICT in Education, Research and Industrial Applications. Integration, Harmonization and Knowledge Transfer*, (pp.727-740). Kyiv, Ukraine: Academic Press.

Oleksandr, G., Vyacheslav, K., & Mario, F. (2015). Evolution of software quality models: usability, security and greenness issues. In *Proceedings of the 19-th International Conference on Computers (part of CSCC 15),* (pp. 519-523). Zakynthos Island, Greece: Academic Press.

Oleksandr, G., Vyacheslav, K., Nataliia, F., & Vladimir, S. (2014). Evolution of software Quality Models in Context of the Standard ISO 25010. In Proceedings of Dependability on Complex Systems DepCoS – RELCOMEX (DepCOS) (pp. 223-233). Brunow, Poland: Academic Press.

Pressman, R. S. (1997). *Software Engineering: A Practioner's Approach.* McGraw-Hill Company.

Vilkomir, S., & Kharchenko, V. (2000). *An "asymmetric" approach to the assessment of safety-critical software during certification and licensing*. Paper presented at ESCOM-SCOPE 2000 Conference, Munich, Germany.

Vilkomir, S. A., & Kharchenko, V. S. (1999). Methodology of the review of software for safety important systems. In G. I. Schueller, P. Kafka (Eds.), *Safety and Reliability. Proceedings of ESREL'99 - The Tenth European Conference on Safety and Reliability* (pp. 593-596). Munich-Garching, Germany: Academic Press.

KEY TERMS AND DEFINITIONS

Common-Cause Failure (CCF): Failure of two or more structures, systems, or components due to a single specific event or cause.

Common-Mode Failure (CMF): Failure of two or more structures, systems, and components in the same manner or mode due to a single event or cause.

Diversity: Presence of two or more redundant systems or components to perform an identified function, where the different systems or components have different attributes so as to reduce the possibility of common cause failure, including common mode failure.

Fault Tolerance: Is the ability of software to retain a certain functioning level during the onset of software malfunctions.

Fault Tree Analysis (FTA): Deductive technique that starts by hypothesizing and defining failure events and systematically deduces the events or combinations of events that caused the failure events to occur.

Failure Mode, Effects, and Criticality Analysis (FMECA): Is a reliability evaluation/design technique which examines the potential failure modes within a system and its equipment, in order to determine the effects on equipment and system performance.

Off-the-Shelf (OTS) Software Component: Pre-developed software components, usually developed by other organization and designed for specific solutions.

Chapter 5
Requirements and Life Cycle Model-Based Assessment of NPP I&C Systems Cyber Security and Safety

Andriy Kovalenko
https://orcid.org/0000-0002-2817-9036
Centre for Safety Infrastructure-Oriented Research and Analysis, Kharkiv National University of Radio Electronics, Ukraine

Oleksandr Siora
Research and Production Corporation Radiy, Ukraine

Anton Andrashov
Research and Production Corporation RadICS, Ukraine

ABSTRACT

The chapter discusses the importance of assessment of interference degree for various attributes of safety-critical instrumentation and control (I&C) systems and proposes applicable metrics. An approach to analysis of safety-critical I&C systems is presented. Such approach relies on performance of gap analysis and consideration of influence of human, technique, and tool. The approach is applicable to cyber security assessment for various safety-critical I&C systems, including complex instrumentation and control systems and field-programmable gate arrays (FPGA)-based systems.

INTRODUCTION

Nowadays safety-critical systems are widely used by the world industry in various areas in forms of I&C systems, including those for NPPs, on-board computer-based systems, electronic medical systems, etc. Moreover, FPGA technology is now being trend in safety-critical systems implementation that inevitably

DOI: 10.4018/978-1-7998-3277-5.ch005

leads to new challenges in various aspects of such systems design, operation and maintenance requiring new approaches, techniques and appropriate requirements. First goal of this chapter is to customize the elements of gap analysis (GA), Intrusion Modes and Effects Criticality Analysis (IMECA) technique and analysis of development processes related to the developer (human), technique, and tool (HTT) to develop an approach, which can be used in analysis and assessment of safety-critical systems (Kharchenko, V. et al. (2012,c), NUREG/CR-7006).

Design and development processes for any safety-critical I&C systems, including complex ones, require predetermined and formalized processes for all types of activities, including design, verification and validation (V&V). There are some set of proposed life cycle representations in a form of appropriate models, including waterfall, incremental, spiral, agile, prototyping, star, hybrid, Y, etc. (Isaias, P. et. al. (2015), Bhuvaneswari1, T. et. al. (2013)).

One of the most convenient forms of design life cycle representation for a safety-critical I&C system is so-called "V-model", which involves the phased development of certain artifacts (components of the future system), and each artifact is subject to formal verification intended to prevent unauthorized changes in design or functionality. (Kharchenko, V. et. al. (2017)).

V-model is a conveniently organized formalized sequence of development phases of the final product, typically complex. It implies stating the requirements in the beginning, and allows correlating the design stages with the appropriate tests. The basic principle of a V-model is that the detailing of the project increases when it moves from left to right, as the time passes. Iterations in the project are executed horizontally, between the left (descending) and the right (ascending) branches. This allows for support during the planning and implementation of the project.

In addition, a V-model can be generalized in several directions to meet the requirements and reflect the complexity layers of the product.

The main disadvantages of existing V-models include the following:

- Lack of support for parallel events.
- Lack of support for the introduction of dynamic changes at different stages of the product's life cycle.
- Too late testing of requirements in the life cycle, which makes it impossible to make changes without affecting the schedule of project implementation.
- Some results can only be obtained after the complete performance of the activities described by the left branch.

In order to consider and take into account all the features of modern complex safety-critical I&C systems and their underlying technologies, it is necessary to analyze the attributes of a system. For this purpose, a development life cycle (DLC) of a system can be represented in the form a V-submodels set that are partially superimposed and corresponding to each of the DLCs associated with the specific components of the system, thus covering the development stages of the components along with the appropriate return points (Shamraev, A. et. al. (2017)).

Given the fact that, in a general case, both the starting point and the length of the DLC are specific for the components, so all V-submodels can be separated to perform a comprehensive evaluation. The complete set of all component-related V-submodels for safety-critical I&C forms a component-oriented (two-coordinate) V-model of the DLC.

In order to link the DLC of a specific attribute to each of the components of a safety-critical I&C system into a component-oriented V-model, consider additional attribute-related plane (formed by a set of attributes of the component). Thus, it can be stated that there are already three coordinates due to the addition of the attribute to a component-oriented (two-coordinate) V-model of DLC. In addition, in conjunction with the DLC, it is already possible to analyze a certain aspect in the three-dimensional space determined by the three coordinates associated with certain system's component, its attribute and the stage of the DLC.

The second goal of this chapter is to construct adequate models of the life cycle for modern safety and energy-efficient systems that will allow building a formalized description of their development process, taking into account all their integral properties.

One of the most challenging modern security problems is security of various safety-critical systems considering increasing attack rate on assets by use of vulnerabilities. Such systems can contain wide set of general and technology-specific vulnerabilities. Number of vulnerabilities and threats become more and more owing to application of different types off-the-shelf (OTS), first of all, commercial-OTS (COTS). There are many security features for I&C systems of NPPs, electronic medical systems and industrial applications. FPGA-based I&C systems are modern systems consisting of hardware modules and software components. The different security aspects of FPGA platform and I&C systems as a whole are analyzed by Drimer, S. (2009), Huffmire, T. et al. (2010), Kharchenko, V. et. al. (2015).

Drimer, S. (2009), Huffmire, T. et al. (2010) and Kharchenko, V. et. al. (2015) also discuss the role of FPGAs within a security system, how various solutions to security challenges can be provided on the basis of FPGA technology and presented basic analysis and description of FPGA security issues. Moreover, Kharchenko, V. et. al. (2015) describes security assessment process of safety-critical systems, applicable metrics, and represented an approach to assessment that relies on performance of gap and IMECA analysis.

The third goal of this chapter is to suggest and illustrate application of an approach to establishment of a secure development and operational environment for FPGA-based safety-critical I&C systems via description the process of establishment of a Secure Development and Operational Environment, its features and vulnerabilities, as well as stages of security-oriented analysis of safety-critical I&C systems. Case study introduces into security assessment for RadICS FPGA-based platform by use of IMECA technique.

SECURITY REGULATIONS ANALYSIS

There are various modern safety regulations and standards in the nuclear industry worldwide, and in the USA in particular, that prescribe requirements and limitations for the I&C development processes. Some of them, which are either issued or endorsed by the US NRC, are Regulatory Guide (RG) 1.152-2011, RG 5.71-2010, IEEE Std. 603-1991, IEEE Std. 7-4.3.2-2003, etc. Mainly they pose various aspects, starting from requirements related to secure development/operational environment, up to safety system's security features, including cyber security ones. Integration of appropriate security activities into I&C life cycle model is still challenging problem.

Tile 10 Code of Federal Regulations (CFR) Part 50, Appendix B, "Quality Assurance Criteria for Nuclear Power Plants and Fuel Reprocessing Plants", as well as 10 CFR 73.54, "Protection of digital computer and communication systems and networks," from the US NRC are generic regulations in the area of licensing the production and utilization facilities within the US market.

Security and cyber security aspects are covered by RG 1.152-2011 and RG 5.71-2010, respectively. RG 1.152-2011, "Criteria for use of computers in safety systems of nuclear power plants" describes a method that the US NRC considers acceptable to implement licensing process with regard to the use of computers in safety systems of NPPs. In such a way, it provides regulatory criteria on the establishment of a Secure Development and Operational Environment (SDOE) for digital I&C via appropriate physical, logical and programmatic controls during development phases and appropriate physical, logical and administrative controls during operation phase.

In turn, RG 5.71-2010, "Cyber security programs for nuclear facilities" is a direct guidance for activities implementation under 10 CFR 73.54 "Protection of digital computer and communication systems and networks", based on international and federal standards, and describes the methods and security activities for the operation and maintenance of a NPP, including appropriate I&C systems. Accordingly, cyber security features should be designed and implemented during the development phase of the I&C, before its installation.

During the establishment of a Secure Development and Operational Environment for I&C on the basis of RG 1.152-2011, security controls, which are in compliance with the RG 5.71-2010, additionally should be planned, designed, and implemented.

TAXONOMIES OF SAFETY-CRITICAL I&C SYSTEMS' ATTRIBUTES

Possible Attributes and Taxonomies

One of the most important attributes of safety-critical I&C systems is dependability. Dependability of a system is the ability to deliver required services (or perform functions) that can justifiably be trusted. Dependability is a complex attribute of a safety-critical I&C system that can be represented by a set of primary attributes, including:

- **Reliability:** Continuity of correct (required) services.
- **Availability:** Readiness for correct services.
- **Survivability:** Ability to minimize loss of quality and to keep capacity of fulfilled functions under failures caused by internal and external reasons.
- **Safety:** Absence of catastrophic consequences for the user(s) and the environment.
- **Integrity:** Absence of improper system alternations.
- **Confidentiality:** Absence of unauthorized disclosure of information.
- **High Confidence:** Ability of correct estimation of services quality, i.e. definition of trust level to the service.
- **Maintainability:** Ability to undergo modifications and repairs.
- **Security:** The protection from unauthorized access, use, disclosure, disruption, modification, or destruction in order to provide confidentiality, integrity, and availability.

In turn, safety attribute of a safety-critical I&C system can have some particular (or secondary) attributes depending on exact system, environment and conditions that have influence on the primary attribute. Here, we distinguished the following attributes (see Figure 1): reliability, security and trustworthiness, and we denoted their two-way influence.

We should note that such particular attributes may be defined for each of primary attributes, thus, representing hierarchical structure of a safety-critical I&C system's generic attributes set. Moreover, those secondary and further attributes may turn to be common for different primary attributes due to their incomplete "orthogonality".

Figure 1. Taxonomy of safety attribute

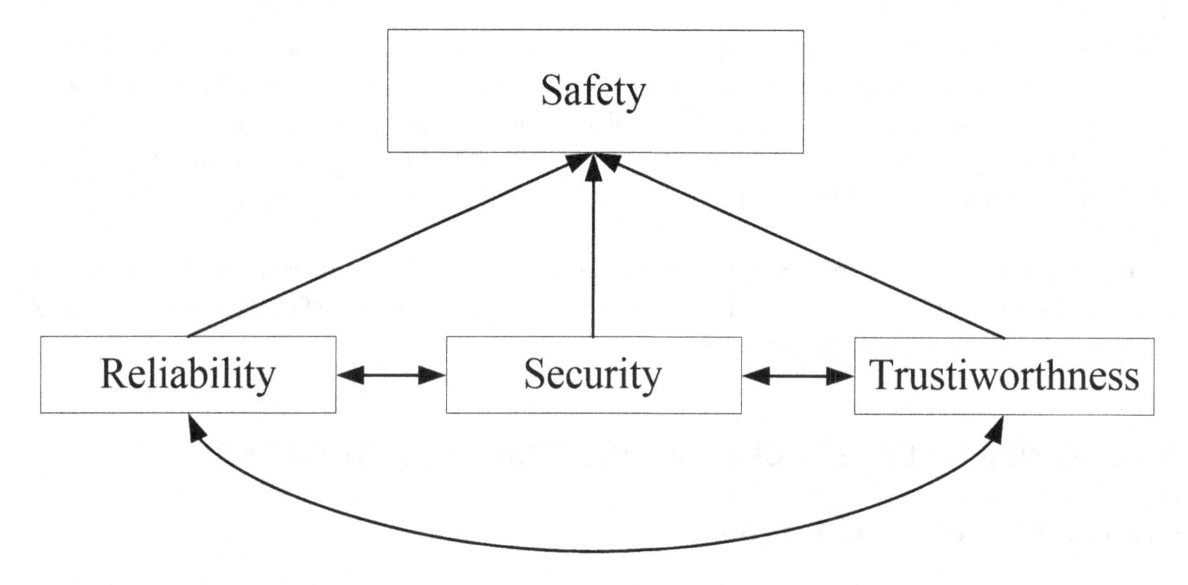

Metrics of Interference

Thus, we can state that a set of a safety-critical I&C system attributes can be represented in a form of i-level hierarchical model, and each of i levels contains k_i attributes. As an example, Figure 2 represents an element of last two levels of an attributes hierarchical model consisting of i levels.

One of the possible ways to reveal criticality of two-way influence for system's attributes, is in creating of attributes influence matrix. Such a problem can be solved, in particular, in the following ways:

1. 1. Create a set of n "local" influence matrixes for i hierarchical levels; each of the matrixes consists of k_i attributes (see Figure 3), and, therefore of k_i rows. Such number n can be calculated using the following equation:

$$n = \sum_{x=1}^{i-1} k_x \qquad (1)$$

The number of rows in each matrix associated with the level m, where m=[1, i-1], is equal to a number of attributes (k_m) at the lower level m+1: for example, the local matrix for a single attribute of i-1 level consists of k_i rows.

Figure 2. Levels of a safety-critical I&C system attributes hierarchy

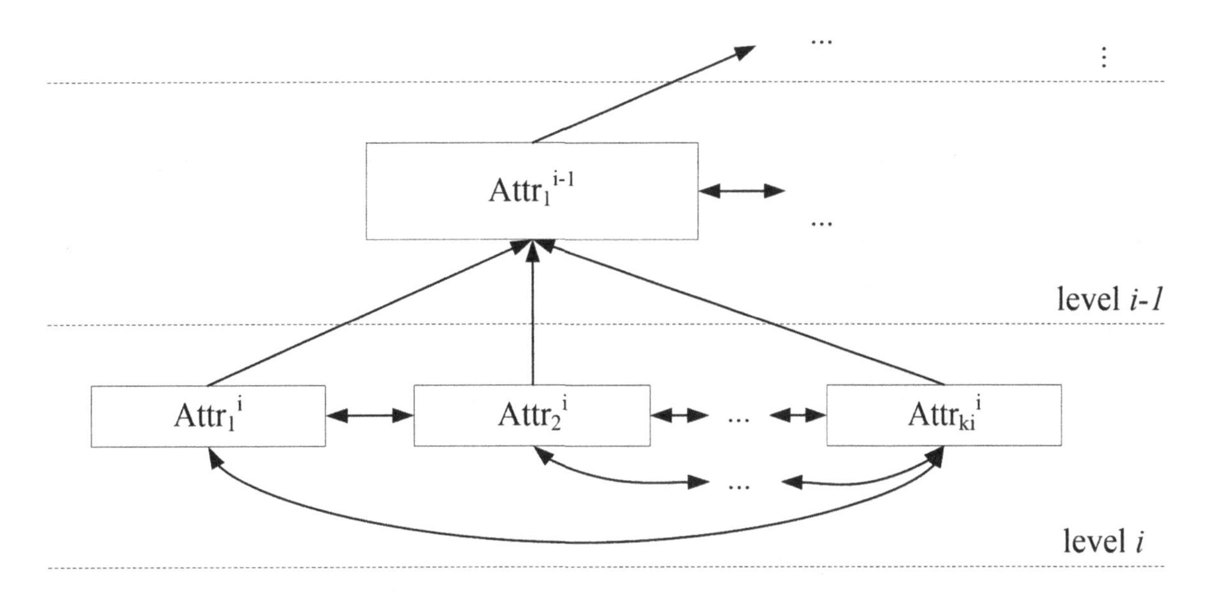

A set of such "local" influence matrixes represents the case of a metric mostly intended for independent assessment of the safety-critical I&C system's attributes within the single level.

Figure 3. Local influence matrix

$Attr_{ki-1}^{i-1}$

	low	*medium*	*high*
$Attr_1^i$		X	
$Attr_2^i$			X
⋮	...	...	...
$Attr_{ki}^i$	X		

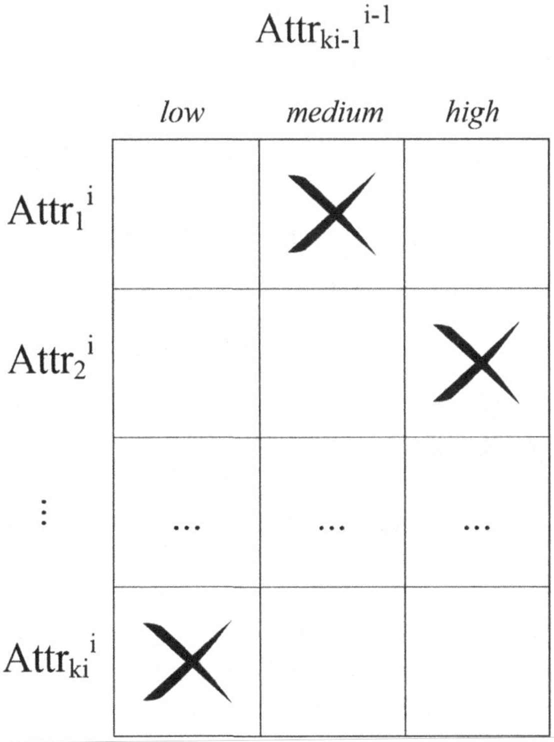

2. Create the single "global" influence matrix where each of all the n attributes (see Eq. (1)) is reflected by a single row and appropriate column (see Figure 4).

"Global" influence matrix can be considered as another metric, which is suitable for assessment of the safety-critical I&C system as a whole.

Thus, on the one hand, such metrics allow sharing system resources in order to assure the required level of security (a vertical related to different levels in Figure 2), on another hand, they allow optimizing the use of the resources (within the same level, see Figure 2).

Figure 4. Global influence matrix

	$Attr_1^i$	...	$Attr_{ki}^i$	...	$Attr_1^1$	...	$Attr_{ki}^1$
$Attr_1^i$		...		...		...	
⋮	...		...	...	...	...	...
$Attr_{ki}^i$	**L**	...		...		...	
⋮	...	...	...		...	...	...
$Attr_1^1$	**M**	...	**L**	...		...	
⋮	...	...	...	...	...		...
$Attr_{ki}^1$	**L**	...	**H**	...	**M**	...	

CLASSIFICATION OF V-MODELS

In this section, we propose a classification of V-models that is based on a number of decomposition attributes (Figure 5):

- Product properties (PPd) that need to be taken into account and for which, in fact, such a V-model is being built (energy efficiency, functional and cyber security, reliability, etc.).
- Components of the product under development – CPd.

- Process components (presence of V-submodels) – CPr.

Figure 5. Attributes used in the classification of V-models

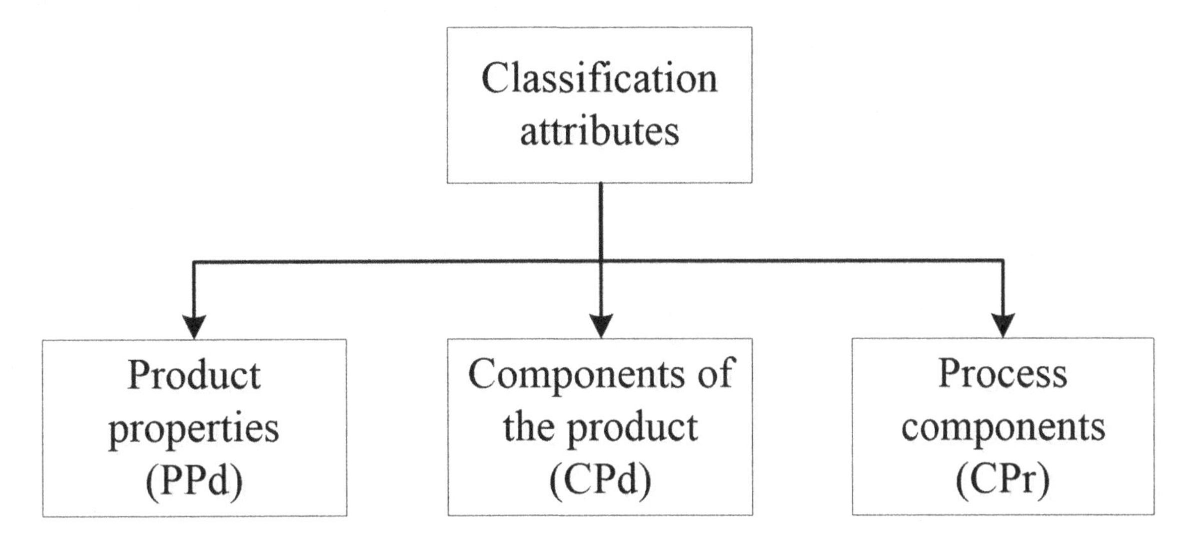

The next level of classification, based on the proposed attributes, allows further detailing the set of different V-models. To do this, we propose the following attributes for classification purposes:

- For the level of PPd: product properties, subproducts of the DLC, stages of the DLC, product properties and products at certain stages of the DLC; when constructing the V-model, such variants can lead to the appearance of classical (linear), plane and three-dimensional variants of the V-model, in different combinations.
- For the CPd level: the presence of intermediate products (subproducts or components) in the final product that is being developed; it leads to the appearance of supporting processes in the implementation of the product DLC that, in turn, leads to appearance of a submodel (incomplete or complete) within the main V-model; incomplete, in this case, means a model in which some stages of the main (complete) model may not be present.
- For the CPr level: the existence of processes for the independent verification for development artifacts result in appearance of an additional branch in the V-model.

Description and Classification of Life Cycle V-Models

It is proposed the following Life Cycle Model (LCM), consisting of stages model (M_s), sets of products (S_{prd}), processes (S_{prc}), process mappings to stages (Φ_{cs}) and products to stages (Φ_{ds}), as well as sets of product converters (MF):

$$LCM = \left\{M_s = \left\{MS, \Phi_{ss}\right\}, S_{prd}, S_{prc}, \Phi = \left\{\Phi_{cs}, \Phi_{ds}\right\}, MF\right\}. \quad (2)$$

The stages model consists of a set of stages of the DLC (a tuple with a fixed placement of the DLC stages) and the mapping of one set of stages to another. In the process of DLC implementation, the products are produced (in compliance with *MF* converters) using the processes; such sets strictly correspond to DLC stages. Moreover, there is an unequivocal representation of the processes set onto a set of stages.

Proposed classification of V-models is presented in Figure 6.

Figure 6. Scheme of the proposed V-models classification

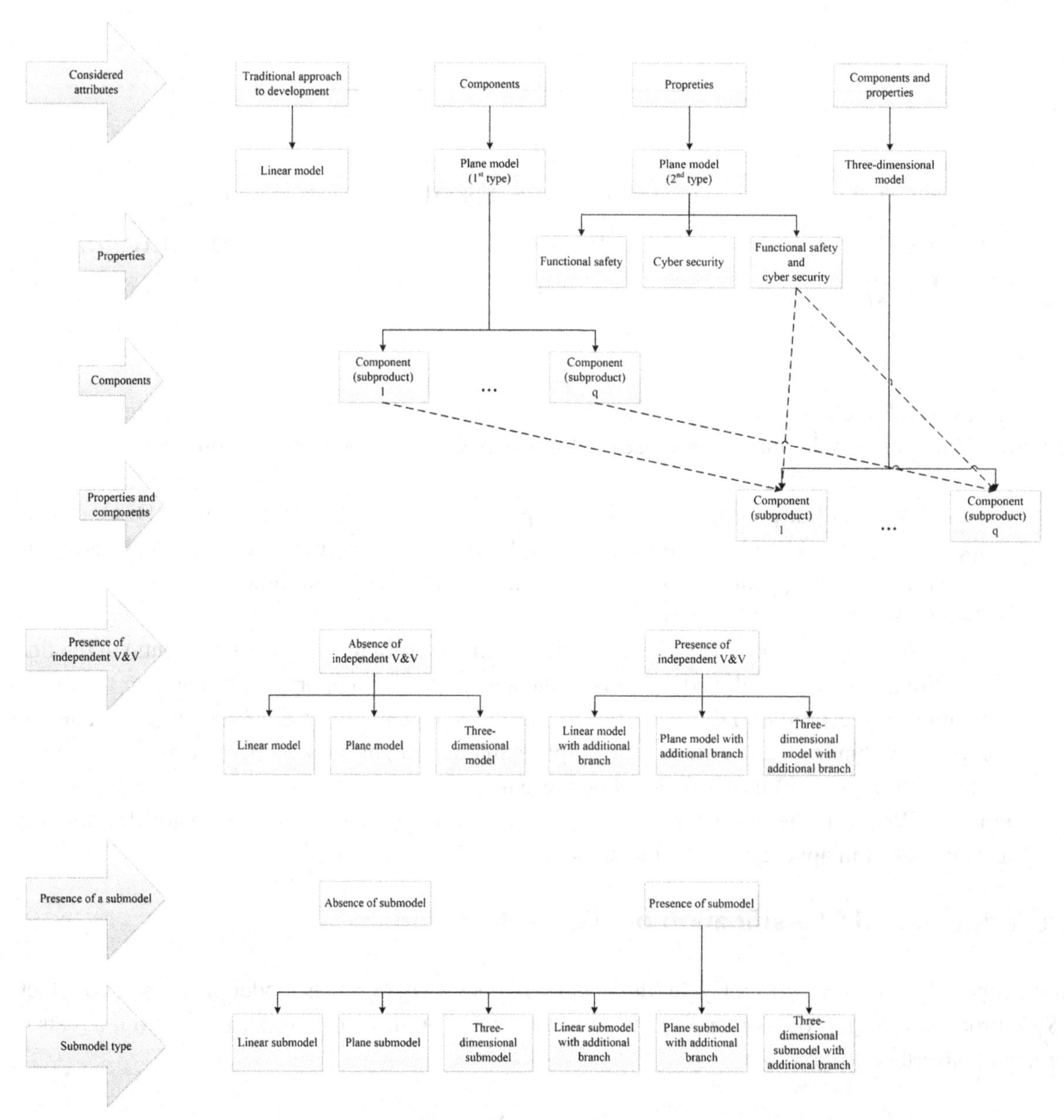

The linear V-model (Figure 7) can be conveniently formalized, and this is true for both of its branches, using two subsets of entities. Such entities can be products that appear during implementation of development activities, as well as processes that lead to the appearance of products at different stages of the final product's life cycle (the result of V-model implementation).

During the construction of descending (left) branch of the V-model, it is being performing the formalization of relationship between the development processes being implemented and the corresponding products, as well as the reflection of the processes onto the products, including the processes of mandatory verification.

During the construction of the ascending (right) branch of the V-model, the formalization of the activities associated with validation by reflection the elements of the right branch to the left is performed.

In a case of the requirements for performing independent verification or validation during the development of a product, an additional branch appears in the V-model, and this can occur for either its left or right main branches.

Figure 7. Linear V-model

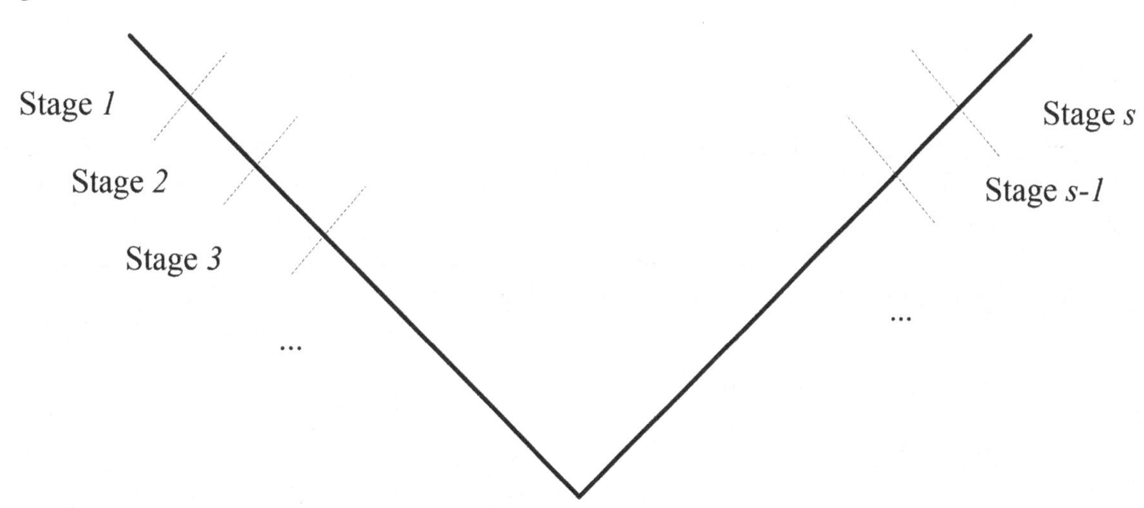

In a case of presence of several intermediate products (subproducts or components) in the final product, which is being developed, appropriate V-model can include several submodels that partially overlap, which, in turn, correspond to each of these products' DLCs.

In a general case, the starting point and the duration of subproduct's DLC differ from those in the product due to various reasons. Consequently, it is possible to separate out all submodels to analyze further the individual attributes of each of the subproducts, if necessary. A complete set of all submodels, associated with each of the subproducts, generates the plane for resulting CPd, V-model of the final product (Figure 8).

Figure 8. Plane for CPd, V-model reflecting the subproducts

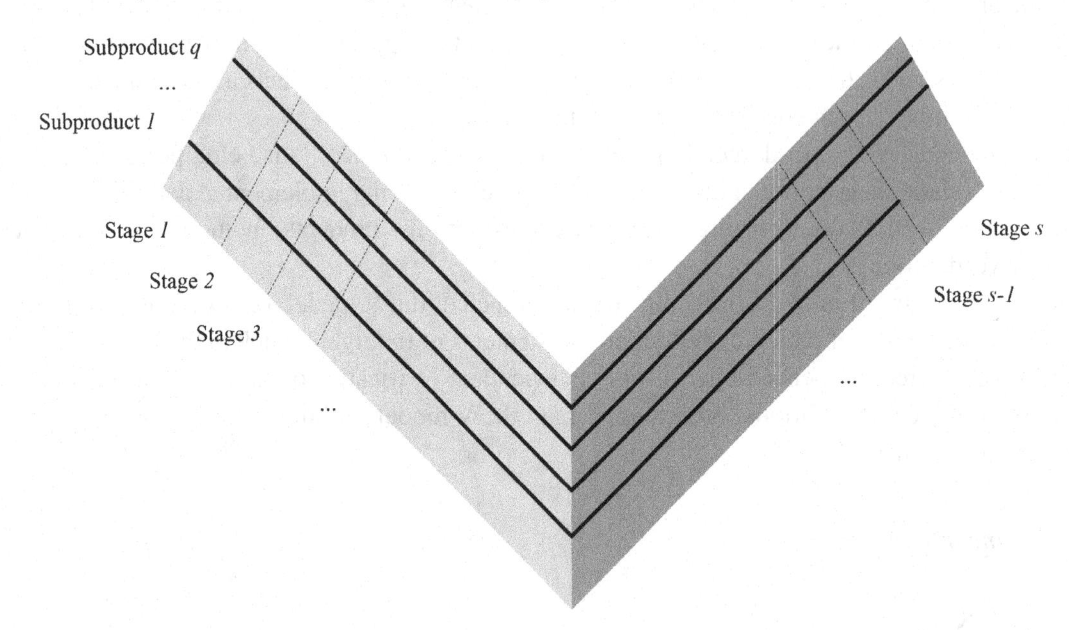

In such V-model, in a general case, the start and end points for each of the submodels are different in time; the products, which were produced at various stages of the DLC, are used in the development of the final product.

Similarly, it is possible to introduce a plane for PPd, V-model, which will reflect the properties of the whole safety-critical I&C or its certain subproduct (Figure 9).

Figure 9. Plane for PPd, V-model reflecting the properties

In a case of adding product properties to a V-model, it is possible to consider the three-dimensional analysis space (Figure 10) and the corresponding three-dimensional V-model (Figure 11): such approach allows for independent analysis of each of the subproducts on the basis of its specific property at certain stage of the DLC.

Figure 10. Three-dimensional analysis space

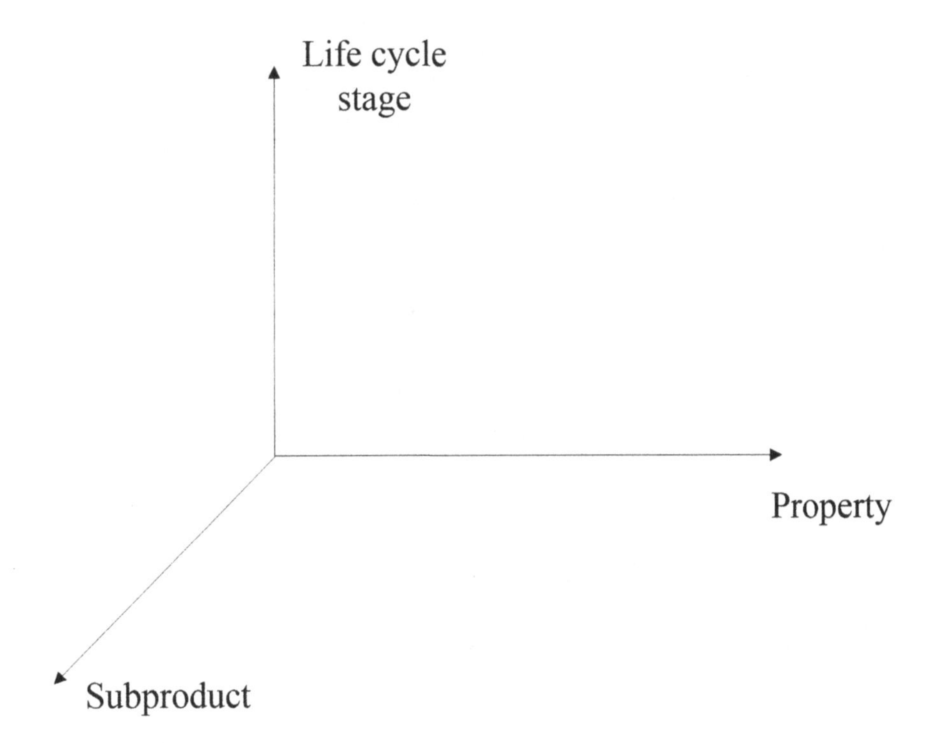

Models with additional branches fall into a separate type of V-models, typically asymmetric, which differ from the linear (traditional) ones by the presence of one or more additional branches. An additional branch in the original V-model can be caused, for example, by the existence of some processes for the independent verification of development artifacts during the product DLC implementation. One of the simplest V-models containing an additional branch is shown in Figure 12.

Case Study: V-Model for FPGA-Based RadICS Platform Used in NPP Safety-Critical I&C Systems

The digital I&C platform RadICS is a product of Research and Production Corporation Radiy. It consists mostly of a set of general-purpose building modules that can be configured and used to implement application-specific functions and systems. The RadICS platform is composed of various standardized modules, each based on the use of FPGA chips as logic solvers.

The Platform was successfully certified under the requirements of IEC 61508 to meet Safety Integrity Level (SIL) 3, where SIL4 would be the most demanding level. The SIL certification process requires that products developed under IEC 61508 requirements applied throughout the product life cycle. The

Figure 11. Three-dimensional PPd, CPd, V-model

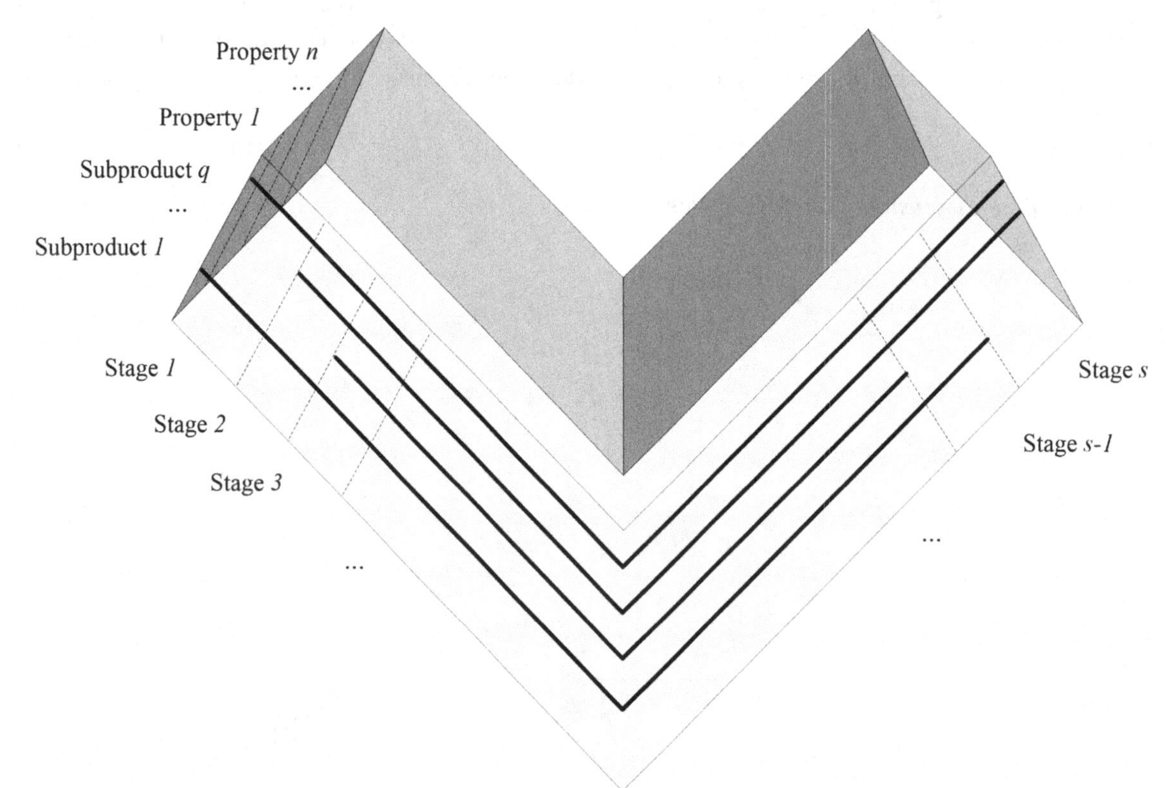

Figure 12. V-model containing an additional branch

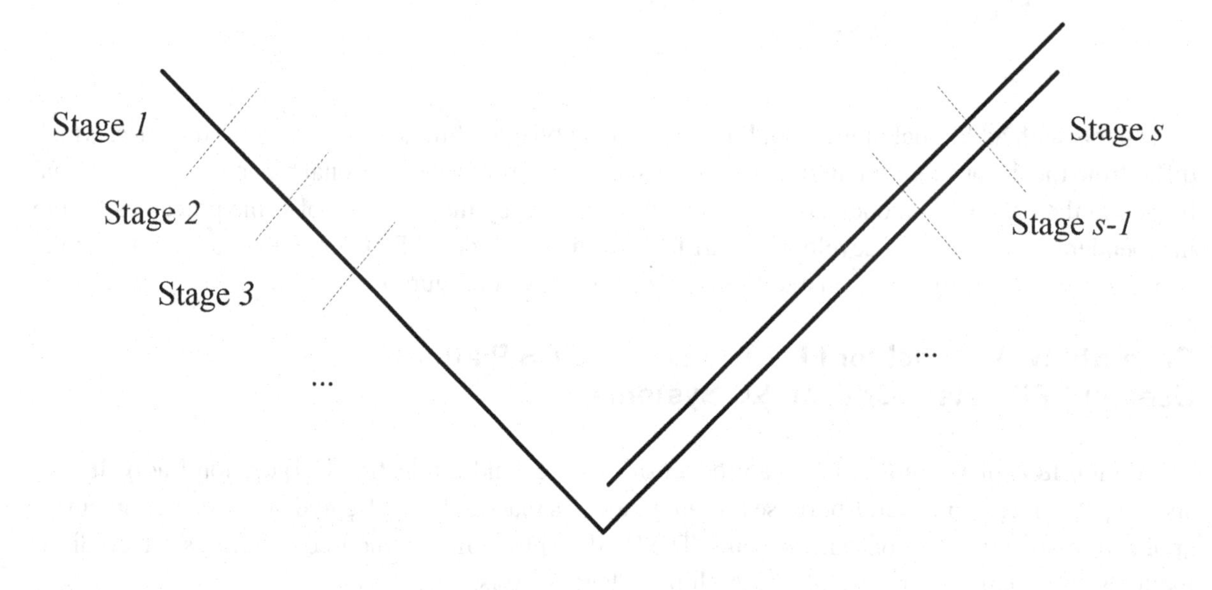

SIL certification process outlined in IEC 61508 requires the preparation of a set of documents specific to each of the phases of the product life cycle. These documents must be subject of an independent auditing process and assessment by a Certification Body.

Thus, life cycle of the RadICS Platform implements specific stages of FPGA design development and verification using formalized DLC. One of the most critical features required for successful SIL3 certification is Requirements Tracing process. The main idea is to achieve complete traceability at all project stages in order to implement all initial requirements and restrict functions to required set only.

For such purpose, it was developed and implemented FPGA technology-specific V-model presented in Figure 13. The RadICS is a complex product, containing specific hardware and software components mostly oriented towards FPGA technology, so its DLC considers all the inherent attributes and features allowing to perform assessments of certain attribute (safety, security, etc.) at certain DLC stage via constructing more complicated DLC models (for example, three-dimensional) described in previous section.

Figure 13. RadICS Platform's DLC simplified model

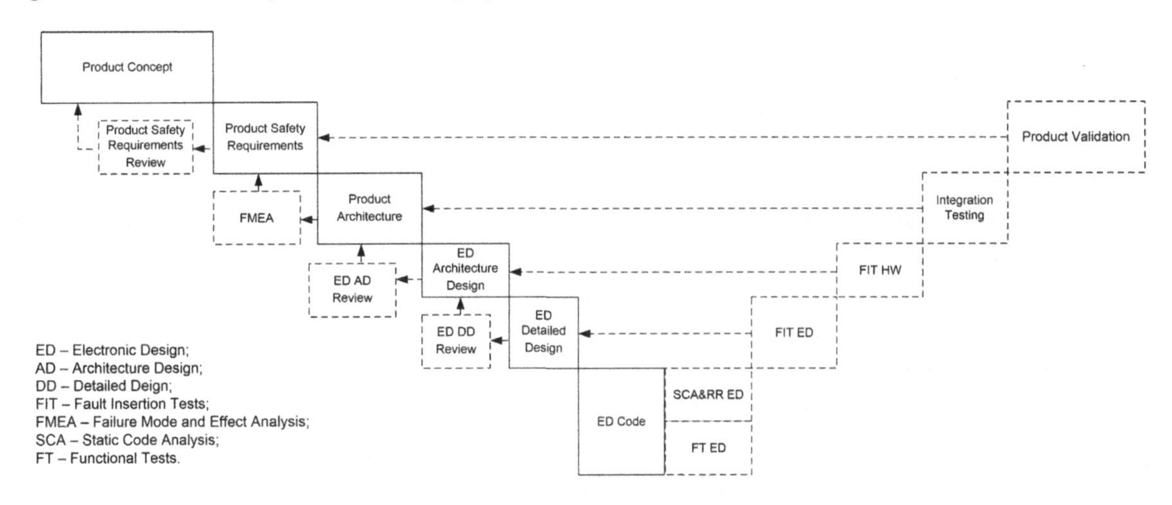

Metrical Assessment of Meeting the Requirements Using V-Models

The requirements presented in the form of a taxonomic structure can be projected onto the stages that form the life cycle model. For example, they can be decomposed within the framework of a V-shaped model of the life cycle depending on the stage on its left or right branches (Figure 14). Decomposed requirements in this way must undergo a verification procedure, i.e. confirmation of their implementation. Verification of requirements at earlier stages reduces the costs of changes in the developed project if they are not fulfilled. Imagine possible "point-stages" of verification of requirements within the framework of a V-shaped model:

- A "point" at which the requirement should be checked; to denote it, the term "point of necessary verification – TNR" is introduced.
- A "point" at which the requirement is verified; this is applicable for a project when the process of verification of requirements is uncontrollable; to denote it, the term "real verification point – TRV" is introduced.

- A "point" at which verification can be carried out or carried out at an earlier stage than at which it should be verified; to denote it, the term "point of earlier verification – TBRV" is introduced.

Consider an example that graphically displays options for the "points" of requirements verification (Figure 14). For a more accurate representation of the options for verification points, we introduce the legend, which are presented in Table 1. Figure 14 illustrates the arrangement of taxa that are part of taxonomic structures according to the stages of the V-shaped model of LC. Taxa are distributed by life cycle stages and software components as follows: P1.1 -> S12C4, P1.2 -> S21C1, P1.3 -> S11C1, P1.2.1 -> S31C1, P1.2.2 -> S32C3, P2.1 -> S42C1, P2.2 -> S33C4, P2.3 -> S33C2, P2.1.1 -> S52C6, P2.1.2 -> S41C5, P3.1 -> S51C5, P3.2 -> S54C2. Next, we consider the formation of sets of types of "points" for verification of requirements, the elements of which will be a pair of "stages-components":

- Necessary verification set of points includes the following:
  THB = {S12C4, S21C1, S11C1, S31C1, S32C3, S42C1, S33C4, S33C2, S52C6, S41C5, S51C5, S54C2}.
- Real verification set of points includes the following:
  TPB = {S52K6, C42P5}.
- Early verification set of points includes the following:
  TPB = {S42C2, S21C4, S22C5, S32C6, S23C5}.

Verification of compliance with requirements at earlier stages of the software life cycle saves project resources. To estimate such savings, quantitative metrics are needed first.

Figure 14. Decomposition of the requirements represented by the taxonomic structure and their projection onto the V-shaped model of the life cycle

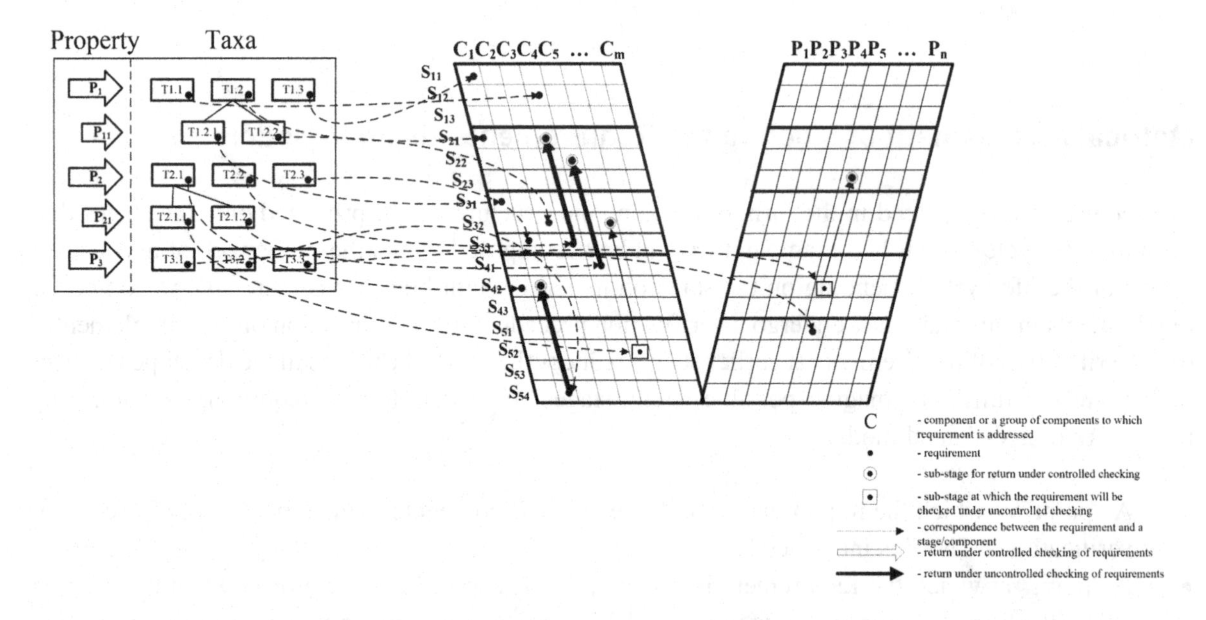

The following metrics can be used as the main metrics that can be used in the processes of evaluating and managing the requirements of NPP I&C software:

1. A metric of verified requirements rate (or completeness of verification), which is equal to the ratio of the number of verified requirements to the total number of requirements:

$$MetCR = DR / GR. \quad (3)$$

2. A metric of verified requirements rate for the completion of a certain stage Ei, which is equal to the ratio of the number of requirements verified by the end of stage Ei to the total number of requirements that must be verified by the end of stage Ei:

$$MetCREi = DREi / GDREi. \quad (4)$$

3. A metric of requirements rate that must be met by the end of stage Ei, which is equal to the ratio of the number of requirements that must be met by the end of stage Ei, to the total number of requirements:

$$MetREi = REi / GREi. \quad (5)$$

4. Return metric for failure to fulfill requirements at stage Ei, which is equal to the ratio of the number of stages / sub-steps to which the return should be made upon failure to fulfill requirements at step / sub-stage Ei to the number of stage / sub-stage Ei:

$$MetRREi = NREi / NEi. \quad (6)$$

5. Total maximum return metric, which is equal to the ratio of the sum of return metrics for all requirements to the total number of requirements:

$$MetGRR = MetRRE1 + MetRRE2 + \ldots + MetRREn / GR. \quad (7)$$

6. Cost reduction metric for timely verification of the Reqi requirement at stage Ej, which is equal to the ratio of the difference in costs for timely and later verification of compliance with the Reqi requirement for the cost of fulfilling the requirement for late verification:

$$MetRSRi = \Delta CSi / CS_Hi. \quad (8)$$

7. Cost reduction metric with timely verification of all requirements:

$$MetRSR = (\Delta CS1 + \Delta CS2 + \ldots + \Delta CSGR) / (CS_H1 + CS_H2 + \ldots + CS_HGR). \quad (9)$$

First of all, it is convenient to detail the process of saving resources in a general way. To do this, we will assign conditional quantitative values to each stage of the life cycle (Figures 15, 16). In Figure 16 they are designated as CSi. The later stage is assigned a larger value (CSi), so for stage S41 CS41 = 10, for stage S23 CS23 = 6, for stage S13 CS23 = 3. Such values are determined by the volume (quantity) of resources. We assume that the assigned values are conditional absolute values. Consider the case when the point of real verification goes to the point of earlier verification, that is, the verification of the requirement moves from S41 to S13 (Figure 15).

Figure 15. Transition of points of necessary and real verification to a point of earlier verification

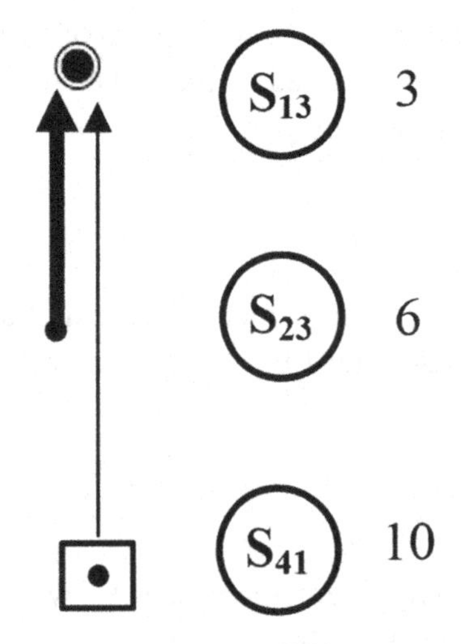

Figure 16. Costing for stages

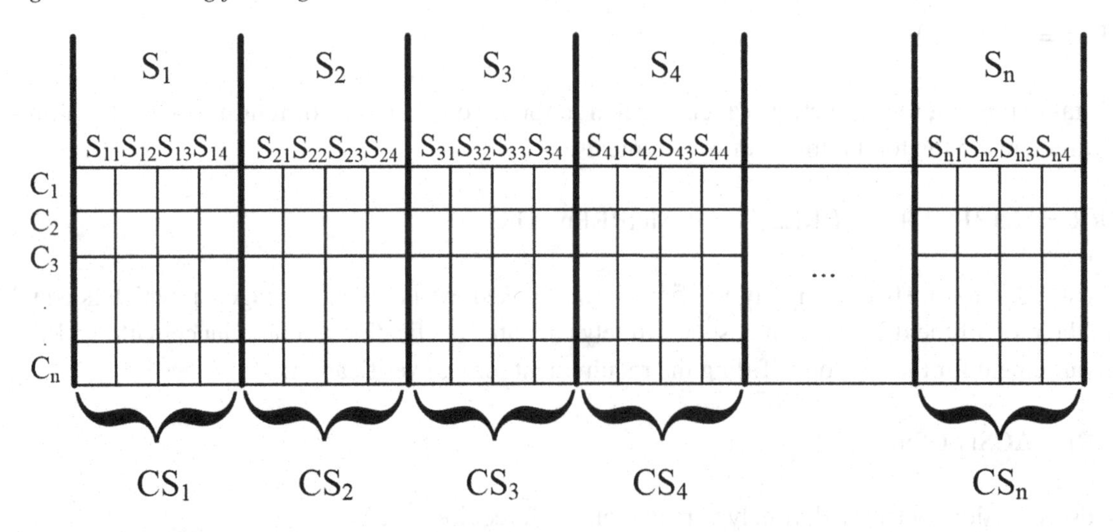

Next, according to formula (9), we calculate the first cost reduction metric for timely verification of requirements (MetRSR1), and then the second cost reduction metric for timely verification of requirements (10) (MetRSR2) and, in conclusion, calculate the difference between the values of the first and second metrics using the formula (11):

$$MetRSR_{1S_i} = \frac{!\,S_{max} - !\,S_{min}}{!\,S_{max}}; \quad (10)$$

$$MetRSR_{2S_i} = \frac{S_i - S_{min}}{S_{max}}; \quad (11)$$

$$\Delta MetRSR_{S_i} = MetRSR_{1S_i} - MetRSR_{2S_i};$$

$$MetRSR_{123} = \frac{10-3}{10} = 0.7;$$

$$MetTRSR_{223} = \frac{6-3}{10} = 0.3;$$

$$\Delta MetRSR_{23} = 0.7 - 0.3 = 0.4.$$

Thus, the savings in the case of transition of the point of real verification to the point of necessary verification is 0.4.

Consider a special case of a development project. Table 1 presents a fragment of the specification of the requirements of the project for the development of a system for NPP.

Table 1. Fragment of requirements specification

Requirement	Verification stage
The worst-case response time for a change of state of a contact input signal through the complete system to the completion of change of state of a contact output shall be 10 ms or less	Validation Testing

Below are the results of calculating two metrics for such a fragment.

Cost reduction metric for timely verification of requirements, at step 12 (Integration Testing) instead 13 (Validation Testing).

MetRSR12 = (230 - 200) / 230 = 0.13.

Return metric for non-compliance at the stage 13 (Validation Testing):

MetRRE13 = (13-1) / 13 = 0.923.

GENERAL CONCEPTION OF GAP-AND-IMECA-BASED APPROACH TO ANALYSIS OF SAFETY-CRITICAL I&C SYSTEMS

In this section, as one of the possible solutions for safety-critical I&C system analysis problem, an approach, which is based on IMECA technique described by Babeshko, E., et al. (2008), is proposed.

One of the fundamental concepts behind the underlying idea of the approach is the concept of gap. Here we can define gap as a set of discrepancies of any single process (which, in a general case, consists of a set of sub-processes) within the system's DLC that can introduce some anomalies in a product and/or cannot reveal (and eliminate) existing anomalies in a product. In particular, such anomalies can be caused by imperfection of product specification (or even representation), implementation, verification, and/or other non-compliances.

For example, in terms of cyber security, some of the anomalies can be vulnerabilities of the product. Vulnerabilities, in turn, can be exploited by an adversary during intrusion into the product to implement an attack in order to introduce some unintended functionality into the product.

In this way, we propose a process-based approach to GA, because "non-ideal" processes, which contain discrepancies, can produce various problems in the corresponding products, and the following statements are true:

1. Presence of gaps in $Process_j$ results in anomalies in $Product_p$ even if $Product_{p-1}$ is "ideal".
2. Presence of anomalies within $Product_{p-1}$ can be eliminated by "ideal" Processj in many cases. This may be true in case of verification and validation processes; however, it does not apply to design processes. For example, anomaly in the technical specification is not eliminated by an "ideal" direct translation process (since it may not include verification).

As an illustrative example for the proposed definition of gap, let us consider a development process within the system's DLC model, where the input of $Process_j$ is represented by $Product_{p-1}$, and the output (result of process implementation) – is $Product_p$ (see Figure 17). Such a situation represents, for example, results of implementation of *i*-1 and *i* stages of system's DLC, respectively.

The transition from the previous product (*p*-1) to next one (*p*) is accomplished by the implementation of a prescribed process (*j*) by developers, using certain tools in compliance with certain prescribed techniques. Thus, this process can be represented as a set of sub-processes, which are related to the HTT, respectively. Such sub-processes are being implemented in serial and/or parallel ways, and each of them may contain problems (or discrepancies towards appropriate "ideal" sub-process) due to various reasons caused by the developer, the used technique or the tool. Therefore, the problems in sub-processes lead to problems in processes, which are implemented in order to produce a new product and can result in product anomalies.

The activities, required to implement the approach, comprise several consequent steps intended for a comprehensive analysis and assessment of a safety-critical I&C system.

The key idea of assessment is in the application of the process-product approach. Therefore, the DLC model of a safety-critical I&C system should include detailed representation of DLC processes and appropriate products. Then, it is possible to identify problems (or discrepancies) within the model, i.e. gaps. In general, such gaps may reflect various aspects of the system, depending on what system attributes are assessed (for example, safety and security).

Figure 17. Development process in the safety-critical I&C system DLC model

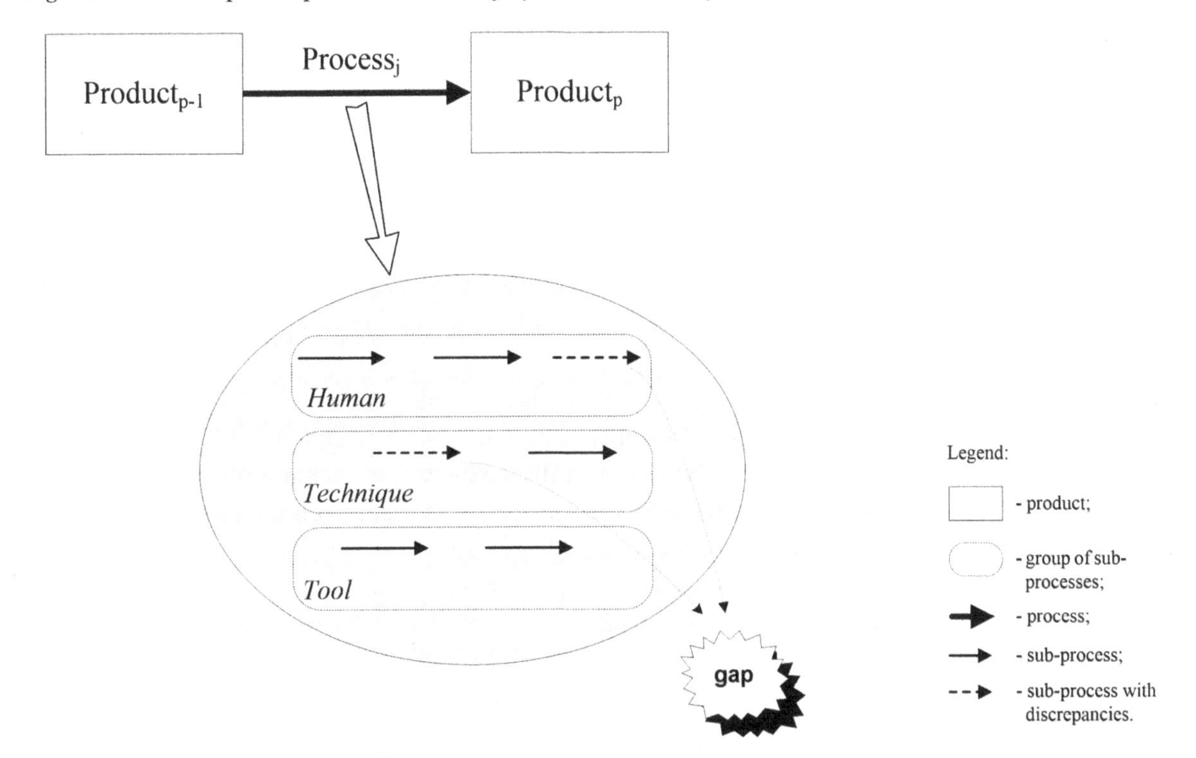

Hence, depending on the safety-critical I&C system aspects under assessment, each gap should be represented in a form of a formal description; such formal description should be made for a set of discrepancies identified within the gap. The IMECA technique is the most convenient, in our opinion, to perform such description: each identified gap can be represented by a single local IMECA table and each discrepancy inside the gap can be represented by a single row in that local IMECA table. In this way, complete traceability of life cycle processes, appropriate products and inherent properties of corresponding discrepancies can be achieved. As a result, the number of local IMECA tables would correspond to the number of identified gaps, and the number of rows within each local IMECA table would correspond to the number of identified discrepancies within the appropriate gap.

After completing the appropriate columns, for example on the basis of expert assessment, for all local IMECA tables, each gap being represented by a set of discrepancies with appropriate numerical values. Data within each row of local IMECA tables reveal, in explicit form, the weaknesses of the safety-critical I&C system aspect under assessment: for example, in terms of safety – system faults and failures, in terms of security – intrusion probability and severity.

Further, in order to implement the approach, the following cases are possible, depending on the scope of the assessment:

1. Assessment of the system as a whole. Then, a set of particular IMECA tables (which represent all the identified gaps by a set of discrepancies) should be integrated into the single global IMECA table that reflects the whole system. In this case, each row of the global IMECA table forms the basis for creating a global criticality matrix.

2. Assessment of particular (sub-)systems within the safety-critical I&C system. In this case, it is possible to create an appropriate set of local criticality matrixes that correspond to certain (sub-) systems, based on a set of local IMECA tables.

Integration of local criticality matrixes into a global one is carried out in accordance with the following rule:

$$e_{yz}^{G} = \bigcup_{r=1}^{g} e_{yz}^{L_r}, \tag{12}$$

where e^G is an element of the global criticality matrix, e^{L_r} is the corresponding element of the r-th local criticality matrix, and g is the total number of local criticality matrixes (equal to total number of gaps).

Moreover, the scales for the numerical values of a discrepancy (for example, its probability and severity) for local criticality matrixes can be set to the same value in order to eliminate the necessity of additional analysis during the creation of a global criticality matrix.

In both cases, the highest risk of the selected aspect corresponds to the highest row in the criticality matrix. In a case of independent gaps and discrepancies, the total risk of R can be calculated using the following equation:

$$R = \sum_{t=1}^{g} \sum_{w=1}^{m} p_{tw} D_{tw}, \tag{13}$$

where g is the total number of gaps, m is the total number of rows in the IMECA table, p is the occurrence probability, and D is the corresponding damage.

Moreover, the criticality matrix can be extended to be K-dimensional (where $K>2$) that allows us to consider, for example, the amount of time required to implement the appropriate countermeasures for the assessed system.

For example, during the assessment of security, the prioritization of vulnerabilities identified on the basis of process-product approach, should be performed according to their criticality and severity, representing their corresponding stages in the cyber security assurance of the given system. The main goal of this step is to identify the most critical security problems within the given set. Prioritization may require the creation of a criticality matrix, where each of the vulnerabilities is represented within single rows. In such cases, it is possible to manage the security risks of the whole system via changing the positions of the appropriate rows within the matrix (the smallest row number in the matrix corresponds to the smallest risk of occurrence).

During the performance of GA, the identification of discrepancies (and the corresponding vulnerabilities in case of security assessment), can be implemented via separate detection/analysis of problems caused by human factors, techniques and tools, taking into account the influence of the development environment.

Then, after all identified vulnerabilities are prioritized, it is possible to assure security of the safety-critical I&C system by implementing of appropriate countermeasures. Such countermeasures should be selected based on their effectiveness (also, in context of assured coverage), technical feasibility, and cost-effectiveness. However, there is an inevitable trade-off between a set of identified vulnerabilities and a minimal number of appropriate countermeasures, which allows us to eliminate vulnerabilities or to make them difficult to be exploited by an adversary. The problem of choosing such appropriate countermeasures is an optimization problem and is still challenging.

ESTABLISHMENT OF A SECURE DEVELOPMENT AND OPERATIONAL ENVIRONMENT

Secure Operational Environment is defined as the condition of having appropriate physical, logical and administrative controls within a facility to ensure that the reliable operation of I&C systems are not degraded by undesirable behavior of connected systems and events initiated by inadvertent access to the I&C system.

The establishment of a SDOE in the context of US NRC's RG 1.152, refers to the following aspects:

- Measures and controls taken to establish a secure environment for development of the safety I&C against undocumented, unneeded and unwanted modifications.
- Protective actions taken against a predictable set of undesirable acts that could challenge the integrity, reliability, or functionality of an I&C system during operations.

Phases of the waterfall life cycle model (WLCM) form a framework for describing specific guidance(s) for the protection of digital safety systems and the establishment of an SDOE via identification and mitigation of potential weakness or vulnerabilities in each of the phases that may degrade the SDOE or degrade the reliability of the system.

WLCM includes: concepts; requirements; design; implementation; test; installation, checkout, and acceptance testing; operation; maintenance; retirement. Each of the phases consists of some prescribed activities performed in order to establish and maintain a SDOE. One of the most important activities during concept phase is assessment of vulnerabilities for both development and operational environments. Such assessment forms framework of security requirements to implementation of further life cycle activities and additional secure design solutions to the system under development.

Figure 18. I&C security vulnerabilities classification

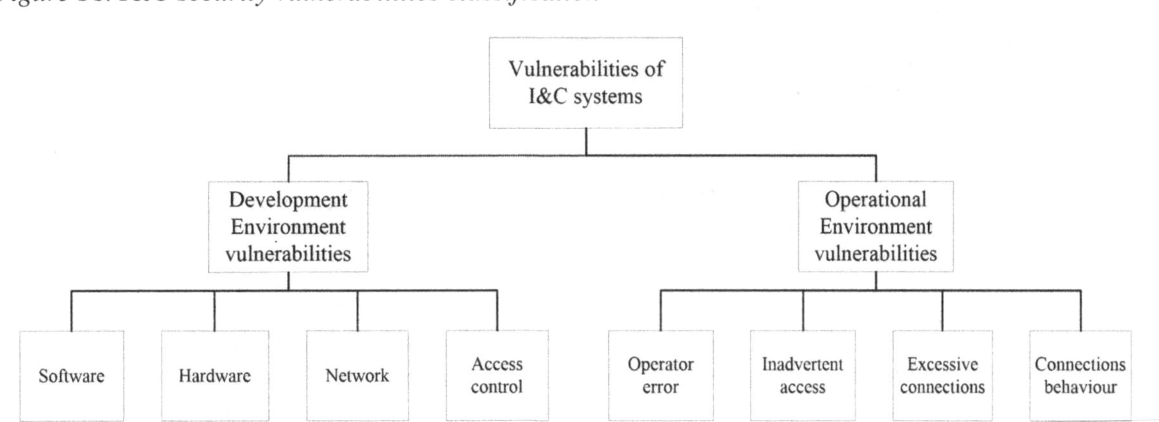

Typical output for vulnerabilities assessment activity is appropriate report describing all the identified vulnerabilities related to both development and operational environments that, in turn, forms the basis for implementation of specific security assurance processes for the I&C system. They should be followed in life cycle phase: for example, in development phase, the set of activities, including measures and controls taken to establish a secure environment for development of the I&C system against undocumented, unneeded and unwanted modifications

I&C system security vulnerabilities classification is presented in Figure 18. SDOE establishment process requires that the development process should identify and mitigate potential vulnerabilities in each phase of the life cycle.

Security-Oriented Analysis of Safety-Critical I&C Systems

The proposed approach is based on IMECA technique (see also Figure 14), as one of the modification of Failure Mode, Effects, and Criticality Analysis (FMECA) which is usually applied to assess reliability and safety. Yastrebenetsky, M. et. al. (2014) analyzed "non-ideal" development processes which can result in various problems in the corresponding products. Each transition between two consequent (*p*-1, *p*) products is accomplished by the implementation of a prescribed process (*j*) using specific tools under prescribed techniques. Thus, process is a set of sub-processes related to the HTT and some of them may contain problems (Figure 19). Such problems can result in product anomalies.

Figure 19. Development process in the I&C development life cycle model

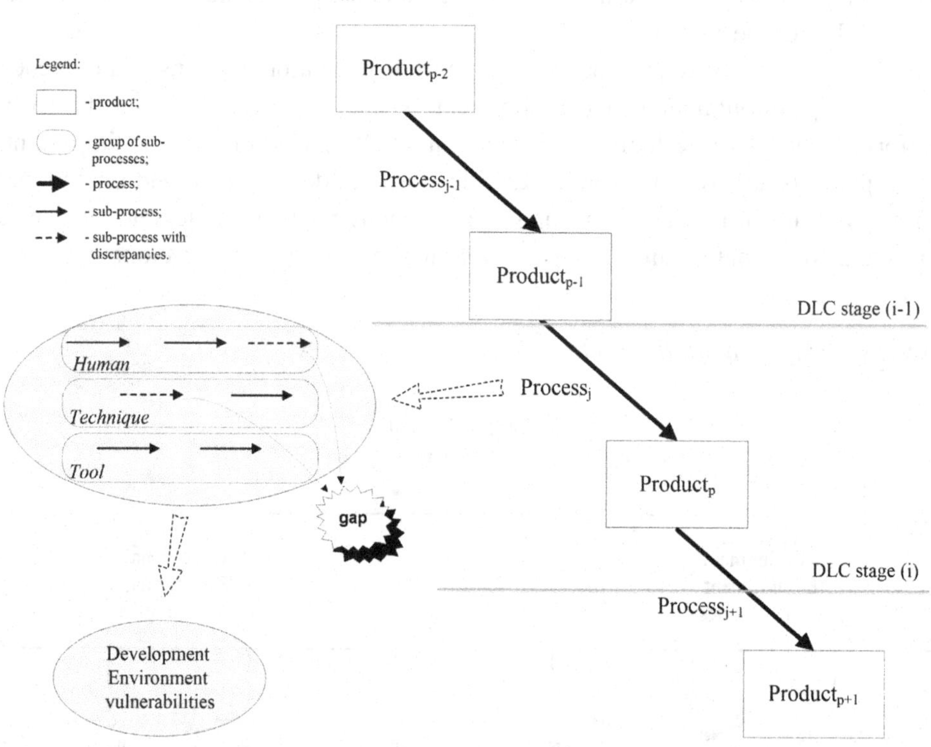

All transition scenarios between two consequent products due to implementation of a process, possibly containing gaps, are presented in Figure 20.

Figure 20. Possible transition scenarios between two consequent products

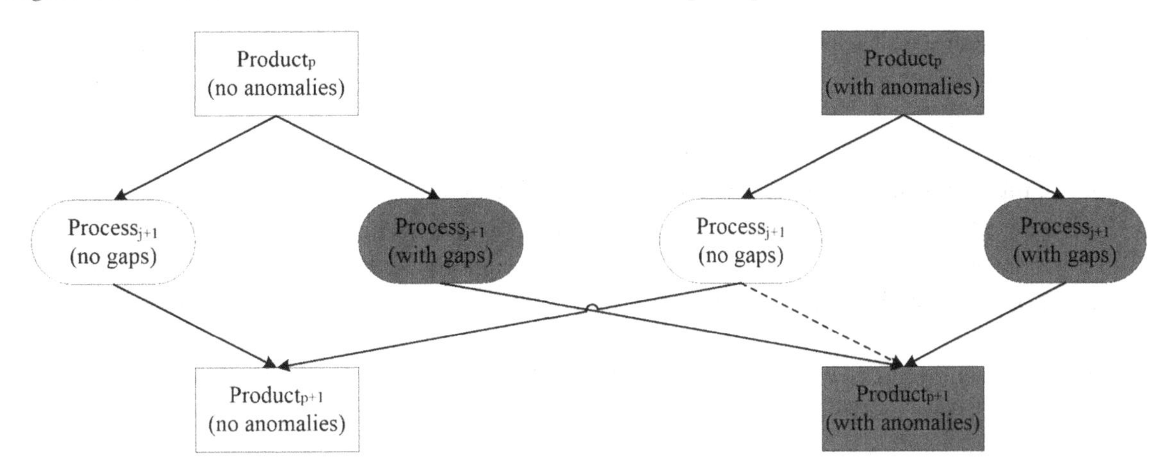

In terms of security, such process gaps represent sources of security threats, which, in turn, can exploit certain vulnerabilities of the development process (see Figures 15 and 16) in order to implement successful attack resulting in introduction some anomalies into the product. Table 2 represents interrelations between the threat sources and possible types of vulnerabilities related to the development environment. In this way, vulnerabilities types that can be exploited by certain threat(s) are designated by "+" symbol. Such interrelations should be considered during choice of appropriate countermeasures in order to reduce security risks.

Table 2. Interrelations between the threat sources and types of vulnerabilities

		Vulnerabilities			
		Hardware	**Software**	**Network**	**Access control**
Threat source	**Human**	+	+	+	+
	Technique	+	+	+	
	Tool	+	+	+	

Columns of IMECA table(s) are being completed on the basis of expert assessment, and each gap is represented by a set of discrepancies with appropriate numerical values.

Then, on the basis of IMECA table(s), criticality matrix should be created to assess the security risks: the highest risk corresponds to the highest row in the criticality matrix. During the assessment of security, the prioritization of identified vulnerabilities (on the basis of process-product approach) should be performed according to their criticality and severity in the security assurance context of the given I&C system, in order to identify the most critical security problems within the given set.

I&C system security can be assured by implementing of appropriate countermeasures. Their selection should be done on the basis of their effectiveness/coverage, technical feasibility, and cost-effectiveness.

Case Study: Security Analysis of FPGA-Based Platform

RG 1.152-2011 states that the establishment of a Secure Development and Operational Environment for I&C refers to:

- Measures and controls taken to establish a secure environment for development of the I&C system against undocumented, unneeded and unwanted modifications.
- Protective actions taken against a predictable set of undesirable acts (e.g., inadvertent operator actions or the undesirable behavior of connected systems) that could challenge the integrity, reliability, or functionality of an I&C system during operations.

Security aspects should be considered at each life cycle stage of I&C system from concept development to system retirement. To decrease the risks of successful cyber attacks, the following procedure can be used:

- Creation of criticality matrix based on results of proposed approach.
- Selection of a set of applicable appropriate countermeasures based on recommendations of the specific regulations according with criteria security (safety) and cost.

Research and Production Corporation Radiy is developer and manufacturer of proprietary FPGA-based platform RadICS intended for application in modern safety-critical applications, including those related to NPP sector. In order to illustrate IMECA-based assessment, we present some of results for attacks modes in operation environment for FPGA-based platform RadICS (see Table 3).

Table 3. Results of IMECA for some of FPGA attacks

Row number	Stage	Attack mode	Attack nature	Attack cause	Type of effects	Occurrence probability	Effect severity	Countermeasures
1	Operation	Black Box Attack	Active	Simple logic of electronic design	Reverse engineering of logic by adversary	Very low	Very low	Complication of electronic design logic
2	Operation	Read-back Attack	Active	Absence of chip security bit and/or availability of physical access to chip interface	Obtaining of secret information by adversary	Moderate	High	The use of security bit Application of physical security controls
3	Operation	Cloning Attack	Active	Storing of decoded configuration	Obtaining of configuration data by adversary	Moderate	High	Checking of chip's internal ID before powering up an electronic design Encoding of configuration file Storing of configuration file within FPGA chip

Basing on IMECA table, criticality matrix can be created. There is a possibility to manage security risks via decreasing of attacks' occurrence probability, since related damage is constant. In terms of FPGA-based NPP I&C systems, such decreasing of the probability can be achieved, for example, by implementation of certain countermeasures in development environment or specific countermeasures in operational environment.

FUTURE RESEARCH DIRECTIONS

Future steps will be dedicated to joint security assessment of the I&C systems based on Kharchenko, V. et. al. (2015), results of this chapter and more detailed product-oriented analysis of FPGA platform-based systems.

CONCLUSION

A problem of NPP I&C systems cyber security analysis and assessment is still challenging due to the fact that such systems consist of interconnected complex components with different functions and different nature. The majority of modern safety-critical I&C systems are being FPGA-based; hence, it is impossible to perform their assessment without consideration of all specific details, including interference of various system's attributes and the special features for all the technologies used. In this chapter we discussed some problems related to assessment of various aspects of safety-critical I&C systems.

The proposed approach is based on gap conception, IMECA technique and HTT. Such an approach is applicable in assessment of various aspects of safety-critical I&C systems, since it considers process-product model to reveal all the process discrepancies that can potentially result in product anomalies.

Gap-and-IMECA-based technique was applied in development of a company standard in Research and Production Corporation Radiy that is harmonized with international standards. This standard is used during implementation of development and verification activities for safety-critical I&C systems for nuclear power plants.

Next steps of research and development activities may be connected with creation and implementation of tool-based support for the proposed approach, taking into account results of qualitative and quantitative assessment.

The description and classification of life cycle models for safety-critical I&C systems, including several applicable metrics, allowing for the decomposition, verification and management of the implementation of non-functional requirements, ensuring the completeness of audits and minimizing subsequent costs for their creation and implementation are proposed in the chapter.

A case study reflecting results of V-model creation for FPGA-based RadICS Platform used in NPP safety-critical I&C systems was also presented. The model allowed assurance of Requirements Tracing process and assessment of various attributes.

Also, problems and features of security environment establishment process are discovered, describing in sufficient details its particular stages. The problems of security-oriented analysis and assessment of complex safety-critical systems are also analyzed. The proposed approach to establishment of secure development and operational environment for FPGA-based safety-critical I&C systems is based on gap conception, IMECA technique and comprehensive analysis of the development processes. The approach can be applied during security-oriented analysis of safety-critical I&C systems. The elements of the approach were applied in security-oriented analysis of RadICS FPGA-based I&C platform.

REFERENCES

Abrial, J.-R. (2010). *Modeling in Event-B*. Cambridge University Press. doi:10.1017/CBO9781139195881

Andrashov, A., Bakhmach, I., Kharchenko, V., & Kovalenko, A. (2019). Equipment qualification of FPGA-based platform RadICS to meet US NRC requirements. *Proceeding of the 11th International Topical Meeting on Nuclear Plant Instrumentation, Control, and Human-Machine Interface Technologies*, 327-335.

Andrashov, A., Bakhmach, I., Leontiiev, K., Babeshko, E., Kovalenko, A., & Kharchenko, V. (2019). Diversity in FPGA-based platform and platform based I&CS applications: Strategy and implementation. *Proceeding of the 11th International Topical Meeting on Nuclear Plant Instrumentation, Control, and Human-Machine Interface Technologies*, 174-182.

ANSI/ISA-99.00.01-2007. (2007). *Security for Industrial Automation and Control Systems: Terminology, Concepts, and Models.*

ANSI/ISA-99.00.02-2007. (2007). *Establishing an Industrial Automation and Control Systems Security Program.*

ANSI/ISA-99.00.03-2007. (2007). *Operating an Industrial Automation and Control Systems Security Program.*

ANSI/ISA-99.00.04-2007. (2007). *Specific Security Requirements for Industrial Automation and Control Systems.*

ANSI/ISA-99.02.01-2009. (2009). *Security for Industrial Automation and Control Systems: Establishing an Industrial Automation and Control Systems Security Program.*

ANSI/ISA-TR99.00.01-2007. (2007). *Security Technologies for Industrial Automation and Control Systems.*

Babeshko, E. (2008). Applying F(I)MEA-technique for SCADA-based Industrial Control Systems Dependability Assessment and Ensuring. *Proceedings of International Conference on Dependability of Computer Systems DepCoS–RELCOMEX 2008*, 309-315. 10.1109/DepCoS-RELCOMEX.2008.23

Badrignans, B. (2011). *Security Trends for FPGAS. From Secured to Secure Reconfigurable Systems.* Springer. doi:10.1007/978-94-007-1338-3

Bhuvaneswari, T. (2013). A Survey on Software Development Life Cycle Models. *Monthly Journal of Computer Science and Information Technology, 2*(5), 262–267.

Christiansen, B. (2006). *Active FPGA Security through decoy circuits* (MS Thesis). Air Force Institute of Technology.

Drimer, S. (2009). *Security for volatile FPGAs*. Technical Report N 763, University of Cambridge Computer Laboratory.

EPRI TR1019181, *Guidelines on the Use of Field Programmable Gate Arrays (FPGAs) in Nuclear Power Plant I&C Systems*, Electric Power Research Institute, (2009).

EPRI TR1022983, *Recommended Approaches and Design Criteria for Application of Field Programmable Gate Arrays in Nuclear Power Plant I&C Systems*, Electric Power Research Institute, (2011).

GAO-04-321, *Cybersecurity for Critical Infrastructure Protection*, U.S. General Accounting Office, (2004).

Grand, J. (2004, March). Practical Secure Hardware Design for Embedded Systems. *Proc. of the 2004 Embedded Systems Conference*.

Huffmire, T. (2010). *Handbook of FPGA Design Security*. Springer. doi:10.1007/978-90-481-9157-4

IEC 60880. (2006). *Nuclear power plants – Instrumentation and control systems important to safety – Software aspects for computer-based systems performing category A functions.*

IEC 61508: Edition 2. (2010). *Functional Safety of Electrical/Electronic/Programmable Electronic Safety-related Systems.*

IEC 61513. (2011). *Nuclear power plants – instrumentation and control important to safety – General requirements for systems*. Ed. 2.

IEC 62138. (2004). *Nuclear power plants – Instrumentation and control important for safety – Software aspects for computer-based systems performing category B or C functions.*

IEC 62566. (2010). *Nuclear Power Plants – Instrumentation and control important to safety – Hardware language aspects for systems performing category A functions*. Ed.1.

IEC 62645. (2011). *Nuclear power plants - Instrumentation and control systems - Requirements for security programmes for computer-based systems.* Ed. 1.

Illiashenko, O., Kharchenko, V., & Kovalenko, A. (2012). Cyber security lifecycle and assessment technique for FPGA-based I&C systems. *Proceedings of IEEE East-West Design & Test Symposium (EWDTS) 2012*, 432-436.

Isaias, P. (2015). High Level Models and Methodologies for Information Systems. Springer Science+Business Media. doi:10.1007/978-1-4614-9254-2

ISO/IEC 15408. (2009). *Information technology – Security techniques – Evaluation criteria for IT security*. Ed. 3.

ISO/IEC 17799. (2005). *Information technology – Security techniques – Code of practice for information security management.*

ISO/IEC 27000. (2009). *Information technology – Security techniques – Information security management systems – Overview and vocabulary.*

ISO/IEC 27001. (2005). *Information technology – Security techniques – Information security management systems – Requirements.*

ISO/IEC 27002. (2005). *Information technology – Security techniques – Code of practice for information security management.*

ISO/IEC 27003. (2010). *Information technology – Security techniques – Information security management system implementation guidance.*

ISO/IEC 27004. (2009). *Information technology – Security techniques – Information security management – Measurement.*

ISO/IEC 27005. (2011). *Information technology – Security techniques – Information security risk management.* Ed. 2.

ISO/IEC 27006. (2011). *Information technology – Security techniques – Requirements for bodies providing audit and certification of information security management systems.*

ISO/IEC 27007. (2011). *Information technology – Security techniques – Guidelines for information security management systems auditing.*

ISO/IEC 27008. (2011). *Information technology – Security techniques – Guidelines for auditors on information security management systems controls.*

Karry, R. (2010, October). Trustworthy Hardware: Identifying and Classifying Hardware Trojans. *Computing. The Magazine*, *43*(10), 39–46. doi:10.1109/MC.2010.299

Kharchenko, V. (2012). GAP- and HTT-based analysis of safety-critical systems. *Radioelectronic and Computer Systems*, *7*(59), 198–204.

Kharchenko, V. (2012c). Gap-and-IMECA-based Assessment of I&C Systems Cyber Security. In *Complex Systems and Dependability*. Springer-Verlag.

Kharchenko, V. (2012d). Cyber Security Lifecycle and Assessment Technique for FPGA-based I&C Systems. *Proceeding of IEEE East-West Design & Test Symposium (EWDTS'2012)*, 432-436.

Kharchenko, V. (2015). Security Assessment of FPGA-based Safety-Critical Systems: US NRC Requirements Context. *Proceedings of the International Conference on Information and Digital Technologies (IDT 2015)*, 117-123. 10.1109/DT.2015.7222963

Kharchenko, V. (2017). Green IT Engineering: Concepts, Models, Complex Systems Architectures. In Studies in Systems, Decision and Control, (vol. 74). Berlin: Springer International Publishing. doi:10.1007/978-3-319-44162-7

Kharchenko, V., Siora, A., Andrashov, A., & Kovalenko, A. (2012). Cyber Security of FPGA-Based NPP I&C Systems: Challenges and Solutions. *Proceeding of the 8th International Conference on Nuclear Plant Instrumentation, Control, and Human-Machine Interface Technologies (NPIC & HMIT 2012)*, 1338-1349.

Kharchenko, V. S., Illiashenko, O. A., Kovalenko, A. A., Sklyar, V. V., & Boyarchuk, A. V. (2014). Security informed safety assessment of NPP I&C systems: Gap-IMECA technique. *Proceedings of the 2014 22nd International Conference on Nuclear Engineering ICONE22*. 10.1115/ICONE22-31175

Kovalenko, A., Kuchuk, G., Kharchenko, V., & Shamraev, A. (2017). *Resource-Oriented Approaches to Implementation of Traffic Control Technologies in Safety-Critical I&C Systems. Springer.*

NEI 08-09, *Cyber Security Plan for Nuclear Power Reactors*, Rev. 6, Nuclear Energy Institute, (2010).

NIST SP 800-30, *Risk Management Guide for Information Technology Systems*, National Institute of Standards and Technology, (2002).

NIST SP 800-53, *Recommended Security Controls for Federal Information Systems and Organizations*, Rev. 3, National Institute of Standards and Technology, (2009).

Nuclear Security Series No, I. A. E. A. (2011). *17, Computer security at nuclear facilities: reference manual: technical guidance*. Vienna: IAEA.

NUREG/CR-7006, *Review Guidelines for Field-Programmable Gate Arrays in Nuclear Power Plant Safety Systems,* U.S. Nuclear Regulatory Commission, (2010).

Ravi, S., Raghunathan, A., Kocher, P., & Hattangady, S. (2004). Security in Embedded Systems: Design Challenges. *ACM Transactions on Embedded Computing Systems*, *3*(3), 461–491. doi:10.1145/1015047.1015049

RG 5.71, *Cyber security programs for nuclear facilities,* U.S. Nuclear Regulatory Commission, (2010).

Sadeghi, A.-R. (2011). *Towards Hardware-Intrinsic Security: Foundations and Practice*. Springer.

Shamraev, A. (2017). Green Microcontrollers in Control Systems for Magnetic Elements of Linear Electron Accelerators. In Green IT Engineering: Concepts, Models, Complex Systems Architectures. doi:10.1007/978-3-319-44162-7_15

Tehranipoor, M. (2010). A Survey of Hardware Trojan Taxonomy and Detection. *Proc. of IEEE Design & Test of Computers*, 10-25. 10.1109/MDT.2010.7

Yastrebenetsky, M. (2014). Nuclear Power Plant Instrumentation and Control Systems for Safety and Security. IGI Global. doi:10.4018/978-1-4666-5133-3

ADDITIONAL READING

Tehranipoor, M. (2011). *Introduction to Hardware Security and Trust*. Springer.

KEY TERMS AND DEFINITIONS

Identification: The process of verifying the identity of a user, process, or device, usually as a prerequisite for granting access to resources in an IT system.

Regulatory Requirement: Requirement, which is established by National Regulatory Authority (authority designated by government for regulatory purposes for safety assurance).

Risk: The level of impact on agency operations (including mission, functions, image, or reputation), agency assets, or individuals resulting from the operation of an information system, given the potential impact of a threat and the likelihood of that threat occurring.

Security: Avoidance of dangerous situation due to malicious threats.

Security Controls: The management, operational, and technical controls (i.e., safeguards or countermeasures) prescribed for an information system to protect the confidentiality, integrity, and availability of the system and its information.

Threat: Any circumstance or event with the potential to adversely impact agency operations (including mission, functions, image, or reputation), agency assets, or individuals through an information system via unauthorized access, destruction, disclosure, modification of information, and/or denial of service.

Vulnerability: Weakness in an information system, system security procedures, internal controls, or implementation that could be exploited or triggered by a threat source.

Chapter 6
Cyber Robust Systems:
The Vulnerability of the Current Approach to Cyber Security

Gary Johnson
Independent Researcher, USA

ABSTRACT

Analysis of the STUXNET attacks on the Natanz gas centrifuge plant illustrated the hazards of a cyber physical attack. STUXNET demonstrated that a cyber-attack can introduce new and malicious function into I&C systems. Cyber robust nuclear power plant systems may be able to provide a truly independent level of defense in depth against cyber-attacks. The development of cyber robust systems involves identifying the plant's vulnerabilities and using non-digital means where such features can defeat malicious attacks. This is a relatively new idea, so a complete roadmap for this is not available. Nevertheless, some principles can be stated, and some methodologies can be discussed.

INTRODUCTION

Cyber security for nuclear power plants tends to be the domain of Information Technology (IT) professionals. The IT approach to cyber security is to protect the instrumentation and control systems from cyber-attacks using methods such as anti-virus software, network segregation, intrusion detection, and security patches. It is an important activity that reduces the hazard to the plant by attempting to prevent the introduction or execution of malicious features into the plant instrumentation and control systems. Most of this book deals with the IT approach.

The IT approach reduces the frequency of attacks, but it has several limitations:

- It depends to a great extent on people faithfully following the cyber security requirements. People are pone to making mistakes and also prone to taking shortcuts that can open pathways for introduction of cyber threats.
- It is vulnerable to zero day attacks.

DOI: 10.4018/978-1-7998-3277-5.ch006

- It cannot deal with cyber hazards that existed before the IT features were applied to the systems and equipment that make up a instrumentation and control system. Experts who are familiar with intelligence community information claim that hazardous cyber threats are already in many of our important instrumentation and control systems- for example (Clark, 2010; Sanger, 2018).
- It is vulnerable to insider threats.

We should expect that some cyber-attacks will be successful. Denial of service attacks and attacks that cause minor damage to the plants are serious problems and will degrade the industry's credibility with the public, but, as long as they are relatively rare they will be survivable.

On the other hand, attacks that cause significant release of radioactivity or core damage will be catastrophic to the entire industry. A defense in depth approach is needed to deal with such possibilities. To deal with these cases we must understand the attacks that may result in radiological releases or core damage and take appropriate action to prevent such attacks or mitigate their consequences.

It is expected that such attacks can come from highly competent and motivated attackers who are employed by nation states or very sophisticated terrorist organizations. Some would argue that nation state attacks are acts of war, and acts of war do not need to be considered in a nuclear power plants safety case. But when such a cybersecurity attack comes it will be very difficult to confirm the origin of the attack. We need only to think of the MH370 shoot down to understand the difficulty of proving the source of mysterious attacks. We should understand that if a serious reactor accident is caused by a mysterious cyber-attack, we will not be forgiven for the consequences.

It is also possible that serious radiological releases might be caused by relatively unsophisticated cyber-attacks. We won't know until we examine the possibilities.

Before discussing the needed analytical process, the next section will discuss a real such event.

THE STUXNET ATTACKS

This section gives a brief overview of the STUXNET attacks on the Natanz gas centrifuge plant to illustrate the hazards of a cyber physical attack. The material in this section is largely derived from Ralph Langner's report "To Kill a Centrifuge" (Langner, 2013). Readers are encouraged to read the full report which gives much more complete discussion of the events than is given here.

In 2010 malware was discovered that attacked Siemens Step 7 programmable logic controllers.

Originally, Siemens concluded that the malware had not affected any Siemens users, but investigation work by Langner and others determined the attack was designed only to attack Natanz gas centrifuge plant.

Natanz is a gas centrifuge gaseous diffusion plant in Iran. To understand the attacks it is necessary to know how these plants operate. The gas centrifuges at Natanz were intended to enrich uranium. This is done by flowing low enriched uranium through a series of centrifuges. In each centrifuge action tends to cause the uranium with higher mass to flow in one direction and the uranium with lower mass in the other direction. This process is very inefficient so a large number of centrifuges are needed to produce significant enrichment. This process can be simply modeled as shown in Figure 1.

Figure 1. A simple model of a gaseous enrichment plant

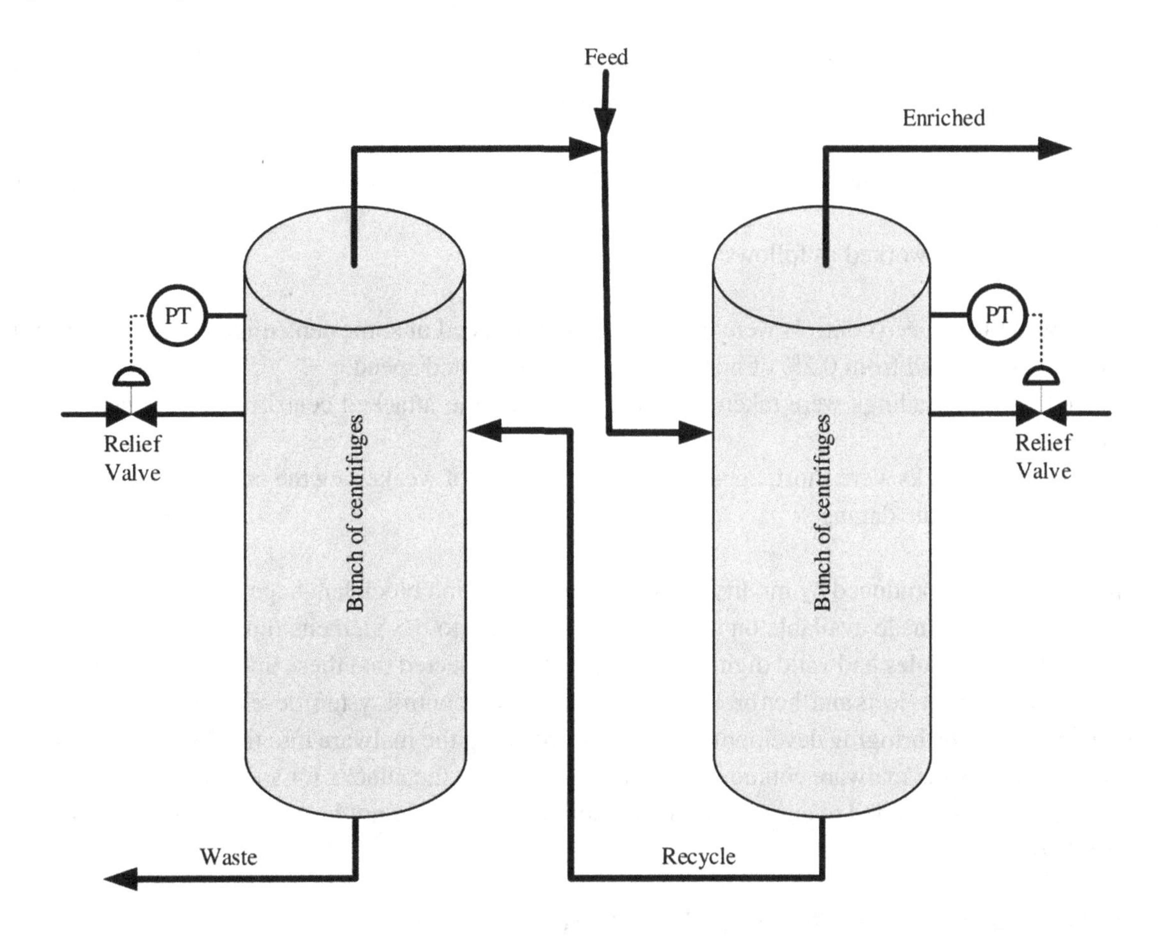

Typically, there are many hundreds of centrifuges. Low enriched uranium is fed in at some point within the string of centrifuges. The centrifuges tend to separate the high enriched uranium from low enriched. The higher enriched material flows in one direction of the system and is finally taken out at the "top end". The lower enriched uranium flows down through centrifuges and is taken off as waste at the "bottom end". By running this process through a large number of centrifuges the material at the top end has a much higher enrichment (e.g., more U235) than the original feed stock.

The system at Natanz was attacked in two different ways at two different times.

The first attack was operated periodically after 2007 and worked as follows.

- Normal plant operating conditions were recorded. When an attack was to be made the malware replaced the data going to the control room displays with the pre-recorded normal data.
- This ensured that control room operators could not detect the changes that the malware made to plant conditions.
- The malware increased the set points for some of the centrifuge high pressure relief valves.
- Valves closed off the lines taking high enriched uranium and low enriched waste, but the feed flow stayed open causing high pressures in the centrifuges.

- This arrangement caused higher than normal stresses in the centrifuges. Short attacks were repeated multiple times. It is believed that the attackers did not want to cause catastrophic damage to the system, instead aimed to cause increased system failure rates in the hopes that the operators would believe that there were flaws in the centrifuges.

The second attack operated periodically depended upon the fact that the gas centrifuges are rather fragile devices.

The second attack worked as follows:

- The centrifuge speed controls were taken over and the speed of some centrifuges were rapidly and repeatedly ramped from 0.2% of normal speed to 130% rated speed.
- Speed control readings were taken over to appear that the attacked centrifuges were running at normal speed.
- Again these attacks were short, apparently with the intent of weakening the centrifuges rather than causing immediate damage.

The malware was produced by modifying the Siemens function block language modules. These corrected modules were made available on the internet as updates to the Siemens function block modules and the modified modules had valid digital certificates It is suspected that these updates were originally downloaded for other projects and then brought into the centrifuge control system development environment on thumb drives or by bringing development laptops containing the malware into the development area.

For both attacks the malware contained instructions to delay the attacks for some weeks or months. Apparently this was intended to ensure that the malicious code would not be exposed during validation and installation testing.

Lessons Learned From the STUXNET Attacks

STUXNET demonstrated that a cyber-attack can introduce new and malicious function into I&C systems.

The attack demonstrated that multiple systems can be attacked to create desired malicious events. The first attack, for example, affected the control room displays, the operation of the centrifuge and the over pressure safety system. Using the defense in depth model given in INSAG 10 we would say that

- Faking the data on the control room displays was an attack on defense in depth level 1 (prevention of abnormal operation and failures)
- Closing the waste and enriched outlet valves was an attack on defense in depth level 2 (control of abnormal operations and failures), and
- Resetting the relief valve set points was an attack on defense in depth level 3 (Control of accidents within the design basis).

Most likely the degree of independence between these levels at Natanz was rather low by nuclear power plant standards. But, consider this, when we are dealing with failures in nuclear power plants we expect the independence between different levels of defense to give us a reduction of risk on the order of 100 to 1000. In the case of malicious acts it's likely that cost of attacking any layer of defense in depth is about the same. Therefore, the difficulty of cyber threat attacking 3 levels defense in depth is closer to a factor of 3 than a factor of 100. The defense in depth strategy that we use for safety is not very effective for protecting against cyber-attack.

Malware can remain dormant for a long time, waiting for completion of validation, and installation tests so that the attack can be made after the target system had become trusted.

The IT procedures meant to prevent insertion of malware will be effective most of the time, but such human procedures will never be perfect. Mistakes happen, procedural requirements may sometimes be bypassed for good or for malicious reasons, the malware may already exist in equipment arrives at the plant, equipment may have unrecognized back doors put there for good or malicious reasons. We must have a method of defense in depth against cyber-attack that is truly diverse from the current approach.

We probably cannot afford to provide defense in depth agains all malicious software, it is necessary establish strong protections against the subset of malware that pose a real risk of core damage or significant radiological release.

CYBER ROBUST SYSTEMS

Cyber robust nuclear power plant systems may be able to provide truly independent level of defense in depth against cyber-attack. The development of cyber robust systems involves identifying the plant's vulnerabilities and using non-digital means where such features can defeat malicious attacks. The idea is not to regress to non-digital systems but to identify the key places where non-digital features can thwart a cyber-attack.

This is a relatively new idea, so a complete roadmap for this is not available. Nevertheless, some principles can be stated and some methodologies can be discussed.

Identifying the Cyber Hazards

A plant system analysis is needed to identify the possible consequences of cyber-attack on systems that are controlled by digital I&C. The analysis must first be a plant analysis to understand the possible consequences of an attack.

The analysis needs to consider the consequence of an attack on the plant equipment itself. At this stage the instrumentation and control systems don't matter. Most instrument and control systems drive analog actuation systems such as solenoid valves, motor operated valves, or analog circuit breakers. No matter what the I&C system does the actuation systems will always act within the bounds of the actuation device design. There are exceptions of course, such as rod control systems and motor drives, but even in these cases it is worthwhile to know the worst case consequences of an attack.

Often the analysis may conclude that the attack cannot cause the plant equipment to create a hazardous response. We often put constraints on the design of equipment for safety reasons. Consider, for example, control rod drives where mechanical design limits the maximum rate of control rod movement, and reactor core feedback characteristics further limit power increases. Such features also protect against cyber-attacks.

The analysis needed here is not the standard plant safety analysis for several reasons. The standard plant safety analysis:

- Considers only the effects of equipment failures. Cyber-attacks may create a whole range of hazardous non-failure states that cause worse consequences than component failures.
- Considers the effect of single failures. There is no reason to believe that cyber-attacks will be limited to single failures If a cyber-attack can attack one system, it can attack all similar or elements of a system.
- Focuses on the response of safety systems to accident conditions. Serious events might be accomplished by attacking non-safety systems.

Once the hazards of concern are identified then the role of instrumentation and control systems can be considered.

It seems likely that the analysis needed will be something like hazards analysis rather than strictly probabilistic studies or traditional plant safety analysis. Proposed methodologies for such studies are just now emerging. These include:

- Security Process Hazards Analysis. This methodology modifies a traditional process hazard analysis to identify hackable components and identifies the possible results if hacks occur. Once the consequences are understood, alternative methods to ensure system safety in the presence of the hacks can be identified. See https://www.kenexis.com/security-pha-review-example-video-from-kaspersky-industrial-cybersecurity-2018/ for a description. This is a technique that was produced for industrial processes. A text book on the methodology is available from the ISA at https://www.isa.org/store/security-pha-review-for-consequence-based-cybersecurity/65832391. Readers are encouraged to watch the video about the Security Process Hazards Analysis as it discusses a simple example that clearly describes the methodology.
- EPRI HAZCADS (2018): Hazards and Consequences Analysis for Digital Systems. This approach uses a Systems-Theoretic Process Analysis model to identify potential system hazards, and identify the unsafe control actions that can trigger the hazards. The probability of systematic unsafe control actions (e.g., cyber-attack) are typically treated as yes/no occurrences and fault trees are used to estimate the probability of random unsafe control actions. Risk reduction worth and achievement worth are used to characterize the importance of failure scenarios. EPRI considers this to be a universal too that can be used for the study of all digital I&C hazards including cyber security, single failures, common cause failures and EMI/RFI effects. The HASCADS report is available from EPRI at https://www.epri.com/#/pages/product/000000003002012755/?lang=en-US. HAZCADS (2018) the implementation of the EPRI digital systems engineering framework which is available at https://www.epri.com/#/pages/product/3002002989/?lang=en-US.

The underlying principles of these two methodologies are the same, but the EPRI method provides a more sophisticated methodology that can help identify existing system elements that can be credited for defeating a postulated attack.

These are relatively new methods and analysis is still needed to identify the most useful approach.

Cyber Robust Design

Where serious cyber hazards are identified corrective action should be considered to implement features that will make the hazardous features more robust against cyber-attack. There are at least four ways to create robustness.

- Change the process or equipment such that the system's physics prevents the hazardous consequences. This may be the strongest protection against cyber-attack, but it may also be the most difficult to implement, especially for existing plants.
- Insert mechanical systems in place of certain components, or limit the range over which the digital I&C system can control the problematic function.
- Replace problematic digital I&C systems or components with analog devices.
- Provide robust administrative controls that protect against cyber security attacks

This list is more or less in the order of most effective to least effective.
How could these protections have been applied to protect the Natanz plant?

- Providing analog information about key parameters such as system inlet and outlet status and abnormal centrifuge speeds would have indicated to the operators that they should be suspicious of the control room displays. (replaces a digital function with an analog function)
- Making the centrifuge shells stiffer could have reduced the damage from rapid speed changes. (a change in the physics of the centrifuges)
- Requiring operator action to change pressure relief set points would have prevented the malware from changing the setpoints. (replaces a digital i&C function with an administrative control)
- Replace the centrifuge safety valves with mechanical valves (Inserts a mechanical device in place of a digital device)
- Provide a non-digital system to safely shutdown centrifuges when excessive speed cycling is detected. (provides a non-digital I&C system to protect against attacks)

There are probably many other better ideas than those given above. But meaningful action cannot be taken until the vulnerabilities are identified.

Example Applications to Plant Modernization and New Plants

Studies of cyber robustness should be part of any new digital system in a nuclear power plant be it for an I&C upgrade or new plant. Ignorance of the cyber hazards in a new design risks introduction of unknown vulnerabilities. There is already evidence that this has happened. Two examples are given below.

Residual Heat Removal Suction Line in a Gen 3 PWR

Most PWR designs have piping to connect the reactor coolant system (RCS) to the residual heat removal (RHR) pumps for reactor coolant circulation during outages. The reactor coolant systems are designed for pressures up to about 17 MPa, but the RHR systems are designed for pressures about 3.4 MPa. Opening the connection between the RCS and the RHR systems when the RCS is highly pressurized would result in a loss of coolant accident that bypasses the containment.

The older plants of the type in question provided redundant isolation valves to isolate the high pressure systems from the RHR during normal operation. Each isolation valve was controlled by a pressure sensor that closed the valves or prevented them from opening if the RCS pressure was more than 3 MPa. The same arrangement was used in the Gen 3 plant except that each valve had two pressure sensors that closed the valve on 1 out of 2 logic and the two valves used diverse pressure sensors. Figure 2 shows this arrangement.

Figure 2. LOCA protection for RCS/RHR interface in the Gen 3 plant

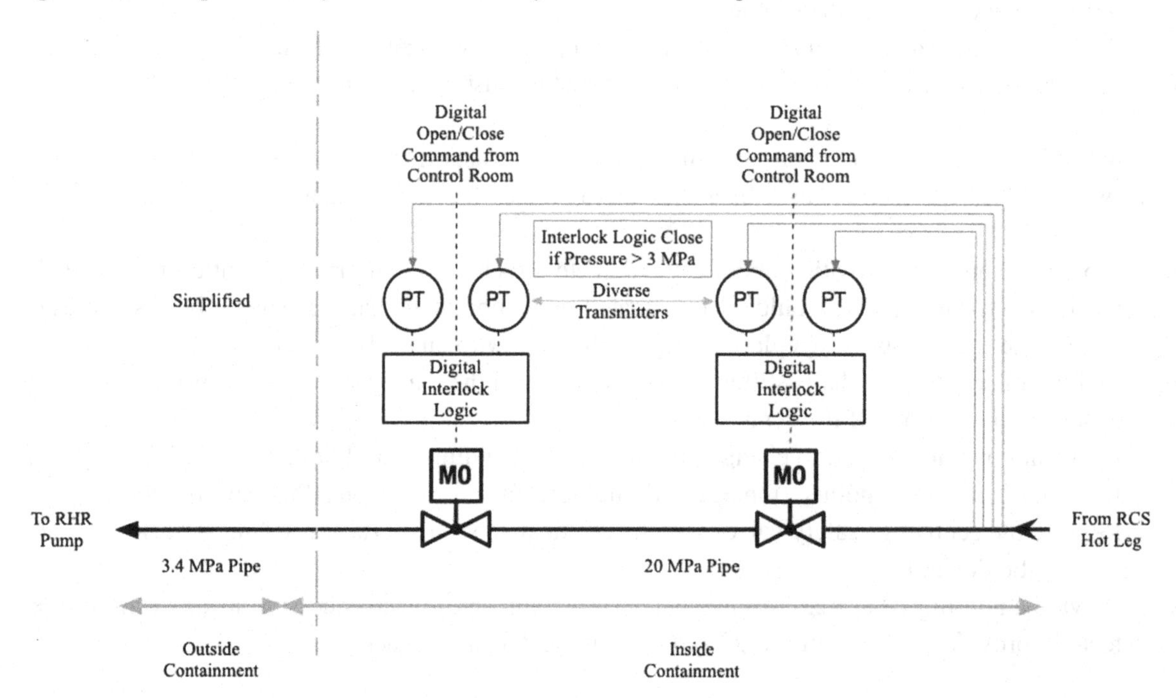

But in the Gen. 3 plant the valve interlock controls are digital. STUXNET demonstrated that attacking even redundant and diverse attacking systems is possible. The Gen 3 plant designers protected against software common cause failure, but they failed to protect against the sophisticated cyber-attacks.

These valves in question are only opened and closed once every refueling cycle. There is no real benefit to put the valve controls on the digital network. It is not much of a problem to send a field operator to the valve motor control center and close the valves once during an outage.

But there is more to this story. There are other protections against opening the valves at when the RCS piping is at high pressure. If that wasn't true, it would be unwise to point out the attack method in open literature.

In addition to the valve logic, there is an administrative requirement to de-energize the valves' motor control centers before reactor coolant system pressure is increased above 3 MPa. Furthermore there is a safety valve to direct coolant back to the containment sump if high pressure water gets past the two motor operated valves, see Figure 3 These features are also in place in the older plants.

Figure 3. LOCA protection showing additional safety features

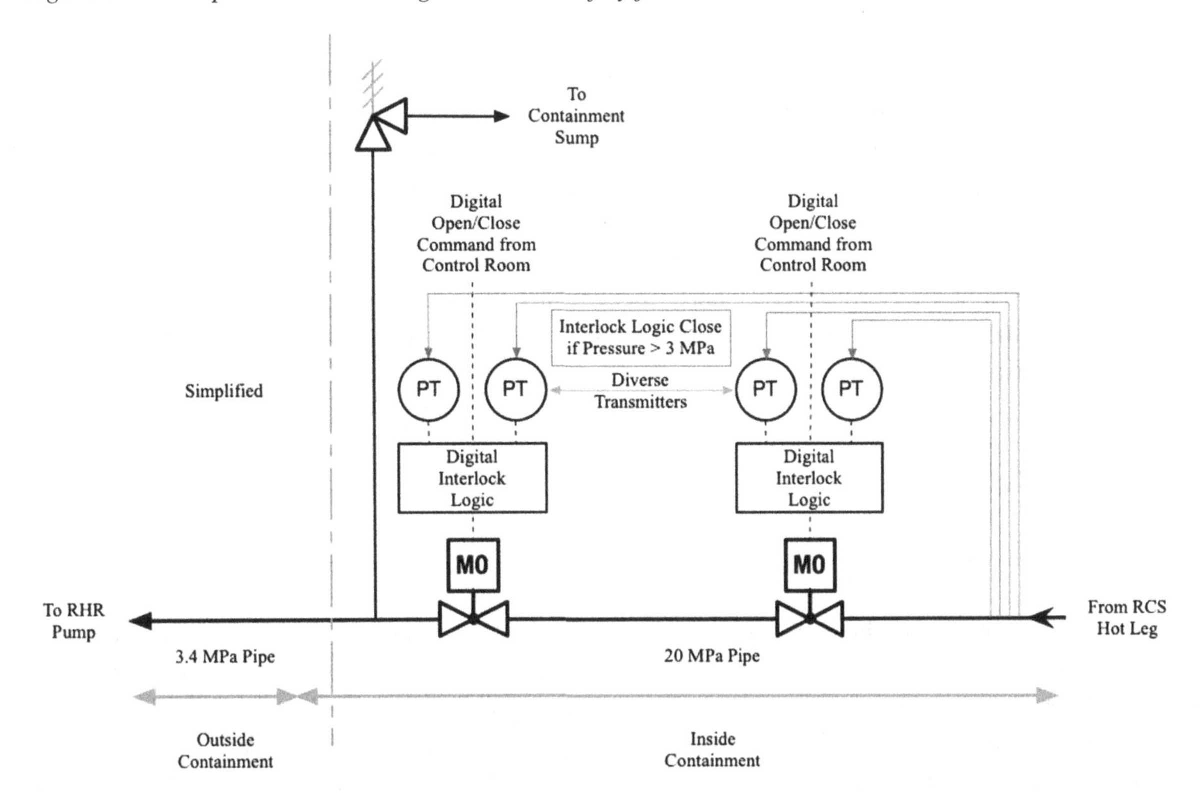

The new design is not so bad, but digitizing the isolation valve interlocks creates very little benefit but causes a cyber vulnerability that does not exist in the older plants.

Boiling Water Reactor Control Rod Drive Upgrade

During the late 2000's certain US boiling Water Reactor plants proposed a license amendment that would allow them to withdraw up to four control rods in assigned gangs during startups in order to minimize the time required to reach full power operation during a startup. This change involved replacing a number of analog rod control subsystem with digital control subsystems.

This proposed modification created a possibility of a new accident a "Multiple Rod Withdrawal Error on Startup". The applicants provided a traditional safety analysis, but such analyses consider only single failures. We know from the STUXNET event that cyber-attacks can cause errors that go beyond single failures. A cyber security analysis was performed and concluded that the system was well protected against cyber intrusions. But again the STUXNET event showed it is possible for threats to bypass even strong barriers to intrusion. Changes to control rod drive systems can have serious consequences. Before installing new digital systems for such critical systems the potential consequences of cyber-attacks should be understood. In the end, NRC and the applicant concluded that the existing safety analysis codes were not adequate to evaluate even the analysis of single failures on the new rod control system. The applicants were faced with the need to develop improved safety analysis codes. This would have been both very expensive and time consuming. In the end the modification was not made.

FUTURE RESEARCH DIRECTIONS

There is a critical need to identify, and provide strong protections, where cyber-attacks on digital functions in nuclear power plants and other high hazard facilities might result in events that risk public safety. Until recently such methods have not been available. We now have two methods that approach this problem by using the existing methods of hazard analysis and probabilistic risk assessment. It is time now to begin using these methods in as part of the cyber security analysis for specific plants. From the first uses we should understand how to improve the methods and to understand the relative merits of the two different methods. More importantly, though, such methods already provide a means to avoid unknowingly introducing significant cyber hazards into nuclear power plants through naive application of digital systems.

CONCLUSION

The current approach to nuclear power plant cyber security protects against the insertion of malware into digital systems in nuclear power plants. This is an important role in reducing plant vulnerabilities, but the effectiveness of the existing approach is highly dependent upon humans that implement the cyber security features. Humans tend to be fallible. Also there is insufficient protection against attacks that are already incorporated into I&C systems either before cyber security programs were put in place or by way of purchased equipment that was insufficiently protected during manufacture.

The current existing methods are sufficient to deal with attacks that do not pose a serious threat of core damage or unacceptable radiological release. The STUXNET attacks illustrate the risk of cyber-attacks creating significant risks to plant safety. To protect the public it is necessary to have a clear understanding of the possible consequences of cyber-attacks so that plant designers can take steps to prevent such attacks or mitigate their consequences.

Understanding the possible consequences of cyber-attacks cannot be accomplished using the normal plant safety analysis methods. For several years, work has been underway to understand the hazards and analytical methods are now emerging. It is time to start using these methods to create cyber robust systems and identify improvements to these emerging methods.

The traditional cyber security activities that protect against cyber-attacks and plant systems that are robust to cyber-attacks should work together as an overall defense in-depth concept as summarized in Figure 4.

Figure 4. The relationship between the traditional cyber protections and plant design that is robust to cyber hazards

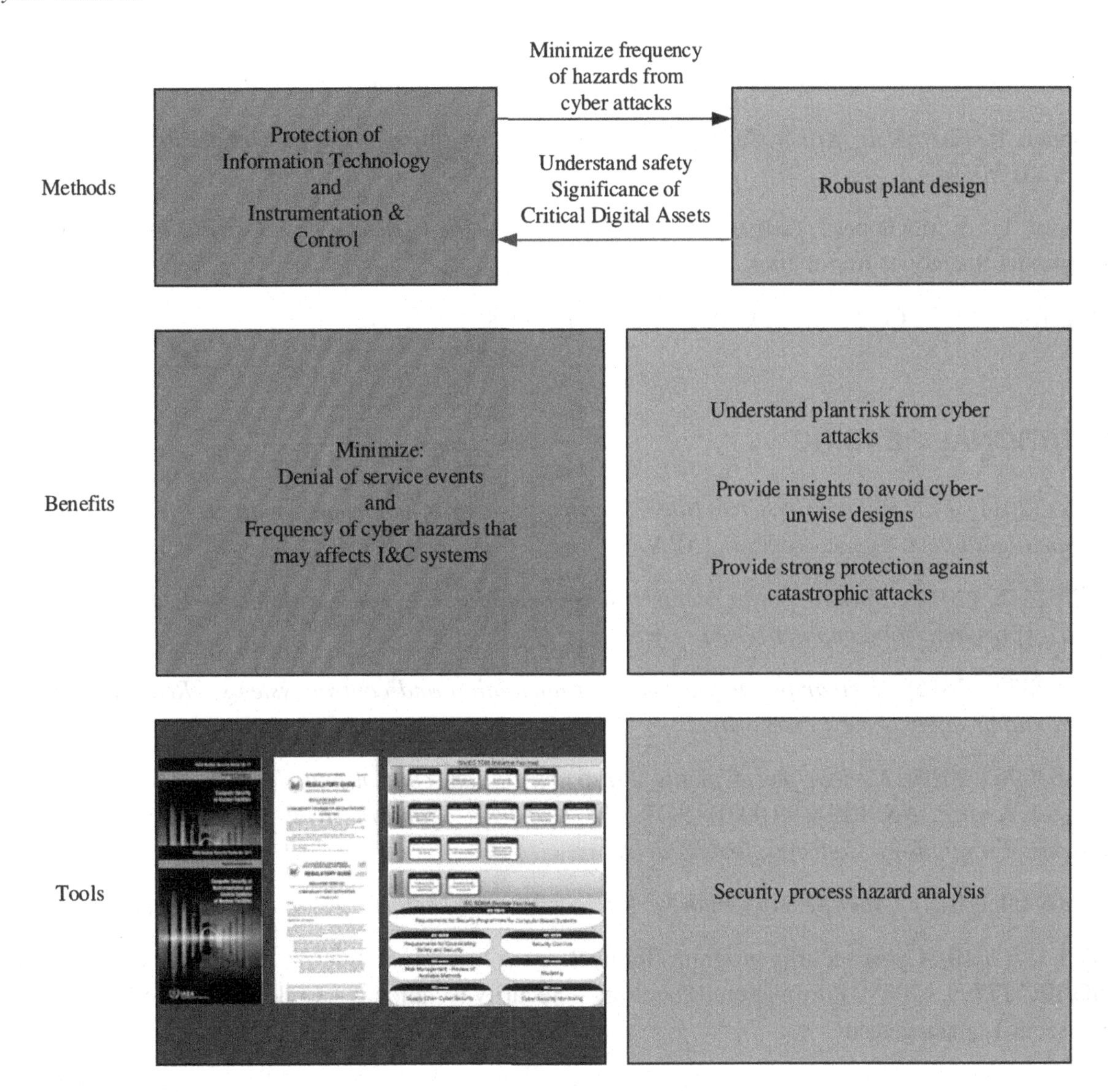

The IT approach to cyber security can minimize denial of service events and reduce the probability that cyber hazards are inserted it the plant systems but robust plant design is necessary to deal with the very low probability events that may pose a risk of creating events that may result in radiological hazards to the public.

REFERENCES

Clark, R. (2010). *Cyber War*. The Perfect Weapon.

HAZCADS. (2018). *Hazards and Consequences Analysis for Digital Systems*. Palo Alto, CA: EPRI.

Langner, R. (2013). *To Kill a Centrifuge*. Retrieved from https://www.langner.com/wp-content/uploads/2017/03/

Marszal, E., & McGlone, J. (2019). *Security PHA Review for Consequence-Based Cybersecurity*. International Society of Automation.

Sanger, D. (2018). *Confront and Conceal*. Gardner Books.

ADDITIONAL READING

IAEA. (2011). *Nuclear Security Series No. 17, Computer security at nuclear facilities: reference manual: technical guidance*. Vienna, Austria: IAEA.

IEC 62645. (2014). *Nuclear power plants - Instrumentation and control systems - Requirements for security programs for computer-based systems. Ed. 1.*

IEC 62859. (2016). *Nuclear power plants - Instrumentation and control systems - Requirements for coordination safety and cybersecurity.*

IEC 63096. (2020). *Nuclear power plants – Instrumentation, control and electrical power systems – Security control.* ANSI/ISA-99.00.01-2007. (2007). *Security for Industrial Automation and Control Systems: Terminology, Concepts, and Models.*

NEI 08-09. (2010). *Cyber Security Plan for Nuclear Power Reactors, Rev. 6, Nuclear Energy Institute.*

RG 5.71. (2010). Cyber security programs for nuclear facilities, U.S. Nuclear Regulatory Commission, ISO/IEC 17799. (2005). Information technology – Security techniques – Code of practice for information security management.

Chapter 7
Tool-Based Assessment of Reactor Trip Systems Availability and Safety Using Markov Modeling

Oleg Odarushchenko
Research and Production Corporation RadICS, Ukraine

Valentyna Butenko
National Aerospace University KhAI, Ukraine

Elena Odarushchenko
Poltava State Agrarian Academy, Ukraine

Evgene Ruchkov
Research and Production Corporation Radiy, Ukraine

ABSTRACT

The accurate availability and safety assessment of a reactor trip system for nuclear power plants instrumentation and control systems (NPP I&C) application is an important task in the development and certification process. It can be conducted through probabilistic model-based evaluation with variety of tools and techniques (T&T). As each T&T is bounded by its application area, the careful selection of the appropriate one is highly important. This chapter presents the gap-analysis of well-known modeling approach—Markov modeling (MM), mainly for T&T selection and application procedures—and how one of the leading safety standards, IEC 61508, tracks those gaps. The authors discuss how main assessment risks can be eliminated or minimized using metric-based approach and present the safety assessment of typical NPP I&C system. The results analysis determines the feasibility of introducing new regulatory requirements for selection and application of T&T, which are used for MM-based assessment of availability and safety.

DOI: 10.4018/978-1-7998-3277-5.ch007

INTRODUCTION

Availability and safety assessment of systems for critical applications, such as NPP I&Cs, is an essential part of the development and certification processes as it demonstrates that relevant regulations have been met and for making an informed decisions about the risks and consequences of inaccurate assessment results.

Such model-based assessment can be performed through discrete-event simulation (DES), analytic models or combining simulation and analytical approaches. The main advantage of DES is ability to consider detailed system behavior in the model, while the drawback is long execution time, when accurate solution is needed. Analytical models tend to be more abstract, but are easier to develop and faster to solve than DES models. The main difficulty is a necessity to set additional assumptions to make the models tractable (Trivedi, K.S. et al. 2000). Analytical modeling techniques can be split into two groups: state space (Markov models, Petri nets, etc.) and non-state space (RBD, FTA, etc.) techniques. Selection of the appropriate technique is provided based on measurements of interest, level of components detalization, etc.

The state space models are always preferred to non-space models, if it is important to model such complex situations as failure/repair dependencies, shared repair facilities (Trivedi, K.S. et al. 2000) or provide the detailed presentation of system behavior for communication with engineering teams (Malhotra, M. et al 1994).

BACKGROUND

System modelers are often interested in transient measures, which provide more useful information than steady-state measures. Modeling components interaction and interdependencies expands the model significantly, thus making the precise computation of system transient measures almost infeasible. Use of Markov models (MM) for modeling the transient behavior of the complex system can lead to the number of description and computational difficulties.

One of the main computational difficulties is model *largeness* (i.e. structural complexity) (Buchholz, P. 1996), which leads to problems in its construction, storage and solution. Adding the software component behavior into the MM increases its size rapidly. Because of models size the closed-form solutions of transient measures become infeasible, in this case, modeler can rely on the numerical methods or imitation modeling. Modeling components interaction enlarge the state space significantly, and results in sparse matrices of differential equations (DE) coefficients.

Sparsity (Hurleyб N. et al. 2009) corresponds to systems, which are loosely coupled. In the subfield of numerical analysis, a sparse matrix is a matrix populated primarily with zeros (Press, W.H. et al. 2007). If the MM is large it becomes wasteful to reserve storage for zero elements, thus solution methods that do not preserve sparsity, is unacceptable for most large problems (Press, W.H. et al. 2007).

The next complexity in solving large models that effect on numerical solution results is the model *stiffness* (Bobbio, A. et al. 1986). It is an undesirable property of many practical MMs as it poses difficulties in finding transient solutions. In practice, stiffness in models of complex computer systems is caused by (Bobbio, A. et al. 1986):

- In case of repairable systems the rates of failure and repair differ by several orders of magnitude;

- Fault-tolerant computer systems (CS) use redundancy. The rates of simultaneous failure of redundant components are typically significantly lower than the rates of the individual components;
- In models of reliability of modular software the modules' failure rates are significantly lower than the rates of passing the control from a module to a module.

Several approaches were developed to deal efficiently with MM largeness and stiffness (Malhotra, M. et al 1994, Bobbio, A. et al. 1986), Arushanyan, O. et al. 1990). In both cases, they can be split into two main groups – "avoidance" and "tolerance" approaches.

The avoidance approach overcomes largeness by exploiting the certain properties of the model to reduce the size of underlying MM (Malhotra, M. et al 1994). In the largeness tolerance approach the new algorithms are designed to manipulate large MM, and special data structures are used to reduce state transition matrix, iteration vector, etc. (Sanders, W. H. et al. 1998).

For stiff models, with tolerance approach (STA) the *specialized numerical methods* (Arushanyan, O. et al. 1990, Reibman, A. et al. 1988) are used to provide highly accurate results despite stiffness. The limitations of STA are:

- STA cannot deal effectively with large models;
- Computational efficiency is difficult to achieve when highly accurate solutions are sought.

The stiffness avoidance (SAA) solution, on the other hand, is based on an *approximation algorithm* which converts a stiff MM to a non-stiff chain first, which typically has a significantly smaller state space [6]. An advantage of this approach is that it can deal effectively with large stiff MMs.

Availability and safety assessment using MM for such complex systems as NPP I&C system can be assessed through model-based evaluation in specialized tools (λPredict, Möbius, SHARP, etc.), off-the-shelf tools (Maple, Matlab, Mathematica, etc.) and own developed utilities (ODU).

The variety of tools and techniques (T&T) is extremely helpful in the process of system modeling but this also poses a difficulty when it comes to choosing the most appropriate method for a specific assessment. As every tool is limited in its properties and applicability, a careful selection is needed for the tools used to solve large and stiff MMs accurately and efficiently. It should be noted that stiffness usually requires the modeler to focus on a number of math details to avoid the use of inefficient approaches, methods (Hairer, E. et al. 2010), and tools. This view goes against the recommendations of IEC 61508-6 (Standard IEC 61508 2010). This standard asserts that methods for solving MM have been developed long ago and trying to improve these methods does not seem sensible. The previous works show (Kharchenko, V. et al. 2013) that solving a large and/or stiff Markov model requires a careful selection of the solution method/tool. Otherwise, the results can differ in several orders of magnitude (Kharchenko, V. et al. 2013), thus, use of inappropriate method/tool for the solution of a non-trivial MM may lead to significant errors.

One of the leading standards in the safety area IEC 61508-2010 provides no special requirements for T&T, which are used to evaluate the system safety indicators, excepting the strong recommendation, that practitioner must have an understanding of the techniques used by software package to ensure its use is suitable for the specific application.

The absence of special requirements can be explained by long-time use of the mentioned techniques and proved by their detailed description in IEC 61508-2010 (part 6) (Standard IEC 60300-3-3 Ed.2.0 (2008). In contrast to the T&T, many requirements were developed for I&Cs verification and validation

Figure 1. Sequence of stages in MM approach

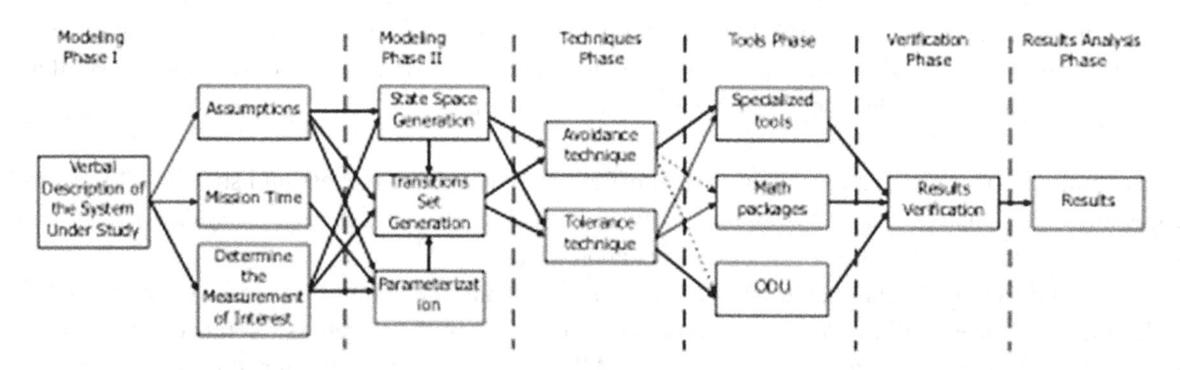

tools are compatible by strength to the requirements of produced software and systems (see standard IEC 60880-2006 (Standard IEC 60800 Ed. 2.0 2006).

This chapter presents the detailed analysis of the main gaps in Markov modeling approach, mainly in T&T selection and application, and how they are traced by few well-known standards in the safety area IEC 61508:2010 (Standard IEC 61508 2010), IEC 60300:2003 (Standard IEC 60300-3-3 Ed.2.0 2008), IEC 61165:2006 (Standard IEC 61165 2006) and ECSS-Q-ST-30-09:2008 (Standard ECSS-Q-ST-30-09 2008). By the "*gap*" we define special risks, which are accompanies the MC application procedure.

We discuss the ways how MM assessment risks and difficulties can be eliminated or minimized and present the case study for safety assessment of a typical NPP instrumentation and control system (I&Cs), the Reactor Trip System (RTS) based on a digital platform produced by RPC "Radiy". This is a two-channel FPGA-based system with three parallel tracks (sub-channels) of voting logic "2-out-of-3" in each of the channels.

GAPS OF THE IEC 61508

In this section, we provide the detailed analysis of the main *gaps* (i.e. risks) of MM approach and how they are tracked by the standards. Overall, application of MM approach can be presented as a sequence of six stages (Figure1), namely: first and second modeling phases, techniques phase, tools phase, verification and results analysis phases.

The first modeling phase presents the processes of verbal system description, definition the measures of interest, development of modeling assumptions and determination of the studied time interval. The most typical gaps of this stage and how they are reviewed by standards are presented in Table 1.

During the second modeling phase, researcher determines the model parameters, using statistic data, etc. and performs generation of system state space and set of transitions taking into account designed modeling assumptions. The main gaps of this stage and their tracking by standards are listed in Table 2. It should be noted, that MC approach described in standards IEC 61508 (part 6) (Standard IEC 61508 2010) and IEC 61165 (Standard IEC 61165 2006) assumes constant time-independent state transition rates, while in practice those rates can vary over time due to physical aging errors, etc.

Table 1. Gaps of the first modeling phase

Gap	Reviewed by Standard
Confirm the applicability of the Markov modeling approach for the specific case	IEC 60300-3-1 "…selection of method for dependability analysis is individual and performed by joint effort of dependability experts and system experts. The selection has to be performed on the early stages of system development and analyzed on applicability… "
Introduce the assumptions, which will keep the required abstraction level with respect to the calculated parameters	The main assumptions used for MC approach are presented in IEC 61165 (section 6) and IEC 61508 – 6 (section B.5.2).

On the techniques phase, using the analysis of constructed state space the researcher choose the assessment technique. If the model appears to be large and/or stiff and/or sparse, a careful selection of the solution approach is required. Using an inappropriate method for the solution of a non-trivial MC may lead to significant errors (Kharchenko, V. et al. 2013). The main gaps of this phase are shown in the Table 3. We also noted the following ambiguity between IEC 61508 and IEC 61165: while the IEC 61508 asserts that modeler should not focus on the underlying mathematical details, the IEC 61165 requires the help of experts in applied mathematics area during model solution.

During the tools phase researcher selects appropriate software package (tool) to obtain the MC solution using the chosen technique. The Table 4 presents the list of main gaps of current phase and how they are reviewed by standards. It is important to note, that the human-based errors can be introduced during the MC manual construction using the means of selected tool. In addition, tool must support portability of a solution project, so researcher can use it in additional calculations, thus the usability-oriented selection of tools is needed.

On the *verification phase* results are checked to ensure confidence and accuracy of the obtained values. We note that standards mainly recommend the manual results verification. In most cases the size of state space makes the manual calculations infeasible in most cases, thus verification can be supported by another tools or approaches.

Table 2. Gaps of the second modeling phase

Gap	Reviewed by Standard
The MM approach requires construction of the full set of all possible system states, which exponentially increase the model size.	IEC 61508 – 6 "…The main problem with Markov graphs is that the number of states increases exponentially when the number of components of the system under study increases…" ECSS-Q-ST-30-09 "…However, the system complexity can generate a high number of expected states that have impact on the calculation aspects (time and accuracy)..." IEC 60300-3-1 "…accounting the additional system components exponentially increases the state space and complicates analysis… "
Analyze the statistics data to determine system parameters	The detailed explanation is presented in IEC 61508 (6 and 7 parts)

Table 3. Gaps of the techniques phase

Gap	Reviewed by Standard
Choose the appropriate solution approach/method based on its applicability for the specific case, accuracy and level of confidence.	IEC 61508 – 7 "… a homogeneous Markov graph is only a simple and common set of linear differential equations with constant coefficients. This has been analyzed for a long time and powerful algorithms have been developed and are available to handle them…"
In case of transient solution, make the analysis of such mathematical details as stiffness, sparsity etc., which can influence on the achieved results.	IEC 61508 – 6 "…Efficient algorithms have been developed and implemented in software packages a long time ago in order to solve above equations. Then, when using this approach, the analyst can focus only on the building of the models and not on the underlying mathematics…" IEC 61165 "...the solution methods can be quiet complex, thus the specialized software packages and/or experts in applied math area are required…"

Table 4. Gaps of the tools phase

Gaps	Reviewed by Standard
Using the selected solution approach/ method choose the efficient and highly trusted software package.	IEC 61508 – 6 "…If software programs are used to perform the calculations then the practitioner shall have an understanding of the formulae/techniques used by the software package to ensure its use is suitable for the specific application…" IEC 60300-3-1 "…m) tools performance. Are the tools user friendly? Do they share the interface with other tools, so the results can be transmitted for multiple use?..."
Construct/generate and solve the MM. For the model manual construction, it is important to verify the resulting MC to detect the human-based errors.	IEC 61508 – 6 "…When dependencies between components cannot be neglected, some tools are available to build automatically the Markov graphs. They are based on models of a higher level than Markov models (e.g. Petri nets, formal language)…"

Table 5. Gaps of the verification phase

Gaps	Reviewed by Standard
Verify the results using manual calculations or another SW package.	IEC 61508 – 6 "…The practitioner should also verify the software package by checking its output with some manual calculated test cases…"
	IEC 61508 – 7 "…when anything but the simplest calculations is performed in floating point, the validity of the calculations must be checked to ensure that the accuracy required be the application is actually achieved…"
	IEC 60300-3-1 "…l) check the trustworthiness. Can we check results manually? If not, the software package is user friendly?..."

The gaps of techniques, tools and verification phases have significant impact on the results accuracy, and observed standards do not exhaustively track all risks. Thus, it is important to know the ways in which the risk of calculating the incorrect resulting value by using inappropriate T&T can be decreased or eliminated.

DIFFICULTIES OF MARKOV MODELS APPLICATION

Stiffness

There is no common definition of "stiffness" because of its complexity. In publication (Malhotra, M. et al 1994) authors introduce the "practical" definitions of stiff problems, based on the interpretation of physical processes in research systems. Here we provide the example of "practical" formation of stiffness problem.

In general there are two ways of how to improve the availability: increase time-to-failure or reduce time-to-recovery. The system failures can be caused by various types of defects (bohrbugs, heisenbugs, aging-related bugs (Nicola, V.F. 1982) the rate of which may vary in orders (more then 10^2). The difference in orders of software – hardware system failure and recovery rates values (Kharchenko, V. et al. 2013) can be shown as an example that system has a feature of *stiffness* (Arushanyan, O. et al. 1990). The given difference appears in matrix of coefficients of Kolmogorov DE and lead to inefficiency of explicit numerical methods use (Malhotra, M. et al 1994).

In research works (Reibman A. et al. 1989, Hayrer, E. et a;. 1999) authors present the definition of stiffness based on the problems of numerical solution: inability or ineffectively use of explicit numerical methods; presence of quick perturbations decay; big Lipschitz constants; big difference of Jacobi matrix eigenvalues etc.

One of the most vides used stiffness definition a method is based on the calculation of stiffness index – s (Hayrer, E. et a;. 1999).

The Cauchy problem $\frac{du}{dx} = F\left(x,u\right)$ is said to be stiff on the interval [x_0, X] if for x from this interval the next condition is fulfilled:

$$s(x) = \frac{\max_{i=1,n} |\mathrm{Re}(\lambda_i)|}{\min_{i=1,n} |\mathrm{Re}(\lambda_i)|} \gg 1,. \quad (1)$$

where: s(x) – denotes the index of stiffness, λ_i – are the eigenvalues of a Jacobi matrix.

In work (Hayrer, E. et a;. 1999)] Ernst Hairer and Gerhard Wanner propose two possible methods of prior detection of stiffness in researches DE system. The implementation of automatically stiffness detection can help to avoid the not accurate, in case of stiffness, numerical methods. The first method is based on the analysis of errors on first steps of system DE solution (not more than 15 steps). The second possibility is based on the estimation directly the dominant eigenvalue of the Jacobian of the problem.

In the last 30 years a lot of approaches have been developed to deal with the problem of stiffness (Arushanyan, O. et al. 1990, Reibman A. et al. 1989). They can be separated into two groups - *stiffness-tolerance* (STA) and *stiffness-avoidance* approaches (SAA).

The basic idea of STA is a model transformation by identifying and eliminating the stiffness from the model, which would bring two benefits: a reduction of the largeness of the initial MM; efficiency in solving a non-stiff model using standard numerical methods. The approach was named an aggregation/ disaggregation technique for transient solution of stiff MCs. The technique, developed by K. S. Trivedi, A. Bobbio and A. Reibmann (Bobbio, A. et al. 1986), can be applied to any MC with transition rates that can be grouped into two separate sets of values – the set of *slow* and the set of *fast* states. While the transformation of the initial stiff MC brings benefits in terms of efficiency, to the best of our knowledge, no systematic study has been undertaken of the impact of the transformation (from a stiff to a non-stiff MC on the accuracy of the solution.

The main idea of SAA is using methods that are stable for solving stiff models. These methods can be split broadly into two classes: "classical" numerical methods for solution of stiff differential equations (DEs) and "modified" numerical methods used for finding a solution in special cases.

Detailed analysis of given classification attribute can determine the type of *stiffness* and according to it choose the optimal (as combination of estimation time, resource cost and accuracy) computational method (Bobbio, A. et al. 1986, Arushanyan, O. et al. 1990).

Based on the analysis of earlier research works (Kharchenko, V. et al. 2013, Butenko, V. 2014) and conducted empirical tests for each group of *s(x)* were selected the main MM solution technique and technique for results verification (Kharchenko, V., et. al. 2014). The Fig.2 shows normalized scale of *s(x)* with corresponding recommendations for method selection, where:

- *STA* – stiffness-tolerance approach;
- *SAA* – stiffness-avoidance approach;
- *m* – subscript, which denotes the main approach;
- *v* – subscript, which denoted the verification approach.

Figure 2. Normalized scale for stiffness metric

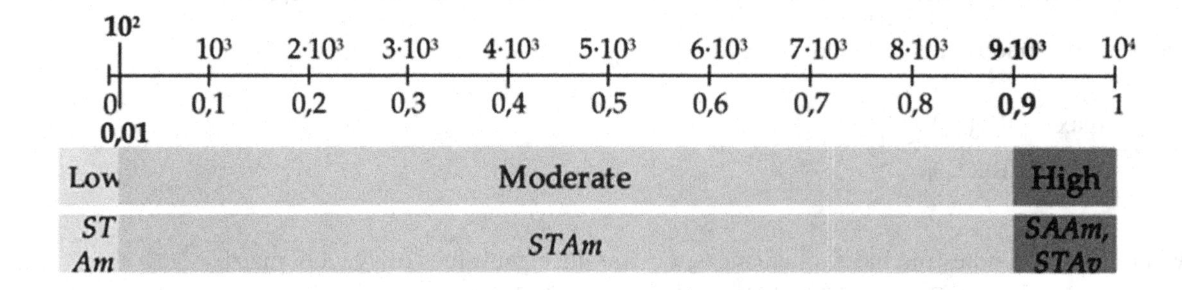

The top values on scale (Figure2) are actual s(x) that are gained using (1) and lower symbols shows the corresponding to them normalized values on interval [0; 1]. Normalization can be performed using:

$$m_i = \frac{x_i - x_i^{max}}{x_i^{max} - x_i^{min}},. \quad (2)$$

where m_i – normalized value, x_i – initial value, x_i^{max} - maximum on scale, for *s(x)* $x^{max} = 10^4$, x_i^{min} – minimum on scale, for *s(x)* $x^{min} = 0$.

Largeness (Decomposability, Irreducibility)

As a second classification attribute of researched system model the term of *largeness* can be used.

In the modeling process the real object is presented with some level of specification. Determine the level of specification at different stages of the modeling process is unique for each system. The nowadays RTS are complex hardware-software systems. High requirements to the reliability of such systems operating process force the modeler to decompose the system to the elementary parts to provide the accurate in-depth analysis. The process of including more details in model makes it larger and more complicated so its analysis will be more difficult or even intractable.

Methods of large MM solution can be divided into two types: largeness-avoidance and largeness-tolerance.

The main idea of largeness-avoidance approach (LAA) is to avoid generation of the large MC from the beginning. Using LAA approach the certain properties of model representation are exploited to reduce the size of the MC to obtain the measures of interest (Sanders, W. H., et al. 2003). The state-level and model-level (Sanders, W. H., et al. 2003) lumping techniques are well-known methods of LAA approach. A state-level lumping technique is a technique that exploits the certain properties on the MM level, while the model-level lumping denotes the lumping properties on the high-level formalism and directly construct lumped MM. Another LAA technique is an aggregation, which set a condition for partition of the state space, and replacing the formed sub-sets by a single state. The aggregation in contrast to lumping gives approximate results, with or without bounds, but may result the smaller MM then a lumping technique.

The largeness-tolerance approach (LTA) is designed to manipulate large MM using special algorithms and data structures to reduce and store transition probabilities matrix (Srinivasan, A., et al. 1990). The numerous works (Srinivasan, A., et al. 1990, Bryant, R. E 1986) present the ideas of using binary and multi-valued decision diagrams (BDD and MDD), matrix diagrams (MD), Kronecker products, etc. to deal with state space size. The disk-based approach for steady-state and path-based approach for transient solutions are also considered in (Sanders, W. H. et al. 1998, Gail, H. R., et al. 1989). Analysis of MM irreducibility and decomposability properties can help to make a prior selection between described techniques (Bryant, R. E 1986).

The analysis of MM using classification attribute largeness will reduce the time for system assessment using special algorithms and amount of computing resources.

The aggregation techniques are mainly based on the decomposability approach. In this case the degree of coupling can be taken as measure of matrix largeness (decomposability) property.

For example, considering a nearly completely decomposable (NCD) MM, which has a matrix with non-zero elements in off-diagonal blocks are small compared with those in the diagonal blocks (Courtois, P. J. 1977):

Figure 3. Normalized scale for decomposability metric

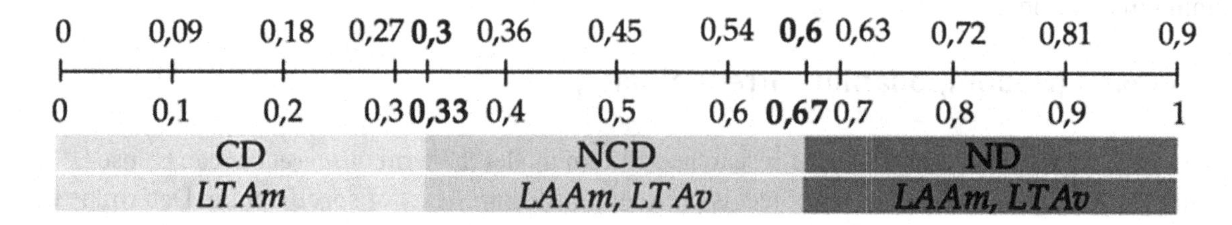

$$A = \begin{pmatrix} A_{11} & A_{12} & \dots & A_{1n} \\ A_{21} & A_{22} & \dots & A_{2n} \\ \dots & \dots & \dots & \dots \\ A_{n1} & A_{n2} & \dots & A_{nn} \end{pmatrix}, \tag{3}$$

$A_{11}, A_{12}, \dots, A_{nn}$ are square diagonal subblocks. The stationary distribution of π can be partitioned such as $\pi = (\pi_1, \pi_2, \dots, \pi_n)$. Assuming that A is of form (4), where E contains all of-diagonal blocks. The quantity (5) is referred to as degree of decomposability (Courtois, P. J. 1977). If $E = 0$ then MM is said to be completely decomposable (CD).

$$A = diag(A_{11}, A_{22}, \dots, A_{nn}) + E, \tag{4}$$

$$\|E\|_{\infty} = \max_{1 \le i \le n} \sum_{j=1}^{n} |e_{ij}| \tag{5}$$

An irreducible MM is presented by a direct graph that is a single strongly connected component. The algorithm for determining strongly connected components is a known graph algorithm (Cormen, T. 2001). Detection of such components (irreducibility property) can help in determining sub-sets for approximate aggregation technique, but it naturally applied after selecting the avoidance approach.

The value (5) can help in deciding between avoidance and tolerance approaches, thus we refer to (5) as a main largeness metric, further decomposability metric. The E values can be split into three groups (Kharchenko, V., et. al. 2014):

- completely decomposable (CD): $E < 0.3$;
- nearly completely decomposable (NCD): $0.3 \le E < 0.6$;
- non-decomposable (ND): $E \ge 0.6$.

As in the previous test, Figure3 presents normalized scale for E with recommendations for approach selection, where:

- *LTA* – largeness-tolerance approach;

- *LAA* – largeness-avoidance approach;
- *m* and *v* denote the main and verification approach, respectively;
- $x_{max} = 0.9$ and $x_{min} = 0$.

Sparsity

Conceptually, sparsity (Barge, W. S., et al. 2002) corresponds to systems which are loosely coupled. In the subfield of numerical analysis, a sparse matrix is a matrix populated primarily with zeros (Press, W.H. et al. 2007).

The analysis of classification attribute sparsity is important part for the special class of problems. As an example of such class we can use the solution of Kolmogorov DE, which describes the MM of system under research. As the matrix of DE coefficients is presented in mostly diagonal form so given attribute can be accompanying in case of using the apparatus of Markov modeling. If research MM is large the sparsity can cause additional assessment difficulties.

Storing and manipulating sparse matrices on a computer is beneficial and often necessary to use specialized algorithms and data structures that take advantage of the sparse structure of the matrix. Operations using standard dense-matrix structures and algorithms are relatively slow and consume large amounts of memory when applied to large sparse matrices.

Most of the approaches are developed to reduce the size of the transition matrix representation and form the dense matrix, by using structured analysis (Reibman, A. et al. 1988) or symbolic data structures analysis (Cormen, T. 2001) and solving them using lumping algorithms or iterative techniques (Cormen, T. 2001).

Paper (Barge, W. S., et al. 2002) presents the formula for evaluation of the heuristic measure of sparsity – matrix score. It gives a measure of how the matrix elements are dispersed from the main diagonal. The analysis of this score in the process of system research will reduce the time for system assessment by using specialized techniques, algorithms and data structures that take advantage of the sparse structure of the matrix.

Let q_i be the number of matrix elements that are a distance *i* from the diagonal. The histogram is weighted and then scaled by n^2 where *n* is matrix order. The matrix score *ms* can be evaluated using:

$$ms = \left(\sum_{i=1}^{n-1} i \cdot q_i\right) / n^2 . \quad (6)$$

In (Barge, W. S., et al. 2002) authors studied the influence of ms value on accuracy of the tolerance techniques for MC solution, and recommended to give additional attention on fill-in amount for matrices with $n \geq 500$ and $ms > 0.8$. The fill-in is a property when initially zero matrix elements become nonzero during solution process and for which storage must be reserved.

The value ms can be classified into three groups (Kharchenko, V., et. al. 2014):

- high sparsity: ms < 0.3;
- moderate sparsity: $0.3 \leq$ ms<0.72;
- low sparsity: ms ≥ 0.72.

Figure 4 presents normalized scale of ms with recommendations for approach selection, where x_{max} = 0.9 and x_{min} = 0.

Figure 4. Normalized scale for sparsity metric

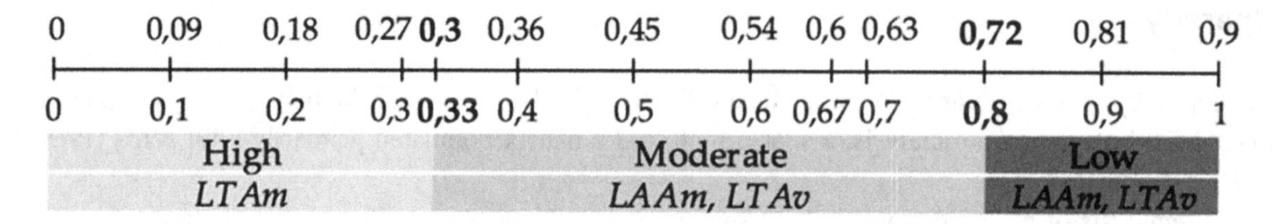

Fragmentedness

The MC are widely applied to analyze the dependability of physical components of safety-critical systems. Assessing the dependability of software components depends on how well the component has been tested, is it available and whether it is a reused or new component (Kharchenko, V. et al. 2011). The verification and validation phases are strongly managed by requirements and recommendations of international standards (see standards IEC 60800-2006, IEC 61508 - 2010). There are lots of required procedures to test the software component, such as documentation analysis, problem review, static code analysis, etc (Butenko, V. 2014). Nevertheless, the residual software bugs can appear during system operation. In this case, to predict the general system dependability we need to observe not only hardware and software components separately, but also analyze their interconnection and total influence on system dependability.

In the previous research works (Butenko, V. 2014, Kharchenko, V., et. al. 2014) we applied the *multi-fragmentation principle* to present such complex interconnection between system hardware and software. The main idea of this approach is to capture and represent using MC the plausible phenomenon – variation of software failure – that is well accepted on practice (Kharchenko, V. et al. 2013). Using this principle the model can be divided into N_{fr} fragments that are with the same structure but may differ in one or more parameters. The use of fragments in MM will increase the modeling clarity and take into account some properties of operating system modes. It is necessary to understand that introduction of parameters change assumption can increase the system size (direct affect on the feature *largeness*) and as a result the increase the sparsity of system transitions matrix (affect on the feature *sparsity*).

The number of fragments N_{fr} in MM depends on the number of expected undetected software faults n_i in *i*-different software versions:

$$N_{fr} = \prod_{i=1}^{m} (n_i + 1). \qquad (7)$$

The structure of fragment and number of such fragments can help to determine the complexity of resulting multi-fragmental model (MFM) and thus help in making decision between avoidance and tolerance approaches. In this paper, we use N_{fr} as a metric of fragmentedness.

Based on N_{fr} value the MM can be classified as follows:

- low-fragmented: $N_{fr} < 6$;
- moderately fragmented: $6 \leq N_{fr} < 15$;
- highly fragmented: $N_{fr} \geq 15$.

The normalized scale ($x^{max} = 30$ and $x^{min} = 0$) with recommendations for approach selection is presented on Figure 5.

Figure 5. Scale for fragmentedness metric

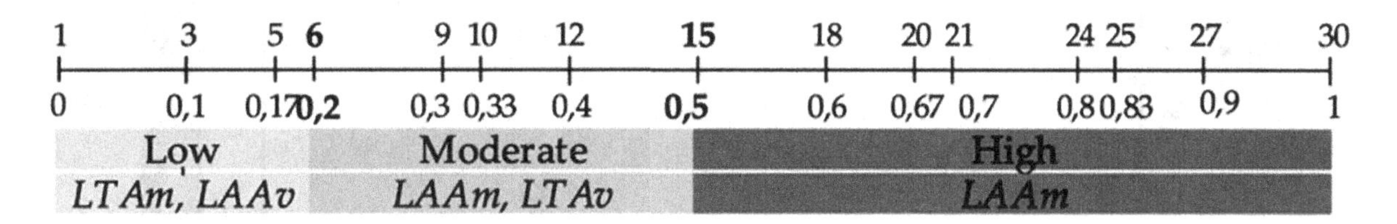

Metric-Based Diagram

The recommendations for selecting the solution approach presented on Figure2 – 5 were received separately for each characteristic. However, it is important to consider each MM feature, while making decision on the most efficient technique.

In this section we present the metric-based diagram (Figure 6), that incorporates all MM characteristics and supports the approach selection based on some specific combination of *s(x)*, *E*, *ms* and N_{fr} values. The diagram passed verification on the wide range of MCs.

The diagram is applied in three main stages.

1. Calculation of stiffness, decomposability, sparsity and fragmentedness metrics using (1), (5) – (8).
2. Normalization of the received values using (2).
3. Marking of the normalized metrics values on the appropriate scale and creating the intersection by drawing perpendicular to the opposite side. As a result, we receive the rectangle, placed in one of the internal zones cij, $i \in (\overline{1,4})$, $j \in (\overline{1,5})$. Each zone provides the recommendation for selection of avoidance or tolerance approach, *AA* or *TA* respectively. If rectangle is placed in two or more zones, selecting the recommendations from zone that contains the larger area of the rectangle. The colored inner zones define the SAA or STA use.

MSMC Tool

The "MSMC – Method selector for Markov chains" was developed to help user, unsophisticated in Markov modeling to select the most effective solution approach automatically, which will give an accurate result. The previously described metric-based approach was implemented in MSMC. This is an alpha version, which runs on Windows platform (XP and later).

Figure 6. Metric-based diagram

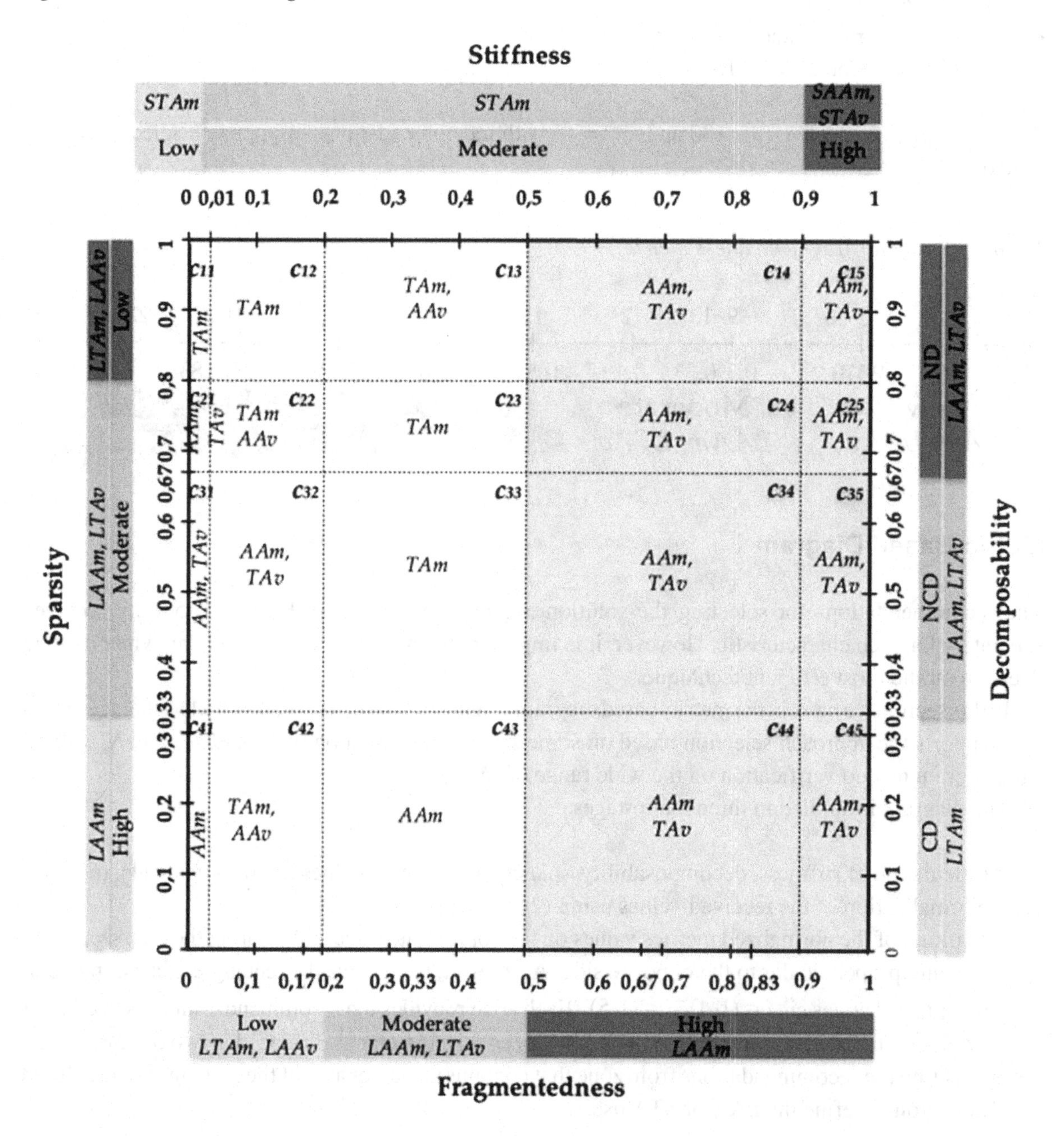

MSMC supports:

1. Two types of MM construction: graphical and analytical.
2. Stiffness, sparsity, decomposability and fragmentedness testing.
3. Selection of the solution approach using the tests results.
4. State probabilities calculation using selected approach.
5. Conversion of transition probabilities matrix and DES into the MATLAB, Mathematica and Maple syntaxes in verification purpose.

Figure 7. MM graphical construction

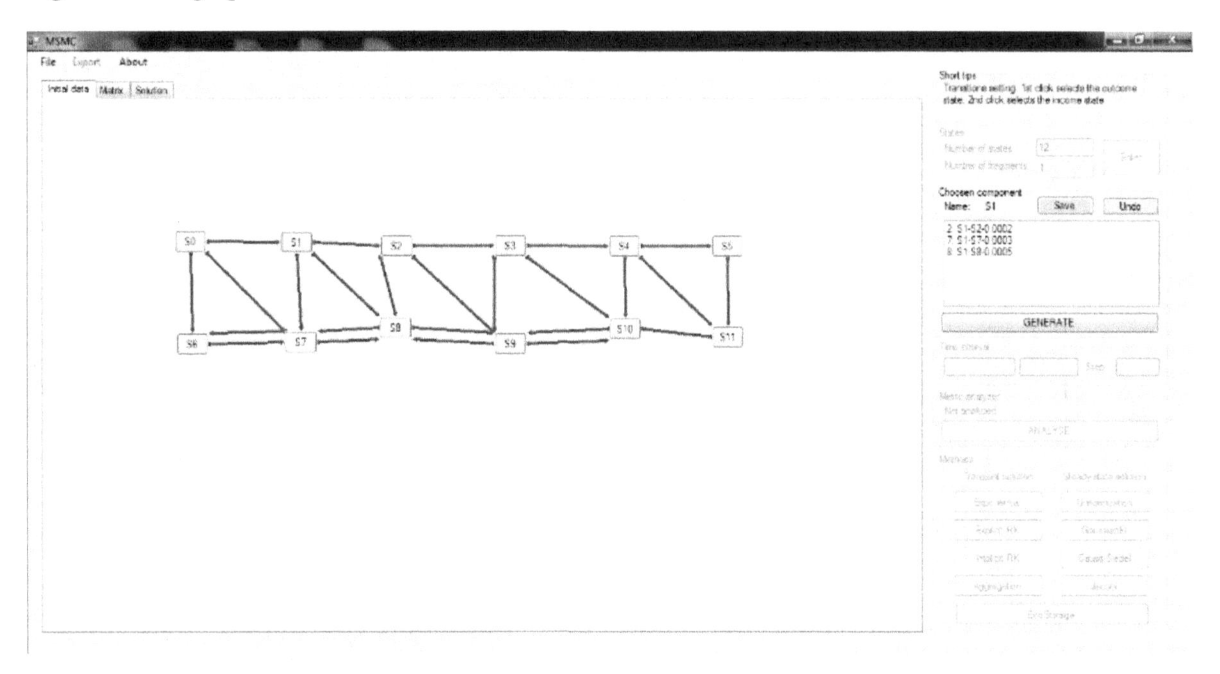

Figure 8. Generated transition probabilities matrix

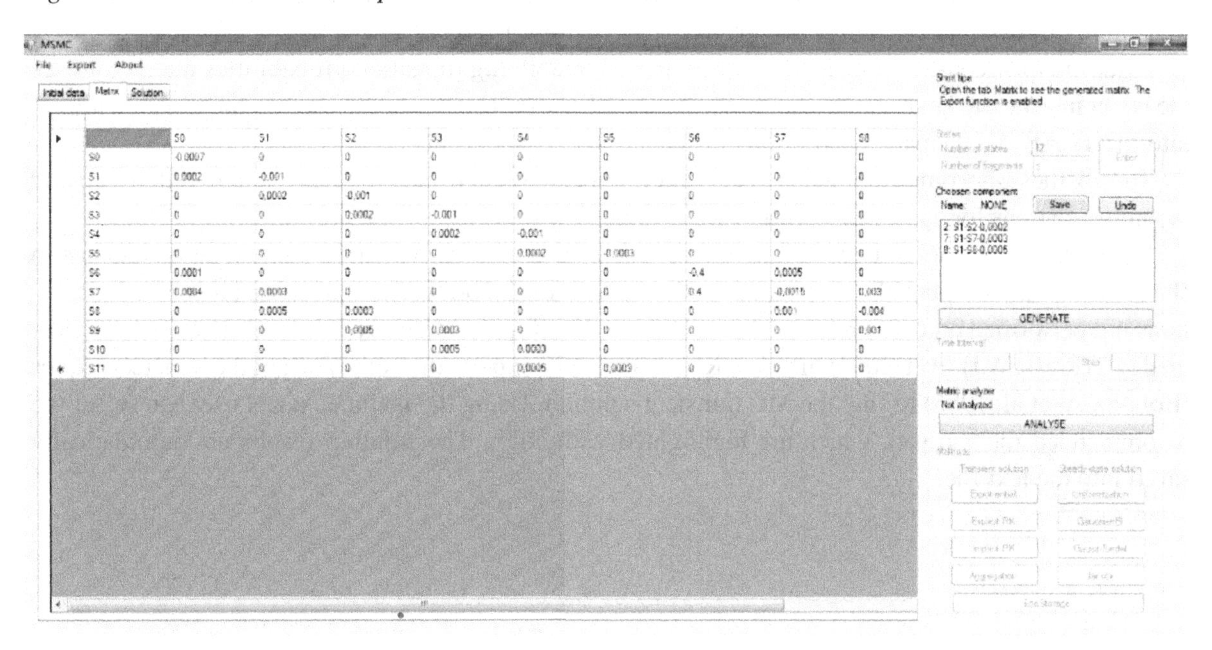

	S0	S1	S2	S3	S4	S5	S6	S7	S8
S0	-0.0007	0	0	0	0	0	0	0	0
S1	0.0002	-0.001	0	0	0	0	0	0	0
S2	0	0,0002	-0.001	0	0	0	0	0	0
S3	0	0	0,0002	-0.001	0	0	0	0	0
S4	0	0	0	0.0002	-0.001	0	0	0	0
S5	0	0	0	0	0.0002	-0.0003	0	0	0
S6	0.0001	0	0	0	0	0	-0.4	0,0005	0
S7	0.0004	0,0003	0	0	0	0	0.4	-0,0015	0,003
S8	0	0.0005	0.0003	0	0	0	0	0.001	-0.004
S9	0	0	0,0005	0.0003	0	0	0	0	0,001
S10	0	0	0	0.0005	0.0003	0	0	0	0
S11	0	0	0	0	0,0005	0,0003	0	0	0

Figure 9. Metric-based approach application

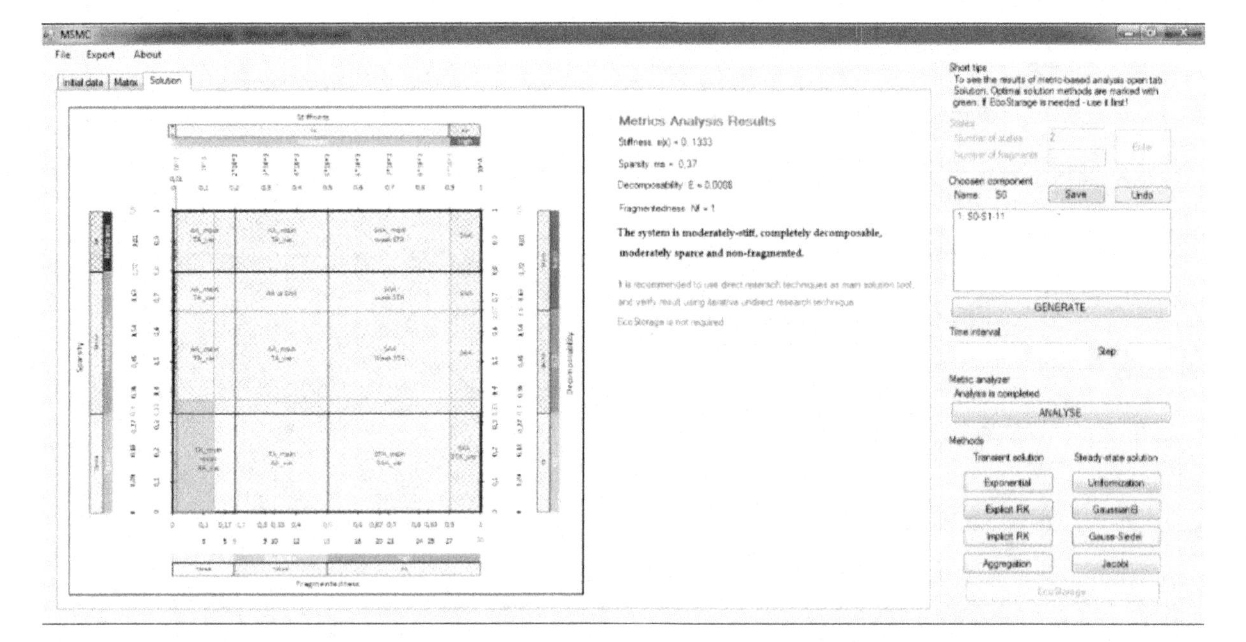

The major components of a GUI are model editor, transitions editor, generated matrix tab, solution tab and methods panel. The model editor (Figure 7) allows a graphical construction of the MM using MSMC primitives – places (states) and arcs (transitions). If MM already exists user can upload the transition probabilities matrix as an Excel spreadsheet (.xls or .xslx) or text (.txt) document. With the transitions editor user can change already assigned transition values.

After the MM construction, MSMC generates an underlying transition probabilities matrix and see the result in generated matrix tab (Figure 8). With the Export option user can convert the received matrix into MATLAB, Mathematica and Maple form and save it as text file.

The stiffness, decomposability, sparsity and fragmentedness test results, metric-based diagram and recommendations for approach selection are displayed in solution tab (Figure 9).

If the matrix appears to be moderately or highly sparse MSMC will store the non-zero elements in flat format using the algorithm described in (Press, W.H. et al. 2007). Thus, all subsequences of calculations are performed on the compressed matrix form. We have implemented the well-known numerical methods (Press, W.H. et al. 2007) (implicit RK, explicit RK, exponential, etc.) and aggregation techniques (Bobbio, A. et al. 1986) to find the MC transient solution. Using the methods panel user can select the recommended method (-s), which are highlighted with color. The solution results are automatically saved into the text file.

Table 6. Feature-based comparison of tools applicable during Markov modeling

Features	Tools								
	SHARP	**Tool Kit Markov**	**MARCA**	**Möbius**	**ASNA**	**MSMC**	**Mathematica**	**Matlab**	**Maple**
MM Construction									
Manually (graphic)	+	+				+			
Manually (text)	+		+				+	+	+
State-space generation			+	+	+				
Importing matrix of DE coefficients									
Import from text file	+		+			+	+	+	+
Import from table editor						+			
Metric-based analysis									
Detection of stiffness						+	+		+
Detection of largeness						+	+	+	+
Detection of sparsity			+			+			
Detection of periodicity			+			+	+		
Numerical methods									
For steady-state analysis	+	+	+	+		+	+	+	+
For transient analysis	+	+	+	+		+	+	+	+
Reports									
Reports generation	+	+	+	+	+	+			
Results export to the text or table editors		+	+			+	+	+	+

ANALYSIS AND SELECTION OF TOOLS APPLICABLE DURING MARKOV MODELING

The IEC 60300-3-1:2003 states a list of questions, which helps to choose an appropriate method as well as software tool (ST), that can be applied during complex systems safety and reliability and parameters assessment. The main questions consider software usability concept and ability to reuse resulting data.

The Table 6 presents comparison analysis of several tools which can be used during MM application – SHARP (Duke University), ToolKit Markov (ITEM), MARCA, MÖBIUS (University of Illinois Urbana Champaign), ASNA (Lviv Polytechnic National University), MSMC, Mathematica (Wolfram), Matlab (Mathworks) and Maple (Maplesoft).

Based on comparison of ST features for construction, analysis, editing, solution end exporting of the MM analysis results we have developed an algorithm of software tools selection and application (Figure 10).

The Figure 10 uses following abbreviations: *Ap* – approaches for MM analysis (8), *TA* – tolerance approaches (9), *AA* – avoidance approaches (10), *STA* – stiffness tolerance approach, *LTA* – largeness tolerance approach, *SAA* – stiffness avoidance approach, *LAA* – largeness avoidance approach, *T* – ST applicable during MM (11), *Build* – ST for MM construction (12), *Gen* – ST for MM generation, *BuildGr* – ST for MM graphical construction, *BuildTxt* – ST for MM text-based construction, *Analyze* – ST for metric-based analysis, *Sol* – ST for numerical MM analysis.

$$AA = \{TA, AA\} \tag{8}$$

$$AA = \{SAA, LAA\} \tag{9}$$

$$AA = \{SAA, LAA\} \tag{10}$$

$$T = \{Build, Gen, Analyze, Sol\} \tag{11}$$

$$Build = \{BuildGr, BuildTxt, Gen\} \tag{12}$$

The algorithm can be divided into following logical parts:

1. Blocks 1 – 4: Initial data. The set of ST (11) and approaches (8), requirements to the accuracy, modeling time interval, acceptable error and required algorithm complexity – are the initial data for presented algorithm.
2. Blocks 5 – 13: MM construction. Based on data gathered after technical documentation analysis, assumptions on how deep the modeled system behavior should be presented during modeling process we can get the initial understanding of MM complexity. In case if resulting MM can be classified as large and complex the most appropriate way is to use automatic state-space generation instead of creating it manually. To do so, researcher can use various high-level modeling formalisms, such as stochastic activity networks, stochastic Petri nets of even verbal presentation. In other case the MM can be created manually with graphical or text editors. As the result researcher should obtain MM and Kolmogorov DE system that describes it.
3. Block 14: Export of the matrix of DE coefficients into text editor. This stage is required to perform further metric-based analysis of MM.
4. Block 15 – 16: Metric-based analysis of the MM. This stage includes application of MSMC methods or several tools which allow user to perform stiffness, largeness, sparsity and periodicity (fragmentedness) analysis. The result of block 16 is applied to make initial recommendations for approach selection.
5. Block 17 – 25: Numerical MM analysis. Based on results of metric-based analysis we can further perform selection of ST that includes the recommended approach. This stage also consider the required accuracy level – if it exceeds 10^{-6} the researcher is recommended to verified obtained numerical results.
6. Block 26: Results analysis. The obtained result are further analyzed by researcher to make informed decisions on modeled system behavior and properties.

Figure 10. Software selection and application during MM

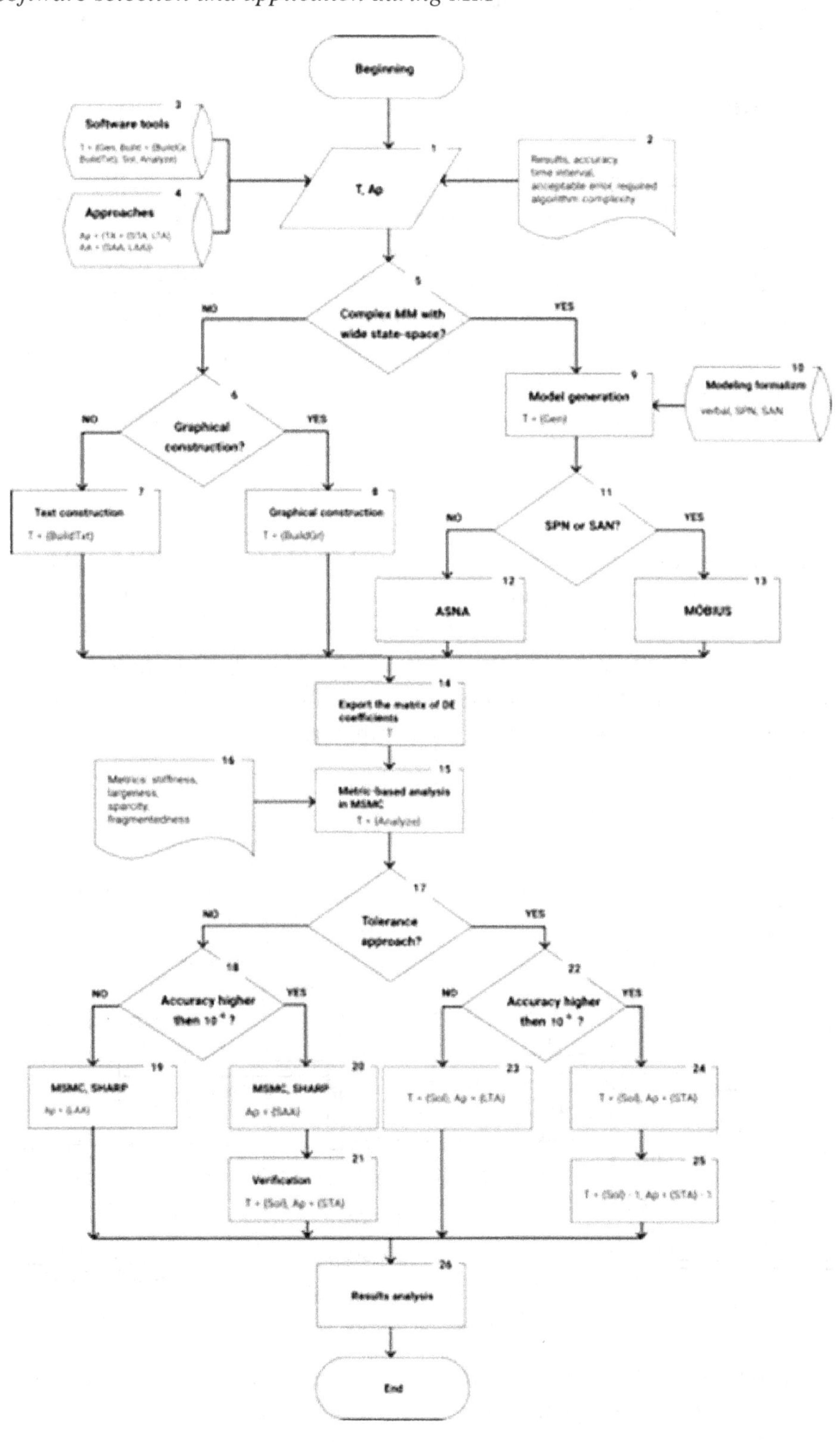

MULTI-FRAGMENTARY MARKOV MODELING FOR THE REACTOR TRIP SYSTEM: CASE STUDY

RadICS Based Reactor Trip System

This section presents the description of a studied NPP I&C system produced by RPC Radiy. This is Reactor Trip System (RTS) with two-channel, three-track architecture, on voting logic "2-out-of-3" for tracks in each channel and "1-out-of-2" between channels. The FPGA-based track is a basic component of observed RTS. Generally, each track can contain up to 7 module types: analogue and digital input modules (AIM, DIM); analogue and digital output modules (AOM, DOM); logic module (LM); optical communication module (OCM); and analogue input for neutron flux measurement module (AIFM). The modules can be placed in 16 different positions on the track (two reserved positions for LM), using LVDS and fiber optical lines for internal/external communications. Such flexible redundancy management helps to ensure the high availability of the system. Each channel independently receives information from sensors and other NPP systems. The channels, each being capable of forming a reactor trip signal, are independent.

Each track, observed in this paper, consists of five modules: LM, DIM, DOM, AIM and AOM. The Figure 11 presents the structure diagram of a typical track. It is assumed that the corresponding components of all the tracks in the channels are identical, i.e. DIM on the 1st track is identical to the same module on other tracks in the channels, etc. The failure of the LM leads to the failure of the whole track, and failures of the DIM, DOM, AIM, AOM result in track malfunction. Therefore, it was assumed that failure of any module implies the general failed state of the track.

Figure 11. The structure diagram of a typical track

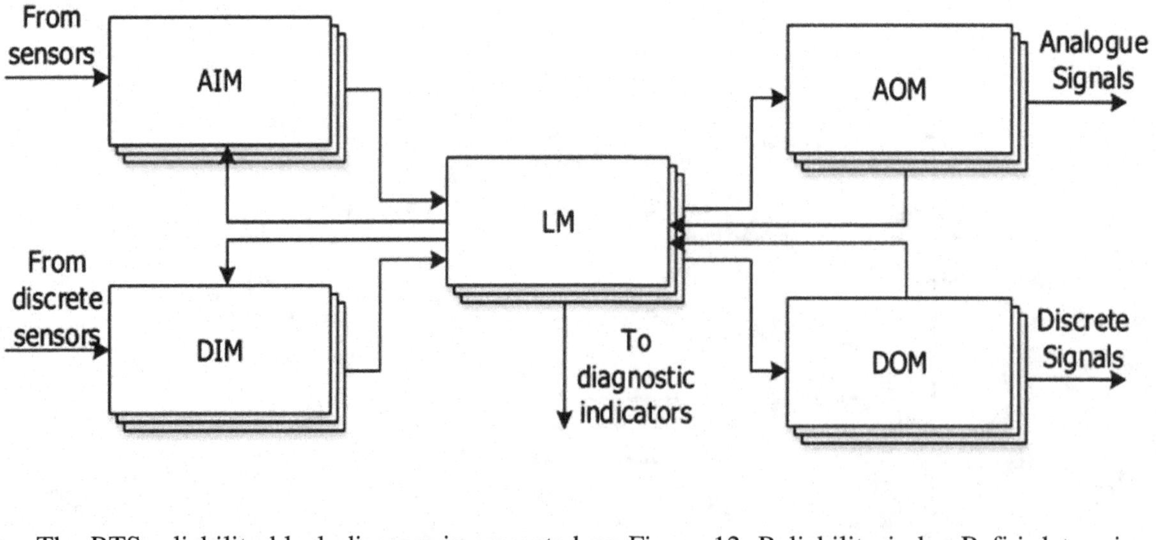

The RTS reliability-block diagram is presented on Figure 12. Reliability index Ppfi.j determines hardware reliability of the track Ti.j (defined by physical faults), where i indicates main (T1.j) or diverse (T2.j) channels, and j indicates the number of the track. Reliability index Pdfi determines software reliability of the main or diverse channels (defined by software faults), where i indicates channel. Reliability index P_{mi}, determines reliability of the majority element mi, where $i \in (\overline{1,3})$.

Figure 12. Reliability-block diagram of two-channel three-track system

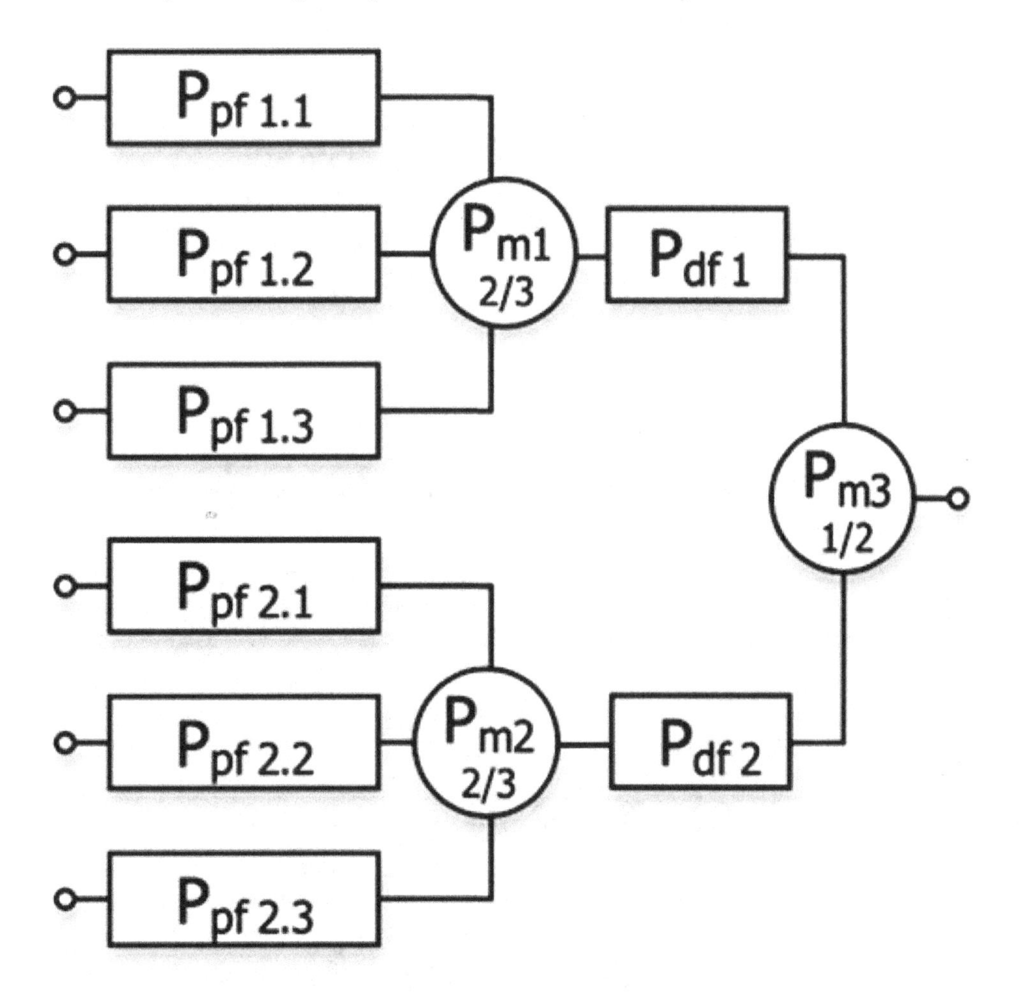

Informally, the system operates as follows. Initially, all components (tracks) are working correctly and deliver the service as expected. If one of the tracks in main channel has failed, by majority voting the failure of the failed track is detected and while the failed track is being repaired, the operation will continue with the second and third tracks in channel. If before the first track has been repaired another one fails, the majority voting component will detect the output disagreement and main channel will stop. The diverse channel operates identically to the main. The operation process continues until one of channels works correctly. Clearly, the reliability of the majority components affects significantly the system reliability. In the observed RTS architecture the failure of majority components (P_{m1} or P_{m2}) leads to the failure of the main or diverse channels, and to the general RTS failure (P_{m3}).

Figure 13. Multi-fragmented Markov model of RTS

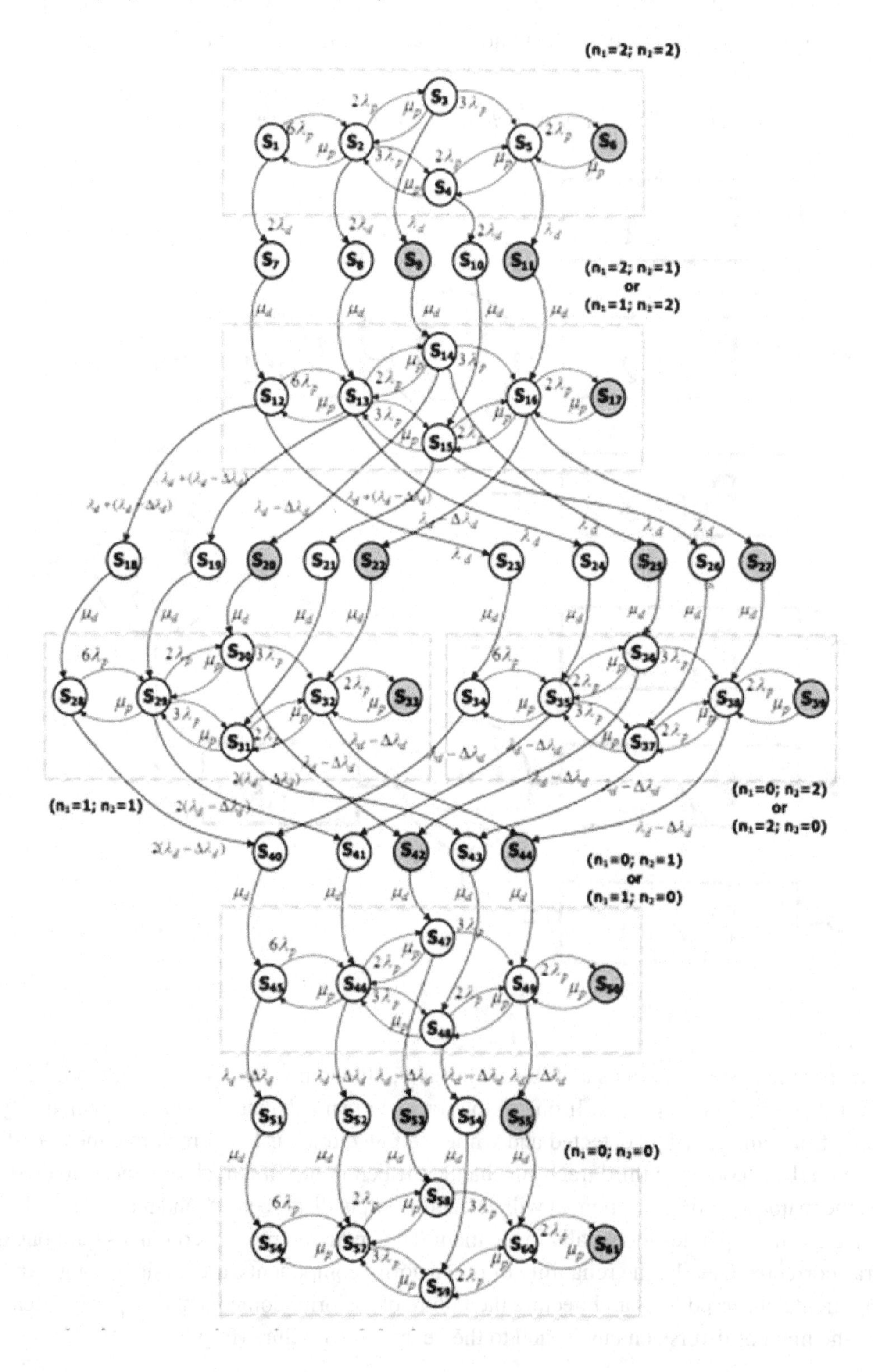

Table 7. MM parameters value

	λ_d *(1/h)*	$\Delta\lambda_d$ *(1/h)*	λ_p *(1/h)*	μ_d *(1/h)*	μ_p *(1/h)*	*t (h)*
1	10^{-5}	$5{\cdot}10^{-6}$	10^{-4}	0.01	1	[0; 30 000]
2	$2.5{\cdot}10^{-5}$	$1.25{\cdot}10^{-5}$				
3	$5{\cdot}10^{-5}$	$2.5{\cdot}10^{-5}$				
4	$7.5{\cdot}10^{-5}$	$3.75{\cdot}10^{-5}$				

Development of RTS Markov Model and Selection of the Tool

Let us further denote by λd and λp the design and physical failure rates, respectively, and by μ_d and μ_p the repair rates after design and physical failures. The MM for RTS system is shown on Figure 13.

The detailed model construction and description is shown in (Butenko, V. 2014). We use the following assumptions to create the MM for observed RTS:

1. Each element of the research system in random moment of time can be only in two states – working and failure.
2. The systems control and majority elements provide unstoppable correct functioning.
3. The system maintenance is performed by one group of engineers, thus failed chassis are repaired sequentially. It should be noted, that recovering strategy use case of two working channels in a priority.
4. All detected defects are eliminated instantaneously and no new defects are introduced. The mean time between failures and mean time to repair are exponentially distributed (Butenko, V. 2014).
5. Software testing datasets are updated after each test. The testing is performed on the complete body of input data.
6. The observed RTS is FPGA-based, thus investigated software faults are such kinds of faults, which are typical for VHDL coding process that were not covered by V&V procedure. The architecture-level MC shows the rare kind of design faults that can cause a general system failure, thus we expect that not more than two undetected design faults on each software version (Butenko, V. 2014, Ehrlich, W., et al. 1990).
7. The failure rate of the design faults $\lambda_{d(i)}$ is proportional to their residual amount n_i in i – different software versions (Butenko, V. 2014). This assumption uses an incremental change of the software failure rate after detected design fault elimination ($\lambda_{d(i)}$ vary on a constant $\Delta\lambda_{d(i)}$). Such failure rates can be presented using multi-fragmentation approach (Kharchenko, V. et al. 2011).
8. 8. The design failures on diverse software versions are independent events, but equal in severity. Thus, we assume that failure and repair rates for the failures caused by design faults are equal:

$$\lambda_{d1} = \lambda_{d2} \Rightarrow \lambda_d = \lambda_{d1} + \lambda_{d2},. \tag{13}$$

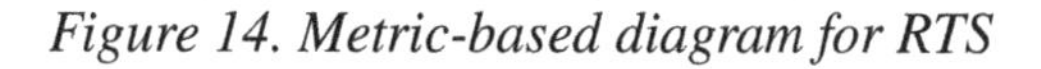

Figure 14. Metric-based diagram for RTS

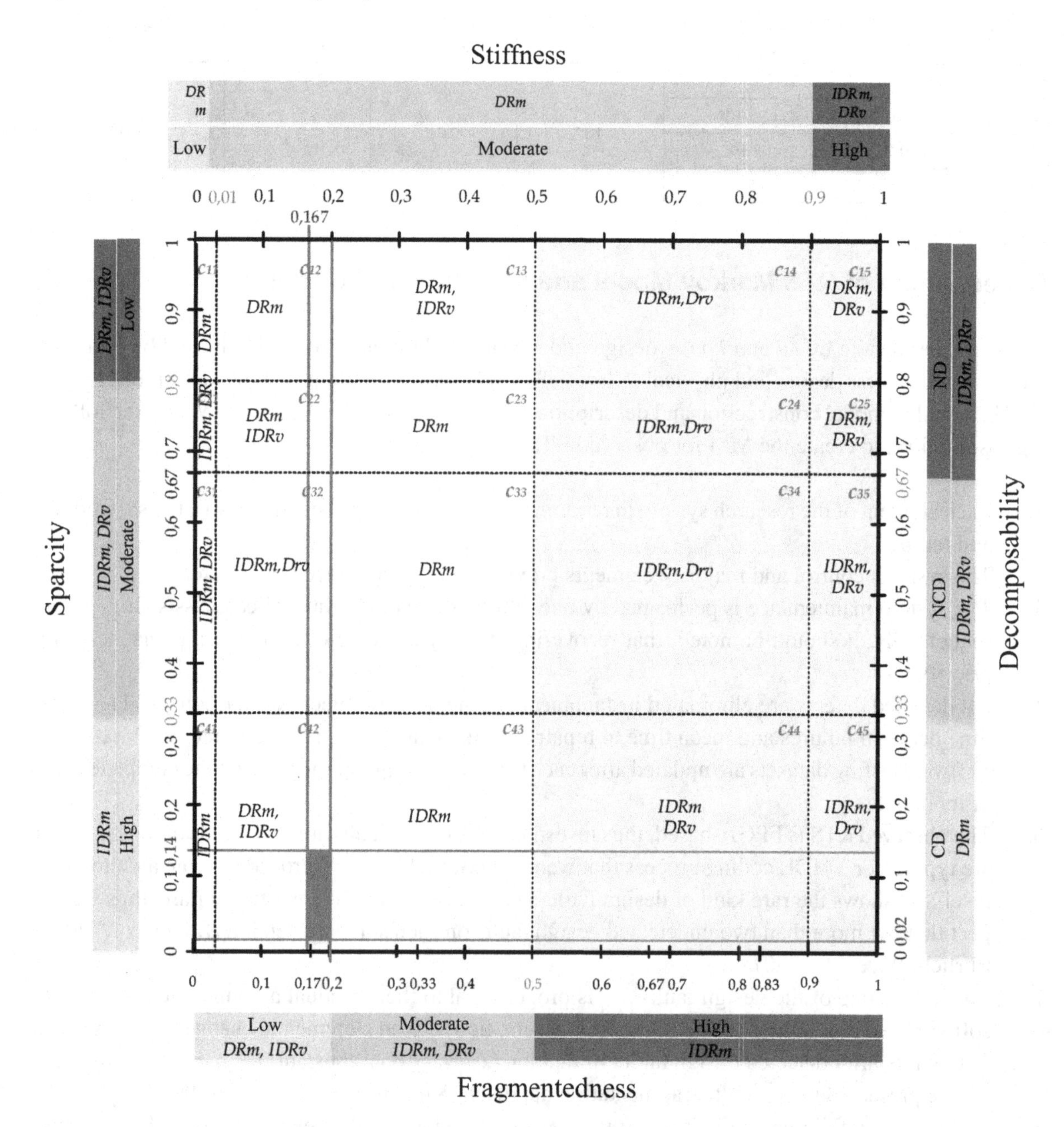

$$\mu_{d1} = \mu_{d2}; \mu_d = \lambda_d / (\sum_{i=1}^{2} \frac{\lambda_{d(i)}}{\mu_{d(i)}}) \quad (14)$$

The neglecting of this assumption will increase the resulting MC state-space, but still we can apply the MFM principle for it construction. Thereby, the assumption was used in a purpose of reducing the model size.

Figure 15. U(t) calculated using STA

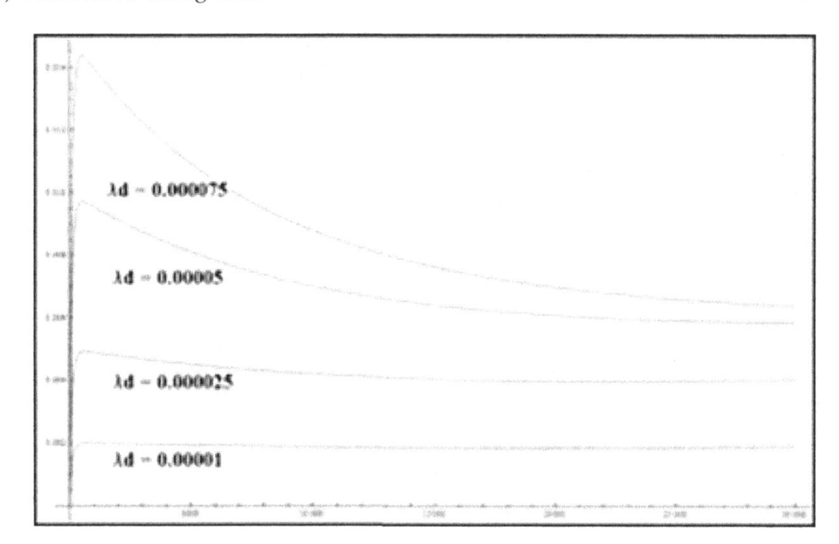

To check the developed model sensitivity we use four sets of parameters values, which are presented in Table 7.

Analysis of MM stiffness, largeness, sparsity and fragmentedness as well as a MM manual graphical construction was performed in MSMC tool. The result of metric-based diagram application is presented on Figure 14.

1. Stiffness: moderately stiff with s(x) = 0.167.
2. Decomposability: completely decomposable (CD) with E = 0.02.
3. Sparsity: highly sparse with ms = 0.2.
4. Fragmentedness: moderately fragmented with Nfr=6.

With the metric-based diagram (Figure 10), we receive the recommendation to use tolerance approach as the main solution technique and verify obtained result with avoidance approach. As the model appears to be moderately stiff, we need to use stiffness-stable numerical methods (STA).

MODELING RESULTS

We use the recommended approaches to assess the RTS unavailability function, which also defines the PFH measure [11].

The unavailability function U(t) is defined as a sum of failed states probabilities, with initial condition U(0) = 0:

$$U(t) = 1 - A(t) = \sum_{i=1}^{n} P_i(t), i \in N_r. \quad (15)$$

where, *A(t)* is RTS availability function.

Due to recommendations from metric-based approach (Fig . 14) and application of tools selection algorithm (Figure 10) we use the STA as a main techniques, thus the *U(t)* for four sets of RTS parameter (Table 7) was calculated using build-in function of implicit Runge-Kutta (RK) method in Mathematica. The results are shown on Figure 15.

The *U(t)* result were verified using MSMC inner function, that implements the implicit RK algorithm and SAA, particularly aggregation/ disaggregation technique (Bobbio, A. et al. 1986). The result of *U(t)* verification for $\lambda_d = 5 \cdot 10^{-5}$ is presented on Figure 16.

Figure 16. U(t) results verification

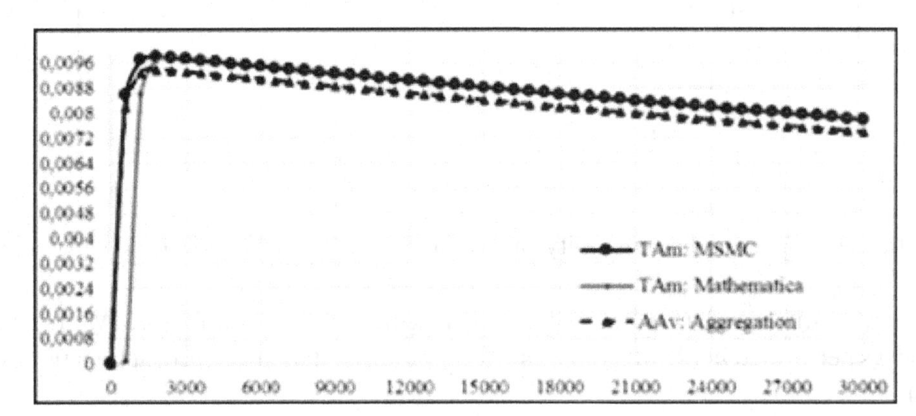

FUTURE RESEARCH DIRECTIONS

The future research work will consider the development of a metric-based algorithm with a self-improvement ability to generate more accurate recommendation for techniques selection.

CONCLUSION

There are some gaps in standards including IEC 61508 regarding techniques and tools choice and accuracy of indicators evaluation on MM-based assessment of safety-critical systems. We conclude that such difficulties as MM largeness, stiffness and sparsity affects significantly on the accuracy of the solution methods, and it is important to take into account such features while selecting the T&T. Analysis of the obtained results allows us to formulate a few problems regarding application of assessment tools: usability-oriented selection of tool in case of solving large MM; results accuracy and level of confidence; high quality verification of the obtained results.

These circumstances determine the feasibility of introducing the additional regulatory requirements to the choice of appropriate T&T and to verification of accuracy of the MC-based safety assessment using these T&T.

This chapter presents the metric-based approach for selection of the applicable solution approach, based on the analysis of such MMs characteristics as stiffness, largeness (decomposability, irreducibility), sparsity and fragmentedness. The metric-based approach was implemented into MSMC tool that support the graphical construction of MM, with further generation of underlying matrix and testing its properties. Based on the test results MSMC tool automatically presents the recommendations for approach selection and provides the transient solution using inner numerical methods.

We present the features-based comparison and discuss the way how software tools can be selected and applied during the modeling of complex systems using MM.

This chapter also presents the case study for assessment the unavailability parameter of NPP I&Cs, in particular, RTS. The system model was build using the MM, taking into account the failures caused by physical and design faults.

REFERENCES

Arushanyan, O. (1990). *Numerical Solution of Ordinary Differential Equations using FORTRAN*. Moscow: Moscow State University.

Barge, W. S. (2002). Autonomous Solution Methods for Large Markov Chains. *Pennsylvania State University CiteSeerX Archives*, 17.

Bobbio, A., & Trivedi. (1986). An Aggregation Technique for Transient Analysis of Stiff Markov Chains. *IEEE Transactions on Computers*, *C-35*(9), 803–814. doi:10.1109/TC.1986.1676840

Bryant, R. E. (1986). Graph-based Algorithms for Boolean Function Manipulation. *IEEE Transactions on Computers*, *35*(8), 677–691. doi:10.1109/TC.1986.1676819

Buchholz, P. (1996). Structured analysis approach for Large Markov Chains. A Tutorial. Proc. of Performane'96.

Butenko, V. (2014). Modeling of a Reactor Trip System Using Markov Chains: Case Study. *Proc. of 2014 22nd ICONE*, 5. 10.1115/ICONE22-31156

Cormen, T. (2001). *Introduction to Algorithms*. MIT Press.

Courtois, P. J. (1977). *Decomposability: Queueing and Computer Applications*. New York: Academic Press.

ECSS (2008). ECSS-Q-ST-30-09. *European Cooperation for Space Standardization (ECSS): Availability analysis*.

Ehrlich, W., Lee, S. K., & Molisani, R. H. (1990). Applying Reliability Measurement: A Case Study. *IEEE Software*, *1990*(2), 56–64. doi:10.1109/52.50774

Gail, H. R. (1989). Calculating Availability and Performability Measures of Repairable Computer Systems. *Journal of the Association for Computing Machinery*, *36*(1), 171–193. doi:10.1145/58562.59307

Hairer, E. (2010). Solving Ordinary Differential Equations II: Stiff and Differential-Algebraic Problems. Springer.

Hayrer, E. (1999). *Solution of Ordinary Differential Equations. Stiff and Differential-Algebraic Poblems*. Mir.

Hurley, N., & Rickard, S. (2009). Comparing measures of sparsity. *IEEE Transactions on Information Theory*, *55*(10), 4723–4741. doi:10.1109/TIT.2009.2027527

IEC (2006, a). IEC 60800. Nuclear power plants – Instrumentation and control system important for safety – Software aspects for computer-based systems performing category A functions.

IEC (2006, b). IEC 61165. *Application of Markov technqiues.*

IEC (2008). IEC 60300-3-3. Application guide – Analysis techniques for dependability – Guide on methodology.

IEC (2010). IEC 61508. *Functional Safety of Electrical/Electronic/ Programmable Electronic Safety-Related Systems.*

Kharchenko, V. (2014). Markov's Model and Tool-Based Assessment of Safety-Critical I&C Systems: Gaps of the IEC 61508. *Proc. 12th Int. Conf. PSAM*, 16.

Kharchenko, V., Odarushchenko, O., & Odarushchenko, V. (2011). Multi-fragmental Availability Models of Critical Infrastructures with Variable Parameters of System Depend*ability. International Journal of Information Security*, *28*, 248–265. doi:10.11610/isij.2820

Kharchenko, V., Odarushchenko, O., Odarushchenko, V., & Popov, P. (2013). Availability Assessment of Computer Systems Described by Stiff Markov Chains: Case Study. *Springer, CCIS*, *412*, 112–135. doi:10.1007/978-3-319-03998-5_7

Malhotra, M., Muppala, J. K., & Trivedi, K. S. (1994). Stiffness-Tolerant Methods for Transient Analysis of Stiff Markov Chains. *Microelectronics and Reliability*, *34*(11), 1825–1841. doi:10.1016/0026-2714(94)90137-6

Nicola, V. F. (1982). *Markovian Models of Transactional System Supported by Check Pointing and Recovery Strategies, Part 1: a Model with State-Dependent Parameters*. Eindhoven Univ. Technol., Eindhoven, The Netherlands, EUT Rep. 82-E-128.

Press, W. H. (2007). *Numerical Recipes. The Art of Scientific Computing*. Cambridge University Press.

Reibman, A. (1989). Analysis of Stiff Markov Chains. ORSA Journal on Computing, 1(2), 126-133. doi:10.1287/ijoc.1.2.126

Reibman, A., & Trivedi, K. (1988). Numerical Transient Analysis of Markov models. *Comput. Opns. Res.*, *15*(1), 19–36. doi:10.1016/0305-0548(88)90026-3

Sanders, W. H. (1998). An Efficient Disk-based Tool for Solving Large Markov Models. *Performance Evaluation*, *33*(1), 67–84. doi:10.1016/S0166-5316(98)00010-8

Sanders, W. H. (2003). Optimal State-space Lumping in Markov Chains. *Information Processing Letters*, *87*(6), 309–315. doi:10.1016/S0020-0190(03)00343-0

Srinivasan, A. (1990). Algorithms for Discrete Functions Manipulation. *Proc. Int'l Conf. on CAD (ICCAD'90)*, 92-95.

Trivedi, K. S. (2000). Availability Models in Practice. *Proc. Int. Workshop on Fault-Tolerant Control and Computing (FTCC-1)*.

ADDITIONAL READING

Smith, W. E., Trivedi, K. S., Tomek, L. A., & Ackaret, J. (2008). Availability Analysis of Blade Server Systems. *IBM Systems Journal*, *47*(4), 1–20. doi:10.1147/SJ.2008.5386524

KEY TERMS AND DEFINITIONS

Architecture: Specific configuration of hardware and software elements in a system.

Channel: Element or group of elements that independently implement an element safety function.

Largeness: Structural complexity of the model.

Multi-Fragment Markov Model: Set of repetitive macromodels (fragments), the internal structure of which and the external relations depend on the sets of selected basic parameters of the model.

Safety Integrity Level (SIL): Discrete level (one out of a possible four), corresponding to a range of safety integrity values, where safety integrity level 4 has the highest level of safety integrity and safety integrity level 1 has the lowest.

Sparsity: Sparse matrix is a matrix populated primarily with zeros.

Stiffness: Value of that in models of complex computer systems is caused by failure and recovery rates varying by several orders of magnitude.

Chapter 8
Composition of Safety and Cyber Security Analysis Techniques and Tools for NPP I&C System Assessment

Ievgen Babeshko
https://orcid.org/0000-0002-4667-2393
National Aerospace University KhAI, Ukraine & Centre for Safety Infrastructure-Oriented Research and Analysis, Ukraine

Kostiantyn Leontiiev
Research and Production Corporation Radiy, Ukraine

ABSTRACT

Safety assessment of nuclear power plant instrumentation and control systems (NPP I&Cs) is a complicated and resource-consuming process that is required to be done so as to ensure the required safety level and comply to normative regulations. A lot of work has been performed in the field of application of different assessment methods and techniques, modifying them, and using their combinations so as to provide a unified approach in comprehensive safety assessment. Performed research has shown that there are still challenges to overcome, including rationale and choice of the safety assessment method, verification of assessment results, choosing and applying techniques that support safety assessment process, especially in the nuclear field. This chapter presents a developed framework that aggregates the most appropriate safety assessment methods typically used for NPP I&Cs.

INTRODUCTION

Modern instrumentation and control systems used at NPPs are mainly complex digital systems responsible for safe plant operation. Such systems include thousands or even tens of thousands of different electronic components. Although digital I&Cs are utilized at NPPs already a long time, the develop-

DOI: 10.4018/978-1-7998-3277-5.ch008

ment of methodologies for the NPP I&C safety assessment is still a critical issue. Safety assessment methodologies include probabilistic safety analysis, operating reliability and safety assessment, static and dynamic analysis of NPP I&C software (IEC, 2004; 2006; 2009; 2010).

Issues on operating reliability assessments we covered in our previous work (Babeshko et al, 2017), in this chapter we focus on probabilistic safety analysis and software assessment and testing issues.

There are several challenges regarding application of the safety assessment techniques:

- Firstly, number of such techniques is very large and choice of appropriate technique (optimal in point of view solved task and assessed system) is a challenging problem: systematization and orchestration of these techniques is required to assure solving of the assessment task;
- Secondly, required accuracy and trustworthiness of safety assessment must be provided. In this case several approaches are possible: to select one technique or to choose and configure several techniques which could be complemented by each other. Question to be answered is how to select and how to configure?

In our previous researches (Babeshko et al, 2011, 2015, 2017, Illiashenko et al, 2011) we have provided assessment idea based on combined assessment techniques usage. But obtained results don't provide formal notations of techniques and, therefore, don't allow to construct models and perform calculations.

St John-Green et al have confirmed necessity of combined security and safety assessment, but integrated approach is stated as to be provided.

Misztal has shown possible combination of FMEA and FTA, but other methods are not considered and assessment of possible benefit is not estimated.

Performed work related analysis has confirmed that researches in this field are in demand in different critical industries, including nuclear.

BACKGROUND

Reliability Block Diagrams

A reliability block diagram (RBD) is a graphical representation of a system's reliability. It shows the logical interconnection of (functioning) components required for successful operation of the system.

RBD allows performing system reliability (no-failure operation) calculation basing on known reliability of its elements.

Probability of no-failure operation in case of series reliability block diagram can be calculated as product of probabilities of no-failure operation of its elements:

$$P_{sys}(t) = \prod_{k=1}^{n} p_k(t), \tag{1}$$

where p_k – probability of no-failure operation of k-th element, n-number of elements in system.

The relation between failure rate and probability of no-failure operation is the following:

$$p(t_0, t) = e^{-\int_{t_0}^{t} \lambda(t)dt}, \tag{2}$$

Basing on formulas (1) and (2) the following expression for failure rate can be obtained:

$$\lambda_{sys}(t) = \sum_{k=1}^{n} \lambda_k(t), \tag{3}$$

where λ_k – failure rate of k-th element, n-number of elements in system.

Fault Tree Analysis

Fault tree analysis is a method to model the chain of causes that lead to an undesired event or effect.

An undesired event is chosen as the top event, e.g., a function event from the event tree. Situations or combination of events that could lead to the top event is connected by logical gates. These second level situations are in turn evaluated and their possible causes determined and connected by logical gates. In this way a tree is built between the top event and a number of basic events and every possible sequence that result in a failing top node is identified.

The basic events are not developed further, they are instead assigned appropriate probability measure that describe their failure probability.

FTA analysis provides both qualitative and quantitative results. Qualitative analysis can be obtained from identification of cut set. While the quantitative analysis of the calculation of failure probability of the system based on the failure probability of each component that compiled it. The qualitative analysis of FTA is based on the cut set that can be easily seen based on the number of components that compose them. Cut set that is only compiled by one component means that if that component failure, then will cause a certain failure on the entire system. While the cut set composed of two or more components implies that there is a redundancy so that the failure of one of them will only increase the failure probability but will not cause a failure of the entire system. Cut set that constructed by many components mean that they have better redundancy. While the quantitative analysis on the FTA is done by calculating the failure probability of the system and/or cut set based on the failure probability of each component by following Boolean algebra rules.

XMEA

FMEA (Failure Modes and Effects Analysis) is a structured, qualitative analysis of a system, subsystem, module, design or function, in order to identify potential failure modes, their causes and their effects on (system) operation, with the objective of improving the design. FMEA is widely used as reliability analysis technique in the initial stage of system development. There are some basic consent of FMEA, such as; how each part can conceivably fail, what mechanisms might produce those modes of failure, what is the effects of the failures, is the failure in the safe or unsafe condition, how the failure can be detected, and what inherent provisions are provided in the design to compensate for the failure. FMEA analysis is done through weighting and ranking. At the end of the analysis will be obtained which component has the greatest weight that means require greater attention and which components have a low weight which means no need to be prioritised. FMEA is a very famous hazard identification in the nuclear industry and still become an important tool to design an NPP.

We are applying this method not only to failures, but also to other domains like possible intrusions. This generic approach we call XMEA.

Markov Models

Markov models provide memoryless modelling and a continuous time stochastic process. There are at less four kinds of Markov models used in different situations depending on sequential state observation and adjustment of the system on the observation, such as Markov chain, Hidden Markov model, Markov decision process, Partially observable Markov decision process. Those Markov models have been used widely for mechanical modelling, especially for reliability modelling. Markov models also very popular for maintenance purposes especially modelling for the single failure of a component. Markov models are very powerful because it can illustrate the failure of the process in detail to present a good quantitative analysis.

NPP I&C SAFETY ASSESSMENT FRAMEWORK

Concept

To assess NPP I&C safety, typically only one assessment technique is being used. Choice of particular assessment technique is usually based on requirements of national or international normative documents, recommendations of regulatory bodies etc. Using only one assessment technique for NPP I&C safety assessment has the following challenges:

- set of safety indicators limited to particular technique possibilities;
- necessity to provide particular set of input information;
- very difficult or even not possible to verify obtained results so as to ground their credibility;
- not possible to see and choose possible alternatives of assessment techniques.

Using several safety assessment techniques is typically not considered due to significant cost and time increase.

Main idea of the proposed framework is to provide possibility to aggregate different formal techniques so as to assess NPP I&C safety. To do this, as a first step we propose to present formal description of techniques. Each technique can be presented as set of input information X, set of output information Z

Figure 1. Framework concept

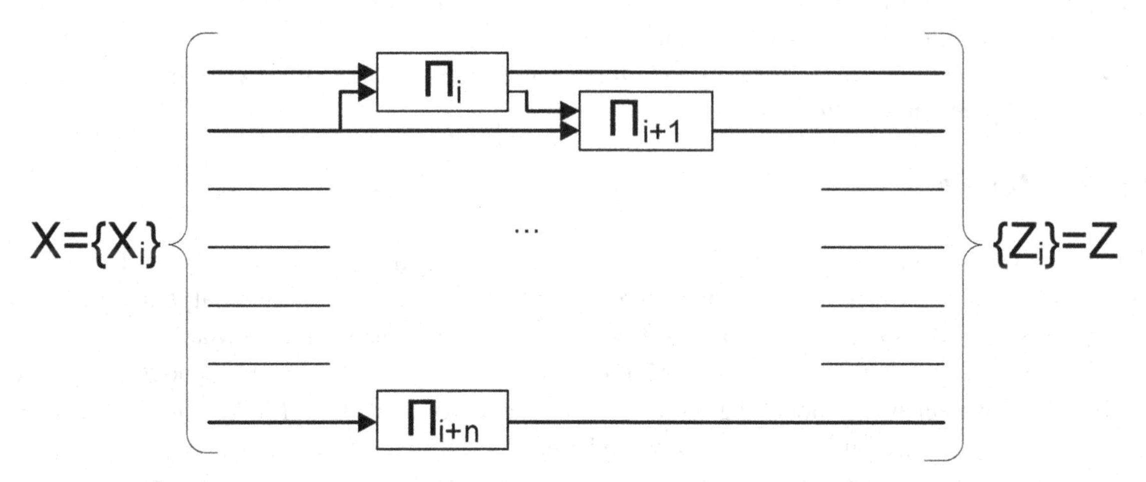

and set of transformers Π (see Figure 1). By transformers we mean steps of particular techniques that have their own inputs and outputs, intermediate or resulting.

Set of input information that is required for the particular transformer and is being obtained from

Figure 2. Sets of input, intermediate and output information

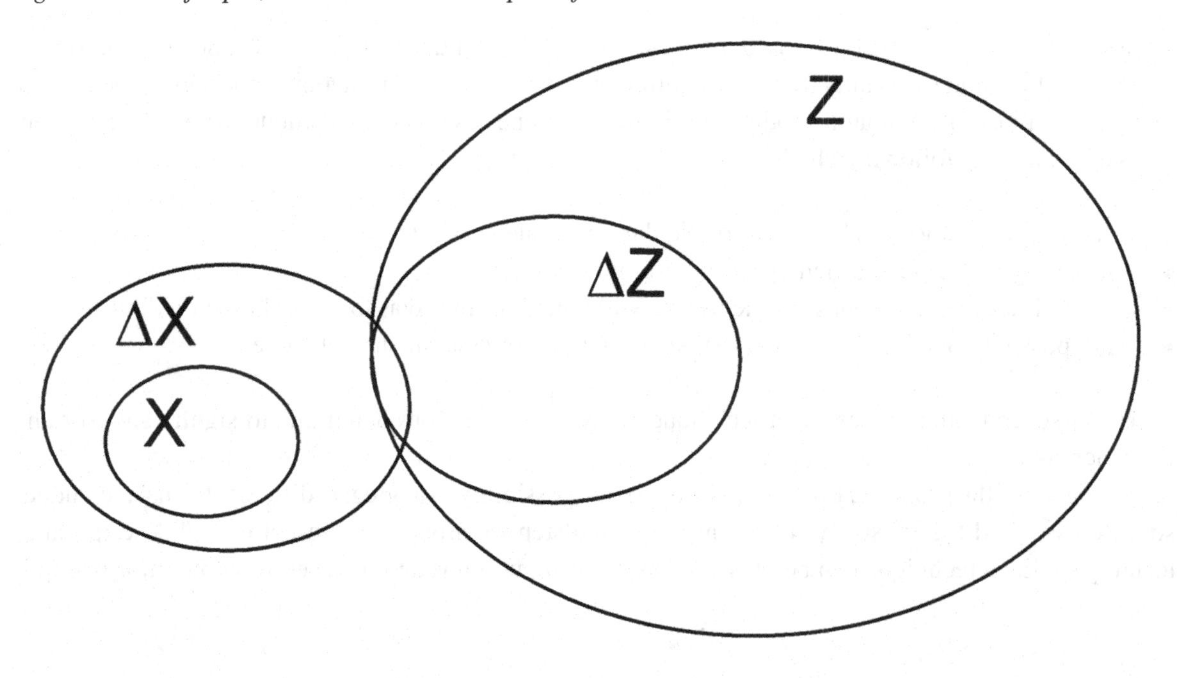

another transformer as output (i.e. intermediate input information), we define as ΔX. Set of intermediate output information we define as ΔZ. Figure 2 shows relations between sets X, Z, ΔX and ΔZ.

Figure 3. Representation in a form of graph

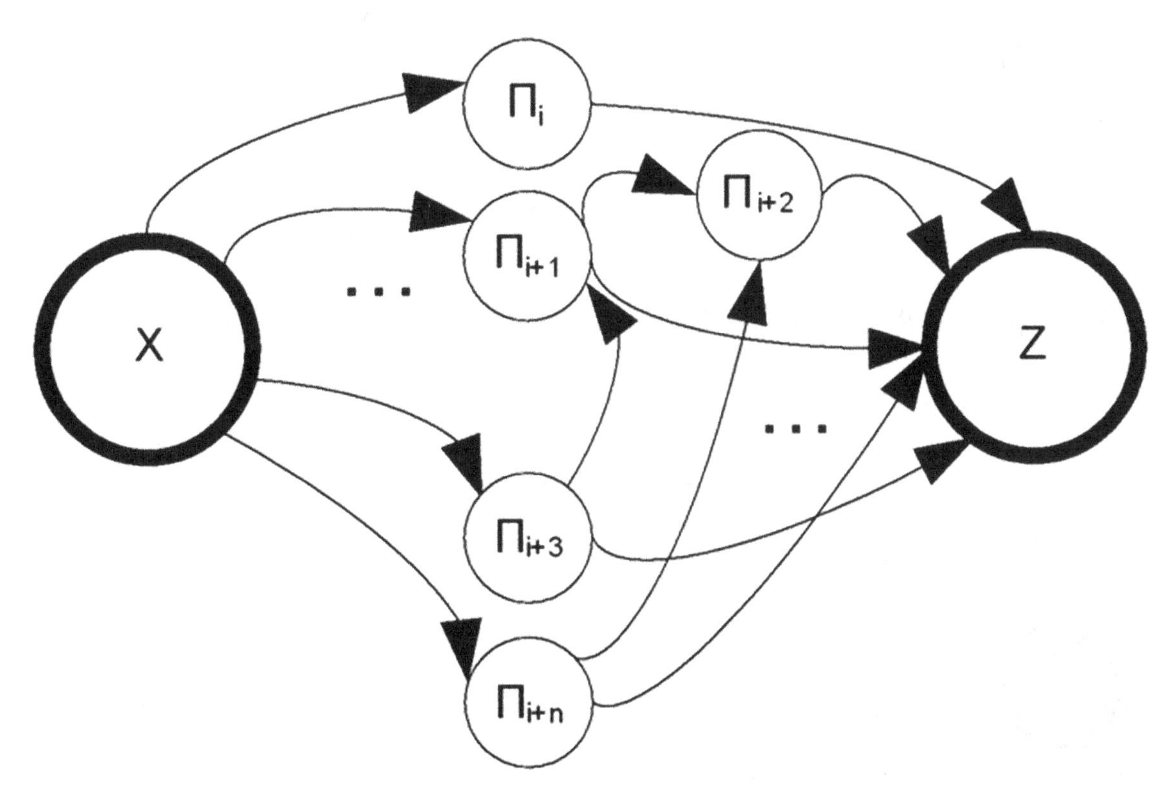

To model we propose to use directed graph that has meta-nodes that correspond to input and output information, nodes that represent transformers and directed edges that represent possible paths that allow obtaining required output information providing available input information (see Figure 3).

Figure 4. Used notation

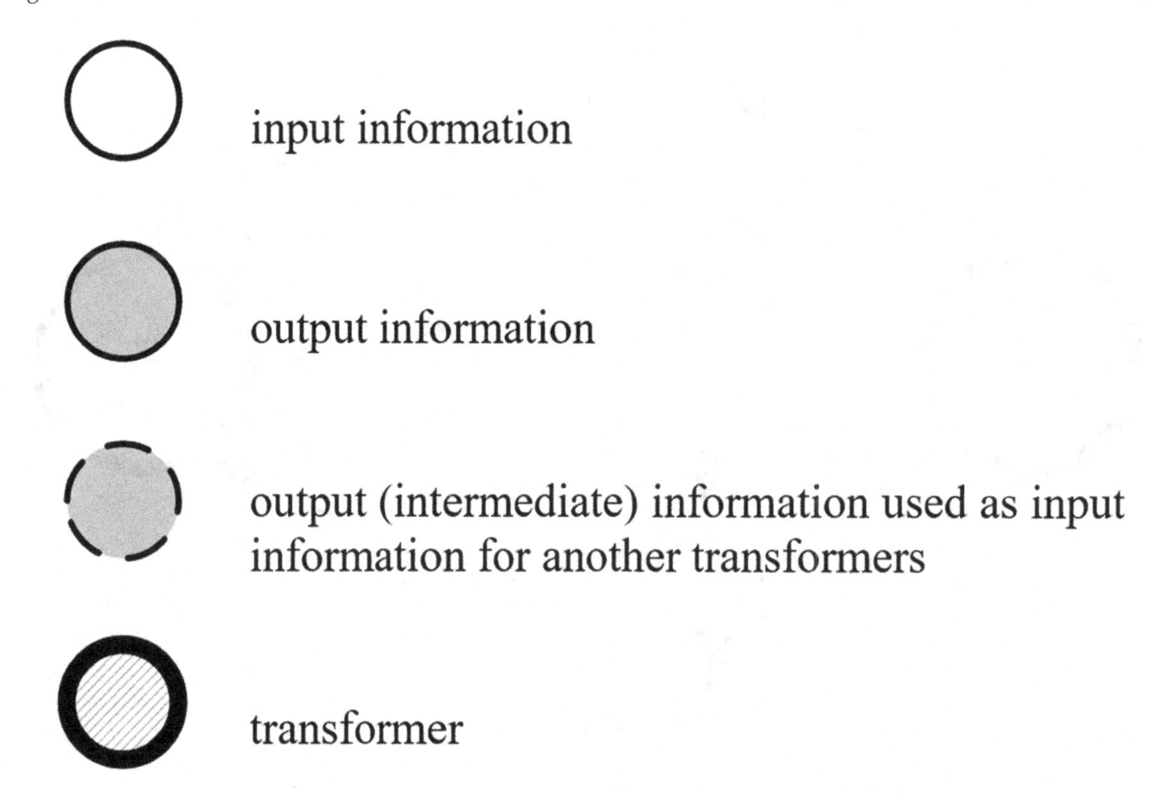

Notation

Next step is to provide formal description of different assessment technique. To do this, we use notation presented on figure 4.

Representation of Techniques

Using this notation, we have described the most popular safety assessment techniques used for NPP I&C evaluations and endorsed in international standards. Figure 5 shows representation of Failure Modes, Effects and Diagnostics Analysis (FMEDA).

The following symbols are used on Figure 5:

- $\{\lambda C\}$ – set of component failure rates;
- dBOM – bills of material;
- dCD – circuit diagrams;
- dSC – safety concept;
- dDS – datasheet (used to find out component failure rates if they are provided by component vendor);
- dMIL – MIL217F (used to find out conservative component failure rates if they are not provided by component vendor);

Figure 5. Representation of FMEDA

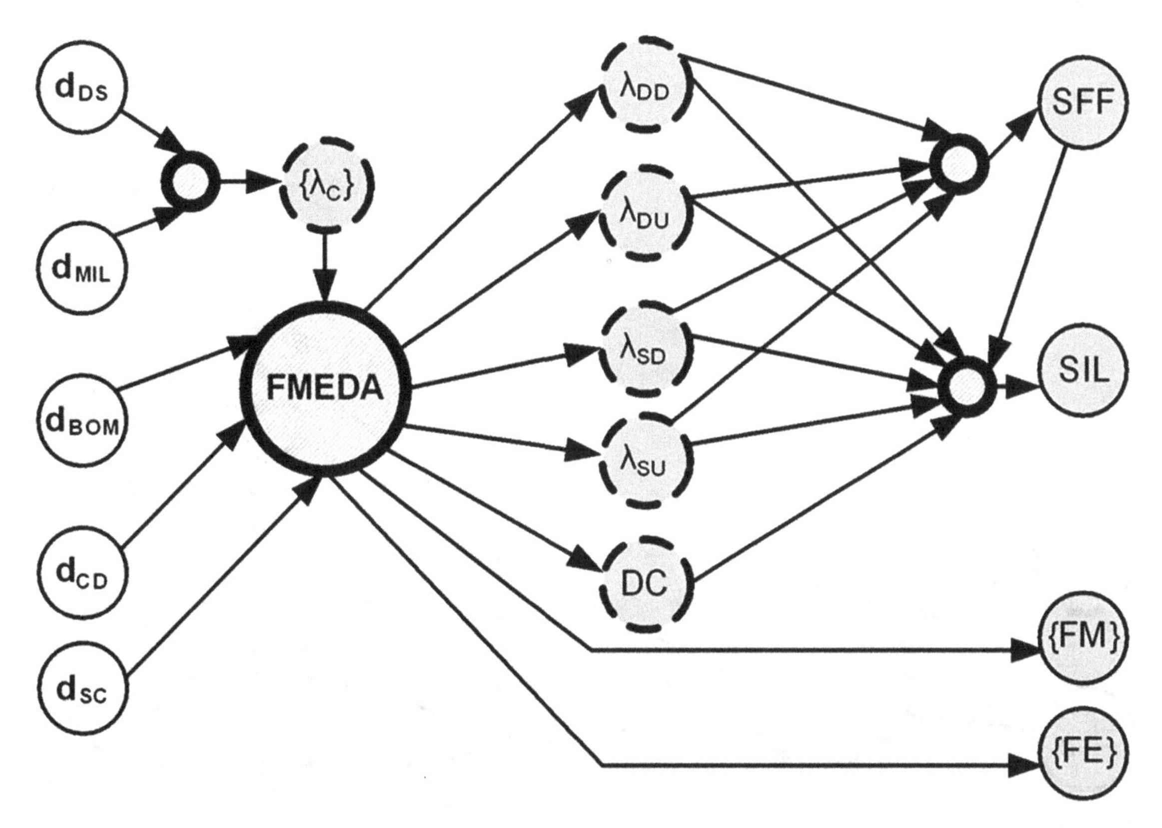

Figure 6. Representation of RBD

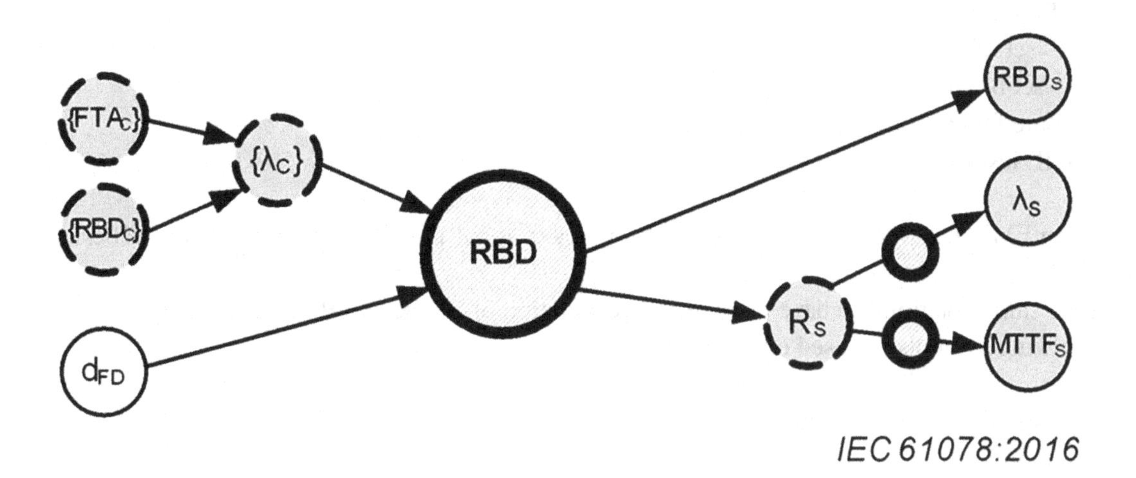

- λDD, λDU, λSD, λSU – failure rates (dangerous detected, dangerous undetected, safe detected and safe undetected);

Figure 7. Representation of FTA

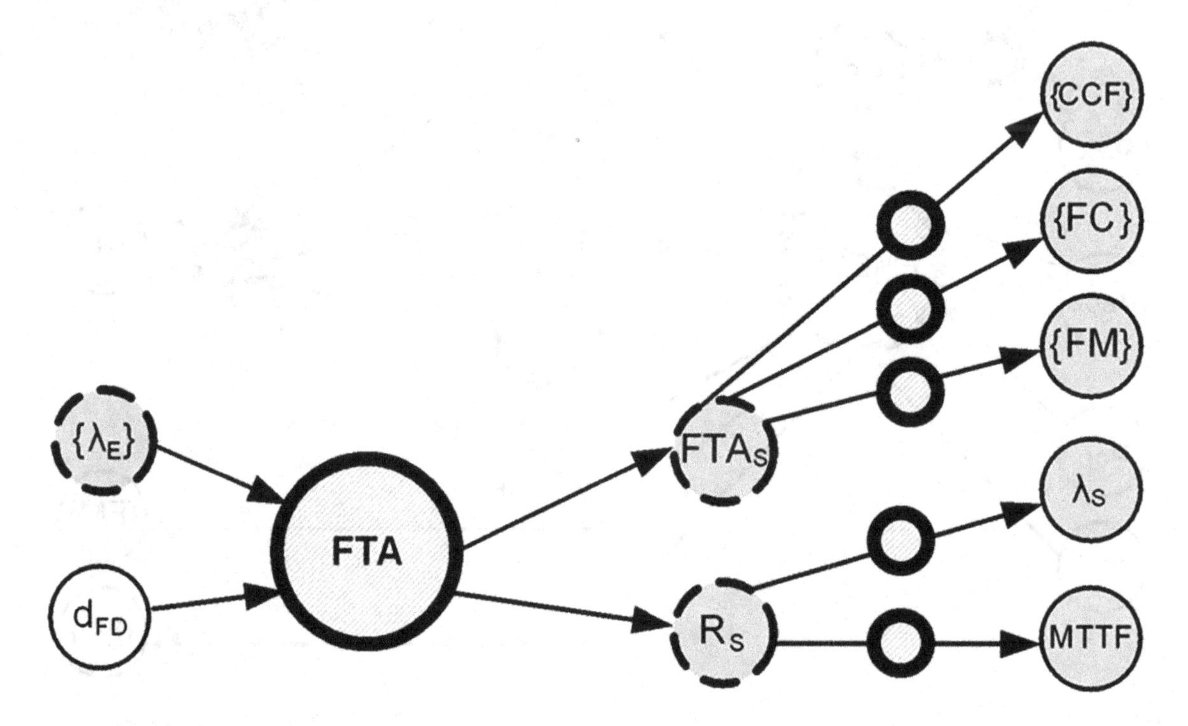

- DC – diagnostic coverage;
- {FM} – set of failure modes;
- {FE} – set of failure effects;
- SFF – safe failure fraction;
- SIL –SIL level.

Details on FMEDA application, its steps and sources of input information are given in (Babeshko, 2015).

Using the same notation, we have prepared representation of Reliability Block Diagram (RBD).

The following symbols are used on Figure 6:

- {λC} – set of component failure rates;
- dFD – functional description;

- {RBDC} – set of RBDs of components;
- {FTAC} – set of fault trees of components;
- RBDS – system (subsystem) RBD;
- RS –reliability;
- λS –failure rate;
- MTTFS – mean time to failure.

Using reliability block diagrams to assess safety is a possible but conservative approach because failure of some component could possibly have effect on reliability but not on safety. A good overcome is to use safety block diagrams instead, although this technique is still not endorsed by any international normative document.

Figure 7 shows representation of Fault Tree Analysis (FTA) using specified notation.

The following symbols are used on Figure 7:

- {λE} – set of initial event probability rates;
- dFD – functional description;

Figure 8. Representation of HAZOP

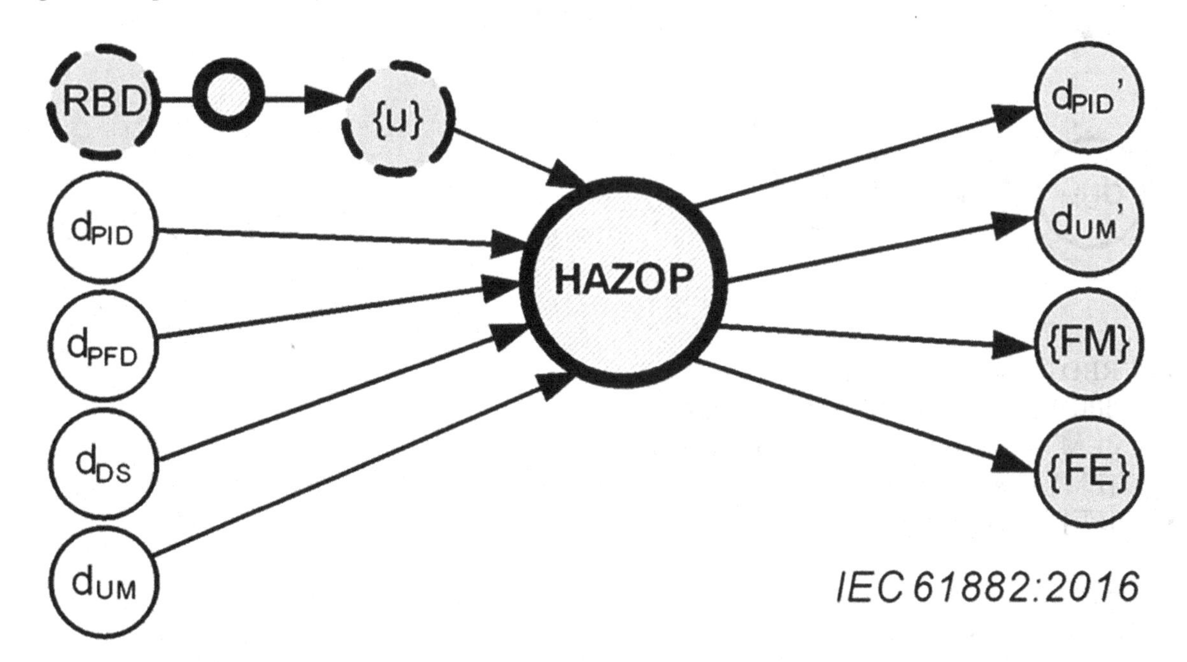

- RS –reliability;
- FTAs –fault tree;
- {CCF} – common cause failures;
- {FM} – set of failure modes;
- {FC} – set of failure causes.

Figure 8 shows representation of Hazardous Operations (HAZOP) using specified notation. The following symbols are used on Figure 8:

- dPID –piping and instrumentation diagram (P&ID);
- dPFD –process flow diagram;
- dDS –datasheet;
- dUM –user manual;
- {u} – set of units (i.e. reactor, turbine, vessel);

Figure 9. Representation of Markov Models

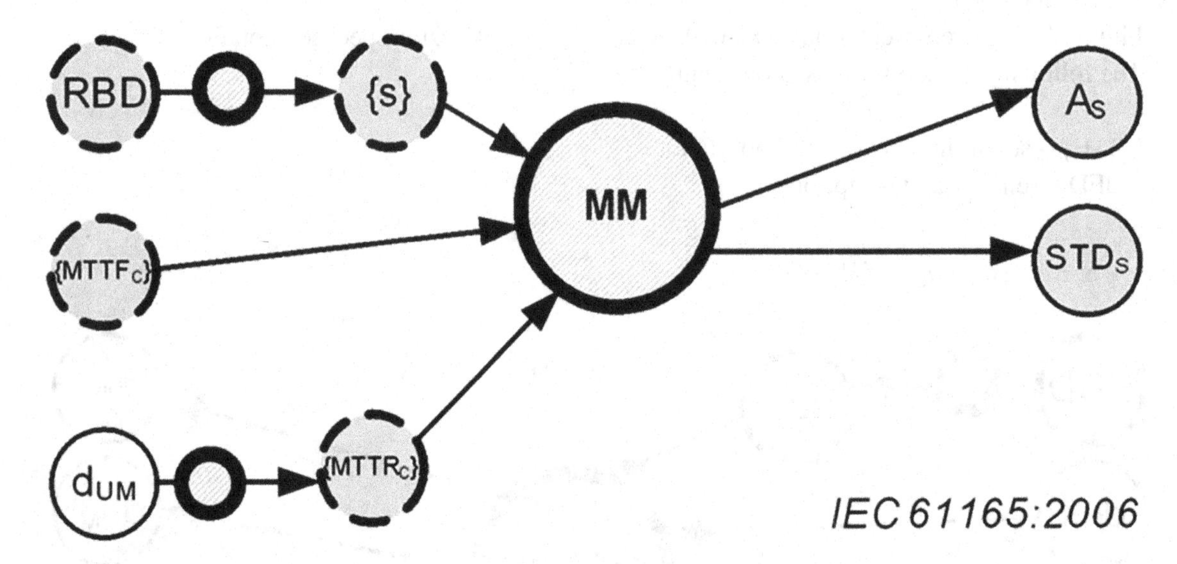

- RBD – results of RBD analysis;
- dPID' – updated P&IDs with markups;
- dUM' – updated user manual with markups;
- {FM} – set of failure modes;
- {FE} – set of failure effects.

Figure 9 shows representation of Markov Models (MM) using specified notation. The following symbols are used on Figure 9:

- {s} – set of possible states;
- RBD – RBD results;
- dUM –user manual;

Figure 10. Graph with several methods

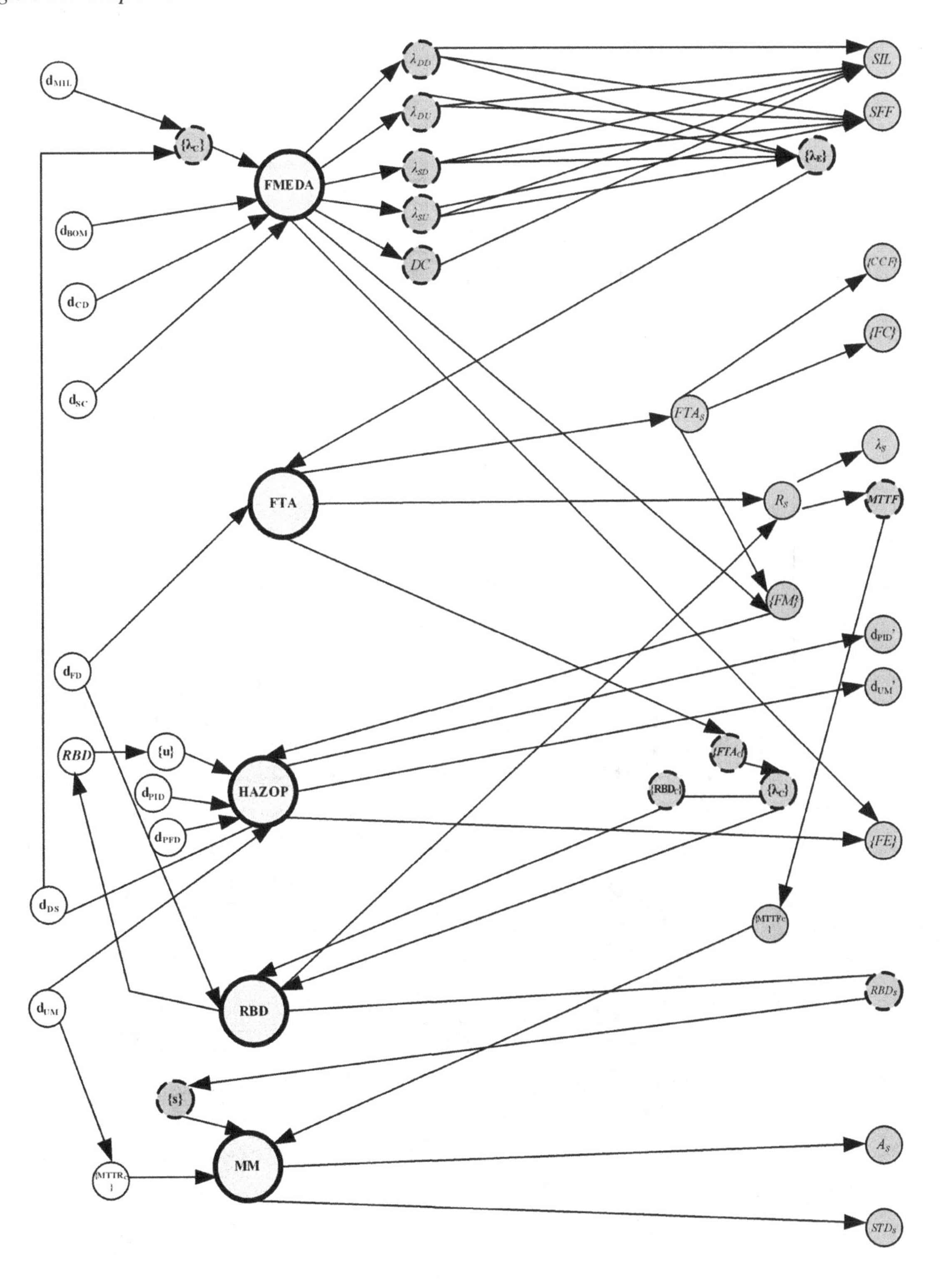

Figure 11. Adding new method to the model

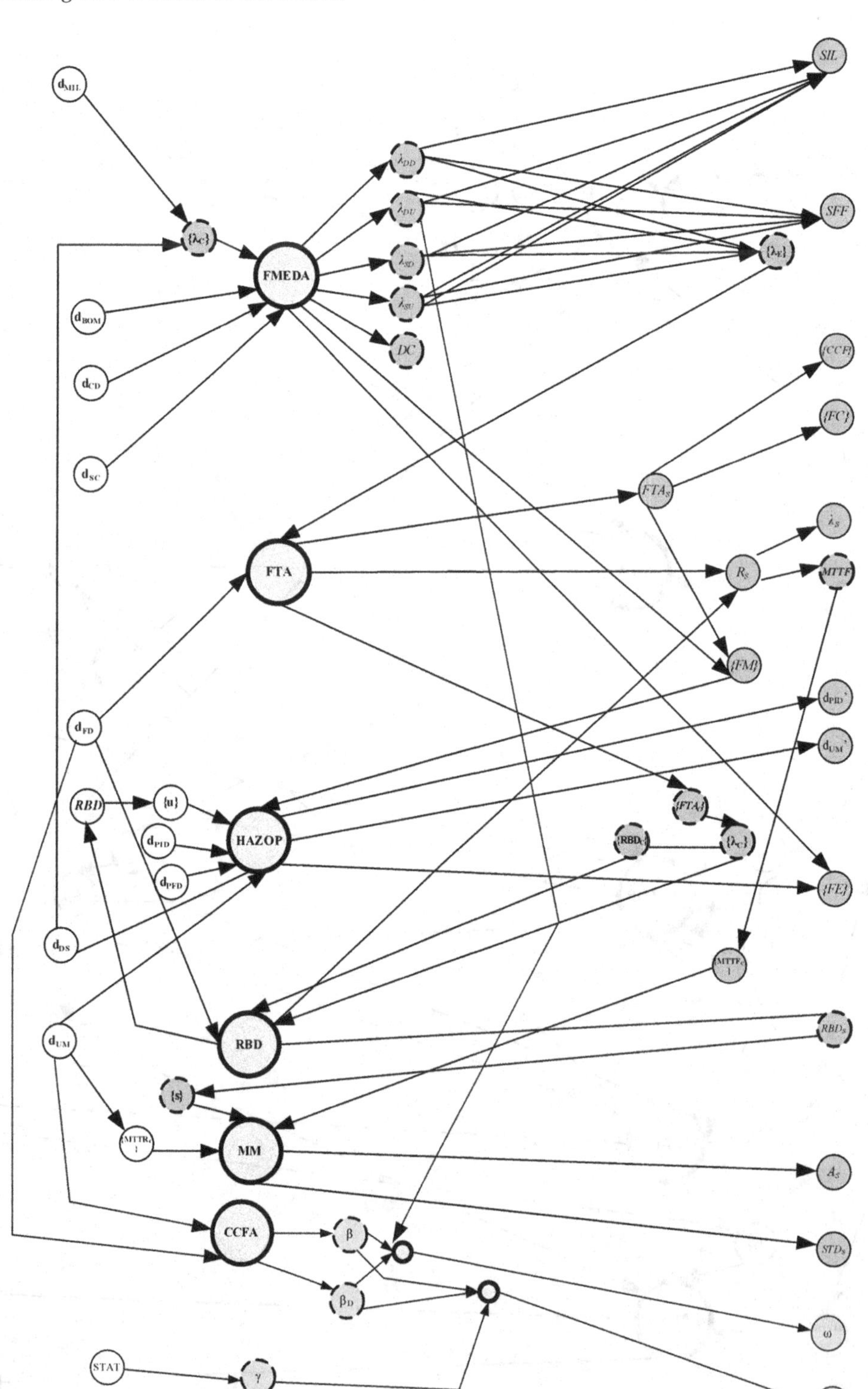

- {MTTFC} – set of components' mean time to failure;
- {MTTRC} – set of components' mean time to repair;
- AS – availability;
- STDS – state transition diagram.

Figures 5-9 have several common inputs and/or outputs. Connecting them together, we obtain general graph model shown on figure 10.

This model could be further updated by adding to it different assessment techniques. For example, there is no universally agreed-upon standard for performing software FMEA (SFMEA) (Goddard, 2000), although it is being widely used during NPP I&C safety assessments. Members of verification teams performing software FMEA projects are encouraged to use the software standards that are relevant for their projects and document their own SFMEA technique. Such technique could be added to our model

Figure 12. Representation of software FMEA

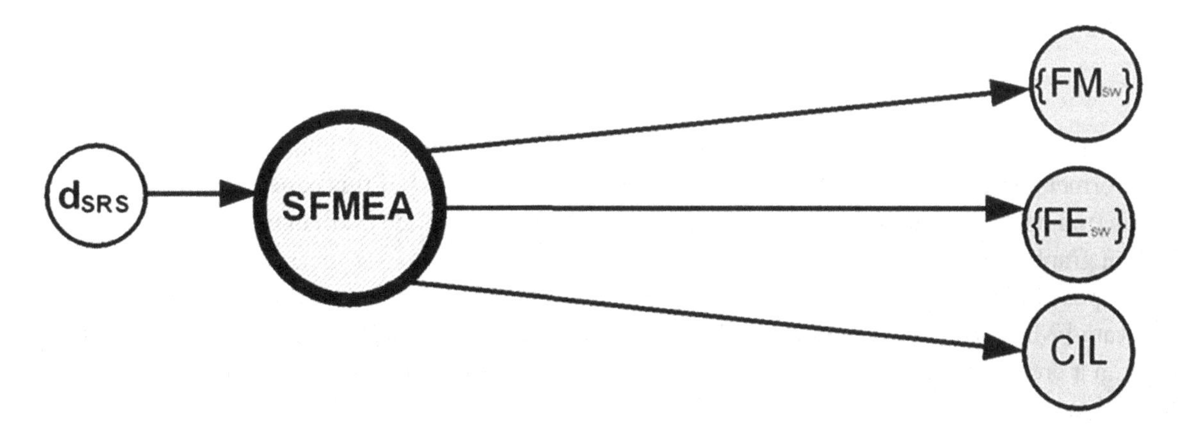

Figure 13. Representation of Software FIT

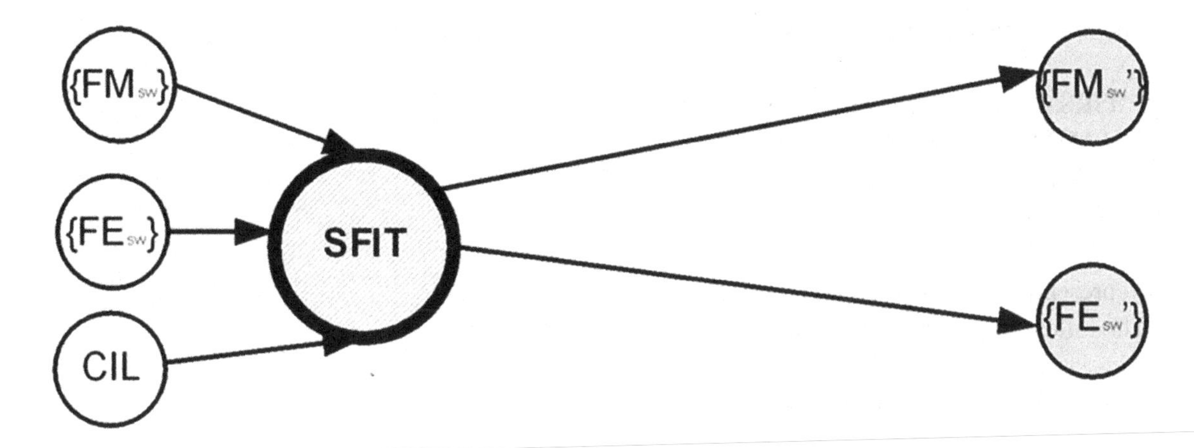

Figure 14. Combination of software FMEA and software FIT

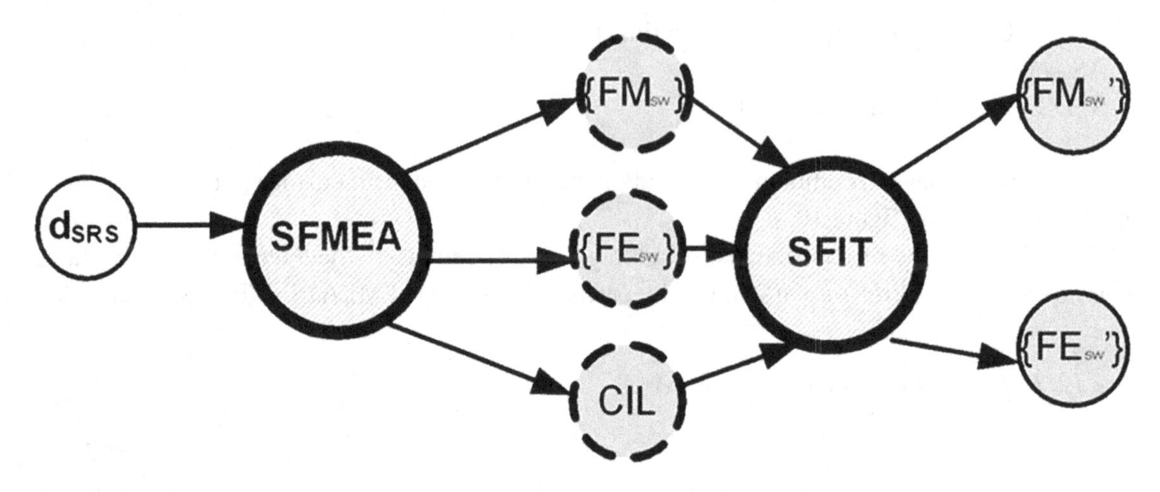

by using the same approach: providing formal description of input and output information, as well as transformers. Example of SFMEA is given in the case study section of this paper.

To find optimal techniques (or sets of techniques) required to obtain particular safety indicator, well-known graph theory approaches could be used.

Proposed idea could also be used for assessment of NPP I&C software. One of possible methods, software FMEA assesses the ability of the system design, as expressed through its software design, to react in a predictable manner to ensure system safety. Issues to be checked during software FMEA are the following:

- Identifying missing software requirements;
- Check correct usage of different data types in different blocks (boolean, signed and unsigned integer, floating point);
- Check possible loopbacks;
- Identifying software response to hardware anomalies;
- Identifying software response to incorrect user actions.

Representation of software FMEA using notation presented in previous section is given on figure 12. The following symbols are used on Figure 12:

- dSRS – software requirements specification;
- {FMSW} – set of software failure modes;
- {FESW} – set of software failure effects;
- CIL –critical items list.

Representation of software FIT is given on figure 13.
The following symbols are used on Figure 13:

- {FMSW} – set of software failure modes;

- {FESW} – set of software failure effects;
- CIL –critical items list;
- {FMSW'} – set of software failure modes didn't detected by diagnostics;
- {FESW'} – set of software failure effects that weren't expected.

Representation of software FMEA and software FIT combination is given on figure 14.

SOLUTION AND RECOMMENDATIONS

A problem of assessment and assurance for safety important I&C systems is still challenging due to the fact that such systems consist of interconnected complex components with different functions and different nature; moreover, the majority of modern I&C systems are being FPGA-based, hence, it is impossible to perform their assessment without consideration of all the special features for all the technologies used.

This approach implies identification of all possible discrepancies, on the basis of product and life cycle processes, and their assessment via application of FMECA, FTA, IMECA and other techniques.

FUTURE RESEARCH DIRECTIONS

The future research and development directions are the following:

- Development of a tool that supports joint application of the different techniques (RBD, FMECA, gap analysis);
- Implementation of tool-based calculation of metrics for choosing the optimal set of applicable methods to ensure reliability and safety of FPGA-based I&C systems.

CONCLUSION

Elements of assessment techniques used for NPP I&C safety assessment were presented. Systematization and orchestration of safety assessment techniques allows to improve solving of the assessment task by providing possibility to achieve higher accuracy and trustworthiness.

Next steps are devoted to further development and clarification of techniques, models and tool support of suggested notation-based composition and assessment of safety.

REFERENCES

Babeshko, E., Bakhmach, I., Kharchenko, V., Ruchkov, E., & Siora, O. (2017). Operating Reliability Assessment of FPGA-Based NPP I&C Systems: Approach, Technique and Implementation. *Icon (London, England)*, E25–E66862. doi:10.1115/ICONE25-66862

Babeshko, E., Kharchenko, V., Odarushchenko, O., & Sklyar, V. (2015) Toward automated FMEDA for complex electronic product. *Proceedings of the International Conference on Information and Digital Technologies (IDT 2015)*, 22-27. 10.1109/DT.2015.7222945

Babeshko, E., Kharchenko, V., Sklyar, V., Siora, A., Tokarev, V. (2011). Combined Implementation of Dependability Analysis Techniques for NPP I&C Systems Assessment. *Journal of Energy and Power Engineering, 5*(42), 411-418.

Goddard, P. L. (2000) Software FMEA techniques. *Proceedings of 2000 Reliability and Maintainability Symposium.*

IEC (2004) IEC 62138, Nuclear power plants – Instrumentation and control important for safety – Software aspects for computer-based systems performing category B or C functions.

IEC (2006) IEC 60880, Nuclear power plants – Instrumentation and control systems important to safety – Software aspects for computer-based systems performing category A functions.

IEC (2009) IEC 61226, Nuclear power plants - Instrumentation and control important to safety - Classification of instrumentation and control functions.

IEC (2010) IEC 61508, Electric / Electronic / Programmable Electronic safety-related systems, parts 1-7.

Illiashenko, O., & Babeshko, E. (2011). Choice and Complexation of Techniques and Tools for Assessment of NPP I&C Systems Safety. *Icon (London, England), 2011*(19), E19–E43484. doi:10.1299/jsmeicone.2011.19._ICONE1943_194

Lee, S. J., Choi, J. G., Kang, H. G., & Jang, S.-C. (2010). Reliability assessment method for NPP digital I&C systems considering the effect of automatic periodic tests. *Annals of Nuclear Energy, 37*(11), 1527–1533. doi:10.1016/j.anucene.2010.06.009

MIL-HDBK-217F, Military Handbook: Reliability Prediction of Electronic Equipment

Misztal, A. (n.d.). Connecting and applying the FTA and FMEA methods together. *Some problems and methods of ergonomics and quality management,* 153-163.

StJohn-Green, M., Piggin, R., McDermid, J. A., & Oates, R. (2015) Combined security and safety risk assessment - What needs to be done for ICS and the IoT. *Proceedings of 10th IET System Safety and Cyber-Security Conference.* 10.1049/cp.2015.0284

Chapter 9
Cyber Security Assessment of NPP I&C Systems

Oleksandr Klevtsov
https://orcid.org/0000-0001-5665-5039
State Scientific and Technical Center for Nuclear and Radiation Safety, Ukraine

Artem Symonov
https://orcid.org/0000-0001-6971-523X
State Scientific and Technical Center for Nuclear and Radiation Safety, Ukraine

Serhii Trubchaninov
State Scientific and Technical Center for Nuclear and Radiation Safety, Ukraine

ABSTRACT

The chapter is devoted to the issues of cyber security assessment of instrumentation and control systems (I&C systems) of nuclear power plants (NPP). The authors examined the main types of potential cyber threats at the stages of development and operation of NPP I&C systems. Examples of real incidents at various nuclear facilities caused by intentional cyber-attacks or unintentional computer errors during the maintenance of the software of NPP I&C systems are given. The approaches to vulnerabilities assessment of NPP I&C systems are described. The scope and content of the assessment and periodic reassessment of cyber security of NPP I&C systems are considered. An approach of assessment to cyber security risks is described.

INTRODUCTION

The problem of the information and cyber security assurance of nuclear facilities, including NPP, is becoming more relevant. The basis for the implementation of various practical measures for protection of the NPP I&C systems against cyber threats is the assessment of cyber security.

DOI: 10.4018/978-1-7998-3277-5.ch009

Cyber security assessment allows to identify possible vectors of cyber-attacks and existing weaknesses in the protection of NPP against cyber threats. Based on the results of the assessment, appropriate cyber security measures are realized for increasing the security of the NPP I&C systems and for reducing the probability of a successful cyber-attack with dangerous consequences for the NPP safety.

The goal of the chapter is the consideration of the main components of the cyber security assessment of the NPP I&C systems:

- Assessments of potential cyber threats;
- Vulnerabilities assessments of the NPP I&C systems;
- Assessment of cyber security measures; and
- Assessment of cyber security risks (if using of risk-based approaches).

BACKGROUND

Cyber security assessment of the NPP I&C systems is a complex multicomponent task, the solution of which requires an integrated approach and implementation of measures in several areas:

- Development of legislative and regulatory framework;
- Compliance with the general principles of cyber security assurance (e.g., defense-in-depth, graded approach, cyber security policy, cyber security culture, etc.);
- Creation of cyber security teams at I&C systems development companies and NPPs;
- Cyber security assessment;
- Implementation of cyber security measures (including design measures) during the development, manufacturing, implementation, operation and decommissioning of NPP I&C systems;
- Development of procedures and training for response to cyber security incidents; and
- Reporting and investigating cyber security incidents in order to take appropriate measures and industry decisions for prevention of the spread of such incidents to other NPPs and their recurrence in the future.

Cyber security assessment is an important stage that precedes the development and implementation of specific measures to ensure cyber security assurance of the NPP I&C systems. The assessment also allows determining the adequacy of realized measures for protection against cyber threats.

Many international and national documents contain requirements and a description of the procedure for cyber security assessment.

Requirements to cyber security assessment are contained in the document IAEA, 2018, according to that such assessment should be performed for each phase of I&C system life cycle to identify potential threats as well as vulnerabilities and weaknesses. Also, each organization that is responsible for development, deploying, operation, maintenance or decommissioning of I&C systems or their components should perform periodic cyber security assessment and audit.

IAEA, 2016 provides a detailed description of the procedure, scope and content of a cyber security assessment of nuclear installations, including vulnerabilities assessment.

IEC, 2014 requires the cyber security assessment of the final design of I&C system, as well as periodic reassessment of risks and security controls during the operation of I&C system.

NUREG, 2004 is prohibited for the public access, however, according to its name, it contains provisions for self-assessment of cyber security recommended for US NPP.

NIST, 2008 does not apply directly to the NPP I&C systems, however, it is one of the most detailed manuals containing general recommendations for cyber security assessment. The document includes a description of the methods of analysis, identification of the targets of cyber-attacks, identifying of vulnerabilities, planning and conducting a cyber security assessment (including testing).

Particular interest is the study published in the report by Masood, R., 2016, where cyber risk scenarios and threat modeling for NPPs are considered.

CYBER THREATS

The cyber security assessment starts with the identification of potential cyber threats that may have a negative impact on the functioning of the NPP I&C systems and, as a result, on the safety of the NPP itself.

Cyber-attacks, viruses, software Trojans, unauthorized modifications of software and data are the threats for NPP since they can dangerously affect the operation of NPP and the technological processes that are fulfilled under the control of these I&C systems. Therefore, the enterprises-developers of the NPP I&C systems and operating organizations should apply measures for protection against cyber threats in case of the purchasing of elements and components, during of development of the I&C systems and during the operation of the NPP I&C systems.

The basic classification of malicious acts committed against computer systems that affect physical protection and nuclear and radiation safety is given in IAEA, 2011:

- Information gathering attacks aimed at planning and executing further malicious acts;
- Attacks disabling or compromising the attributes of one or several computers crucial to facility security or safety; and
- Compromise of one or several computers combined with other concurrent modes of attack, such as physical intrusion to target locations.

The presented classification is general. It is advisable to develop more detailed classifications of potential threats for a better understanding of the scope of the solved tasks for cyber security assurance and the subsequent development of measures for protection against these threats.

In general, two large groups of cyber threats can be considered:

- Cyber threats at the stage of development of I&C systems;
- Cyber threats at the stage of implementation and operation of NPP I&C systems.

In each of the indicated groups, several main types of cyber threats can be considered by the type of impact on the I&C systems, important for the safety of nuclear power plants.

Cyber Threats at the I&C Development Stage

This group includes such malicious impacts that occur during the development of I&C systems or their components (i.e., hardware and software), and results of these impacts then adversely affect the operation of the I&C systems at nuclear power plants. Let us consider in more detail the possible types of threats of this group.

Type 1: Trojans in own developed software. The presence of certain malicious software code that can be included into the own developed I&C system software by someone from the development group is a significant danger for the subsequent operation of the I&C system. Therefore, it is necessary to monitor the program code as much as possible at all stages of software development (i.e., forming of software requirements, software design, software coding, integration of software and hardware). One of the means of such monitoring is verification, which is mandatory for I&C system software, which is important for NPP safety. However, verification, increasing the probability of detecting Trojans, does not ensure their complete absence in the software. Therefore, explicit two-way tracing is a necessary part of the verification to ensure needed safety elements are not excluded (forward trace) and that unwanted functionality is not introduced (backward trace).

A single system can include both software important for safety and software that is not important for safety. However, nobody can exclude the possibility of a negative impact of software that is not important for safety on software important for safety. Verification of software that is not important for safety is usually not required in many countries. Thus, this software becomes a convenient object for introducing of certain malicious program code into it, which subsequently can have a negative impact on other software components important for safety used in the same system, and, as a result, on the operation of the I&C system as a whole. It is necessary to give attention to such software from the point of view of cyber security, since this software is the most vulnerable. Therefore, the controls should be based on the classification of the resident system not on the functionality of software module subsets.

Type 2: Trojans in purchased software. Currently, in any I&C system important for NPP safety, some purchased software is used. At the same time, the probability of presence of Trojans in this software is not excluded. When dealing with a finished product, the I&C system developer cannot monitor the software development process at all stages (i.e., forming of requirements, design, coding, integration) and analyze the software source code. The analysis of the executable code available to the developers of the I&C system is a difficult task, which is usually solved only partly with use of scan tools that look for malicious code signatures.

Type 3: The negative impact on I&C system software from software development tools. Various software development tools (e.g., code generators, compilers, translators, etc.) can generate incorrect source or executable program code, both due to unintentional internal errors in these tools and as a result of malicious functions intentionally embedded in them. Therefore, it is necessary to monitor the correct functioning of the development tools themselves and a detailed check of the automatically generated source and executable code.

Type 4: Threats of object-oriented programming. In case of using object-oriented programming for the development of NPP I&C system software, it must be taken into account that, as a rule, a significant number of standard objects offered by the developer of the programming language or by third-party developers will be used in the program code. The problem is that the code of standard objects can theoretically contain malicious commands, which under certain conditions can adversely affect the operation of the system. Code analysis of standard objects is a rather complicated task.

Type 5: Hardware Trojans. Malicious and intentionally hidden modifications of electronic devices (e.g., microcircuit or its internal programmable configuration) can be done. Such a modification can change the functionality of modules that contain programmable components or are based on Field-Programmable Gate Array (FPGA) technology, which leads to disruption of their functioning (e.g., as a result of unpredictable failures or defects) and, thus, can adversely affect functioning of I&C system used this device.

The introduction of Trojans in hardware is possible at the stages of development and production of the electronic devices themselves or the whole system built on the base of these devices.

Hardware Trojans, depending on the location, can directly affect the following components:

- Processor (change the order of execution of commands);
- Memory, including its interfaces (change the values stored in the memory; block read/write operations for certain memory areas);
- Input-output system (affect intra-system communications or communications with external components);
- Power supply circuit (change the voltage and current supplying a microcircuit or module, thereby causing failures); and
- Time loop (affect the synchronization of the functioning of various modules, thereby leading to errors and system failures).

The effect of hardware Trojans can vary from minor violations in the operation of devices containing such a Trojan to dangerous failures of a system that contains such devices (e.g., if hardware Trojan is embedded in the rector trip system). Possible consequences can be divided into the following groups:

- Changing the functionality of the device (by adding extra logic or deleting part of the existing logic), which usually leads to subtle errors that are almost unpredictable;
- Decreasing of productivity as a result of deliberate changes of the device technical characteristics using the hardware Trojans;
- Information leakage through open or hidden channels;
- Failure of performing of operations by a device that impedes the performance of I&C system functions and is usually caused by an unexpected lack of bandwidth, processing power, or power supply; and
- Physical damage, turning off the device or changing its configuration.

Cyber Threats at the I&C Operation Stage

This group of threats includes harmful effects that are realized during the implementation and operation of I&C system at NPP and adversely affect its operation. We list the main types of threats at this stage.

Type 1: Negative impact through the data networks (e.g., Internet or local networks). Such impacts may include the intrusion of ordinary viruses that are not aimed at a specific I&C system or targeted planned attacks on a specific I&C system or a specific NPP. The first type of threats is quite successfully neutralized by the use of antivirus software and firewalls. Scan tools must be kept up to date to address new threats. However, this only reduces the risks and does not guarantee complete protection, since it does not protect against penetration using vulnerabilities of the software itself. In particular, new viruses pose a threat, because of the absence of appropriate information in the databases of antivirus software. In addition, these tools are not effective enough for protection against threats when the attack has a specific target, is well planned and exploits software vulnerabilities.

In this regard, the software of I&C systems important for NPP safety should be designed in such a way as to eliminate or minimize vulnerabilities. It is advisable to completely eliminate the connection of I&C systems important to NPP safety with the Internet. For safety systems, only one-way (outgoing) information transfer via data transmission networks to other I&C systems should be implemented. For systems important to safety, it is necessary to limit the data transfer with systems that are not important to safety and apply firewalls. It is advisable to develop special requirements for cyber security of systems that are not important for security, in order to prevent a negative impact on systems important to security.

It is necessary to assess the impact of protection tools (e.g., antivirus programs, firewalls, etc.) on the traffic capacity of communication channels and on the productivity of systems important to NPP safety since the protection tools should not adversely affect the operation of I&C systems.

Note that networks can also be used by attackers to obtain sensitive information or privileged remote access as preparation for subsequent attacks. Prevention of such actions is an important component of the overall protective measures against computer attacks.

Type 2: Intrusion of malicious software or data from portable devices (e.g., mobile phones, tablets, laptops, etc.) or external storage media (e.g., CD, DVD, flash drives, memory cards, portable hard drives, etc.) during operation. Portable devices can be used to damage of information or intrusion of malware into various computer systems of NPP via wireless connections (e.g., Wi-Fi or Bluetooth), as well as by connecting such devices to computers via USB ports. In this regard, the use of portable devices should be limited, implementing, in particular, appropriate protection of wireless connections and their isolation from networks to which I&C systems important for safety are connected. Generally, it also requires devices control programs that secure the device when not in use and scans them prior to use.

The most striking example of the negative impact of using an external data medium is the well-known Stuxnet, which was inadvertently entered into the system from a personal flash drive of a Siemens employee. This is the first known computer worm that intercepts and modifies the information flow between programmable logic controllers of the Simatic S7 brand and workstations of the SCADA-system Simatic WinCC from Siemens. The uniqueness of the virus lies in the fact that it was designed strictly for a certain configuration of hardware (i.e., it was launched in a specific system) and for the first time in the history of cyber-attacks physically destroyed the infrastructure (i.e., disabled uranium enrichment centrifuges by sudden changes in their rotation speed). This (or similar) computer worm can potentially be used as a means of unauthorized data collection (espionage) and sabotage in other process control systems of critical infrastructure (e.g., industrial enterprises, power plants, airports, etc.). The existence of such a pathway for intrusion of malicious software requires organizations to set rules and implement special procedures for use of external storage media and portable devices.

Type 3: Negative impact of the test equipment. Test equipment is connected to I&C systems important for NPP safety during the testing, recovery or maintenance. Interacting with the hardware and software of the I&C system, this equipment is a potential source of or way to distribute harmful influence on I&C system important for NPP safety.

In case of interacting of test equipment with the NPP I&C system, it is necessary to exclude the possibility of distortion of data or I&C system software, as well as the possibility of intrusion of malicious software or data into the I&C system.

Testing equipment may include its own software, which should be monitored for the absence of malicious functions, viruses, or Trojans that could have a negative impact on the I&C system important for NPP safety.

Type 4: Malicious actions carried out directly by personnel of NPP or third-party organizations. Such actions can be either unintentional or deliberate. In turn, deliberate actions can range from sabotage of an individual employee to well-planned diversion.

If third-party specialists are involved in the installation, commissioning, or repair of the NPP I&C systems, their actions should be strictly monitored to avoid violating the confidentiality, integrity, availability, and suitability of hardware, software, and system data.

Type 5: Distortion of information from sensors. Incorrect information from the sensors can ultimately lead to disruption of the technological process. Of course, attention should be paid to checking the sensors themselves. However, besides this, it should be borne in mind that the sensors are usually located at a considerable distance from the central part of the I&C system (e.g., some radiation monitoring sensors are located even outside the NPP), and this simplifies the possibility of unauthorized access to them in order to damage or distort transmitted information. Therefore, it is necessary to provide protective and monitoring measures and means of monitor over unauthorized access to sensors and measures of checking the reliability of data received from sensors.

Type 6: Incorrect update of the NPP I&C system software. Often there are situations when during the operation of the I&C system the refinement, modification or updating of the software is needed. At the same time, there is a risk of introducing malware or unintentional errors into the system. There is a known incident that occurred at Hatch NPP in the state of Georgia (USA) and consisted of the fact that the reactor was stopped for 48 hours after an incorrect software update on one computer. The computer was used for monitoring of the chemical and diagnostic data received from the main control system. The software in the computer was modified in order to synchronize the data stored in it with the data in the control system. When the computer was restarted after a software update, all data was deleted both in it and in the associated control system. In turn, the security system interpreted the lack of data as a loss of cooling water and as a result initiated an emergency shutdown of the reactor.

In this incident, several important shortcomings appeared at once:

- A mistake of the developer who did not foresee the possibility of such a development of events;
- The unpreparedness of the technician directly performing the software update, since he was not aware of the impact of this software on the operation of other systems;
- The negative impact of a less important system on safety important system; and
- Incorrect design of both systems and the implementation of their software without considering the principles of cyber security.

The described case clearly demonstrates what dangers may be caused by software modification during the operation of the I&C system and how errors in one system can adversely affect the operation of other systems. At the same time, such threats can be not only unintentional (as in the considered example), but also be the result of a deliberate malicious attack.

The presented list of types of threats is not exhaustive and can be expanded. However, the performed analysis shows the main possible paths of computer attacks, demonstrates their diversity and allows us to assess the complexity of protection against all possible threats.

ASSESSMENT OF VULNERABILITIES OF I&C SYSTEMS

It is of utmost importance to analyze the cyber security vulnerabilities of I&C systems and software at each stage of the I&C systems life cycle.

First of all, it is necessary to assess the vulnerability and security measures against physical unauthorized access to I&C systems (e.g., the presence or absence of locks and seals on cabinet doors, alarms on opening cabinet doors, etc.). This is important because, in the presence of such vulnerabilities, insiders can gain unauthorized access to I&C system hardware and take malicious action against the system.

Then the potential vulnerability of an I&C system to unauthorized access to the I&C system software through the human-machine interface should be assessed. It should include the assessment of the order and means of software monitoring of users' access to the I&C system. Carrying out such an assessment is intended to confirm that in the I&C system:

- Authentication of users (i.e., for reading of configuration files containing details of user accounts) is implemented and anonymous access to I&C system is not allowed;
- Access is provided only to a limited set of the I&C system functions, data and parts that are necessary and accessible to a particular user or user group;
- Remote access to the I&C system is prevented;
- There are no bypass accounts with administrator privileges to allow vendor access to perform maintenance or upgrade;
- Locking the user in case of several unsuccessful attempts to access the I&C system account and informing the responsible staff about unauthorized access to the I&C system is provided;
- The required length, reliability, complexity and frequency of password change are enforced; and
- Procedures for creating, modifying, blocking, and deleting user accounts are defined and documented.

The vulnerability and security measures against any unauthorized connection (including wireless) of any external devices to the I&C system must be assessed, including service and testing equipment, laptops, mobile devices (e.g., phones, tablets, etc.), removable media (e.g., disks, flash drives, memory cards, or portable HDDs).

Vulnerability of and protection for local networks should be assessed against the potential negative impact on NPP I&C system through these networks. At the same time the following checks are performed:

- Correct determination of the security perimeter;
- Correct configuration of network equipment;
- Ensuring the security of ports on network equipment and limiting access to specific ports of I&C system hardware;
- The presence or absence of appropriate network segmentation (e.g.,use of unmanaged traffic in the control networks, availability or inaccessibility of I&C system in a general NPP local network, placing the services of control network directly in this network, etc.);
- Use of firewalls for separation of local networks and the presence or absence of connections that bypass firewalls and audits of compliance with administrative controls;
- The presence or absence of demilitarized zones;
- Filtering of incoming data packets; and

- Restriction of access.

The availability and adequacy of organizational measures and procedures for cyber security must be evaluated. In particular, it determines the presence or absence of:

- A cyber security staff in the organizational structure of the NPP;
- Cyber security documentation;
- Documented backup and recovery procedures;
- Safety assessments;
- Defined order of users registration;
- The simplest and documented network architecture;
- Monitoring of input and output data flows; and
- Monitoring of cyber security events.

The results of the I&C system threats and vulnerabilities analysis are documented in a relevant report and should be taken into account when developing specific security measures in the I&C system cyber security program, which also defines data and software security and availability requirements. If the analysis shows that the system-level security measures are insufficient, then the requirements for software design security measures and/or additional compensatory measures should be defined.

ASSESSMENT OF MEASURES FOR CYBER SECURITY ASSURANCE OF I&C SYSTEMS

Assessment of cyber security during the modification or implementation of a new I&C system may be carried out by a third-party expert organization specializing in this field or by a specially formed evaluation team comprising representatives of the operating organization and the developer of the I&C system. The evaluation team should include I&C system operation, maintenance, nuclear and radiation professionals, and cyber security professionals. Usually, the evaluation team members sign a confidentiality and non-disclosure agreement because disclosure of sensitive information can be used by attackers to identify vulnerabilities and development of possible attack scenarios.

It should be noted that the assessment of cyber security of I&C systems, which have been implemented earlier and are already in operation at the NPP, is carried out within the framework of a similar separate procedure, but it should be taken into account that the design of existing NPP I&C systems may not have provided adequate protective measures against cyber threats.

Assessment of cyber security for I&C system should be performed using the following methods to obtain the necessary information:

- Analysis of documentation (e.g., cyber security policy, program and plan, cyber security assessment reports, cyber security training materials, I&C system technical documentation, inventory lists of I&C system hardware, access control lists, local network architecture, operating systems logs, cyber security incident reports, risk assessments (if any), etc.);
- Interviewing of staff (including administrative management, operational and maintenance personnel, cyber security professionals); and

- Direct survey of I&C systems, their components and local networks.

During the analysis of documentation, the assessment of the compliance of existing cyber security measures with the requirements of regulations, cyber security policy, program and plan is performed. Additionally, the current status of existing cyber security threats is assessed.

Interviews with staff are aimed at assessing staff's awareness and understanding of cyber security policy, program and plan, the effectiveness of training and cyber security assurance, staff's perception of threats and risks, preparedness to respond to incidents, defining duties and responsibilities, efficiency cyber security culture, confidentiality assurance measures.

The direct survey evaluates cyber security measures (taking into account I&C system security levels), implementation of cyber security zones, control of access to I&C system and its components (including spare parts), actual local network architecture, testing and maintenance of I&C system, configuration management, monitoring and logging of cyber security events.

There are three main steps of assessment of cyber security for I&C system.

Stage 1: Gathering Information

At this stage, the evaluation team conducts a preliminary gathering of the information necessary for further detailed analysis. The following is estimated:

- Cyber security policy, program and plan and reports about their implementation;
- The order, extent and results of the analysis of threats and their possible consequences;
- Applying of a graded approach to cyber security, defining cyber security levels;
- Implementation of defense-in-depth in I&C system and NPP in general; and
- Existence of risk assessment and appropriate security measures.

Stage 2: Detailed Analysis

At this stage, the evaluation team performs a detailed analysis based on the previously collected information:

- Staff awareness of cyber security policy, specialized cyber security training and existence of a cyber security team;
- The responsibilities and procedures for access to I&C system by staff, contractors and third-party users;
- Existence of inventory list of I&C system, hardware, network equipment, software, spare parts, their cyber security classification, list of their physical location, functional scheme of the I&C system and scheme of the zonal model (at the same time the conformity of the zonal model to the physical placement of the I&C system and absence of referring of equipment to more than one zone are assessed);
- Implementation of administrative, technical and software means of protection and monitoring of unauthorized access to I&C system, hardware, network equipment, software, spare parts and effectiveness of their implementation;
- Physical separation of I&C system, hardware, networks with different cyber security levels;

- The procedures for testing, adjusting equipment, portable devices and external data carriers at I&C system locations;
- Procedures for disposal of inoperable or replaced technical equipment and destruction of data carriers;
- Limitation of users and software access only to information and resources that are minimum necessary for the successful performance of relevant functions;
- Possibilities of unauthorized access to I&C system, hardware, software and data through networks, modems, wired or wireless connection points, ports, unlocked hardware and workstations, interconnected I&C system, etc.;
- Implementation of vulnerabilities detection through appropriate analysis and testing;
- the adequacy of the cyber security measures implemented in the I&C system and hardware in Accordance with the cyber security plan;
- Realization of additional corrective measures if the necessary cyber security measures cannot be applied within a specific I&C system or hardware;
- Procedures for upgrade, replacement or implementation of new hardware, modification or installation of new software and assessment of the impact of these changes on cyber security;
- Implementation of cyber security measures on site of developers, manufacturers and suppliers of I&C system, hardware, software and element base;
- Cyber security measures during the installation of I&C system, hardware;
- Programs and methods and results of factory acceptance tests after the manufacturing of I&C system, hardware and at the NPP after the installation of I&C system, hardware;
- Cyber security measures during maintenance fulfilled by third party organizations at the NPP; and
- Existence of documented procedures for response (i.e., personnel actions, informing, countermeasures, recovery, investigation, and corrective action) to cyber security incidents, considering external and internal (insider) threats.

Stage 3: Reporting

At this stage, the evaluation team produces an appropriate report that reflects the most detailed results of the analysis, points potential threats, identified weaknesses and vulnerabilities, recommendations for addressing the identified deficiencies, improving cyber security, implementing compensatory security measures, taking into account and implementation of individual cyber security measures in the design of I&C system in the future.

Each organization responsible for the development, implementation, testing, operation, maintenance of I&C system (including components and/or software performs a periodic reassessment of cyber security. It is recommended to do periodic reassessment at least once every two years.

An additional reassessment of the cyber security of I&C system should also be performed in case of:

- Modifications of I&C system;
- The occurrence of a cyber security incident; and
- Identification of new I&C system vulnerabilities or cyber security threats.

The results of periodic reassessment of cyber security are reflected in a report.

It should be noted that an adequate level of protection of confidential information, including the proper labeling, storage, transmission and destruction of all documents, preparatory materials, technical records, draft reports and the final report, must be ensured during the assessment or periodic reassessment and preparation of the report. Appropriate restrictions apply to the use of electronic devices and storage media when preparing a report.

CYBER SECURITY RISK ASSESSMENT

When using a risk-informed approach to cyber security for I&C system, a risk assessment is conducted to identify vulnerabilities to computer attacks affecting this system and to determine the potential consequences of successful exploitation of these vulnerabilities by attackers. Appropriate cyber security measures are based on the results of such risk analysis (if performed). Risk is measured as a combination of the probability of an event and the severity of its consequences.

In accordance with ISO, 2004 the diagram of the relationships between the concepts of threat, vulnerability, and risk is shown in Figure 1.

Figure 1. Security concepts and relationship (adapted from ISO, 2004)

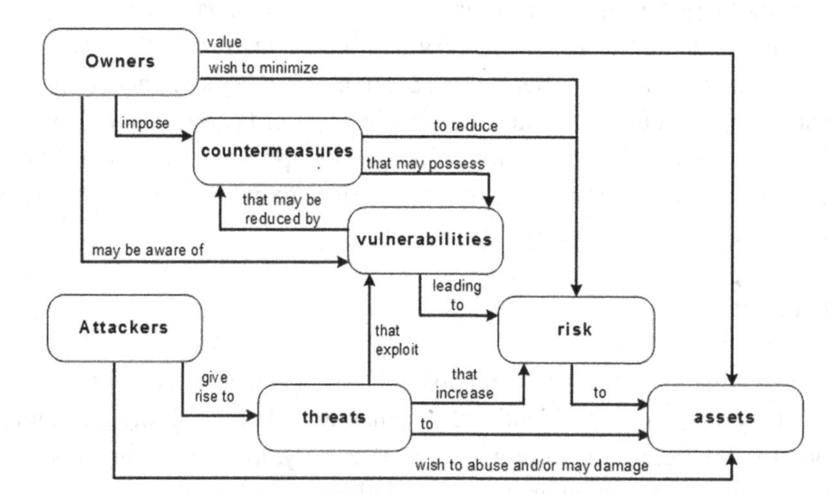

Risk assessment is an important tool for determining the best place to allocation of resources and effort during the analysis of vulnerabilities and the probability of their exploitation. During the risk assessments specific combinations of threats, vulnerabilities and consequences should be identified and documented. Then appropriate additions to the cyber security requirements and measures which are needed to prevent or mitigate the effects of attacks on I&C system should be developed (if it is necessary). In other words, an impact on risk is exercised for risk mitigation.

Risk assessment provides:

- Understanding of potential threats, their sources and their consequences;
- Obtaining information necessary for decision making;

- Identification of vulnerable systems, objects and elements - key factors that create risk;
- The ability to compare risk (with a previously assessed risk; with a risk assessed using a different methodology; with the risk of other similar systems, etc.);
- Sharing information about risk;
- Information necessary for ranking the risk;
- Prevention of new incidents based on the study of the consequences of incidents occurred;
- Selection of risk mitigation methods; and
- The ability of risk assessment at all stages of the I&C system life cycle.

The risk assessment of I&C system involves:

- Determination of perimeter and general conditions (operation of I&C system);
- Identification and characterization of the threat;
- Development of computer attack scenarios;
- Vulnerabilities assessment;
- Assessment of the probability of successful exploitation of vulnerabilities;
- Risk level assessment; and
- Determination of countermeasures.

The I&C system risk assessment should consider the possibility of cyber attacks at each stage of the I&C system life cycle. When assessing, it is necessary to take into account that cyber attacks may affect an individual I&C system or multiple I&C systems and could be combined with other malicious acts causing physical damage. Malicious actions that could change process signals, equipment configuration data, or software should also be considered. In addition, the risk assessment should take into account all attack vectors that could be used for injection of malicious code or data into the I&C systems.

For effective risk assessment of I&C system, it is necessary to assign roles and responsibilities throughout the I&C system life cycle. This process requires focused efforts by multidisciplinary organizations and teams that would be responsible for different stages of the I&C system life cycle and for different directions during the I&C system risk assessment. It is also necessary to fulfil an inventory of the I&C systems, including software, subsystems and components, which are updated and maintained throughout the life cycle of the system. The consequences for safety and security that may result from improper operation or compromise of these components should be identified.

For new nuclear facilities, risk assessment should be performed as part of the design process and accepted before completion of the initial commissioning phase.

Specific risk assessment methods for I&C system are identified in the cyber security program and plan and should be kept up to date.

Over time, the NPP network will be expanded and/or updated, its components changed, and its software applications replaced or updated with newer versions. In addition, personnel and security policies changes will occur. These changes mean that new risks will appear and risks previously mitigated may again become a problem.

Risk reassessments are carried out periodically throughout the life cycle of the I&C system and in case of modification and/or change of threats.

The need for an I&C system risk assessment, the depth of the assessment, and the frequency of risk assessment updates depend on their level of cyber security.

Stages of NPP I&C System Risk Assessment Methodology

In order to understand the threats and vulnerabilities of a particular I&C system, initially it is necessary to analyze this system from functional and technical points of view and identify the relevant factors for dependability assurance that need to be provided. Next, risks associated with these factors should be identified and analyzed.

Stage 1: Determination of Perimeter and General Conditions (Operation of I&C System)

The first step is the determination of the functions and elements of the I&C system, as well as the information that this I&C system operates (e.g. the I&C system which controls the parameters of the coolant uses various kinds of information, such as measures, parameters and computation results, and various functions allowing this computation to be carried out). Based on these data, the aggregate interest in this I&C system for compromise is determined.

It should be taken into account that the main elements of I&C system are linked to a set of various types of objects: hardware, software, networks, human resources, etc.

Stage 2: Identification of Sensitive Elements

In order to ensure proper functioning, it is necessary to determine the sensitivity of each essential element selected in stage 1. The definition of sensitivity is based on various security criteria, such as availability, integrity and confidentiality.

Stage 3: Threat Study

Each object is characterized by a specific threat factors. The threat factor can be characterized by its type (natural, human or environmental) and by its cause (unintentional or intentional). Examples of threat factors include espionage, insiders, hackers, organized crime, terrorists, Nation States, etc.

The threat factor can use various attack methods that should be identified. An attack method is characterized by the security attributes (e.g. availability, integrity, confidentiality) that it can violate and by the likely threat factor. Examples of attack methods: eavesdropping, flooding, software intrusion, login/password attacks, etc.

Each object has vulnerabilities that can be exploited by threat using the relevant attack methods. An example of vulnerabilities associated with the NPP I&C systems:

- The possible existence of hidden functions introduced during the design and development phases (software);
- Use of non-assessed equipment (hardware);
- The possibility of creating or modifying system commands online (networks);
- The network, which can be used for intrusion to system resource software (networks);
- The simplicity of intruding into the site through indirect access routes (premises);
- Operators' failure to comply with instructions (personnel); and
- The absence of security measures during the design, installation and operation phases (organization).

The result of this step should be a formalization of threats, including attack scenarios.

Stage 4: Definition of Security Objectives

At this stage, it is necessary to determine how the threat factors and their attack methods can affect the essential elements identified in stage 1: this is the risk. The risk represents possible damage. It arises from the fact that a threat factor can affect the essential elements by using a given attack method to exploit the vulnerabilities of the objects on which these elements depend.

The security objectives consist mainly in decreasing the vulnerabilities of the objects representing all the remaining risks. Clearly, there is no sense in protecting what is not exposed. However, if the risk potential increases, the strength of the security objectives must also increase. Therefore, these objectives are a completely adapted set of technical requirements.

Stage 5: Determination of Security Requirements

At the final stage, it is necessary to determine the requirements that can be achieved and then demonstrate the necessary level of confidence in ensuring security.

One of the requirements to security assurance may be that the developer must perform an analysis of the stability of the system security functions under a certain impact.

SOLUTIONS AND RECOMMENDATIONS

The cyber security assessment of NPP I&C systems should be required part of complex of measures for NPP I&C systems cyber security assurance. Therefore, it is recommended:

- To develop the regulatory requirements for cyber security assessment of NPP I&C systems;
- To fulfill the cyber security assessment for new and existed NPP I&C systems; and
- to Develop the detailed methods of cyber security risk assessment.

FUTURE RESEARCH DIRECTIONS

The future research directions can be as follows:

- Ongoing studying of cyber security threats;
- Sharing of information about new cyber security threats, attacks and detected vulnerabilities;
- Development of unified reports about cyber security assessment; and
- Creation of expert teams for cyber security assessment and developing knowledge and skills for a sufficient number of experts to support the assessments.

CONCLUSION

Cyber security assurance of the NPP I&C systems is relevant in many countries, including Ukraine. The required preliminary step is a comprehensive assessment of various aspects of cyber security.

Cyber security assessment includes the study of possible cyber threats, assessment of vulnerabilities of I&C system, assessment of adequacy of applied cyber security measures and cyber security risk assessment.

The analysis of the types of cyber threats shows that cyber security assurance is a complex task and solving this task involves many different factors and the various types of cyber-attacks should be taken into account.

Assessment of vulnerabilities in I&C system and confidentiality of appropriate documentation (e.g., cyber security policy, program, plan, etc.) is required to identify the weaknesses in the I&C system and possible ways for potential cyber-attacks.

Identified cyber threats, vulnerabilities of I&C system, and possible weaknesses of confidentiality of appropriate documentation allows to develop appropriate measures for protection of I&C system against cyber-attacks. Adequate protective measures should cover all stages of the life cycle of the NPP I&C systems.

The initial assessment and further periodical reassessments of adequacy and completeness of applied cyber security measures are very important stages of cyber security assessment of NPP I&C systems which allows to keep protection measures on appropriately high level.

In case of use of risk-informed approach the cyber security risk assessment also should be performed.

Competent cyber security assessment is a good fundament for efficient cyber security assurance.

REFERENCES

IAEA. (2011). *Nuclear Security Series No. 17, Computer security at nuclear facilities: reference manual: technical guidance*. Vienna, Austria: IAEA.

IAEA. (2016). *Conducting Computer Security Assessment at Nuclear Facilities*. Vienna, Austria: IAEA.

IAEA. (2017). *NST047. Computer security techniques for nuclear facilities*. Draft.

IAEA. (2018). *Nuclear Security Series No. 33-T, Computer security of instrumentation and control systems at nuclear facilities*. Vienna, Austria: IAEA.

IEC. (2019). *62645, Nuclear power plants — Instrumentation, control and electrical power systems — Cybersecurity requirements*. Geneva, Switzerland: IEC.

ISO/IEC. (2004). *13335-1, Information technology – Security techniques – Management of information and communications technology security – Part 1: Concepts and models for information and communications technology security management*. Geneva, Switzerland: ISO/IEC.

ISO/IEC. (2018). *27005, Information technology — Security techniques — Information security risk management*. Geneva, Switzerland: ISO/IEC.

Masood, R. (2016). *Report GW-CSPRI-2016-03. Assessment of Cyber Security Challenges in Nuclear Power Plants.* Washington, DC: The George Washington University.

NIST. (2002). *Special Publication 800-30, Risk Management Guide for Information Technology Systems.* Gaithersburg, MD: NIST.

NIST. (2008). *Special Publication 800-115, Technical Guide to Information Security Testing and Assessment.* Gaithersburg, MD: NIST.

NUREG. (2004). *CR-6847, Cyber Security Self-Assessment Method for U.S. Nuclear Power Plants.* Richland: PNNL.

ADDITIONAL READING

Baylon, C., Brunt, R., Livingstone, D. (2015). *Cyber Security at Civil Nuclear Facilities. Understanding the Risks*. Chatham House Report.

EPRI. (2018). *3002012752, Cyber Security Technical Assessment Methodology: Risk Informed Exploit Sequence Identification and Mitigation, Revision 1*. Palo Alto, CA, USA: Electric Power Research Institute.

EPRI. (2019). *3002015760, Risk Informed Target Level Topical Guide*. Palo Alto, CA, USA: Electric Power Research Institute.

NEI. (2012). *Identifying Systems and Assets Subject to the Cyber Security Rule* (pp. 10–04). Nuclear Energy Institute. [Revision 2]

NEI. (2017). *Cyber Security Control Assessments* (pp. 13–10). Nuclear Energy Institute. [Revision 6]

NRC (2019). *OIG-19-A-13. Audit of NRC's Cyber Security Inspections at Nuclear Power Plants.*

Park, J. W., & Lee, S. J. (2017). Probabilistic Risk Evaluation of Cyber-Attacks on a Nuclear Power Plant Safety. NPIC&HMIT 2017, San Francisco, CA, USA.

Varuttamaseni, A., Bari, R. A., & Youngblood, R. (2017). *Construction of a Cyber-attack Model for Nuclear Power Plants. 10th International Topical Meeting on Nuclear Plant Instrumentation, Control and Human Machine Interface Technologies.*

KEY TERMS AND DEFINITIONS

Attack: An attempt to destroy, expose, alter, disable, steal or gain unauthorized access to or make unauthorized use of an asset.

Cyber Security: A particular aspect of information security that is concerned with computer-based systems, networks and digital systems.

Cyber Security Incident: An occurrence that actually or potentially jeopardizes the confidentiality, integrity, or availability of a computer based, networked or digital information system or the information that the system processes, stores, or transmits or that constitutes a violation or imminent risk of violation of security policies, security procedures, or acceptable use policies.

Demilitarized Zone (DMZ): Physical or logical subnetwork that contains and exposes an organization's external-facing services to an untrusted network, usually a larger network such as the Internet. The purpose of a DMZ is to add an additional layer of security to an organization's local area network: an external network node can access only what is exposed in the DMZ, while the rest of the organization's network is firewalled. The DMZ functions as a small, isolated network positioned between the Internet and the private network and, if its design is effective, allows the organization extra time to detect and address breaches before they would further penetrate into the internal networks.

Threat: Potential cause of an unwanted incident, which may result in harm to a system or organization.

Vulnerability: Weakness of an asset or control that can be exploited by a threat.

Chapter 10
Diversity for NPP I&C Systems Safety and Cyber Security

Ievgen Babeshko
https://orcid.org/0000-0002-4667-2393
National Aerospace University KhAI, Ukraine

Vyacheslav Duzhiy
National Aerospace University KhAI, Ukraine

Oleg Illiashenko
National Aerospace University KhAI, Ukraine

Alexander Siora
Research and Production Corporation Radiy, Ukraine

Vladimir Sklyar
Research and Production Corporation Radiy, Ukraine

Artem Panarin
Research and Production Corporation Radiy, Ukraine

Eugene Brezhniev
National Aerospace University KhAI, Ukraine & Research and Production Corporation Radiy, Ukraine

ABSTRACT

This chapter presents a cost-effective approach to selection of the most diverse NPP Reactor Trip System (RTS) under uncertainty. The selection of a pair of primary and secondary RTS is named a diversity strategy. All possible strategies are evaluated on an ordinal scale with linguistic values provided by experts. These values express the expert's degree of confidence that evaluated variants of secondary RTS are different from primary RTS. All diversity strategies are evaluated on a set of linguistic diversity criteria, which are included in a corresponding diversity attribute. The generic fuzzy diversity score is an aggregation of the linguistic values provided by the experts to obtain a collective assessment of the

DOI: 10.4018/978-1-7998-3277-5.ch010

secondary RTS's similarity (difference) with a primary one. This most rational diversity strategy is found during the exploitation stage, taking into consideration the fuzzy diversity score and cost of each strategy.

INTRODUCTION

An important task in the development of safety-critical computer systems is achieving a high level of reliability and safety. To protect safety-critical systems from common-cause failures (CCFs) that can lead to potentially dangerous outcomes, special methods are applied, including multi-version technologies operating at different levels of diversity. Diversity is the general approach used for decreasing CCF risks of instrumentation and control (I&C) systems. Differences in equipment, development and verification technologies, implemented functions, etc. can mitigate the potential for common faults. Diversity and defense-in-depth (D3) is the required attribute of Nuclear Power Plant (NPP) I&C systems important for safety. One of the key theoretical and practical problems is the diversity estimation and optimization of used version redundancy capacity. Existing NUREG guidelines published in reports CR-7007 and CR-6303 present the technical basis for establishing acceptable mitigating strategies that resolve D3 assessment findings. These approaches work in the terms of diversity attributes and associated criteria aimed at the potential for CCF vulnerabilities and make possible to choose I&C system architecture based on combinations of diversity criteria. But they do not provide measures (diversity indexes or metrics) to calculate reliability of such a system in the context of CCF. In the report some other techniques for diversity assessment are analyzed, as well as advantages and disadvantages of these techniques in comprising with NUREG-based method are described. Possibilities of their joint applications and tool support are considered.

Although safety-critical computer systems that perform safety and security functions are required to be isolated from the external networks, including the Internet, and, therefore, are protected from many cyber threats, cyber security issues still are to be addressed. Diversity can be used in this process so as to reduce repetition of a single vulnerability and prevent possible cyber-attacks or impair their consequences. NUREG guideline CR-7141 states that deployment of diversity could be used to reduce the risk of a successful exploitation because of a common flaw or vulnerability.

The combined use of reliability, safety and cyber security models, diversity metrics and reliability indexes of system components allows enhancement of estimation sensitivity, making sufficiency criteria for diversity and redundancy more concrete and choice of technical solutions more informed and confident at the early stages of NPP I&C system design. To guarantee required level of dependability, safety and cyber security of computer-based systems for critical (safety-critical, mission-critical and business-critical) applications a diversity approach is used. This approach implies development, choice and implementation of a few diverse design options of redundant channels for created system. Probability of CCF of safety-critical systems may be essentially decreased due to selection and deployment of different diversity types on the assumption of maximal independence of redundant channels realizing software-hardware versions.

Risk of CCF is the main factor of reducing redundant I&C systems dependability. Diversity and defense-in-depth is the required principle of development for NPP I&C systems important for safety, first of all, reactor trip systems (Jonson, G., 2010). Diversity is the general approach used for decreasing CCF risks of I&C systems, because differences in hardware and software components, development and verification technologies, implemented functions, etc. can mitigate the potential for common faults (Jonson, G., 2010, NUREG/CR-6303, 1994).

One of the key theoretical and practical problems is diversity estimation and optimization of used version redundancy capacity. Diversity related decisions should be made at the first design stages, because ones affect safety and cost of NPP I&C system. There are risks of the inaccurate or untrustworthy assessment of diversity and I&C system safety as a whole.

If diversity indicator is overstated, it causes increasing risks of CCF. If result of assessment is understated, it increases costs unreasonably at the production, implementation and operation stages.

This circumstance calls for that a lot of international and national standards and guides contain the requirements to use diversity in safety-critical systems, first of all, in NPP I&Cs (RTS), aerospace on-board equipment (automatic/robot pilot, flight control systems), railway automatics (signalling and blocking systems), service oriented architecture (SOA)-based web-systems (e-science) etc. (Pullum, 2001; Wood et al., 2009; Gorbenko et al., 2009; Kharchenko et al., 2018; Sommerville, 2011, Kemikem et al., 2018).

Objectives of the chapter are: (a) analysis of the challenges caused by use of the diversity approach in NPP I&C systems in context of FPGA and other modern technologies application; (b) development of multi-version NPP I&C systems assessment technique and tool based on a check-list and metric-oriented approach; (c) case-study of the technique: assessment of multi-version FPGA-based NPP I&C developed by use of RadICS Platform including safety and security issues.

BACKGROUND

In a modern world, there are many various regulations, which, in general case, cover the most important areas widely used by the mankind. It is possible to distinguish those related (in some way) to safety important I&C systems, grouped into several sets to cover general issues of critical I&C systems at various lifecycle stages (including their development, operation and maintenance), cyber security, as well as covering various technology-related aspects.

Application of the modern information and electronic technologies and component-based approaches to development in critical areas, on the one hand, improve reliability, availability, maintainability and safety characteristics of digital I&Cs. On the other hand, these technologies cause additional risks or so-called safety deficits. Microprocessor (software)-based systems are typical example in that sense. Advantages of this technology are well-known, however a program realization may increase CCF probability of complex software-based I&Cs. Software faults and design faults as a whole are the most probable reason of CCFs. These faults are replicated in redundant channels and cause a fatal failure of computer-based systems. It allows to conclude that "fault-tolerant" system with identical channels may be "non-tolerant" or "not enough tolerant" to design faults. For example, software design faults caused more than 80% failures of computer-based rocket-space systems, which were fatal in 1990 years (Kharchenko et al., 2003) and caused 13% emergencies of space systems and 22% emergencies of carrier rockets (Tarasyuk et al., 2011).

The CCF risks may be essential for diversity-oriented or so-called multi-version systems (MVSs) (Kharchenko, 1999) as well if choice of a version redundancy type and development of channel versions are fulfilled without thorough analysis of their independence and assessment of real diversity degree assessed by special metrics, for example, β-factor (Bukowsky&Goble, 1994).

COMMON EVENT AND COMMON CAUSE FAILURES

CCF is an event, when e_f (two or more) channels (versions) of redundant e-channel (e-version) system fail simultaneously, and there is a common reason caused this event. Thus, CCF is a multiple failure (MF). It is an alternative of a single failure (SF). On the other hand, multiple failures occur as a result of not only one (common) cause. Multiple failures may be caused by an influence of a few different reasons if these reasons concur or spread of influence time value is less than a speed of on-line testing and reconfiguration means. In this case MF may be called a common time failure (CTF). Hence, CCF and CTF are multiple failures or common event failures (CEF).

Attributes of the classification form simple hierarchy. CCFs and CTFs may be additionally divided in two groups in accordance with a number of failures (partial and full CCFs, i.e. PCCFs and FCCFs, and partial and full CTFs, i.e. PCTFs and FCTFs) and distinguishability of channel output data on failures, i.e. distinguishable (DCCFs, DCTFs) and undistinguishable (UDCCFs, UDCTFs) failures.

Authors of works related to NPP safety problems, first of all, attend to CCFs analysis. However, CTFs are the important objective of a research, as there are examples of serial failures caused by attacks on vulnerabilities of redundant channels and other reasons. Besides, a very important problem, in our opinion, is the analysis of distinguishability of effect failures, because it allows determining the moment of partial or full CCFs (or CTFs) by simple means of channel output data comparison.

ANALYSIS OF DIVERSITY RELATED STANDARDS

There are the following standards and guides contained requirements to diversity:

- IEC 61513: 2011. NPPs - I&Cs important to safety – general requirements for systems;
- IEC 60880: 2006. NPPs - I&Cs important to safety - SW aspects for computer-based systems performing category A functions;
- IAEA Safety Standards Series No. SSG-39. 2016. Design of Instrumentation and Control Systems for Nuclear Power Plants;
- IAEA Nuclear Energy Series NP-T-3.17. 2016. Application of Field Programmable Gate Arrays in Instrumentation and Control Systems of Nuclear Power Plants;
- IEEE std.7-4.3.2:2016. IEEE Standard Criteria for Programmable Digital Devices in Safety Systems of Nuclear Power Generating Stations;
- IEEE std. 603:2018. IEEE Standard Criteria for Safety Systems for Nuclear Power Generating Stations;
- NUREG/CR-6303:1993. Method for Performing Diversity and Defense-in-Depth Analyses of Reactor Protection Systems;
- DI&C-ISG-02, Diversity and Defense-in-Depth Issues, Interim Staff Guidance, BTP 7-19, Guidance for Evaluation of D&DiD In Digital I&C Systems (USA);
- NP 306.2.202-2015 Nuclear and Radiation Safety Requirements for Instrumentation and Control Systems Important to NPP Safety (Ukraine), etc.

These standards contain general requirements concerning: systems which must/should be developed using the diversity approach (RTSs); types of diversity used to develop NPP I&Cs and to decrease CCF probability; features of the diversity implementation, determination of types and volume of the diversity; assessment (justification) of real level of the diversity in developed systems; drawbacks and benefits connected with the use of the diversity.

The standards are not enough detailed to make all necessary decisions concerning the diversity. It's important to develop additional detailed techniques of assessing diversity and choosing optimal kinds and volume of the diversity according to criterion "safety-cyber security-reliability-cost".

NEW CHALLENGES OF DIVERSITY IMPLEMENTATION IN NPP I&C SYSTEM

Technology and Risks

Modern software/microprocessor (MP)-based and hardware/mixed FPGA-based technologies ensure new possibilities for implementation of the diversity approach (DA), because their application allows to use two additional kinds of the diversity:

- FPGA vs MP (main system is developed using FPGAs, diverse system is developed using MPs);
- FPGA1 vs FPGA2 (different manufacturers Intel, Xilinx, Microsemi, etc., subtechnologies SRAM, Flash, Antifuse, development techniques are used to develop main and diverse systems) (Kharchenko, V., et al, 2011, Kharchenko, V., et al, 2008).

The technologies of FPGA projects development, in particular graphical scheme and library blocks in CAD environment, special hardware describing languages (VHDL, Verilog, Java HDL, etc), microprocessor emulators, which are implemented as IP-cores allow increasing a number of possible options of different project versions and multi-version I&Cs. But they can create additional risks and deficits of safety or transform pre-existed ones caused by features of FPGA technology. Hence, they stipulate necessity to analyze and decrease such risks, to use positive features of new technologies.

Uniqueness of Multi-Version Systems

There are a lot of DA implementations in critical domains (Kharchenko, V., et al, 2011) but:

- MVS component failures occur rarely; it does not allow to use statistical methods to evaluate reliability indicators;
- comparative analysis of failures for different applications is not enough.

It concerns both MP-based and FPGA-based MVSs, but MP-based NPP I&C systems are operated more than forty years, when FPGA-based are operated during last ten-twelve years and are more unique.

Key questions are:

- how we should collect, compare experience of different domains and take into consideration features of DA applications?
- is long time of non-failure operation reliable proof?

Standards Related to D3 Principle

A lot of standards and technical reports contain requirements to diversity and recommendations regarding to assessment of MVSs: IEC and IAEA standards (IEC 61513: 2011, IEC 60880: 2006, IAEA SSG-39: 2016, IAEA NP-T-3.17: 2016 etc), IEEE standards and NUREG guides (IEEE std.7-4.3.2:2016, IEEE std. 603:2018, NUREG/CR-6303:1994, NUREG/CR-7007:2009, etc.), EPRI reports (EPRI 1019183:2009, EPRI 1019181:2009, EPRI 1019182:2010, EPRI 1022983:2011), some national guides, for example (NP 306.2.202: 2015).

The requirements of these documents concern:

- NPP I&C systems which must/should be developed and produced using diversity approach;
- diversity types to decrease a common cause failure probability of NPP I&Cs;
- features, benefits and limitations of DA implementation;
- postulation of necessity regarding: determination of the required diversity volume; assessment (justification) of the real diversity level; risks associated with the use of the diversity.

Existed standards are not enough detailed to make the assessment procedure. The most representative document is NUREG 7007. The main questions are the following:

- what should be specification and severity of regulation for DA implementation?
- how regulated should be requirements and procedures of assessment and development of FPGA-based NPP I&Cs?

Safety Assessment

There is a problem of CCF risks assessment and MVS safety assessment as a whole. Inaccurate assessment either increases risk of a fatal failure (understated assessment) or increases risk of unreasonable costs.

The main question is the following: what indicators (metrics), techniques and tools we should use:

- To assess the actual diversity level and MVS safety,
- To assess cost and limitations of developing and implementing such structures,
- To compare different structures of MVS according to a criterion "safety-cost" and make optimal decision?

CCF Risk Decreasing and MVS Safety

There is a problem of decreasing number of common version faults (CVF). The CVF number (and probability of CCF) may be decreased using several types of the diversity (multi-diversity or "diversity of diversity").

There are problems of a compatibility and dependence of diversity types. Main questions are the following:

- What type (types) of diversity should be used?
- How much versions developers should use to ensure required level of the MVS safety?
- How to take into account dependencies of diversity types and to search regularized set of decisions (sets of diversity types)?

Challenges: Some Conclusions

There are two main theoretical and practical problems of the diversity approach application in NPP I&C systems. Firstly, a problem of the actual diversity level assessment for developed MVSs, reliability safety and taking into account:

- Product/process technologies (types, rate of physical, design and interaction faults);
- System architectures (type and capacity of the applied diversity and redundancy).

Second problem is a choice of product-process diversity types, MVS architecting and configuration of diverse components, etc.

WORK RELATED ANALYSIS

Known works, related to the current problem and taking into account features of NPP I&C systems, are divided into three groups: (1) classification and analysis of version redundancy types and diversity-oriented decisions; (2) methods and techniques of the diversity level assessment and evaluation of multi-version systems safety in context of CCFs; (3) multi-version technologies of safety critical systems development.

1. A set of diversity classification schemes (general, software and FPGA-based) was analyzed in (Kharchenko et al., 2009). First one is based on NUREG technical reports and guides, samples two-level hierarchy and includes seven main groups of version redundancy (Wood et al., 2009): a signal diversity (different sensed reactor or process parameters, different physical effects, different set of sensors); a equipment manufacture diversity (different manufacturers, different versions of design, different CEC versions, etc); a functional diversity (different underlying mechanisms, logics, actuation means, etc); a logic processing equipment or architecture diversity (different processing architectures, different component integration architectures, different communication architectures, etc); a logic or software diversity (different algorithms, operating system, computer languages, etc); a design diversity (different technologies (González et al., 2019), approaches, etc); a human or life

cycle diversity (different design organizations/companies, management teams, designers, programmers, testers and other personnel). Software diversity types are classified in according to following attributes (Pullum, 2001; Volkoviy et al., 2008): life cycle models and processes of development (for example, V-model for main version and waterfall model with a minimum set of processes for duplicate version); resources and means (different human resources, languages and notations, tools); project decisions (different architectures and platforms, protocols, data formats, etc). Next one FPGA-based classification includes the following types of the diversity (Kharchenko&Sklyar, 2008; Siora et al., 2009): the diversity of electronic elements (different electronic elements manufactures, technologies of production, electronic elements families, etc); the diversity of CASE-tools (different developers, kinds and configurations of CASE-tools); the diversity of projects development languages (different graphical scheme languages, hardware description languages and IP-cores); the diversity of specifications (specification languages) and others.

2. There are following methods of the diversity level assessment and evaluation of the MVS dependability and safety (Kharchenko et al, 2009). Theoretical-set and metric-oriented methods are based on: an Eiler's diagram for sets of version design, physical and interaction faults (including vulnerabilities for assessment intrusion-tolerance); a matrix of diversity metrics for sets of different faults (individual, group and absolute faults of versions); calculation of diversity metrics by use of Eiler's diagrams or other data about results of testing and faults of different versions. Probabilistic methods use reliability block-diagrams (RBDs), their modifications (survivability and safety block-diagrams), Markovian chains, Bayesian method, etc. Statistical methods include the following procedures: receiving and normalization of version fault trends using testing data; choice of software reliability growth model (SRGM) taking into account features of version development and verification processes and fitting SRGM parameters; metrics diversity assessment; calculation of reliability and safety indicators. Fault injection-based assessment consists of: receiving project-oriented fault profiles; performing of faults injection procedure; proceeding of data and metrics diversity calculation; calculation of reliability and safety indicators. Expert-oriented methods use two groups of metrics: diversity metrics for direct assessment of versions and MVS reliability and safety (direct diversity metrics); indirect diversity metrics (product complexity metrics and process metrics); values of these metrics may be used to assess direct diversity metrics. Expert methods are added to other techniques founded on interval mathematics-based assessment of diversity metrics and MVS indicators, soft computing-based assessment (fussy logic, genetic algorithms), risk-oriented approach and so on.
3. Multi-version technologies (MVTs) of diversity types selection and application, development of MVSs as a whole are based on (Siora et al., 2009; Wood et al., 2009) use of diversity types and strategies table, a model of multi-version life cycle (MVLC), a special graph of diversity types and their modifications, and procedures of diversity type and volume choice according to different criteria. The set of developed diversity strategies (Wood et al., 2009) consists of three families of strategies: different technologies—Strategy A (digital vs analog), different approaches within the same technology—Strategy B (MP vs FPGA) and different architectures within the same technology—Strategy C (IP-based vs VHDL). Each of the strategy families is characterized by combinations of diversity criteria that may provide adequate mitigation of potential CCF vulnerabilities according to metrics determined in an expert way.

There are a lot of examples of multi-version systems and multi-version technologies application in different safety critical areas. Generalized results of MVS application analysis are presented by the matrix “types of diversity – areas of multi-version I&Cs application” in Table 1 (Wood et al., 2009; Kharchenko et al., 2010).

Table 1. Matrix “types of diversity – areas of multi-version I&Cs application”

Diversity types	**Multi-version I&C systems application**												
	Space		*Aviation*				*Rail. ways*	**Che-mic. indu-stry**	**Defense**	**Power Plants**	**NPPs**		**e-Com-merce**
	Shut-tle	**ISS**	**MC JVC**	**FAA FCS**	**Air-bus A320**	**Boeng 777**	**SCB**	**CCPS**	**MICS**	**Electr. Grid**	**RTS**	**ESFAS**	**WSOA**
Design													
Equipment													
Function													
Human													
Signal													
Software													
Others													

Types of diversity (diversity redundancy) are classified according to NUREG 6303 and painted by different colors. Last row of the matrix corresponds to other types of diversity. MVSs are used in space systems (Shuttle, ISS), aviation equipment (MC JVC, FAA FCS, Airbus and Boeing on-board systems), railway automatics (signaling, centralization and blocking systems SCB), chemical industry (CCPS), defense systems, power plants (electricity grid), NPPs (RTS and ESFAS), e-commerce and e-science (web-systems with diverse target web-services).

A LAW “NEGATION OF NEGATION”: STAGES OF DIVERSITY APPROACH IMPLEMENTATION EVOLUTION IN NPP I&CS

Interesting are the results of transformation of multi-version I&Cs for the last decades in context of hardware-software-FPGA technologies development. There are a few diversity implementation evolution stages in safety-critical NPP I&Cs, in particular, reactor trip systems. Analysis of these stages allows formulating (or demonstrating truth) a law “negation of negation” (Kharchenko et al., 2009) (Figure 1):

- **Stage 1 (1970-1980s):** Use of hardware (hard logic, HL)-based one-version systems and transition from hardware (HW)-based systems with identical subsystems to systems with hardware (HL)-based primary subsystem and software (MP)-based secondary subsystem; it was the first “negation”;

Figure 1. Stages of diversity approach implementation evolution in safety-critical NPP I&Cs

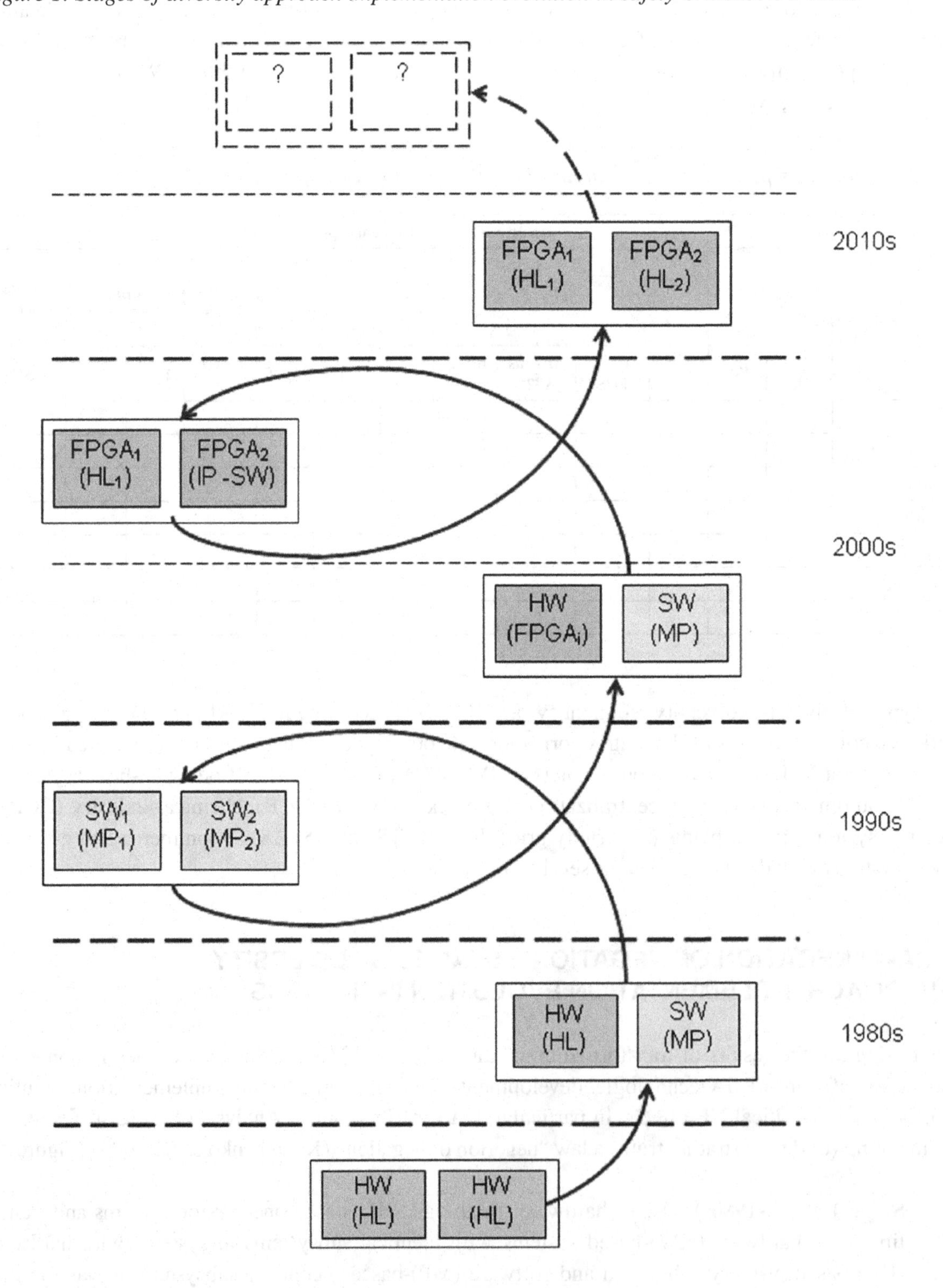

- **Stage 2 (1990s):** Use of primary and secondary subsystems with software (SW) diversity (I&C platforms produced by Siemens, WH and other companies); example of multi-version systems with software diversity is two-version system consisting of subsystems developed using microprocessors Intel and Motorola (languages C and Ada); it completed the first cycle of "negation of negation";
- **Stage 3 (2000s, First Half):** Transition to FPGA-based primary and software-based secondary subsystems with equipment, design and software diversity (first generation of the I&C platforms produced by RPC Radiy); it was next "negation";
- **Stage 4 (2000s, Second Half):** Application of FPGA-oriented soft processors for a primary subsystem and FPGA project developed using HDL-oriented language (hard logic) for creation of a secondary subsystem (next generation of the I&C platform produced by RPC Radiy); it completed the second cycle of "negation of negation";
- **Stage 5 (Beginning Of 2010s):** Application of different FPGAs (hard logic) produced by different manufacturers (and other types of diversity) for primary and secondary subsystems correspondingly; it is next "negation".

What will be the next step? Probably, advancement of electronic technologies, in particular, nanotechnologies, naturally dependable, safe and secure chips will create new perspectives and possibilities for development of diversity-oriented decisions. Microsemi, Intel and other companies inform about creating first chips called nano FPGAs allowing to develop fault-tolerant projects using large-scale means.

AN EVOLUTION OF FPGA TECHNOLOGY AND DIVERSITY APPLICATION IN NPP I&CS

Complex Electronic Components and FPGA Technology for NPP I&Cs Development

An analysis of development and introduction trends of computer technologies to NPP I&Cs has specified a number of important aspects affecting their safety, peculiarities of development, update and licensing. Such trends include, among others (Yastrebenetsky, 2004): introduction of novel complex electronic components (CECs); expanded nomenclature of a software applied and increased effect of its quality to I&Cs safety; realization of novel principles and technologies in I&Cs development; advent of a large number of novel standards regulating the processes of I&Cs development and safety assessment. During recent decades the application of microprocessor techniques in NPP I&Cs design has substantially expanded. Microprocessors are used both in a system computer core and in realization of intellectual peripherals – various sensors, drives and other devices with built-in programmable controllers.

Another contemporary trend is dynamically growing application of programmable logic technologies, particularly, FPGA in NPP I&Cs, onboard aerospace systems and other critical areas. FPGA as a kind of CECs is a convenient mean not only in realization of auxiliary functions of transformation and logical processing of information, but also in execution of basic monitoring and control functions inherent in NPP I&Cs. This approach in some cases is more reasonable than application of software-controlled microprocessors (Kharchenko&Sklyar, 2008). In assessment of FPGA-based I&Cs it should be taken into consideration that application of this technologies somewhat levels the difference between hardware and software, whereas obtained solutions are an example of a peculiar realization of so called heterosystems – systems with a "fuzzy" software-hardware architecture and mixed execution of functions. This circumstance and other features of FPGA technology increase a number of diversity types and enlarge a set of possible diversity-oriented decisions for NPP I&Cs.

FPGA Peculiarities in Context of Dependability, Safety and Cyber Security

FPGA architecture topologically originates from channeled Gates Arrays (GA). In FPGA internal area a set of configurable logic units is disposed in a regular order with routing channels there between and I/O units at the periphery. Transistor couples, logic gates NAND, NOR (Simple Logic Cell), multiplexer-based logic modules, logic modules based on programmable Look-Up Tables (LUT) are used as configurable logic blocks. All those have segmented architecture of internal connections.

System-On-Chip architecture appeared due to two factors: a high level of integration permitting to arrange a very complicated circuit on a single crystal, and an introduction of specialized hardcores into FPGA. Additional hardcores may be: additional Random Access Memory (RAM) units; JTAG interface for testing and configurating; Phase-Locked Loop (PLL) – a frequency control system to correct timing relations of clock pulses as well as for generation of additional frequencies; processor cores enabling creation of devices with a control processor and a peripheral.

An analysis of dependability assurance possibilities in FPGA-based systems permit to determine the following FPGA peculiarities (Kharchenko & Sklyar, 2008; Bobrek et al., 2009, Kharchenko & Illiashenko, 2016).

1. Simplification of development and verification processes: an apparatus parallelism in control algorithms execution and realization of different functions by different FPGA elements; an absence of cyclical structures in FPGA projects; an identity of FPGA project presentation to initial data; advanced testbeds and tools; verified libraries and Intellectual Properties (IP) - cores in FPGA development tools.
2. There are three technologies of FPGA-projects development: development of a graphical scheme by means of library blocks in CAD environment; development of a software model by means of especial hardware describing languages (VHDL, Verilog, Java HDL, etc); development of a program code for operation in the environment of microprocessor emulators, which are implemented in FPGA as IP-cores. It does allow increasing a number of options of different project versions and multi-version I&Cs.
3. **Assurance of Fault-Tolerance, Data Validation and Maintainability Due to Use Of:** redundancy for intra- and inter-crystal levels; diversity implementation; reconfiguration and recovery in the case of component failures; improved means of diagnostic.
4. **Cyber Security Assurance:** FPGA reprogramming is possible only with use of especial equipment. Stability and survivability assurance due to: tolerance to external impacts (electromagnetic, climatic, radiation); possibilities of implementation of multi-step degradation with different types of adaptation. Also, FPGA reprogramming requires physical access to the system, and, therefore, cyber security is assured by physical security which is to be ensured anyway.

FPGA Technology Application in Safety-Critical Systems and NPP I&Cs

Due to these peculiarities area of FPGA technology application has essentially expanded. We can say about an affirmative answer to question "Expansion of FPGA-technology application in safety-critical systems for the last decades: evolution or revolution?" It is confirmed by (Bakhmach et al., 2009):

- A substantial increase of applying the technologies based on programmable logic (FPGA, CPLD, ASIC);

Figure 2. Application of FPGA technology in the NPP I&Cs produced by RPC Radiy

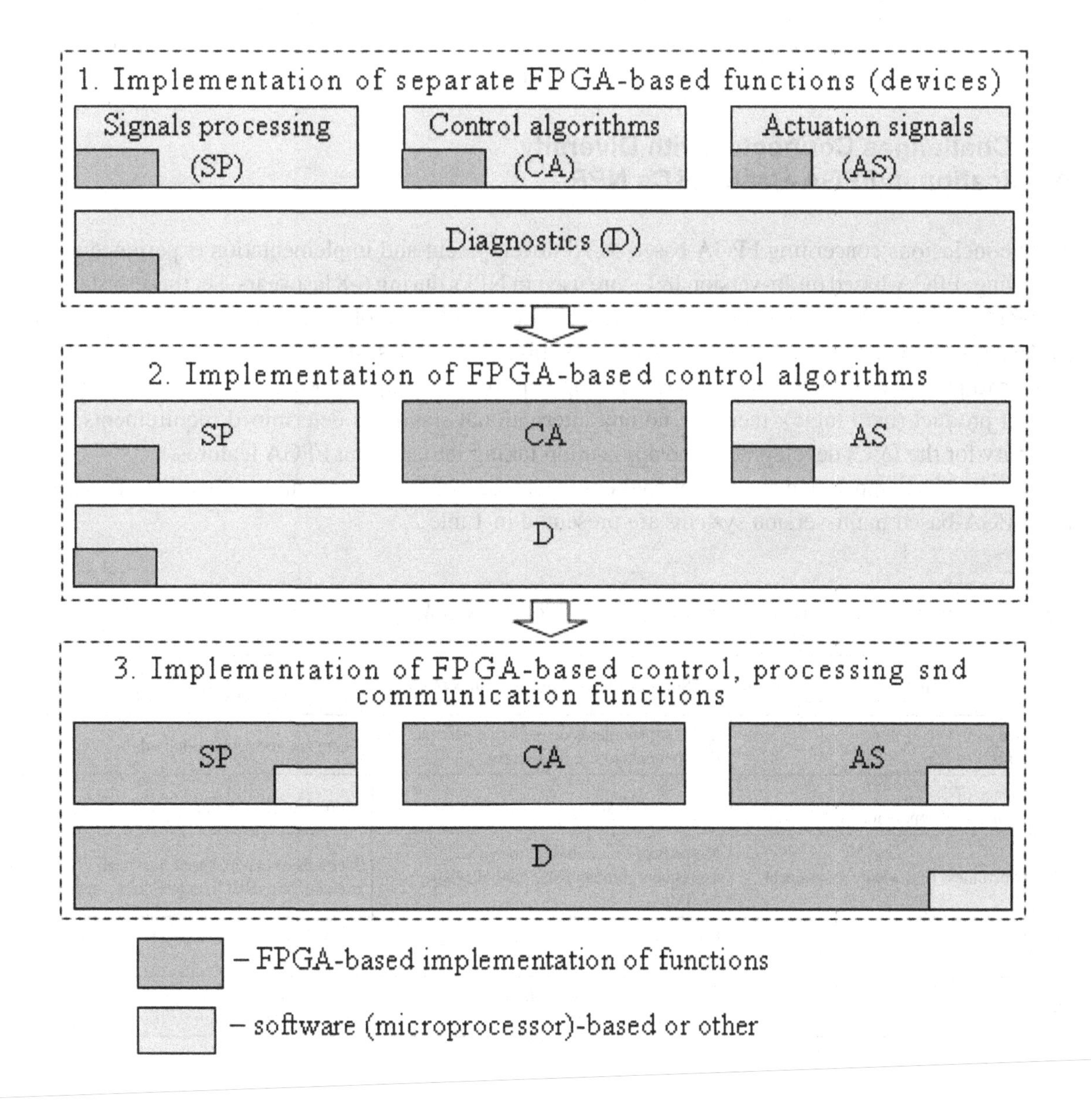

- The FPGA technology is improved and ensures new possibilities to develop more reliable and effective systems; application of the FPGA technology for development of military (B-1B, F-16, etc) and civil aircraft control systems (Boeing 737, 777, AN70, 140), space control systems (satellites FedSat, WIRE; the Mars-vehicle Spirit), etc;
- The application of FPGAs in NPP I&Cs (Ukraine, Russia, Bulgaria: 1999-start, 2002 – 1000, 2006 – 6000, 2008-2010 – more than 8000 chips every year).

Besides, the illustration of FPGA expansion is an evolution of the NPP I&Cs produced by RPC Radiy during 2000-2008 years (Kharchenko&Sklyar, 2008).

There are three stages of the evolution (Figure 2): from implementation of separate FPGA-based functions in I&Cs (signals processing (SP), control algorithms (CA), actuation signals formation (AS) and diagnostics (D), stage 1, and implementation of FPGA-based CA, stage 2, to preferred implementation of FPGA-based SP-, CA-, AS-, D- and communication functions, stage 3.

An analysis of industrial application experience of FPGAs in NPP I&Cs is described in a technical report prepared by EPRI (Naser, 2009, Fink et al, 2011).

Key Challenges Connected with Diversity Application in FPGA-Based I&Cs NPP

Main conclusions concerning FPGA-based MVS development and implementation experience are the following: FPGA-based multi-version I&Cs are used in NPPs during 6-8 last years, i.e. these systems are a new object of analysis and still more unique one; the FPGA technology gives additional possibilities to develop MVSs and ensure high safety and reliability; processes of FPGA project development are similar to processes of SW-based project development. FPGA project product is similar to HW-based project product (hard logic); there are no any international standards determined requirements to use diversity for the I&Cs development and application taking into account FPGA features.

Results of a comparative analysis of challenges caused by development and application of software- and FPGA-based multi-version systems are presented in Table 2.

Table 2. Key challenges for software-based and FPGA-based MVSs

Challenges	Software-based multi-version I&C	FPGA-based multi-version I&C
Detailed standards	There are standards determining general requirements to use of diversity	There are no special standards
Experience of development and operation	More than 20 years	More than 10 years
Trustworthiness of diversity assessment	Methods of expert-based, metrical assessment, probabilistic methods using SRGMs	Methods of expert-based, metrical, probabilistic (RBD), deterministic methods
Development of MVSs	Choice of diversity types, generation of really diverse software versions	Number of diversity types increases
Verification of MVSs	Verification activities volume are significantly increased	Verification is more simple due to simplifying of version verification

MAIN CONCEPTS AND MODELS OF MULTI-VERSION COMPUTING

Taxonomy Scheme of Multi-Version Computing

A set of concepts concerning the diversity may be united by a general term "multi-version computing" on the analogy with a "dependable computing" (Avižienis et al., 2004). Multi-version computing is a type of dependable computing organization based on use of the diversity approach. The taxonomy scheme of multi-version computing, developed taking into consideration concepts in this area, described in international standards, includes the following elements (Kharchenko et al, 2009) (Figure 3).

Figure 3. Taxonomy scheme of multi-version computing

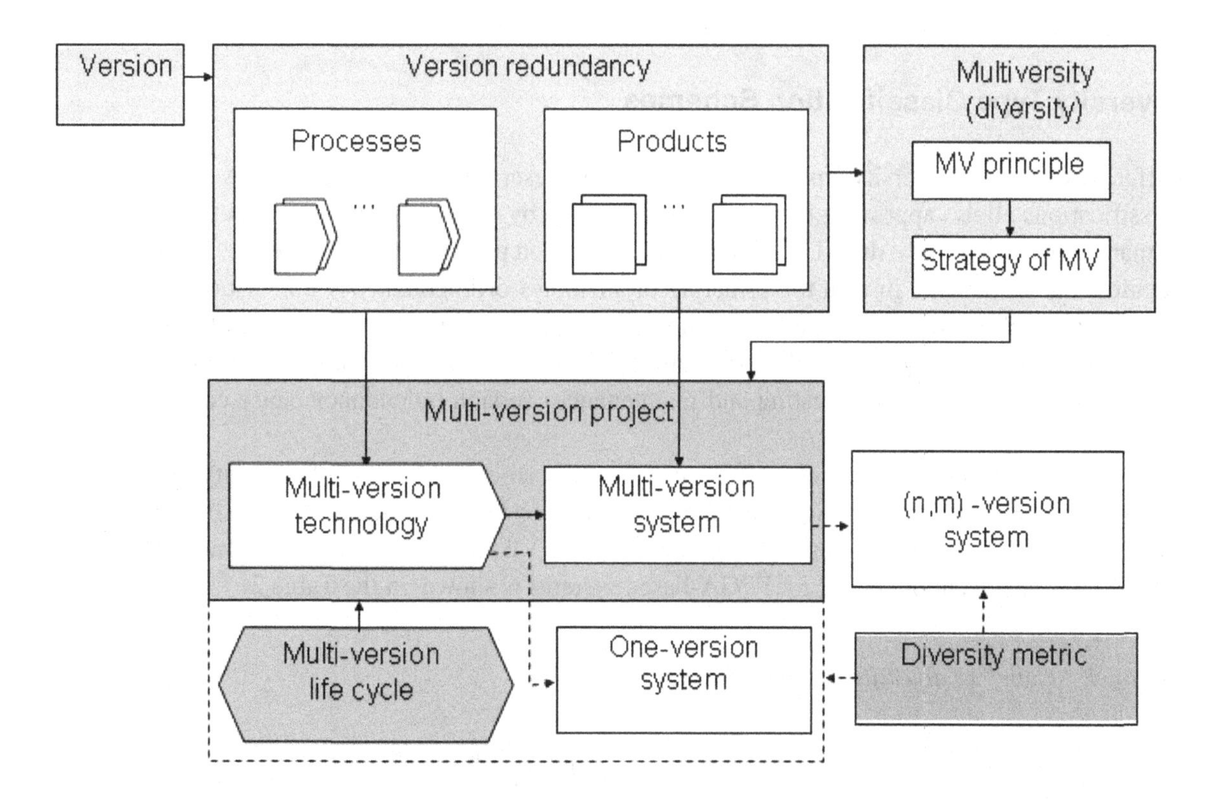

Version is an option of the different realization of an identical task (by use software, hardware or FPGA-based products and life cycle processes); an identical versions of structure redundancy-based system are trivial. Version redundancy (VR) is a type of product and process redundancy allowing to create different (non-trivial) versions; product VR is realized jointly with structure, time and other types of non-version redundancy.

Diversity or multiversity (MV) is a principle providing use of several non-trivial versions; this principle means performance of the same function (realization of products or processes) by two and more options and processing of data received in such ways for checking, choice or formations of final or intermediate results and decision-making on their further use.

Multi-version system (MVS) is a system, in which a few versions-products are used; one-version systems may be redundant but consist of a few trivial versions. Multi-diversion system (MDVS) is MVS, in which two or more VR types are applied. Multi-version technology (MVT) is set of the interconnected rules and design actions, in which in accordance with MV strategy a few versions-processes leading to development of two or more intermediate or end-products are used; thus, for development of MVS MVT should be used, for development of one-version systems multi-version and one-version technology can be used both.

Multi-version project (MVP) is a project, in which the multi-version technology is applied (version redundancy of processes is used) leading to creation of one- or multi-version system (realization of version redundancy of products). Strategy of diversity (MV) is a collection of general criteria and rules defining principles of formation and selection of version redundancy types and a volume or/and choice of MVTs. Besides, important elements of the multi-version computing are concepts "multi-version life cycle", "diversity metric". More detailed interpretation of these concepts will be done below.

Diversity Type Classification Schemes

Different variants of diversity type classifications were described above. The analysis of the considered classifications allows approving that: they are presented by classifications of mixed facet-hierarchical or matrix (network) types; the NUREG-based classification presented in (Wood et al., 2009) is the most detailed and systematic, though the principle of attributes orthogonality is not sustained in full in it; for example, subsets of a design and software, a functional and signal version redundancy are crossed and dependent; a variety of a product (system, hardware and software components) and of a process (technologies of development, testing and maintenance) version redundancy cause complexity of VR selection and MVS development.

More general diversity type classification scheme is so-called "cube" of diversity described by a matrix MVR =|| vrijk|| in three-dimensional space (Figure 4). The scheme has coordinates: a stage of LC (i); a level of project decisions (PD, j) and a type of VR (project decision). Example of two-space matrix presented a cut of "cube" for FPGA-based systems is shown on the Table 3.

Figure 4. "Cube" of diversity-oriented decisions

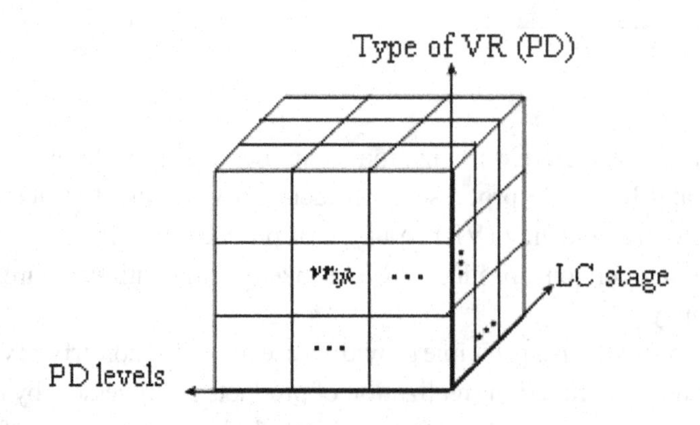

Table 3. Matrix of diversity-oriented FPGA-based decisions

Stages of FPGA-based I&C life cycle	Kinds of version redundancy			
	1 Diversity of electronic elements (EE)	**2 Diversity of CASE-tools**	**3 Diversity of project development languages**	**4 Diversity of scheme specification (SS)**
1 Development of block-diagrams according to signal formation algorithms		1.2.1 Different developers of CASE-tools 1.2.2 Different CASE-tools kinds 1.2.3 Different CASE-tools configurations		1.4.1 Different SSs 1.4.2-1.4.4 Combination of couples of diverse CASE-tools and SSs
2 Development of program models of signal formation algorithms in CASE-tools environment		2.2.1 Different developers of CASE-tools 2.2.2 Different CASE-tools kinds 2.2.3 Different CASE-tools configurations	2.3.1 Joint use of graphical scheme language and HDL 2.3.2 Different HDLs 2.3.3-2.3.8 Combination of diverse CASE-tools and HDLs	
3 Integration of program models of signal formation algorithms in CASE-tools environment		3.2.1 Different developers of CASE-tools 3.2.2 Different CASE-tools kinds 3.2.3 Different CASE-tools configurations	3.3.1 Joint use of graphical schemes and HDL 3.3.2 Different HDLs 3.3.3 – 3.3.8 Combination of couples of diverse CASE-tools and HDLs	
4 Implementation of integrated program model in FPGA	4.1 Different manufacturers of EEs 4.2 Different technologies of EEs production 4.3 Different families of EEs 4.4 Different EEs of family	4.2.1 Different developers of CASE-tools 4.2.2 Different CASE-tools kinds 4.2.3 Different CASE-tools configurations 4.2.4-4.2.15 Combina-tion of diverse CASE-tools and EEs		

This table contains variants of a joint application of one or two diversity types (items 1.4.2-1.4.4, 2.3.3-2.3.8, 3.3.3-3.3.8, 4.2.4-4.2.15; for example, last combinations correspond to 12 = 4 (kinds of EE diversity) x 3 (kinds of CASE-tool diversity)) couples.

DIVERSITY ASSESSMENT TECHNIQUES AND TOOLS

Diversity Assessment Techniques

To analyse diversity assessment techniques it is needed to describe their basic principles and procedures. Further three techniques are analysed: NUREG-A, CLB-A, GMB-A.

NUREG-A Technique and its Extensions

1. Features of NUREG-based assessment technique are the following:
    a. It is based on a diversity classification described in (NUREG/CR-6303, 1994, NUREG/CR-7007, 2009);
    b. It allows to fulfill metric-based assessment of two-version systems;
    c. Diversity is assessed using a value Yes or No (if Yes, there are to subvalues: INT = intentional (X), INH = inherent (i): if Intel (diversity of chip), hence Quartus (diversity of tools));
    d. 2-level analysis procedure is used for types and subtypes of the diversity (attribute and criteria);
    e. Weight of attribute depends on rate of application of the diversity type in I&Cs;
    f. Metric is non-normalized;
    g. Acceptable value of diversity metric equals 1.0.
2. Assessment procedure consists of the following stages:
    a. An expert analyzes design and fills assessment table (X (i) or No);
    b. Diversity metric is calculated ("automatically");
    c. An expert makes decision "accepted/not accepted".

The described technique allows assessing level of diversity using general metrics; values of metrics are determined in advance. But this technique does not permit to calculate safety indicators of MVS safety.

Taking into account the comments on the NUREG-A assessment technique, two approaches to improve the diversity assessment technique can be proposed including modern technological and technical solutions:

- To propose am approach to addition of the NUREG-7007 classification by new subtypes of diversity;
- To improve the existing assessment technique.

The commitment to modernize the NUREG-A technique is based on the belief that its potential is not exhausted, and in addition it is well known to specialists in the design of safe NPP I&Cs.

To avoid possible assessment errors, existing two-level classification could be improved by converting it into a three- or four-level classification. The first two levels are already present in the classification used in the NUREG-A methodology. The third level, an additional level of diversity classification, is designed to take into account new or specific diversity types. The classification elements entering this level will henceforth be called sub-subtypes of the diversity. The fourth level, the level of diversity fragmentation, is intended to uniquely decide on the full or partial implementation of the diversified properties of the given I&Cs.

The expert adds a third or fourth level guided by the following rules:

- If it is necessary to clarify the diversity classification by introducing new properties into it, then they enter the third level;
- If it is impossible to determine unambiguously the presence of diversity, then they enter the fourth level.

Addition of diversity classifications or fragmentation levels is carried out individually for each diversity type or subtype, if necessary.

In general, all functions implemented by I&Cs can be divided into several categories - basic and additional. The basic functions include the implementation of safety functions, and additional functions include monitoring functions and self-diagnostics. If necessary, the number of levels can be increased to three or four, but not more, so as not to reduce the significance of each level. Instead of assessment of the presence of some diversity type or subtype, it is proposed to assess the presence of diversity of basic functions, additional functions or functions of a different level. Such approach will allow to avoid assessment errors when the presence of the diversity of functions of some level is regarded by the expert as the presence of a diversion of the I&Cs as a whole.

After selecting the number of levels, their ranking with the subsequent prioritization is performed. Prioritization of levels consists in assigning weight to each level depending on its number.

Priority can be considered successful if the expert can unequivocally answer the question whether this property is implemented in I&Cs or not implemented diversely.

A part of the table for diversity assessment using the NUREG-A method, in which the level splitting procedure is applied to the "Design" diversity type, is presented in Table 4, where DCE – Diversity Criterion Effectiveness, DCE WT – Diversity Criterion Effectiveness weight.

Tables for other diversity types for the I&Cs assessment using the NUREG-A technique can be obtained in a similar way.

The procedure for improving the diversity classification by "splitting the functions of diversity" includes the following steps:

- Choose the diversity type or subtype that will be divided into levels;
- Split the selected diversity type or subtype into the required number of levels (no more than four);
- Perform ranking of levels;
- Prioritize the levels.

Table 4. Table for assessment of "design" diversity type after the level splitting procedure is applied

Attribute criteria		Rank	DC E WT	Level	Rank	WT
DESIGN	Design					
	Different technologies	1	0.500			
				Base	1	0.667
				Addition	2	0.333
	Different approaches within a technology	2	0.333			
				Base	1	0.500
				Addition	2	0.333
				Other	3	0.167
	Different architectures	3	0.167			
				Base	1	0.667
				Addition	2	0.333

Table 5. Table for assessment of "design" diversity type in accordance with the standard nureg-7007

Attribute criteria		Rank	DCE WT
DESIGN	Design		
	Different technologies (Analog/Digit)	1	0.400
	Different approaches within a technology (MP-MC/FPGA)	2	0.300
	Different architectures (Intel/Xilix)	3	0.200
	Different architecture implementation (Soft/Hard Core)	4	0.100

Soft-processors are the hardware implementation of the processor architecture using FPGA technology, an example of such processor is Nios II by Intel. It is proposed to consider soft processors as one of the possible microprocessors implementations and include it in the assessment table as a diversity subtype for the Design type (Table 5).

Consider the following hardware-software diversity types, which have become widespread in the FPGA- based I&Cs development:

- Diversity implemented using soft processors;
- Diversity of methods and interfaces for information transfer;
- Diversity of self-diagnostics (CRC16 vs CRC32, software implementation using tabular method vs hardware implementation);
- Diversity of the processor architecture (bit depth, architecture, address space);
- Diversity of the development environment (versions, language, hardware bus);
- Diversity of coding templates (pointers vs arrays, while loops vs for loops);
- Hybrid systems that implement one part of the functions programmatically, and the other part using hardware.

In the general case, the data transmission in the diversity classification is negligible. Meanwhile I&Cs is a distributed system consisting of a large number of constructive elements with an intensive data transmission between them. Therefore, it is important to take into account methods and interfaces of the data transfer in the diversity classification, and not just the system bus of the processor. This type of I&Cs properties can include the following: data transfer medium (copper conductor or optical fiber), various types of interfaces (RS485 or others), the implementation of these interfaces (in the form of hardware or software logic). A possible way to account several levels of the data transfer protocol is presented in Table 6.

The "Logic Processing Equipment" level in the NUREG-A technique is well developed, which allows the expert to unambiguously determine the existence of a particular diversity subtype. However, the use of FPGA as an element base provides additional possibilities. So, different system interfaces, which differ in the number of signals, can be used. For example, the Avalon bus in the Quartus 9 and Quartus 16 development environment is different in various parameters, which can be considered as the bus divergence. This specification is included in the diversity classification table (Table 7).

Table 6. Table for assessment of "equipment manufacturer" diversity type after the level splitting procedure is applied

	Attribute criteria	Rank	DC E WT	Leve l	Rank	WT
EQUIP.MANUF.	Equipment Manufacturer					
	Different manufacturers of fundamentally different equipment designs	1	0.400			
				Equipment	1	0.667
				Data transmission medium (copper / optics)	2	0.333
	Same manufacturer of fundamentally different equipment designs	2	0.300			
				Equipment	1	0.667
				Interfaces (RS485/other)	2	0.333
	Different manufacturers of same equipment design	3	0.200			
				Equipment	1	0.667
				Interfaces implementation	2	0.333
	Same manufacturer of different versions of the same equipment design	4	0.100			

Table 7. Table for assessment of "logic processing equipment" diversity type

	Attribute criteria	Rank	DCE WT
LOGIC PROC. EQUIP.	Logic Processing Equipment		
	Different logic processing architectures (Xilinx/Intel)	1	0.400
	Different logic processing versions in same architecture (486/586)	2	0.300
	Different component integration architectures (PCB)	3	0.200
	Different data flow architectures (Bus)(Avalon Quartus 9/16)	4	0.100

The "Logic" diversity type for the FPGA-based systems is proposed to be substantially transformed. The "Different algorithms, logic, and program architecture" diversity subtype should be divided into three diversity sub-subtypes, in order to separately assess the presence of diversity in the architecture, algorithms and logic that implements the algorithms. Further, "Different runtime environments" diversity subtype is proposed to be replace with "Different CAD systems", as the development environment defines many properties of the implemented device. In addition, this level is the second most important. The fourth diversity subtype is proposed to be clarified, noting that it should reflect the diversity of the algorithms coding templates (Table 8).

Examples of coding templates can be understood as a specific implementation of logic, for example, the use of pointers vs arrays, the use of while loops vs for loops, etc.

Table 8. Table for assessment of "logic" diversity type

Attribute criteria		Rank	DCE WT	Level	Rank	WT
LOGIC	Logic					
	Different algorit hms, logic, and program archit ect ure	1	0.400			
				Program architect ure	1	0.500
				Algorit hms	2	0.333
				Logic	3	0.167
	Different CAD systems (Quartus 9/16)	2	0.300			
	Different timing or order of execut ion	3	0.200			
	Different funct ional represent at ions (Coding Pat t erns)	4	0.100			

An important auxiliary function of I&Cs is self- diagnostics, which can be implemented in various ways and at different levels of I&Cs. To implement self- diagnostics, cyclic codes (CRC8, CRC16 or CRC32) are often used, the calculation of which can be implemented both programmatically using table methods and hardware using FPGA. The initial set of cyclic codes, as well as several possible implementation methods, makes it possible to implement diverse self-diagnostics solutions. The type of cyclic code is proposed to be referred to the level of algorithms ("Algorithms"), and the way of their implementation - to the level of logic ("Logic").

In the formation of an improved NUREG-A diversity classification, unintentional errors may occur due to the fact that some property is considered to be diverse and included in the incorrect classification position. The occurrence of classification errors leads to an incorrect definition of the diversity type and subtype, and, respectively, the weight of this property in the assessment.

The following errors in the diversity properties classification are possible:

- "Incorrect diversity status" error – some diversity property is treated as diversity attribute while in fact it is diversity criterion;
- "Incorrect diversity subtype choice" error – some property is included in the incorrect diversity subtype, but it should be included to the diversitysub-subtype;
- "Incorrect diversity sub-subtype status" error – some property is included in incorrect sub- sub-type position within correct diversity subtype; it leads to wrong weight of this property during diversity value evaluation.

CLB-A Technique

1. Features of CLB-based assessment technique are the following:
    a. It is based on a classification of diversity types described in (Kharchenko, V.,2011) and detailing NUREG classification;

   b. A main document is a multilevel check-list (CL) and questionnaires for assessment of diversity type application;
   c. Diversity metric is normalized [0,1];
   d. It may be used as a stage of I&C safety assessment (calculation of β-factor, reliability and safety indicators using RBD or MM).
2. Assessment procedure consists of the following stages:
   a. Analysis of I&C specification, design and development process;
   b. Identification of MVS types, product/process diversity (according to presented CL);
   c. Determination of metric values for different n types of applied diversity (local diversity metrics μ_i for diversity type d_i, metrics μ_{ij} for diversity subtype d_{ij};
   d. Determination (correction) of weight coefficients ω_i (ω_{ij}) of metrics;
   e. 2. e. Calculation of the general diversity metric μ for system:

$\mu=\Sigma\omega_i\ \Sigma\omega_{ij}\ \mu_{ij}$, $i=1,\ldots,n;\ j=1,\ldots,n_i$.

GMB-A technique analysis. This technique is a next step of developing CLB-based one. It is additionally based on a graph model of diversity types for two-version I&C systems (number of joint nodes, k; length of a minimal version, n_{min}). It details CLB-A technique regarding evaluation of metrics and weight coefficients and takes into account features of technological and architectural aspects of applied diversity (sensors, HW, SW, design, etc.).

Besides, an acceptable value of diversity correlates with NUREG-A in this technique.

Diversity Assessment Tools

NUREG-A-Based Tool: NUREG-A-based tool supports corresponding technique and allows calculating the diversity metric ccording to attributes and criteria, values of weights (Kharchenko, V. 1999).

CLB-A-Based Tool: Tool DivA (Diversity Analysis Helper) (Kharchenko, V., et al, 2012, Kharchenko, V., et al, 2012), is based on CLB-A and has main window displays (Figure 5):

- Hierarchy (multi-level and extensible) of diversity types;
- Calculated results (weights, metrics,…);
- Options for metric calculations.

Green colours mean diversity type is included in result of calculation. Gray colours mean diversity type is disabled for managing. User can add new diversity subtype for a selected type.

There are a few metric calculation options:

- Fixed value – user inputs metric manually;
- Value determined by children – metric is calculated as the sum of sub-types metrics;
- pre-defined value – shows an additional window, where a user can select pre-defined metric;
- Value determined by help questions – shows additional window with helper.
- The special window appears after selection, for example, of "Pre-defined value" option on main window. Features of this and other options are the following:

Figure 5. Main window of the DivA tool

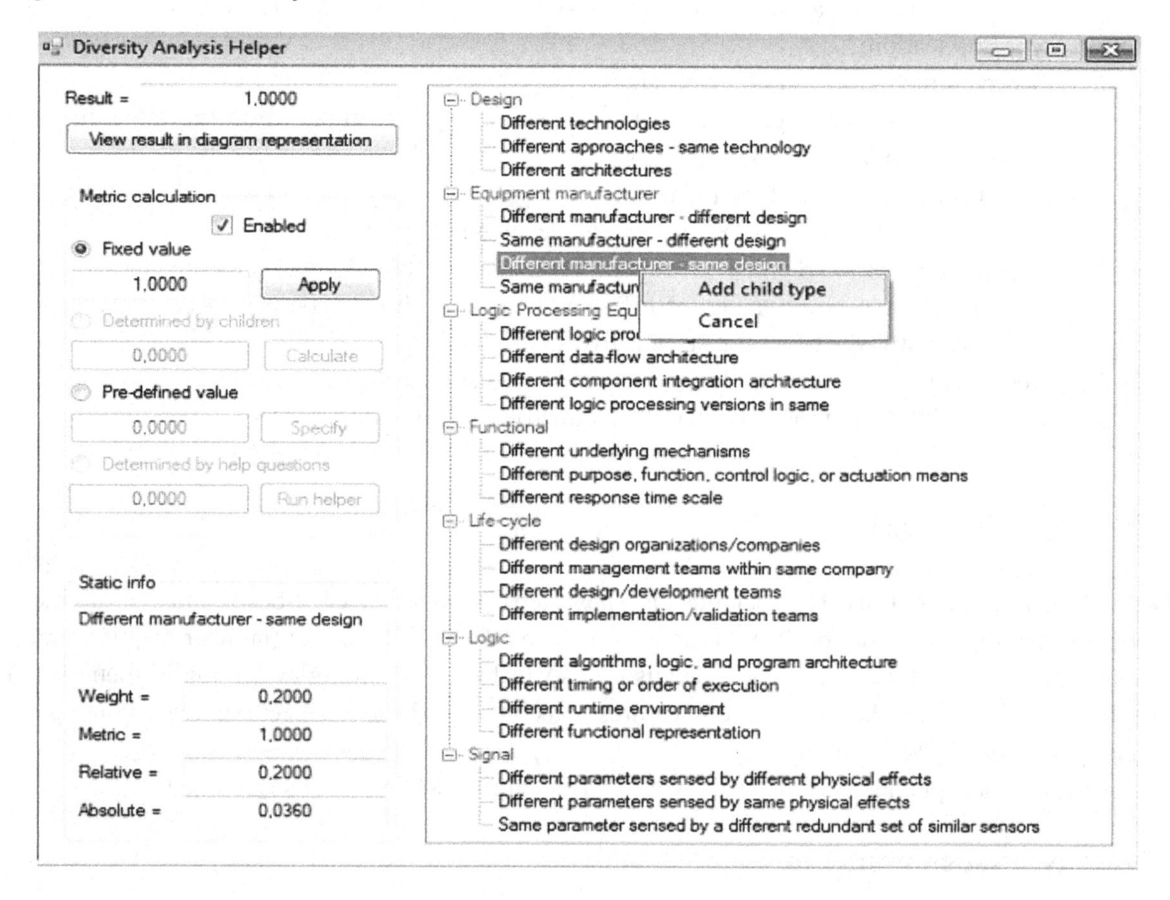

- Only one item can be selected;
- Dependencies between components are shown (for example, selection of Intel and Microsemi manufacturers causes selection of Quartus and Libero tools);
- Helper runs after selection of "Determined by use of questions" (see Figure 6) option on the main window, etc.

Current result represents the metric for a corresponding diversity type; user can choose answers: "YES" (answer value is considered as 1), "NO" (answer value is considered as 0), "Partially" (expected input of a answer value in the range between 0 and 1). The result is represented in the table and by coloured radial diagram. Absolute value and percentage of result are shown for each diversity type (on all levels of a diversity hierarchy).

Comparison of Diversity Assessment Techniques and Tools

Assessment of Different MVSs: To analyze the selected techniques of diversity assessment, five different multi-version projects MVP-1 - MVP-5 (see Figure 7, top part) were evaluated:

Figure 6. A special window for additional questions to calculate local diversity metrics

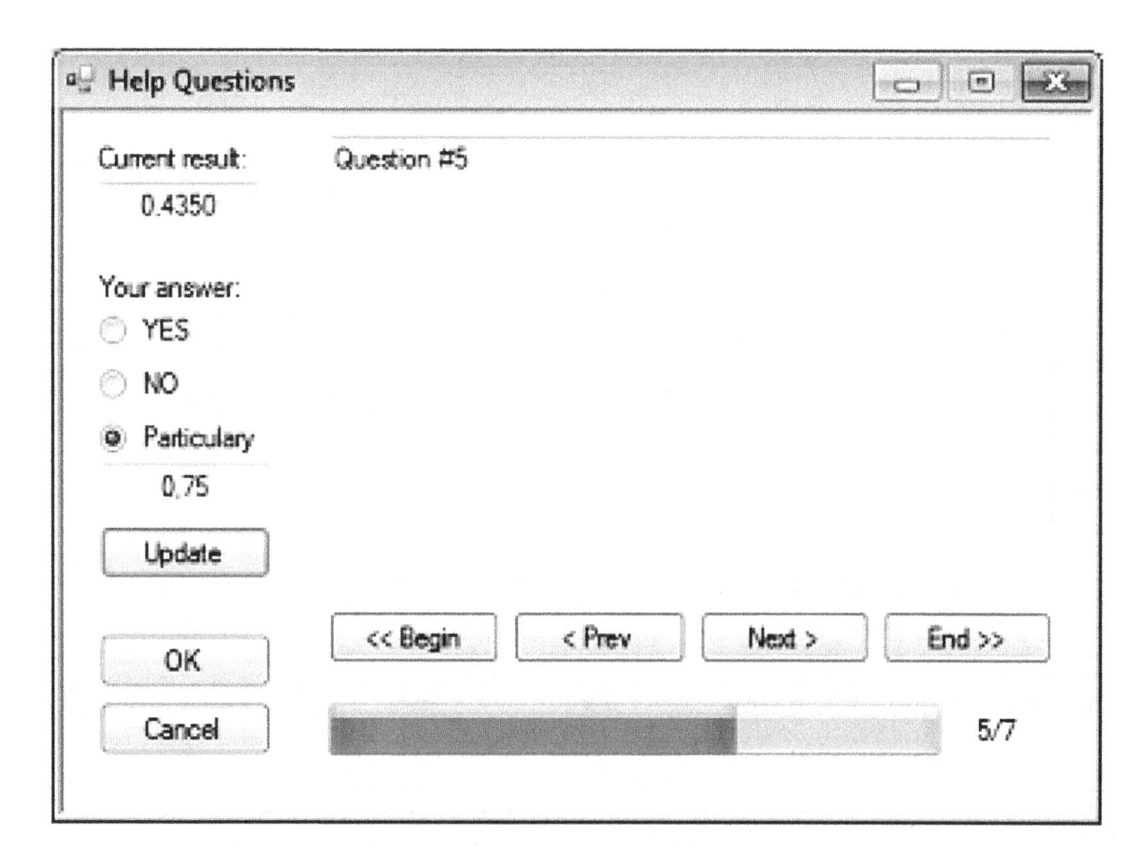

- **MVP-1:** Diversity is implemented by application of different FPGA manufacturers (Intel and Microsemi), technologies and others;
- **MVP-1:** Diversity is implemented by application of different FPGA manufacturers (Intel and Microsemi), technologies (SRAM and Antifuse) and others;
- **MVP-2:** Diversity is implemented by application of different FPGA families, processes and others;
- **MVP-3:** Diversity is implemented by application of different strategies (MP and FPGA) and others;
- **MVP-4:** Diversity is implemented by application of different strategies (MP and FPGA), processes and others;
- **MVP-5:** Diversity is implemented by application digital (FPGA) and analog technologies.

One of systems (main or diverse) of two-version I&C systems for analysed MVP-1, MVP-5 is based on the RadICS platform. Results of MP-1 - MP-5 assessing by use of NUREG-A and GMB-A techniques are shown in the Figures 7 – 9 (ID – total indicator of diversity calculated by use NUREG-A technique, HS-core – ID for hardware/software core of I&C system). Acceptable value of the diversity is defined by NUREG-A as ID = 1.

Besides, results of diversity assessment using these techniques for four I&C systems (PWR, DAS, AWTS which were described in (NUREG/CR-7007, 2009) and a variant of two-version RadICS-based

Figure 7. NUREG-A diversity assessment results for different two-version NPP I&C systems with FPGA-based subsystem

			MVP-1			MVP-2			MVP-3			MVP-4			MVP-5		
			FPGA-FPGA			FPGA-FPGA			FPGA-MP			FPGA-MP			FPGA-AnalogDevice		
			Intel-Microsemi			Intel-Intel			Intel-MP			Intel-MP			FPGA-Analog Device		
	DCE WT	Max %	Score	Norm score	%	Score	Norm score	%	Score	Norm score	%	Score	Norm score	%	Score	Norm score	%
Design	1.000	21	0.167	0.062	6	0.167	0.062	6	0.500	0.185	15	0.500	0.185	14	0.667	0.246	18
Equipment Manufacturer	0.250	5	0.050	0.018	2	0.025	0.009	1	0.075	0.028	2	0.100	0.037	3	0.100	0.037	3
Logic Processing Equipment	0.644	13	0.386	0.143	14	0.386	0.143	14	0.451	0.166	13	0.451	0.166	13	0.451	0.166	12
Function	0.600	13	0.500	0.184	18	0.500	0.184	18	0.500	0.184	15	0.500	0.184	14	0.500	0.184	14
Life-Cycle	0.683	14	0.410	0.151	15	0.478	0.176	17	0.410	0.151	12	0.478	0.176	14	0.478	0.176	13
Signal	0.867	18	0.722	0.266	26	0.722	0.266	26	0.722	0.266	21	0.722	0.266	21	0.722	0.266	20
Logic	0.733	15	0.513	0.189	19	0.513	0.189	18	0.733	0.270	22	0.733	0.270	21	0.733	0.270	20
Total	4.78	100	2.75	1.01	100	2.79	1.03	100	3.39	1.25	100	3.48	1.29	100	3.65	1.35	100
HS-Core				0.41			0.40			0.65			0.66			0.72	
HS-Core, %					41			39			52			51			53

Figure 8. Graphical illustrations of NUREG-A diversity assessment results of the MVP-1 - MVP-5

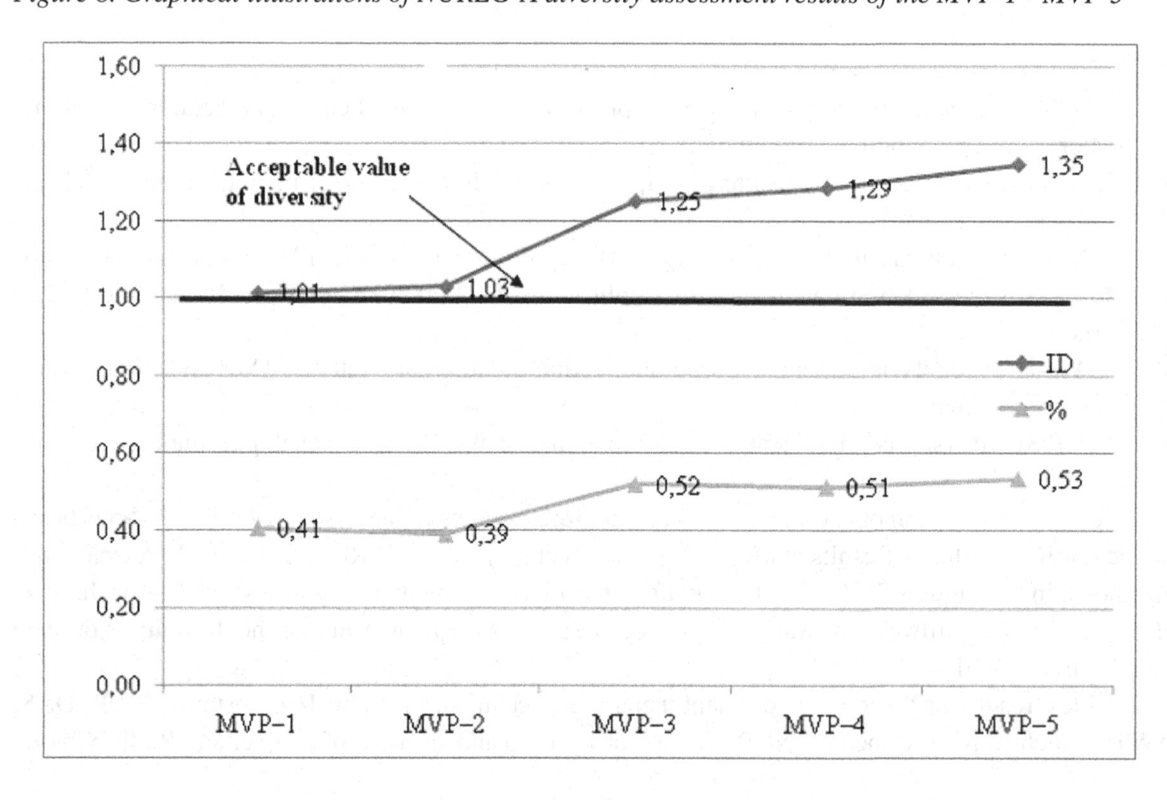

Figure 9. GMB-A diversity assessment results of the MVP-1 - MVP-5

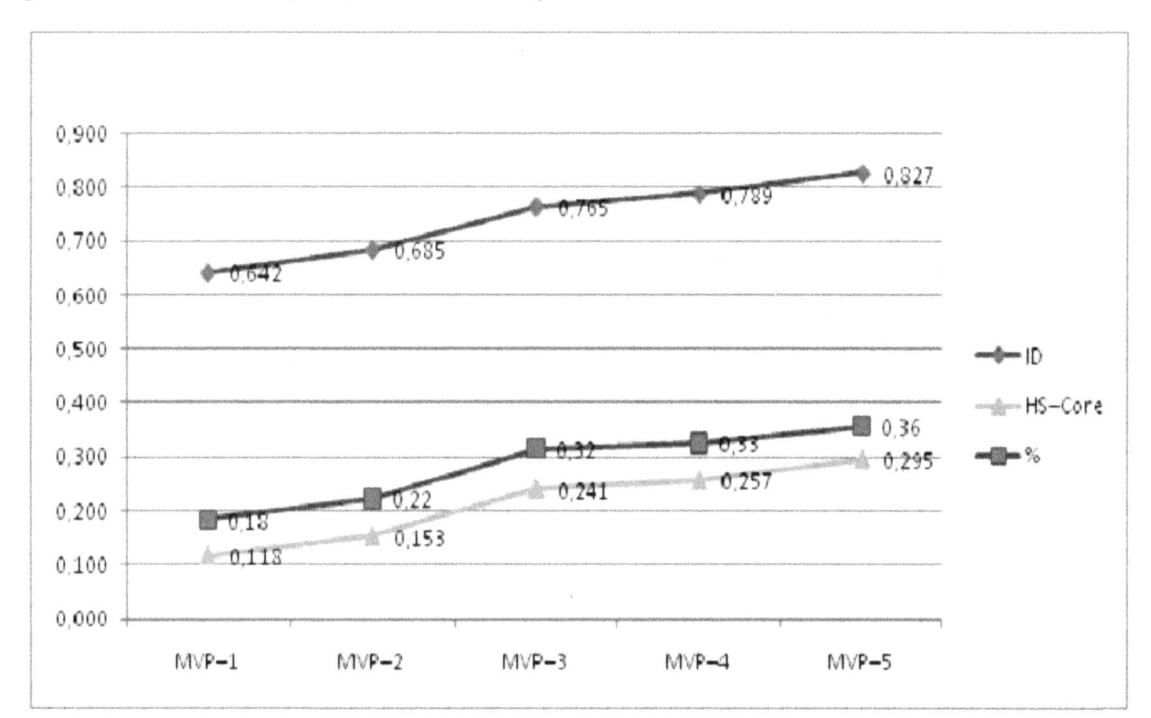

Figure 10. NUREG-A diversity assessment results of two-version I&C systems (examples taken from (NUREG/CR-7007, 2009) and RadICS-based)

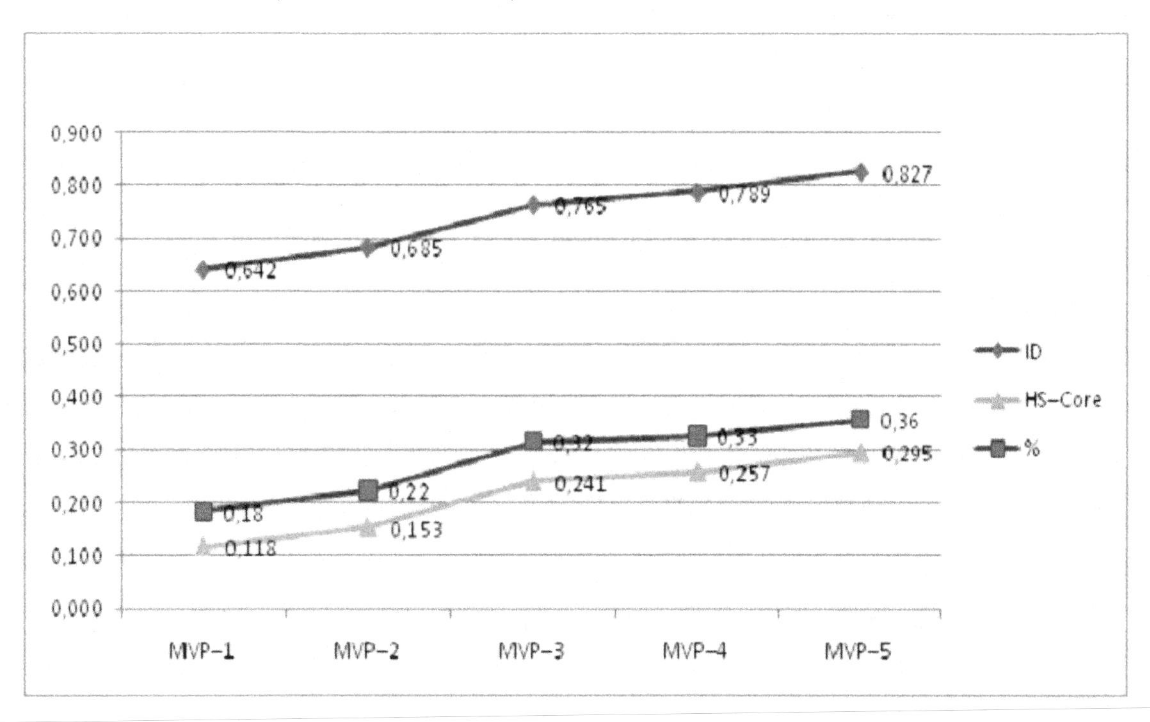

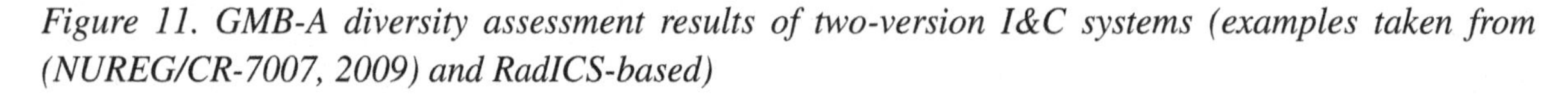

Figure 11. GMB-A diversity assessment results of two-version I&C systems (examples taken from (NUREG/CR-7007, 2009) and RadICS-based)

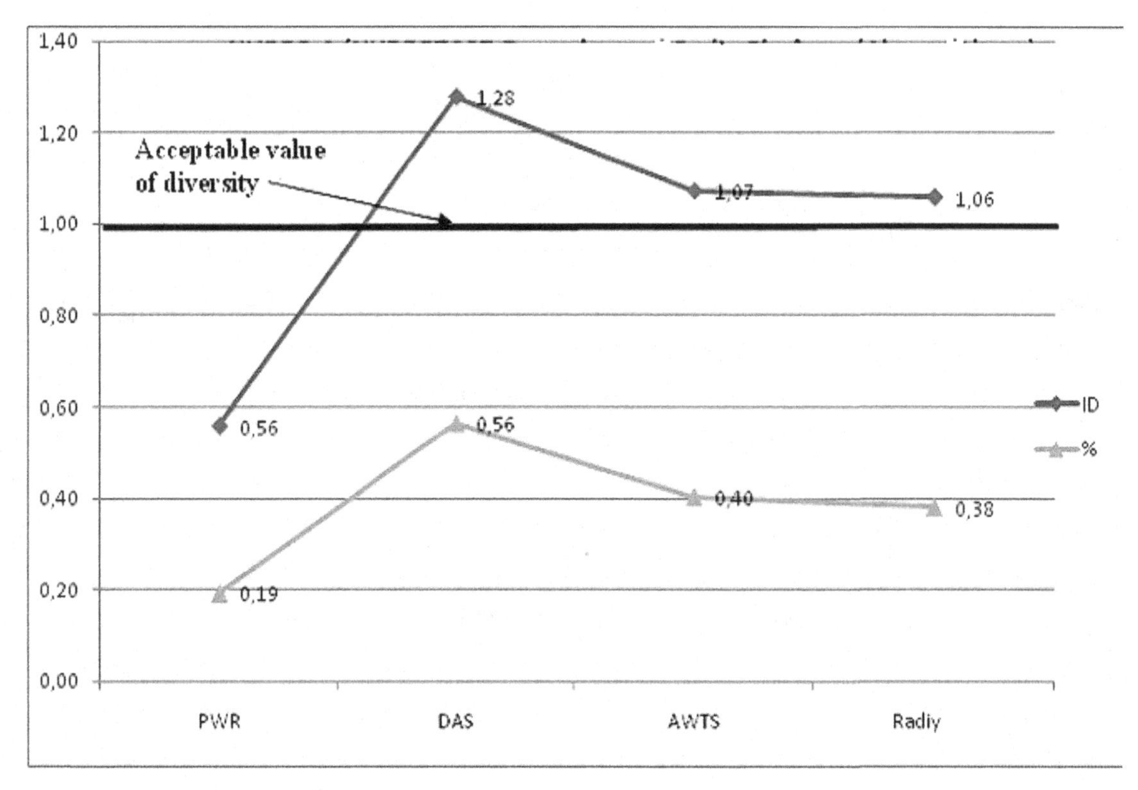

Figure 12 GMB-A diversity assessment results for FPGA-based two-version I&C systems

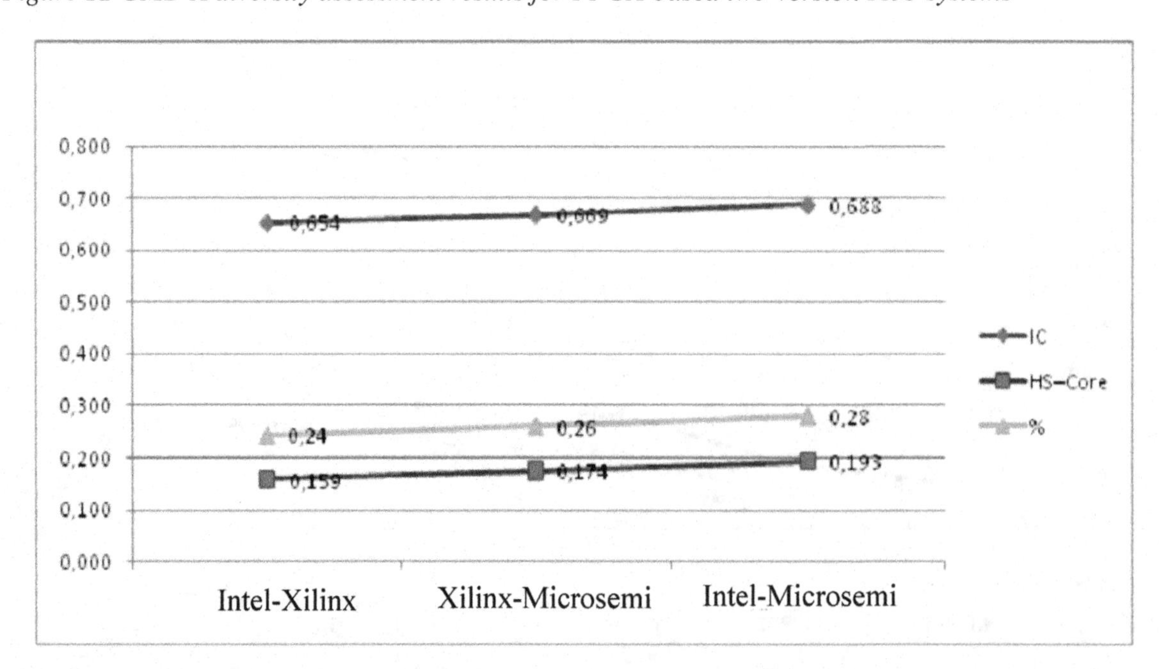

I&C system, Radiy) are illustrated in the Figures 10 and 11 correspondingly. Results of the GMB-A diversity assessment of FPGA-based I&C are shown in the Figure 12.

Results of Diversity Assessment Techniques Comparison: Results of a comparative analysis of NUREG-A, CLB-A and GMB-A techniques are shown in the Table 9. NUREG-A technique is more general. CLB-A and GMB-A techniques are detailed to evaluate the actual diversity level of a MVS. Results of assessing a few I&C systems using the techniques are consistent (the same priority row).

Table 9. Results of Comparative Analysis of Diversity Assessment Techniques

Technique	Diversity classification	Number of level	Diversity metrics	Sensibility to diversity type	Tool
NUREG7007- technique	NUREG6303-based classification	Two-level hierarchy	Non-normalized	Fixed	+
CLB - technique	Extensible (in depth) NUREG7007-based classification	Multi-level	Normalized metric, may be used to calculate β factor	May be increased	+
GMB - technique	Additionally take into account feature technological and architectural aspects	Multi-level	Normalized metric, may be used to calculate β factor	May be increased	+

ASSESSMENT OF MULTI-VERSION FPGA-BASED SYSTEMS SAFETY

General Approach to Assessment

Assessment of a diversity level and MVS safety is based on the following basic procedures analysis and evaluation:

- Check-list-based analysis of applicable diversity types (CLD); initial data for the CLD analysis are I&C design and documentation, a table of diversity types (subtypes) was developed in advance; a result of the CLD analysis is a formalized structured information about used diversity types and subtypes in analyzed I&C system;
- MAD; initial data for the MAD procedure are results of the CLD analysis and values of metrics and weight coefficients for diversity types (subtypes) used in I&C systems; a result of the MAD assessment is a value of general diversity metric;
- RBD and MM-based assessment taking into account results of MAD.

Stages of Assessment

The main stages and operations of diversity analysis and MVS assessment depend on the type of the evaluated system. The first stage is a Check-list-based analysis of MVS design and documentation. This stage contains two operations:

1. Analysis of I&C specification and requirements to a system, definition of system safety class; requirements to diversity (necessary for diversity application);
2. Analysis of I&C design and development process that involves activities: (a) identification of MVS types: which of the subsystems are FPGA-based and which are software and microprocessor-based; (b) identification of a product diversity; for FPGA-based MVSs: manufacturer of chips; FPGA technology; FPGA families; FPGA chips, languages; tools, etc); (c) identification of process diversity types.

Results of the analysis are entered in a check-list in accordance with a rule Yes (if corresponding diversity type is used in a system) / No (in opposite case) and is presented as a n-bit Boolean vector.

The second stage is a metric-based assessment of diversity. This stage contains two operations:

1. Determination of metric values for different types of applied diversity, i.e. performing two activities: (a) determination of metric values (local diversity metrics μ_i for the diversity type d_i and local diversity metrics μ_{ij} for the diversity subtype d_{ij}); the metric values may be predefined; (b) correction of metric values in accordance with development and operation experience.
2. Calculation of general diversity metric μ for a system: (a) determination (correction) of weight coefficients ω_i (ω_{ij}) of metrics (taking into account multi-diversity aspect); sum of weight coefficients ω_i (ω_{ij}) is equal 1; (b) convolution (additive or more complex) of metrics and calculating value of the general diversity metric $\mu = \Sigma \omega_i \Sigma \omega_{ij} \mu_{ij}$, $i = 1,\ldots, n$; $j = 1,\ldots n_i$.

Thus, result of this stage is a value of general diversity metric μ, which is some approximation of β, and can characterize the diversity effect on CCF probability.

The third stage is a probabilistic RBD- or MM-based (RDM) assessment of MVS reliability and safety. Initial data for the RDM procedure are I&C design and documentation, results of the CLD and MAD analysis; results of the RDM procedure are values of safety and dependability indicators. Detailed description of the RDM procedure is given in (Kharchenko et al., 2004).

IMPLEMENTATION OF FPGA-BASED SAFETY-CRITICAL NPP I&CS. UKRAINIAN EXPERIENCE

General Description of the FPGA-Based RADIY Platform

The platform RADIY produced by RPC Radiy is an example of a dependable and scalable FPGA-based I&C platform ensuring possibility of development of multi-version systems. Dependability assurance feature of the I&C platform RADIY is a multi-diversity implementation through the following diversity types: equipment diversity is provided by different electronic components, different programmable components (FPGAs and microcontrollers) and different schemes of units; software diversity is provided by different programming languages and different tools for development and verification; life cycle (human) diversity is provided by different teams of developers.

Scalability of I&C platform RADIY permits to produce different types of safety-critical systems without essential changing of hardware and software components. The I&C platform RADIY provides the following types of scalability: scalability of system functions types, volume and peculiarities by changing quantity and quality of sensors, actuators, input/output signals and control algorithms; scalability of dependability (safety integrity) by changing a number of redundant channel, tiers, diagnostic and reconfiguration procedures; scalability of diversity by changing types, depth and criteria of diversity choice.

The FPGA-based I&C RADIY platform comprises both upper and lower levels (Kharchenko&Sklyar, 2008). The upper level has been created on purchased IBM-compatible industrial workstations. The software for the upper level RADIY platform was developed by RPC Radiy and is loaded on the workstations. The functions of the upper level workstations are the following: receipt of the process and diagnostic information; creation of a man-machine interface in the Control Room; display of process information on each of the control algorithms relating to a control action executed by I&C system components; display of diagnostic information on failures of I&C system components; registration, archiving and visualization of process and diagnostic information.

The lower level of the RADIY platform consists of standard cabinets including standard functional modules blocks). The RADIY platform comprises the following standard cabinets (Bakmach et al., 2009):

- Normalizing Converters Cabinets performs inputting and processing of discrete and analog signals as well as feeding sensors;
- Signal Forming Cabinets performs inputting and processing of discrete and analog signals, processing of control algorithms, and formation of output control signals;
- Cross Output Cabinets receives signals from three control channels (signal formation cabinets) and forms output signals by "two out of three" mode;
- Remote Control Cabinets controls 24 actuators on the basis of Control Room signals, automatic adjustment signals and interlocks from signal formation cabinets;
- Signalling Cabinets forms control signals for process annunciation panel at Control Room and others.

The platform includes the following main modules: chassis and backplanes; power supply modules; analog input modules; normalizing converter modules, thermocouples; normalizing converter modules, resistive temperature detector; discrete input modules; discrete information input modules, pulse; potential signals input modules, high voltage; protection signal forming modules (logic modules); analog output modules, voltage; analog output modules, current; discrete output modules; potential signal output modules; solid-state output modules; relay output modules; actuator control modules; fiber optic communication modules; system diagnostic modules; fan cooling modules etc.

The latest RPC Radiy innovation is the FPGA-based Digital Instrumentation and Control Platform RadICS. This is a new generation product, designed in 2011 on the basis of an earlier RADIY platform having more than 10 years of experience in a platform design, manufacturing, operation, and maintenance. The RadICS platform provides IEC 61508:2010 SIL 3 architecture in an individual chassis, fast response time (less than 5 ms) and a comprehensive set of functional modules.

The RadICS Platform can be treated as an example of internal-diversity features that provide sufficient protection to address CCF vulnerabilities that may be introduced by the FPGA technology:

- **Functionally Independent and Diverse Self-Testing and Diagnostics:** Provides physically separate FPGA logic circuits for self-monitoring features that are independent and functionally diverse from the FPGA logic circuits executing control functions. The self- monitoring features put the Modules in the safe state when critical failures are detected.

- **Functionally Independent and Diverse Power Supply and Watchdog (PSWD) Monitoring:** Provides a functionally and structurally diverse method of monitoring the FPGA logics and power supplies. The PSWD Unit provides an independent method of placing a RadlCS Module in a safe state when critical failures are detected.
- **Separate Clocks for Diverse Functional Domains:** Physically separate clocks are used for safety functions, self-testing, and PSWD monitoring to ensure different timing or order of execution based on the parallel processing of the FPGA and CPLD circuits.
- **Diverse Chip Technologies:** The CPLD (Complex Programmable Logic Device)-based PSWD Unit is separate and inherently diverse from the Module FPGAs.

Opportunities of the RADIY Platform

Application of the RADIY platform with the use of FPGA technology provides the following opportunities:

- To implement control and other safety-critical functions in the form of FPGA with implemented electronic design, without software;
- To use software only for diagnostics, archiving, signal processing, data reception and transfer between I&C systems components; failures of those functions do not affect execution of basic I&C systems control functions, and an operation system is not applied at I&C systems lower levels;
- To process parallel of all control algorithms within one cycle, thus ensuring high performance of the system (for instance, a processing cycle of Reactor Trip System is 20 ms) and proven determined temporal characteristics;
- To develop the software-hardware platform in such a way that it becomes a universal interface to create I&C systems for any type of reactors;
- To assure high reliability and availability due to the application of industrial components as well as using the principles of redundancy, independency, single failure criterion, and diversity;
- To modify the I&C system after commissioning in a quite simple manner, including algorithm alterations, without any interference in I&C systems' hardware structure;
- To reduce by more than 10 times the number of contact and terminal connections, which cause many operational failures of equipment on account of the wide use of integrated solutions and fiber optic communication lines, etc.

Licensing of the RADIY Platform

The RADIY platform has been licensed for NPP application in Ukraine and in Bulgaria. The main idea for licensing FPGA-based NPP I&C systems lays in consideration of FPGA-chip as hardware and FPGA electronic design as a special kind of software with specific development and verification stages (Siora et al., 2009).

Qualification tests of FPGA-based hardware in accordance with International Electrotechnical Commission (IEC) standard requirements include: radiation exposure withstand qualification; environmental (climatic) qualification; seismic and mechanical impacts qualification; electromagnetic compatibility qualification. Results of qualification tests confirmed FPGA-based hardware compliance with IEC safety requirements.

FPGA electronic design has a V-shape life cycle in accordance with requirements of standard IEC 62566 "NPP – I&C important to safety – Selection and use of complex electronic components for systems performing category A functions".

The safety assessments have been conducted by Ukrainian State Scientific Technical Centre on Nuclear and Radiation Safety (SSTC NRS), which is the supporting organization of Ukrainian Regulatory Authority. Experts of SSTC NRS have considerable experience in the area of FPGA-based systems safety assessment, as they have performed reviews of all thirty three FPGA-based safety systems supplied to Ukrainian NPP units since 2003.

Implementation of the RADIY Platform-Based I&Cs in NPPs

The RADIY platform has been applied to the following NPP I&Cs systems, which perform reactor control and protection functions: Reactor Trip Systems (RTS); these I&Cs were developed as two-version systems consisting of two triple module redundant subsystems; It should be noted that this list is not concluded because of an universality of Radiy Platform. PRC Radiy has the ability to build different digital I&C systems for reactors of any type. The example of two-channel RTS is shown at Figure 13.

Figure 13. Radiy platform based reactor trip system with primary and diverse channels

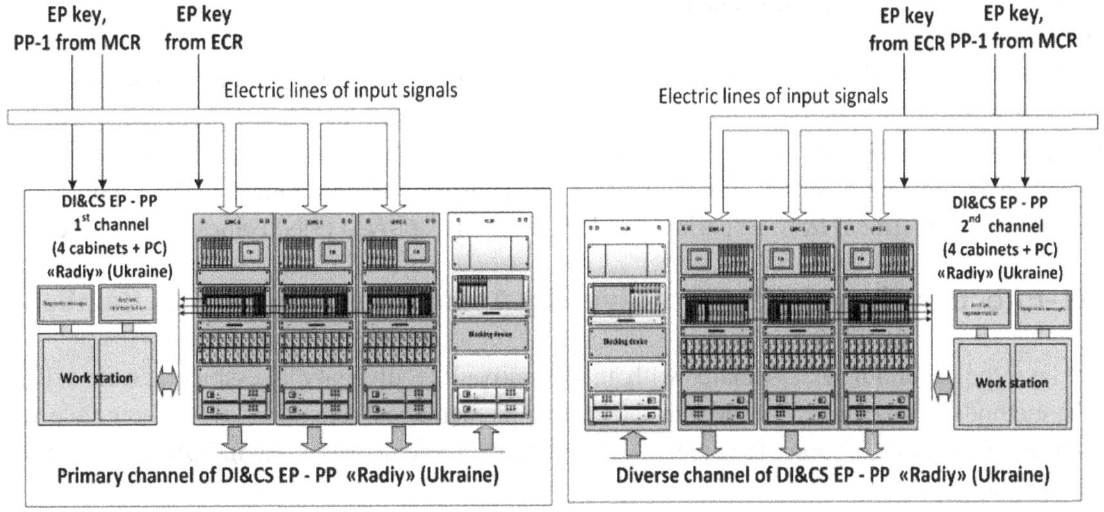

Both channels implemented on Radiy Platform. Reactor Power Control and Limitation System; Engineering Safety Features Actuation System (ESFAS); Control Rods Actuation System; Automatic Regulation, Monitoring, Control, and Protection System for Research Reactors; these I&Cs were developed as one-version systems consisting of triple module redundant subsystems.

The first commissioning of the RADIY platform was done in 2003 for Ukrainian NPP unit Zaporozhe-1. In seven years since that time, more than 80 applications of RPC Radiy systems have been installed in 17 nuclear power units in Ukraine and Bulgaria. These systems are commissioned in pressurized water reactor (PWR) plants known as "WWER" reactors developed by the former Soviet Union.

WWER reactors are used in Armenia, Bulgaria, China, Czech Republic, Finland, Hungary, India, Iran, Russia, Slovakia, and Ukraine.

The largest project implemented by RPC Radiy is the modernization of six ESFASs for Bulgarian NPP Kozloduy (three ESFASs for Kozloduy-Unit 5 and three ESFASs for Kozloduy-Unit 6).

RADIY RTS DIVERSITY ASSESSMENT

Radiy RTS Diversity Assessment Based on General Approach

RadICS Platform can be used for building whole multi-channel systems as well as for building one (primary or diverse) channel of I&C System. For cases using the platform in several channels of a system, diverse solutions should be used and appropriate diversity assessment have to be performed.

The following example shows the results of assessment for two variants of RTS. Both systems are two-channel and have diverse channels with primary channel based on Intel FPGA. The first variant diverse channel is based on Microsemi FPGA (see Table 10). The second variant uses FPGA produced by Xilinx in the diverse channel (see Table 10). Data in Tables 10 and 11 shows that using FPGA chips from different manufacturers allows obtaining the value of the general diversity metric more than 0.7. Increasing this value is possible primarily by increasing the independence of the processes and enhancement of the diversity of languages and models.

Described models of multi-version systems and multi-version technologies (life cycle) may support selecting of cost-effective technique and optimal architecture according to requirements to diversity, safety, reliability and limitation of applied technologies. These theoretical issues were used on development of FPGA-based I&C RadICS platform. Main peculiarities of the platform are realization of control and other safety-related functions without software and ensuring dependability- and diversity-scalable decisions of safety-critical I&C. Experience of RPC Radiy has proved effectiveness of these decisions.

Cost–Effective Approach to RTS Diversity Assessment Under Uncertainties

Nowadays, the uncertainties, associated with an alternative RTS diversity assessment, create a demand for the methods to make possible the translation, to a mathematical language, of the intangible values and human experience, improving the available resources in the decision making process in this complicated area.

Usually, in a quantitative setting, the information is expressed by means of numerical values. However, when we work in a qualitative setting, that is, with a vague or imprecise knowledge, the information cannot be estimated with an exact numerical value. In that case, a more realistic approach may be to use linguistic assessments instead of numerical values, that is, to suppose that the variables, which participate in the problem area, are assessed by means of linguistic terms (Zadeh L. 1999, Mendel, 2002). This approach is appropriate for a lot of problems, since it allows a representation of the information in a more direct and adequate form if we are unable to express it with precision.

A linguistic variable differs from a numerical one in that its values are not numbers, but words or sentences in a natural or artificial language. Since words, in general, are less precise than numbers, the concept of a linguistic variable serves the purpose of providing a means of approximated characteriza-

Table 10. Assessment of the first variant of two-channel system (intel and microsemi FPGA)

Diversity types		Result of analysis		Assessment results		
		Yes/ No	Implementation	Local metric	Ratio	Weighting coefficient
Diversity of programmable components (A)	Diversity of manufacturers of FPGA (A1)	Yes	Intel vs. Microsemi	0.8	0.25	0.25
	Diversity of technologies of FPGA producing (A2)	Yes	SRAM vs. Antifuse	1	0.4	
	Diversity of FPGA families (A3)	Yes	Cyclone vs. ProASIC (based on A1)	1	0.25	
	Diversity of FPGA from the same family (A4)	Yes	different families (based on A1)	1	0.1	
Diversity of printed circuit boards (PCBs) (B)	Diversity of PCB development technologies and manufacturers	Yes	Different manufacturers but the same technology	0.5	1	0.15
Diversity of CASE-tools (C)	Diversity of CASE-tools developers (C1)	Yes	Intel vs. Microsemi	1	0.5	0.15
	Diversity of CASE-tools (C2)	Yes	Quartus II vs. Libero (based on C1)	1	0.3	
	Diversity of CASE-tools configurations (C3)	Yes	Different tools for design but the same for verification	0.7	0.2	
Diversity of languages of FPGA projects development (D)	Diversity of language kinds (D1)	Yes	Graphic Notation and Hardware Description Language are used	0.5	0.4	0.15
	Diversity of hardware description languages (D2)	Yes	VHDL vs. Verilog	0.8	0.6	
Diversity of specification presentation (E)	Diversity of FPGA initial specification languages (E1)	No	The same language	0	0.5	0.1
	Diversity of FPGA specification models (E2)	Yes	B&HDL used for Intel	0.8	0.5	
Diversity of processes (P)	Diversity of development processes (P1)	Yes	Different teams	0.7	0.5	0.2
	Diversity of verification processes (P2)	Yes	Different departments	0.85	0.3	
	Diversity of maintenance (P3)	Yes	Different teams	0.7	0.2	
Overall assessment (general diversity metric)						0.74

tion of phenomena, which are too complex or too ill-defined to be amenable to their description in conventional quantitative terms.

In fact, considering the approach suggested in (NUREG/CR-7007, 2009, NUREG/CR-6003, 1994), it is often difficult to determine the precise values of diversity attributes' weights and rank of all alternatives on diversity criteria. We need to evaluate all appropriate experience of applications of different diversity approaches in all industrial area, take into account all relevant statistics of I&C failures caused by CCFs etc. A part of this information is often represented as linguistic information, being the expert's subjective opinions. The transformation and formalization of this linguistic information into precise

Table 11. Assessment of the second variant of two-channel system (Intel and Xilinx FPGA)

Diversity types		Result of analysis		Assessment results		
		Yes/ No	**Implementation**	**Local metric**	**Ratio**	**Weighting coefficient**
Diversity of programmable components (A)	Diversity of manufacturers of FPGA (A1)	Yes	Intel vs. Xilinx	0.8	0.25	0.25
	Diversity of technologies of FPGA producing (A2)	No	SRAM	0	0.4	
	Diversity of FPGA families (A3)	Yes	Cyclone vs. Virtex (based on A1)	1	0.25	
	Diversity of FPGA from the same family (A4)	Yes	different families (based on A1)	1	0.1	
Diversity of printed circuit boards (PCBs) (B)	Diversity of PCB development technologies and manufacturers	Yes	Different manufacturers and technologies	1	1	0.15
Diversity of CASE-tools (C)	Diversity of CASE-tools developers (C1)	Yes	Intel vs. Xilinx	1	0.5	0.15
	Diversity of CASE-tools (C2)	Yes	Quartus II vs. ISE (based on C1)	1	0.3	
	Diversity of CASE-tools configurations (C3)	Yes	Different tools for design but the same for verification	0.7	0.2	
Diversity of languages of FPGA projects development (D)	Diversity of language kinds (D1)	Yes	Graphic Notation and Hardware Description Language are used	0.5	0.4	0.15
	Diversity of hardware description languages (D2)	Yes	VHDL vs. Verilog	0.8	0.6	
Diversity of specification presentation (E)	Diversity of FPGA initial specification languages (E1)	No	The same language	0	0.5	0.1
	Diversity of FPGA specification models (E2)	Yes	B&HDL used for Intel	0.8	0.5	
Diversity of processes (P)	Diversity of development processes (P1)	Yes	Different teams	0.7	0.5	0.2
	Diversity of verification processes (P2)	Yes	Different departments	0.85	0.3	
	Diversity of maintenance (P3)	Yes	Different teams	0.7	0.2	
Overall assessment (general diversity metric)						0.72

form without application of special methods is characterized by loss of important information. This is another aspect, which increases the difficulties of the I&C diversity assessment.

At the initial stage of selection of secondary (primary) RTS it is more convenient approach for the experts to compare the possible alternatives of primary (secondary) RTS and express their preferences using the natural language expressions.

The experts have to deal with portion of a qualitative information stipulated by several types of the following uncertainties:

- Uncertainties caused by lack of a sufficient and objective information on RTSs, which could be considered as an alternative for given RTS. The lack of required information is stipulated by policies of some I&C company-manufacturer to conceal the part of information related to its possible shortages and defects. In addition, a part of information on RTS features is confidential and not available for objective expert assessment;
- Strategic Uncertainties caused by dependencies on activities of other subjects involved (directly or indirectly) in the process of selection of alternative RTS (partners, suppliers etc.);
- Uncertainties caused by application of an imprecise information (different system parameters) expressed in natural language (for example the linguistic nature of some diversity attributes).

On the one hand, it is possible to neglect all these uncertainties and use deterministic approaches for selection of the most diverse I&C system for a given one. But on the other hand, some of important information might be lost.

We suggest using fuzzy metrics, derived from application of Computing, with words (CW) methodology to form the initial subset of possible alternatives and determine the most diverse I&C system under uncertainties.

Diversity Strategies Description: According to (NUREG/CR-7007, 2009, NUREG/CR-6003, 1994) the rational choice of a pair of primary and secondary RTS could be named as diversity strategies. Both of (NUREG/CR-7007, 2009, NUREG/CR-6003, 1994) describe three types of diversity strategies.

Strategy S_1 focuses on the use of fundamentally diverse technologies as the basis for RTS diverse systems, redundancies, or subsystems. In this case, the primary RTS is built on an analog (digital) technology, and the diverse RTS is based on a digital (analog) platform. This choice of technology *inherently* contributes notable equipment manufacturer, processing equipment, functional, life-cycle, and logic diversities.

Intentional application of life-cycle and equipment manufacturer diversities is included in the baseline, while the traditional use of functional and signal diversities is also adopted.

Strategy S_2 involves the use of distinctly different technology approaches as the basis for diverse RTS, redundancies, or subsystems. In other words, this approach presumes using some variations inside either digital or analog technologies. In this case, the primary RTS is built on general-purpose microprocessors (MC), and the diverse RTS is based on, for example, FPGA platform.

This choice of technology inherently contributes some measure of equipment manufacturer, processing equipment, functional, life-cycle, and logic diversities. Intentional application of a logic processing equipment, life-cycle, and equipment manufacturer diversities is included in the baseline, while the traditional use of functional and signal diversities is also adopted.

Strategy S_3 represents the use of architectural variations within a technology as the basis for diverse systems, redundancies, or subsystems. In this case, the primary RTS is built on static random access memory (SRAM)-based FPGA, and the diverse RTS is based on, for example, on Flash –based FPGA platform.

This choice of technology inherently contributes some limited degree of equipment manufacturer, life-cycle, and logic diversities. Intentional application of equipment manufacturer, logic processing equipment, life-cycle, and logic diversities is included in the baseline, while the traditional use of functional and signal diversities is also adopted.

Considering the system approach to strategies formulation and representation, represented in NUREGs 6303, 7707 two additional strategies S_4 and S_5 are introduced in this section.

Strategy S_4 represents the variations inside of one SRAM (Flash) FPGA technologies. One family of SRAM (Flash) FPGA is used for the primary RTS, and second (third) family of SRAM (Flash) FPGA is used for secondary RTS. For example, the primary RTS is based on application of Arria family FPGA (Intel), and the secondary RTS is based on application of Stratix family FPGA (Intel).

Strategy S_5 represents the variation inside of SRAM (Flash) FPGA family. The application of this technology supposes using the representatives from one family to provide diversity for both secondary and primary RTS. For example, the primary RTS is based on application the Stratix II FPGA, and the secondary RTS is based on application of Stratix III FPGA.

It is apparently that the lower layer of hierarchy the less diversity for I&C provided. Each strategy is characterized by the set of possible alternatives (secondary RTS) available to provide the diversity between the primary and secondary RTS. The bigger index of strategy the more possible alternatives are available.

The choice of strategies is stipulated by the existence of some restrictions, which could limit the set of possible alternatives for the secondary RTS. The S_1 strategy represents the policy of absence of any restrictions (financial, organizational, political etc) related to the choice of the secondary RTS.

The S_2 strategy is characterized by the freedom "inside" of digital technology. Any of FPGA – based RTS could be chosen as primary RTS and any of MC –based RTS could be selected as a secondary one.

The S_3 strategy is characterized by some freedom "inside" of FPGA technology. In this case, there are no restrictions for selection any FPGA-based RTS either SRAM or Flash.

The S_4 strategy is characterized by some freedom "inside" of FPGA SRAM (Flash) technology. In this case, the second RTS could be selected from FPGA SRAM (Flash) families.

The S_5 strategy is characterized by freedom "inside" of one family of FPGA SRAM (Flash) technology. In this case, the second RTS could be selected from one of family of FPGA SRAM (Flash) families.

Generally, each S_i diversity strategy includes the subset of diversity strategies S_{ij}, where j – a number of possible alternatives classified as a type of S_i strategy. For example, different types of S_3 might be the following strategies: S_{31} – the primary RTS – FPGA SRAM – based RTS (Stratix IV (E,GX,GT)) and the secondary RTS – FPGA FLASH-based RTS (XC3000), S_{32} – the primary RTS – FPGA SRAM – based RTS (Stratix IV (E,GX,GT)) and the secondary RTS – FPGA Flash-based RTS (XC4000), S_{33} – the primary RTS – FPGA SRAM – based RTS (Stratix IV (E,GX,GT)) and the secondary RTS – FPGA Flash-based RTS (Virtex), etc.

The hierarchy of diversity strategies is shown on Figure 14.

The Stages of RTS Diversity Assessment: The linguistic approach for selection of the most diverse RTS (Zadeh L. 1999) deals with qualitative aspects that are represented in qualitative terms by means of linguistic variables. When a problem is solved using linguistic information, it implies the need for computing with the words (CW) (Zadeh L., 1999). Since CW deals with words or sentences defined in a natural or artificial language instead of numbers, it emulates human cognitive processes to improve solving processes of problems dealing with uncertainty. Consequently, CW has been applied as a computational basis to linguistic decision making, because it provides tools close to human beings reasoning processes related to decision making, which improve the resolution of decision making under uncertainty as linguistic decision making. CW is an approximate technique in its essence, which represents qualitative aspects as linguistic values by means of linguistic variables, that is, variables, whose values are not numbers but words or sentences in a natural or artificial language.

To compare the secondary and primary RTS, using the diversity criteria and evaluate the similarity (difference), expert should take into consideration the compelling evidence (i.e., some adequate com-

Figure 14. The hierarchy of diversity strategies

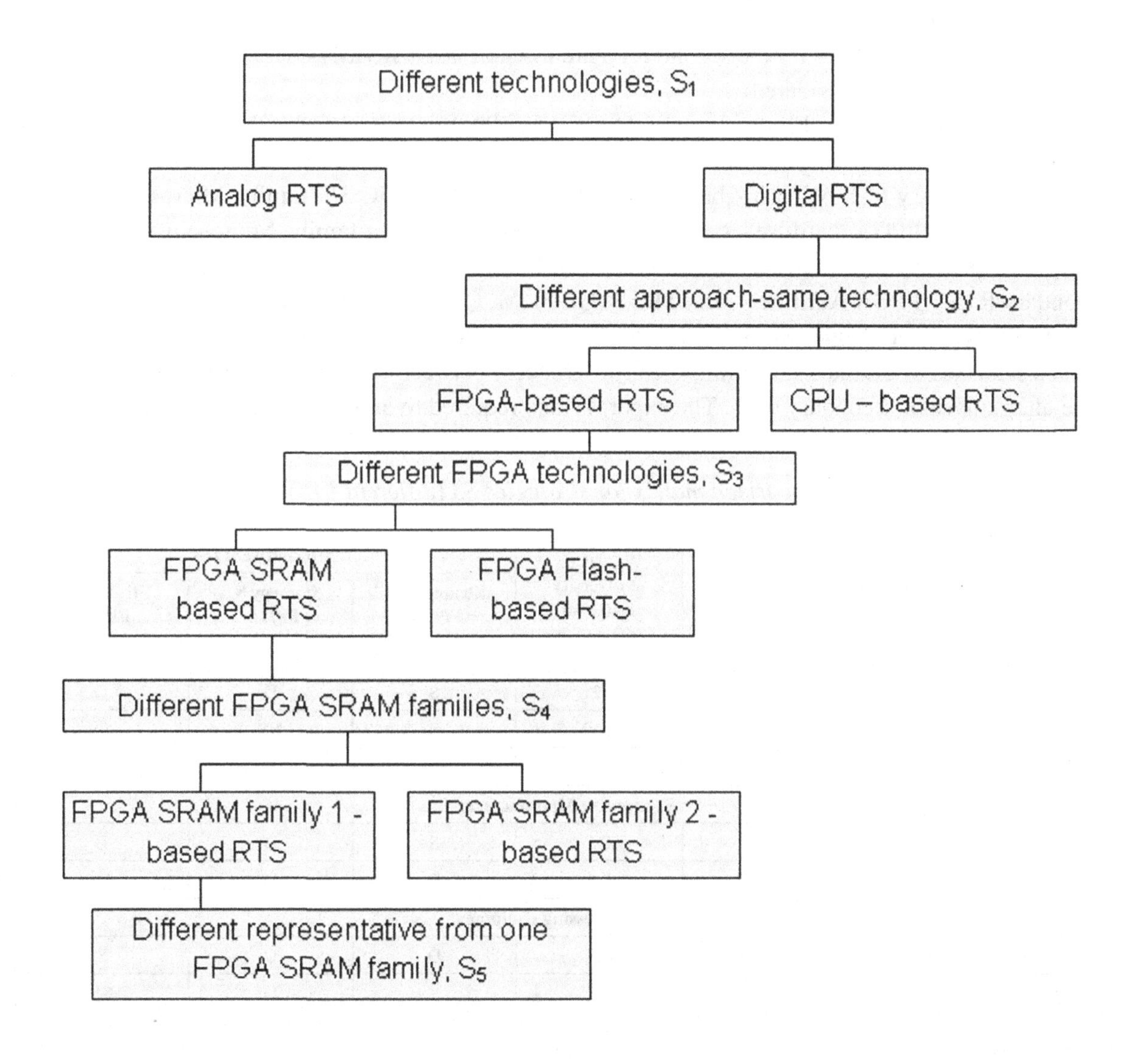

binations of thorough testing, substantial usage history for a comparable application under very similar demands and conditions, extensive formal proofs, detailed hazard/threat analysis, etc.). Based on these evidences experts evaluate the difference (similarity) between the primary RTS and secondary RTS for each diversity strategy using the linguistic terms: SAME (S), NEARLY SAME (NS), DIFFERENT (D).

The stage of cost – effective approach to selection of diverse NPP RTS consists of the following stages.

The Formation of Diversity Strategies Se:. When the primary RTS has been already determined, and a set of possible alternatives for the secondary RTS has also been established, it is suggested to classify the type of a diversity strategy according to the hierarchy of diversity strategies. When the type of a diversity strategy is determined, it is suggested to use the set of diversity criteria predefined in (NUREG/CR-7007, 2009, NUREG/CR-6003, 1994). These diversity criteria are used to complete the comparison matrixes. These comparison matrixes are chosen according to strategies used to provide the required diversity. In this case the primary RTS is FPGA Flash – based RTS (A3PE1500 from ProASIC 3/E family, Microsemi).

The possible alternatives for the second RTS are FPGA Flash (SRAM) – based RTS. The following diversity strategies are considered:

S_{31} - the primary RTS is Flash – based RTS (A3PE1500, ProASIC 3/E family, Microsemi) and the secondary RTS – FPGA SRAM – based RTS (EP1SGX40G, Cyclone IV GX family, Intel);

S_{32} - the primary RTS is Flash – based RTS (A3PE1500, ProASIC 3/E family, Microsemi) and the secondary RTS – FPGA Antifuse – based RTS (AX2000, Axcelerator family, Microsemi);

S_{33} - the primary RTS is Flash – based RTS (A3PE1500, ProASIC 3/E family, Microsemi) and the secondary RTS – FPGA Antifuse – based RTS (QL904M, QuickMIPS family, QuickLogic).

The Diversity Strategies Set's Expertise: During this stage experts are supposed to fill the comparison matrixes to evaluate the similarities (differences) between the primary RTS and each of possible alternatives of secondary ones. The expert is also required to assign the weight of each criterion.

Table 12. The example of comparison matrix for strategies S3 (different FPGA technologies)

Diversity criterion	W_k, weight of diversity criterion	Alternative RTSs		
		Strategy S_{31}, $FPGA_1$	Strategy S_{32}, $FPGA_2$	Strategy S_{33} $FPGA_3$
Design				
Technologies	0,21	S	D	D
Approach (for the same technology)	0,19	S	NS	NS
Architecture	0,6	D	S	D
Equipment Manufacturer				
Design (for different Manufacturer)	0,5	S	D	D
Design (for the same Manufacturer)	0,5	D	D	NS
Logic Processing equipment				
Logic Processing Architecture	0,3	D	D	S
Component integration Architecture	0,7	S	D	NS
Functional				
Purpose, function, control logic, or actuation means	1	S	D	D
Life-cycle				
Design organizations/companies	0,24	S	D	NS
Design/development teams	0,36	D	NS	NS
Implementation/validation teams	0,4	S		D
Logic				
Algorithms, logic, program architecture	0,33	D	D	D
Runtime environment	0,47	D	NS	D
Functional representation	0,2	D	S	S
Signal				
Parameters sensed	0,6	D	NS	S
Physical effects used	0,4	NS	D	D

Generally, the weight might be evaluated on various scales. The criterion's weight might be expressed either as linguistic value (Low, Medium, High) or any numerical values from [0, 1]. The more weight the more criterion influence on diversity RTS. For sake of simplicity the weight of criterion is presented as scalar value.

Table 12 represents the example of diversity assessment for the set of alternative strategies S_3 (different FPGA technologies) and corresponding set of diversity criteria, which could be applicable for diversity evaluation.

Diversity for Security

General diversity classification scheme was presented by "cube of diversity" with three coordinates: "stage of the life cycle" – "level of project decisions" and "type of version redundancy" (Fig. 4). Using this classification we can analyse safety and security issues for FPGA-based systems and traditional SW-based I&Cs, first of all, for NPPs.

Table 13 summarizes variety of diversity attributes from NUREG-CR/7007:2009 for NPP I&Cs and their accordance with kinds of version redundancy of FPGA-based systems.

Table 14 shows results of research on diversity attributes from NUREG-CR/7007 which could be applied to mitigate CCF in diverse SW- and HW/FPGA-based systems with the same vulnerabilities in both versions. Different vulnerabilities in both versions have four grades: VH – very high, H –high, M – medium, L – low.

Gradation is based on risk reduction after appliance of a certain diversity attribute. In this case diversity is considered as a countermeasure for elimination of harmful consequences after successful attacks.

Table 15 summarizes some attacks on FPGA-based MVS and results of security assessment using IMECA-analysis. Countermeasures are employed to thwart such tampering attacks. The table contains countermeasures strategies which could be applied as a requirements from Regulatory Guide 5.71:2010 (Cyber Security Programs For Nuclear Facilities, U.S. NRC) to eliminate the attack causes and, moreover, FPGA-based MV I&Cs diversity kind and its attributes as a countermeasures.

Thus diversity of FPGA-based MVS is reviewed as a countermeasure and mitigation strategy for ensuring of security and safety of systems.

SOLUTION AND RECOMMENDATIONS

Described models of multi-version systems are a base for the development of different architecture variants. The proposed techniques of diversity level and multi-version systems safety assessment are founded on two interconnected approaches. First of them is the metric-based technique allowing to assess a diversity level and to compare multi-version systems on application of different kinds and different volume of diversity. Second one is based on the probabilistic models, which include β calculated using metric analysis.

Development and implementation of multi-version FPGA-based systems is a new stage of the evolution in area of improving safety of NPP I&Cs. In this chapter we discussed basic concepts of diversity as a key approach to decreasing a probability of a common cause failure of safety-critical I&Cs and the taxonomic scheme of multi-version computing as a part of dependable, safe and secure computing.

Table 13. Diversity attributes and correspondent MVS version redundancy kinds

DIVERSITY ATTRIBUTES (NUREG-CR/7007:2009)	KINDS OF VERSION REDUNDANCY (FPGA-BASED MVS)
Design	**Diversity of electronic elements (EE)**
Different technologies	Different manufacturers of EEs; Different technologies of EEs production
Different approaches within a technology	Different technologies of EEs production
Different architectures within a technology	Different families of EEs
Equipment Manufacturer	**Diversity of electronic elements (EE)**
Different manufacturers of fundamentally different equipment designs	Different manufacturers of EEs
Same manufacturer of fundamentally different equipment designs	Different families of EEs
Different manufacturers of same equipment design	Different manufacturers of EEs
Same manufacturer of different versions of the same equipment design	Different EEs of the same family
Logic Processing Equipment	**Diversity of project development languages**
Different logic processing architectures	
Different logic processing versions in same architecture	
Different component integration architectures	Joint use of graphical scheme language and hardware description language (HDL)
Different data flow architectures	Joint use of graphical scheme language and HDL
Function	**Diversity of CASE-tools**
Different underlying mechanisms to accomplish safety function	Combination of couples of diverse CASE tools and SSs
Different purpose, function, control logic, or actuation means of same underlying mechanism	Different SSs
Different response time scale	
Life-Cycle	**Diversity of CASE-tools**
Different design companies	Combination of couples of diverse CASE-tools and HDLs
Different management teams within the same company	Combination of diverse CASE-tools and HDLs
Different designers, engineers, and/or programmers	Different HDLs
Different implementation/validation teams	
Signal	**Diversity of CASE-tools, Diversity of scheme specification (SS)**
Different reactor or process parameters sensed by different physical effect	Combination of couples of diverse CASE tools and SSs
Different reactor or process parameters sensed by the same physical effect	
The same process parameter sensed by a different redundant set of similar sensors	
Logic	**Diversity of CASE-tools, Diversity of scheme specification (SS)**
Different algorithms, logic, and program architecture	Combination of couples of diverse CASE-tools and HDLs
Different timing or order of execution	Different CASE tools configurations
Different runtime environments	Different CASE tools
Different functional representations	Different HDLs

Table 14. Diversity attributes as a countermeasure

DIVERSITY ATTRIBUTES (NUREG-CR/7007:2009)	VULNERABILITIES			
	Software		*Hardware*	
	common vulnerability	*different vulnerabilities*	*common vulnerability*	*different vulnerabilities*
Design				
Different technologies	H	H	H	H
Different approaches within a technology	M	M	M	M
Different architectures within a technology	L	L	L	L
Equipment Manufacturer				
Different manufacturers of fundamentally different equipment designs	H	H	H	H
Same manufacturer of fundamentally different equipment designs	HM	HM	HM	HM
Different manufacturers of same equipment design	M	M	M	M
Same manufacturer of different versions of the same equipment design	L	L	L	L
Logic Processing Equipment				
Different logic processing architectures	H	H	H	H
Different logic processing versions in same architecture	HM	HM	HM	HM
Different component integration architectures	M	M	M	M
Different data flow architectures	L	L	L	L
Function				
Different underlying mechanisms to accomplish safety function	H	H	H	H
Different purpose, function, control logic, or actuation means of same underlying mechanism	M	M	M	M
Different response time scale	L	L	L	L
Life-Cycle				
Different design companies	H	H	H	H
Different management teams within the same company	HM	HM	HM	HM
Different designers, engineers, and/or programmers	M	M	M	M
Different implementation/validation teams	L	L	L	L
Signal				
Different reactor or process parameters sensed by different physical effect	H	H	H	H
Different reactor or process parameters sensed by the same physical effect	M	M	M	M
The same process parameter sensed by a different redundant set of similar sensors	L	L	L	L
Logic				
Different algorithms, logic, and program architecture	H	H	H	H
Different timing or order of execution	HM	HM	HM	HM
Different runtime environments	M	M	M	M
Different functional representations	L	L	L	L

Table 15. IMECA-analysis of attacks on FPGA-based MVS

No	Attack mode	Attack nature	Attack cause	Occurrence probability	Effect severity	Type of effects	Countermeasures (including RG 5.71)	FPGA-based I&C diversity kinds and its attributes
1	Readback	Active	Absence of chip security bit and/ or availability of physical access to chip interface (e.g. Joint Test Automation Group, JTAG)	M	H	Obtaining of secret information by adversary	• The use of security bit ; • Application of physical security controls; (B.1.18 Insecure and Rogue Connections, Appendix B to RG 5.71, Page B-6)	Diversity of (EE): • Different technologies of EEs production;
2	Cloning	Active	Storing of decoded configuration	H	H	Obtaining of configuration data by adversary	• Checking of chip's internal ID before powering up an electronic design; • Encoding of configuration file; • Storing of configuration file within FPGA chip (requires internal power source)	Diversity of EE: • Different technologies of EEs production; • Different element kinds of EE families
3	Brute force	Active	• Search for a valid output attempting all possible key values; • Exhaustion of all possible logic inputs to a device in order; • Gradual variation of the voltage input and other environmental conditions	L	M	Leak of undesirable information	Detecting and documenting unauthorized changes to software and information, (C.3.7, Appendix C to RG 5.71, Page C-7)	Diversity of project development languages • Combination of couples of diverse CASE-tools and HDLs
4	Fault injection (glitch)	Active	• Altering the input clock; • Creating momentary over- or under-shoots to the supplied voltage	M	H	• Device to execute an incorrect operation • Device left in a compromising state • Leak of secret information	• Making sure all states are defined and at the implementation level, verifying that glitches cannot affect the order of operations; • Detection of voltage tampering from within the device; • Clock supervisory circuits to detect glitches	Diversity of EE: • Different manufacturers of EEs; • Different technologies of EEs production; Diversity of SS • Different SSs; • Combination of diverse CASE tools and SSs

Known version redundancy classification schemes were generalized in three-space matrix ("cube of diversity") taking into account features of FPGA technology. This unique technology allows to simplify NPP I&C development and verification, realize multi-reconfiguration (dynamical function- and dependability-oriented architecting, multi-parametrical space-structural adaptation, etc.), to propose decisions with different product-process version redundancy.

Key challenges related to diversity-oriented and FPGA-based systems are the following: existing standards are not enough detailed to make all necessary decisions concerning diversity (all the more FPGA-based decisions); multi-version I&Cs are still unique, failures occurred rarely and information about failures is not enough representative; methods of diversity assessment and kind selection, as a rule, are based on expert approach.

FPGA technology allows developing multi-version systems with different product-process version redundancy, diversity scalable multi-tolerant decisions for safety-critical NPP I&Cs.

FUTURE RESEARCH DIRECTIONS

Future R&D steps may be the following:

- Development of the detailed standards and guides to assess and choice types and capacity of diversity according to requirements and criteria of safety and cost;
- Research of different diversity types application to decrease risks of CCF taking into consideration dependencies of these types including hardware and software diversity for programmable systems;
- Research and implementation of internal diversity to minimize risks of CCF for redundant channels;
- Development of Safety Case-oriented techniques and tools for diversity assessment;
- Research and development of diversity application techniques for cyber security improvement taking into account features of attacks on microprocessors and FPGA technologies.

CONCLUSION

Application of the diversity allows a decrease in the probability of CCFs. A new graphical model is presented in this chapter for different variants of diversity and can be used during the development of safety-critical systems and selection of optimal algorithms for diversity types based on a criterion of safety-reliability-cost. The model addresses diversity types at different levels: complex electronic components (FPGA, etc.), printed circuit boards, manufacturers, specification languages, design and program languages, etc. It takes into consideration the dependencies among diversity types. The graphical model is developed using the subgraph splitting algorithm, which has been previously used for software test generation.

Key challenges related to MP- and FPGA-based multi-version I&C systems concern uniqueness of ones, specific risks of CCFs (including CCFs for different versions of MVS) existing standards (are not enough detailed), approved diversity-oriented assessment techniques. One of the main challenges related to diversity approach is a fact that multi-version I&C systems are still unique, failures occurred very rarely and information about failures is not enough representative.

Analysis of NUREG 7007-and CLB (GMB)-based assessment techniques allows determining advantages/disadvantages of these techniques, possibilities of their joint applications and tool support. It the chapter three-stages CLD-MAD-RMD-technique for the assessment of multi-version NPP I&C systems is proposed. This technique has got an approbation in the analysis of multi-channel FPGA-based I&C Systems based on RadICS Platform and allows to decide the issue of assessment in conditions of lack of the statistical data about CCF.

REFERENCES

Avizienis, A., Laprie, J.-C., Randell, B., & Landwehr, C. (2004). Basic Concepts and Taxonomy of Dependable and Secure Computing. *IEEE Transactions on Dependable and Secure Computing*, *1*(1), 11–33. doi:10.1109/TDSC.2004.2

Bakhmach, E., Kharchenko, V., Siora, A., Sklyar, V., & Tokarev, V. (2009). Advanced I&C Systems for NPPS Based on FPGA Technology: European Experience. *Proceedings of 17th International Conference on Nuclear Engineering (ICONE 17).*

Bobrek, M., Bouldin, D., & Holkomb, D. (2009). *Review Guidelines for FPGAs in Nuclera Power Plants Safety Systems*. NUREG/CR-7006 ORNL/TM-2009/020.

Bukowsky, J., & Goble, W. (1994). An Extended Beta Model to Quantize the Effects of Common Cause Stressors. *Proceedings of ISAFECOMP*.

Dubois, D., & Prade, H. (1980). *Fuzzy Sets and Systems: Theory and Application*. New York: Academic.

Duzhyi, V., Kharchenko, V., Starov, O., & Rusin, D. (2010). Research Sports Programming Services as Multi-version Projects. *Radioelectronic and Computer Systems*, *47*, 29–35.

Fink, R., Killian, C., & Nguyen, T. (Eds.). (2011). Recommended Approaches and Design Criteria for Application of Field Programmable Gate Arrays in Nuclear Power Plant Instrumentation and Control Systems. EPRI.

González, C. J. (2019). Reducing Soft Error Rate of SoCs Analog-to-Digital Interfaces with Design Diversity Redundancy. *IEEE Transactions on Nuclear Science*.

Gorbenko, A., Kharchenko, V., & Romanovsky, A. (2009). Using Inherent Service Redundancy and Diversity to Ensure Web Services Dependability. In M. Butler, C. Jones, A. Romanovsky, & E. Troubitsyna (Eds.), *Methods, Models and Tools for Fault Tolerance* (pp. 324–341). Springer. doi:10.1007/978-3-642-00867-2_15

Jonson, G. (2010). The INSAG Defense in Depth Concept and D-in-D&D in I&C. *Proceedings of 7th ANS Topical Meeting on NPIC-HMIT*.

Kemikem, D. (2018) Quantitative and Qualitative Reliability Assessment of Reparable Electrical Power Supply Systems using Fault Tree Method and Importance Factors. *2018 13th Annual Conference on System of Systems Engineering (SoSE).*

Kharchenko, V. (1999). Multi-version Systems: Models, Reliability, Design Technologies. *Proceeding of 10th ESREL Conference*, 73-77.

Kharchenko, V. (2018). Multi-Diversity for FPGA Platform Based NPP I&C Systems: New Possibilities and Assessment Technique. *Proceedings of the 2018 26th International Conference on Nuclear Engineering*. 10.1115/ICONE26-82377

Kharchenko, V., Bakhmach, E., & Siora, A. (2009). Diversity-scalable decisions for FPGA-based safety-critical I&C systems: From theory to implementation. *Proceedings of the 6th Conference NPIC&HMIT.*

Kharchenko, V., Duzhyi, V., Sklyar, V., & Volkoviy, A. (2012). Safety Assessment of Multi-version FPGA-based NPP I&C Systems: Theoretical and Practical Issues. *Proceedings of the 5th International Workshop on the Applications of FPGA in Nuclear Power Plants.*

Kharchenko, V., & Illiashenko, O. (2016). Diversity for security: case assessment for FPGA-based safety-critical systems. *MATEC Web Conf., 76.* 10.1051/matecconf/20167602051

Kharchenko, V., Siora, A., & Bakhmach, E. (2008). Diversity-scalable decisions for FPGA-based safety-critical I&C systems: from Theory to Implementation. *Proceedings of the 6th Conference NPIC&HMIT.*

Kharchenko, V., Siora, A., & Sklyar, V. (2011). Multi-Version FPGA-Based NPP I&Cs: Evolution of Safety. In NPP – Control, Reliability and Human Factors. InTech.

Kharchenko, V., Siora, A., & Sklyar, V. (2011). Multi-Version FPGA-Based NPP I&Cs: Evolution of Safety. In NPP – Control, Reliability and Human Factors. InTech.

Kharchenko, V., Siora, A., Sklyar, V., & Volkoviy, A. (2012). Defence-in-Depth and Diversity Analysis of FPGA-based NPP I&C Systems: Conception, Technique and Tool. *Proceedings of the ICONE20.* 10.1115/ICONE20-POWER2012-54349

Kharchenko, V., Siora, A., Sklyar, V., Volkoviy, V., & Bezsaliy, V. (2010). Multi-Diversity Versus Common Cause Failures: FPGA-Based Multi-Version NPP I&C Systems. *Proceedings of the 7th Conference NPIC&HMIT, Las-Vegas, USA, November, 2010.*

Kharchenko, V., & Sklyar, V. (Eds.). (2008). *FPGA-based NPP Instrumentation and Control Systems: Development and Safety Assessment. RPC Radiy, National Aerospace University "KhAI"*. State Scientific and Technical Center for Nuclear and Radiation Safety.

Kharchenko, V., Sklyar, V., Siora, A., & Tokarev, V. (2008). Scalable Diversity-oriented Decisions and Technologies for Dependable SoPC-based Safety-Critical Computer Systems and Infrastructures, *Proceeding of IEEE International Conference on Dependability of Computer Systems*, 339-346. 10.1109/DepCoS-RELCOMEX.2008.21

Kharchenko, V., Sklyar, V., & Volkoviy, A. (2007). Multi-Version Information Technologies and Development of Dependable Systems out of Undependable Components. *Proceedings of International Conference on Dependability of Computer Systems*, 43-50. 10.1109/DEPCOS-RELCOMEX.2007.34

Kharchenko, V., Yastrebenetsky, M., & Sklyar, V. (2004). Diversity Assessment of Nuclear Power Plants Instrumentation and Control Systems, *Proceeding of 7th International Conference on PSAM and ESREL Conference*, 1351-1356. 10.1007/978-0-85729-410-4_218

Mendel, J. M. (2002). An architecture of making judgment using computing with words. *International Journal of Applied Mathematics and Computer Science*, *12*(3), 325–335.

Naser. (Ed.). (2009). *Guidelines on the Use of Field Programmable Gate Arrays (FPGAs) in Nuclear Power Plant I&C Systems.* EPRI.

NP 306.2.202-2015. Nuclear and Radiation Safety Requirements for Instrumentation and Control Systems Important to NPP Safety (Ukraine).

NUREG/CR-6003. Method for Performing Diversity and Defense-in-Depth Analyses of Reactor Protection Systems, United States Nuclear Regulatory Commission, 1994.

NUREG/CR-7007. Diversity Strategies for NPP I&Cs, United States Nuclear Regulatory Commission, 2009.

NUREG/CR-7141. Cyber Security Regulatory Framework for Nuclear Power Reactors, United States Nuclear Regulatory Commission, 2014.

Prokhorova, Y., Kharchenko, V., Ostroumov, B., Ostroumov, S., & Sidorenko, N. (2008). Dependable SoPC-Based On-board Ice Protection System: from Research Project to Implementation. *Proceeding of IEEE International Conference on Dependability of Computer Systems*, 312-317. 10.1109/DepCoS-RELCOMEX.2008.43

Pullum, L. (2001). *Software Fault Tolerance Techniques and Implementation.* Artech House Computing Library.

Siora, A., Krasnobaev, V., & Kharchenko, V. (2009). Fault-Tolerance Systems with Version-Information Redundancy. Ministry of Education and Science of Ukraine, National Aerospace University KhAI.

Siora, A., Sklyar, V., Rozen, Yu., Vinogradskaya, S., & Yastrebenetsky, M. (2009). Licensing Principles of FPGA-Based NPP I&C Systems. *Proceedings of 17th International Conference on Nuclear Engineering (ICONE 17).* 10.1115/ICONE17-75270

Sommerville, J. (2011). Software Engineering (9th ed.). Addison-Wesley.

Tarasyuk, O., Gorbenko, A., Kharchenko, V., Ruban, V., & Zasukha, S. (2011). Safety of Rocket-Space Engineering and Reliability of Computer Systems: 2000-2009 Years. Radio-Electronic and Computer Systems, 11, 23-45.

Vilkomir, S. (2009). Statistical testing for NPP I&C system reliability evaluation. *Proceedings of the 6th American Nuclear Society International Topical Meeting on Nuclear Plant Instrumentation, Controls, and Human Machine Interface Technology (ICHMI 2009).*

Vilkomir, S., Swain, T., & Poore, J. (2009). Software Input Space Modeling with Constraints among Parameters. *Proceedings of the 33rd Annual IEEE International Computer Software and Applications Conference COMPSAC*, 136-141. 10.1109/COMPSAC.2009.27

Volkoviy, A., Lysenko, I., Kharchenko, V., & Shurygin, O. (2008). Multi-Version Systems and Technologies for Critical Applications. National Aerospace University KhAI.

Wood, R., Belles, R., & Cetiner, M. (2009). *Diversity Strategies for NPP I&C Systems.* NUREG/CR-7007 ORNL/TM-2009/302.

Yastrebenetsky, M. (Ed.). (2004). Safety of Nuclear Power Plants: Instrumentation and Control Systems. Technika.

Zadeh, L. (2009). From computing with numbers to computing with words-from manipulation of measurements to manipulation of perceptions. IEEE Trans. Circ. Syst, Fund. Theory Applic., 4(1), 105 –119.

Zadeh, L., & Kacprzyk, J. (1999). *Computing with Words in Information/Intelligent Systems – Part 1:Foundation; Part 2: Applications. Physica-Verlag.*

ADDITIONAL READING

Littlewood, B., & Popov, P. (2000). Littlewood B. Modelling the effects of combining diverse software fault removal techniques. *IEEE Transactions on Software Engineering, SE-26*(12), 1157–1167. doi:10.1109/32.888629

Medoff, M., & Faller, R. (2010). *Functional Safety – An IEC 61508 SIL 3Compatible Development Process. Exida.com L.L.C.*

Smith, D., & Simpson, K. (2004). *Functional Safety. A Straightforward Guide to applying IEC 61508 and Related Standards.* Oxford, UK: Elsevier Butterworth-Heinemann.

KEY TERMS AND DEFINITIONS

Diversity or Multiversity (MV): A principle providing use of several non-trivial versions. This principle means performance of the same function by two and more options and processing of data received in such ways for checking, choice or formations of final or intermediate results and decision-making on their further use.

Multi-Diversion System: MVS in which two or more VR types are applied.

Multi-Version Project (MVP): A project, in which the multi-version technology is applied (version redundancy of processes is used) leading to creation of one- or multi-version system (realization of version redundancy of products).

Multi-Version System (MVS): A system, in which a few versions-products are used; one-version systems may be redundant but consist of a few trivial versions.

Multi-Version Technology: (MVT): Set of the interconnected rules and design actions, in which in accordance with MV strategy a few versions-processes leading to development of two or more intermediate or end-products are used.

Strategy of Diversity: A collection of general criteria and rules defining principles of formation and selection of version redundancy types and a volume or/and choice of MVTs.

Version: An option of the different realization of an identical task by use software, hardware or FPGA-based products and life cycle processes.

Version Redundancy: A type of product and process redundancy allowing to create different (non-trivial) versions.

Chapter 11
Safety and Security Management for NPP I&C Systems

Vladimir Sklyar
National Aerospace University KhAI, Ukraine

Vyacheslav S. Kharchenko
National Aerospace University KhAI, Ukraine

ABSTRACT

The main contribution of this study comprises a set of detailed contents for safety and security management. The following aspects of safety and security management requirements for NPP I&C systems are considered: relation between safety and security management, safety and security management plan, human resource management, configuration management, computer tools selection and evaluation, documentation management, planning of safety and security assessment.

INTRODUCTION

International and national standards introduce requirements to safety management of Instrumentation and Control (I&C) systems. Taking into the account increasing role of information security in nuclear safety assurance we have to consider also requirements to security management. So it makes a sense to discuss safety and security management as a common process implemented for I&C systems.

The standards IEC 61508 "Functional safety of electrical/ electronic/ programmable electronic safety-related systems" has been firstly issued in 1990s. It collected the existing experience of risk-oriented approach in safety-critical industries. The second edition has been issued in 2010. The IEC 61508 includes seven parts. The main idea of the IEC 61508 is that I&C system has to produce risk depending on potential damage for people and environment. The IEC 61508 is an umbrella standard for different industrial brunches including nuclear. In the nuclear industry we have the standard IEC 61513 "Nuclear Power Plants – Instrumentation and Control important to safety – General requirements for

DOI: 10.4018/978-1-7998-3277-5.ch011

systems" which complies with the top level requirements of the IEC. Some nuclear authorities consider the IEC 61508 as an important standard which is applicable in many issues for I&C systems. In Canada compliance with the IEC 61508 is mandatory for I&C systems of Nuclear Power Plants. The IEC 61508 introduces the following groups of requirements:

- Requirements to safety management (this set of requirements can be considered as the umbrella part of safety regulations applicable also for the nuclear industry).
- Requirements to safety life cycle.
- Requirements to random (hardware) failures avoidance.
- Requirements to system and software failures avoidance.
- Requirements to safety assessment.

BACKGROUND

Researches in safety and security management area always have had extremely practical focus directed to achievement of successful licensing or certification. This is a reason why consulting companies pay attention to this issue. Smith and Simpson, 2004 represented a concept and interpretation of IEC 61508 requirements. Medoff and Faller, 2010 provided a detailed description for safety management processes including recommendation concerning compliance with IEC 61508 requirements and templates of managerial plans. Sklyar (2016); Sklyar and Kharchenko (2017) harmonized safety management and security management process provided a common licensing framework.

Based on the results of the above works, the aim of the chapter is to give details of safety and security management process based on the IEC 61508 requirements and applicable for Nuclear Power Plants Instrumentation and Control systems. The main contribution of this study comprises a set of detailed contents for safety and security management.

Safety and Security Management Plan

The umbrella part of requirement to safety and security is related with safety and security management. The safety and security management plan (SSMP) is the document, which states the main safety and security issues and requirements for specific Instrumentation and Control (I&C) system or systems development and operation project. The SSMP covers a set of processes which can be described in a view of separated documents. There are the following safety and security processes which have to be reflected in the SSMP (see Figure 1):

- Human resource management;
- Configuration management;
- Tools selection and evaluation;
- Verification and validation;
- Requirements tracing;
- Documentation management;
- Safety and security assessment.

Figure 1. Structure of the safety and security management plan (SSMP)

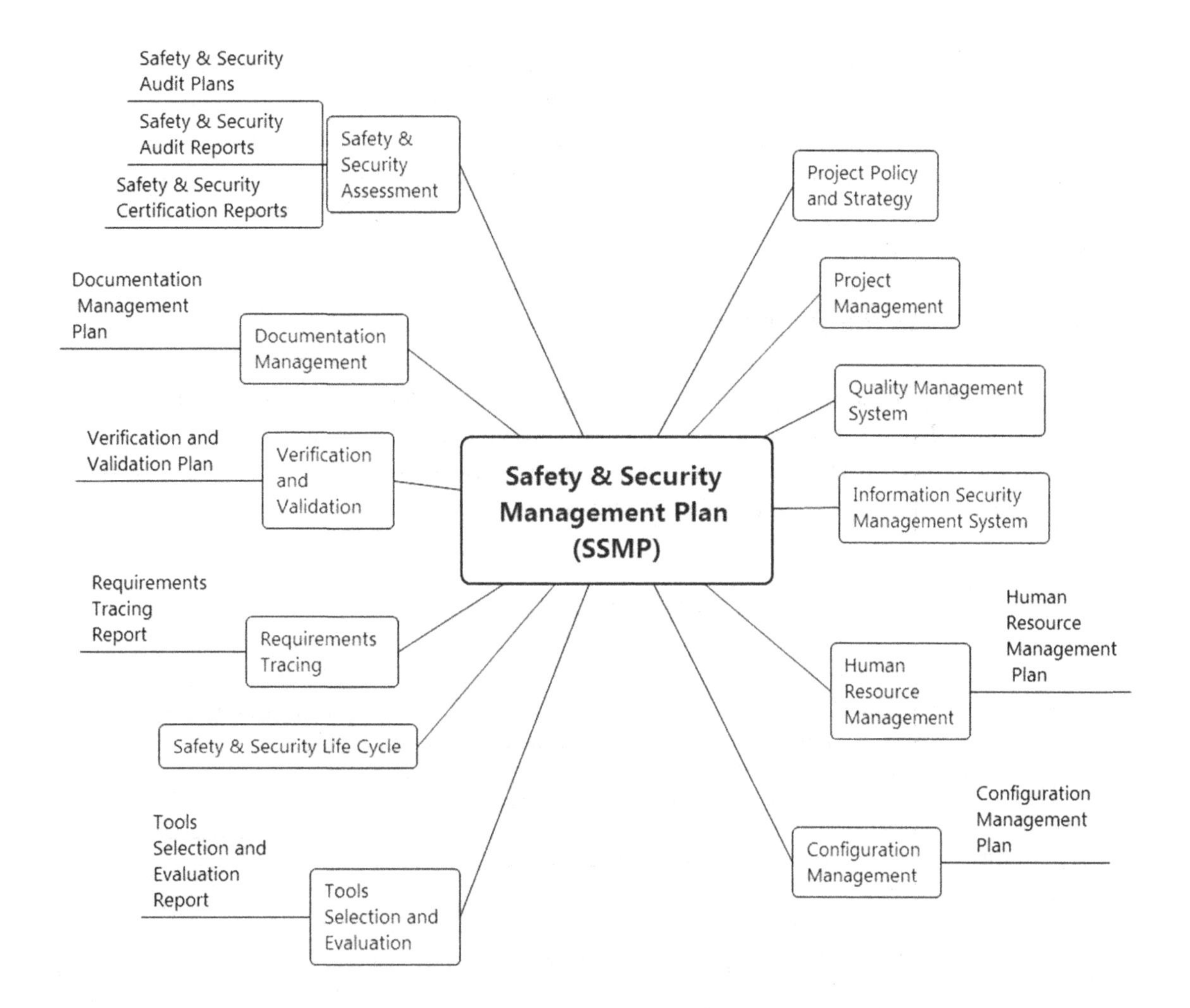

All the mentioned processes are described below in this chapter. Also the SSMP has to cover the following issues (see Figure 1):

- Project policy and strategy is a declarative description of how and why the goals of the project will be achieved;
- Project management is reasonable applicable to project performance since, for example, the IEC 61508-2 (Annex B) requires applying this method to protect the product against systematic failures;
- Quality management system it important to implement quality for all products and processes; special attention is paid to interaction with suppliers of products and services that affect safety and security;
- Information security management system (ISMS) has to cover activities in accordance with requirements of ISO/IEC 27000 “Information technology – Security techniques – Information security management systems” or any other relevant ISMS framework;

- Safety & security life cycle has to be described in SSMP stage by stage.

All the above activities cover both safety and security issues. Additionally ISMS has to cover activities like the following: asset management, identification and authentication, access control, system perimeter protection, work stations, servers, and other devices protection, network and communications protection, cloud infrastructure protection, database protection, cryptography, monitoring and recovery, incidents response and investigation. All appropriate measures and activities have to be implemented for the considered I&C system.

Human Resource Management

For detailed personnel management planning, an appropriate the human resource management plan has to be developed. Note that this plan does not apply to the organization as a whole, but only to the participants in the project of I&C system development and certification against safety and security requirements. The personnel management plan should contain (see Figure 2):

Figure 2. Structure of the human resource management plan

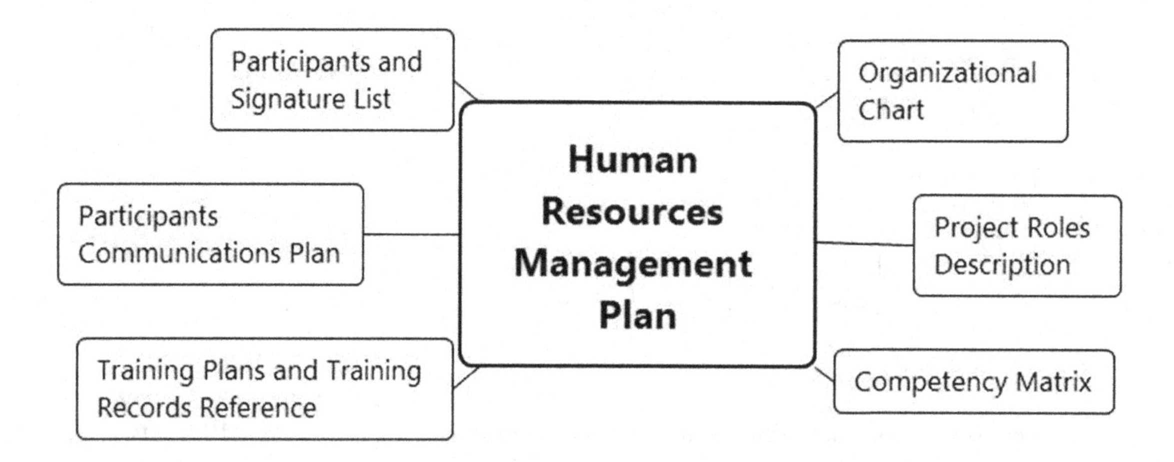

- Organizational chart of the project with a description of project roles;
- A list of project participants indicating project roles and responsibilities for planning and performing work at various stages of the life cycle;
- The competence matrix and the conclusions on the adequacy or lack of competencies of the appointed performers, i.e. what knowledge and skills are required for a particular project role and to what extent a particular employee corresponds to them;
- Personnel training activities aimed at achieving and maintaining the above mentioned competences that are critical for the implementation of the project; training plans and reports should be documented;
- Communication plan for the project participants;
- A list of the signatures of personnel, indicating the familiarization with this plan.

Configuration Management

When defining configuration items in the context of safety and security, it is important to understand that they include not only source codes and program builds, but also development and testing tools, a complete set of design, user, and any other relevant documentation, including design documentation, according to which all mechanical, electrical and electronic components of I&C systems are manufactured (see Figure 3). Such structure can serve as the basis for the project repository.

Figure 3. A set of configuration items of I&C system

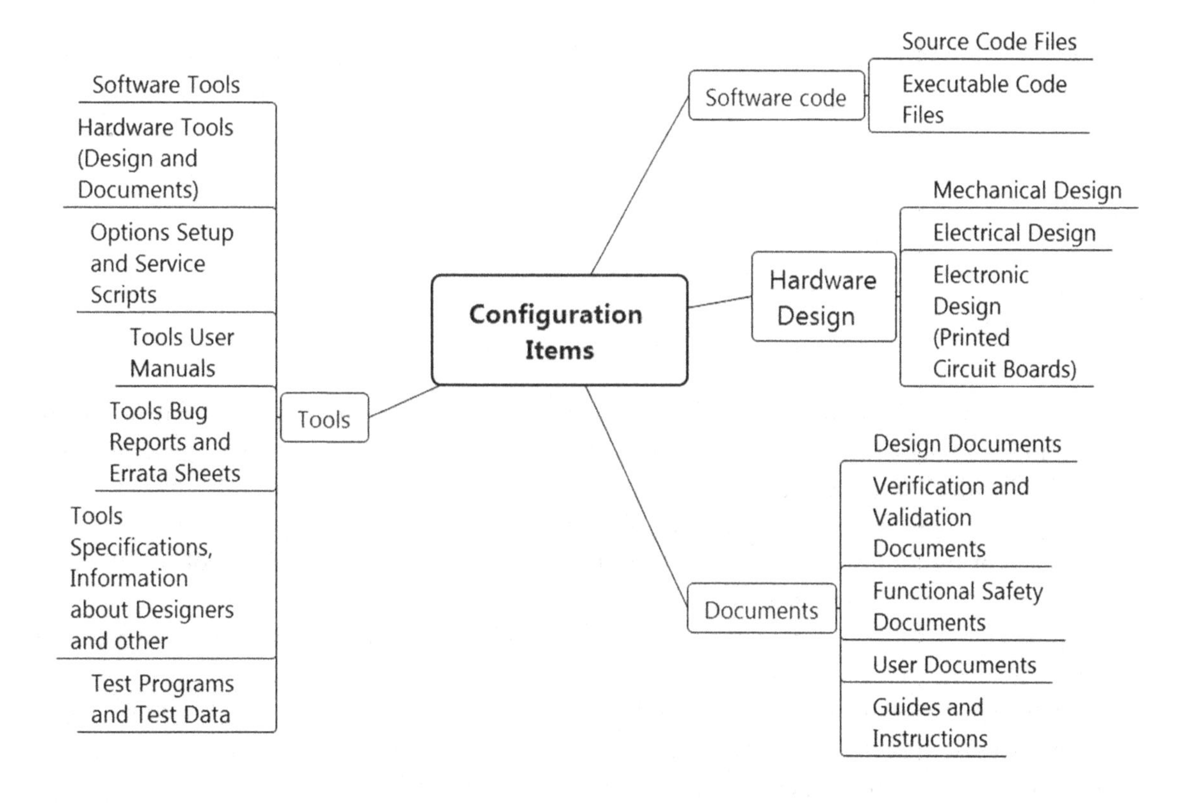

Configuration management directly depends on the used electronic document management tools, however, some the following general points can be included in the configuration management plan (see Figure 4):

Figure 4. Structure of the configuration management plan

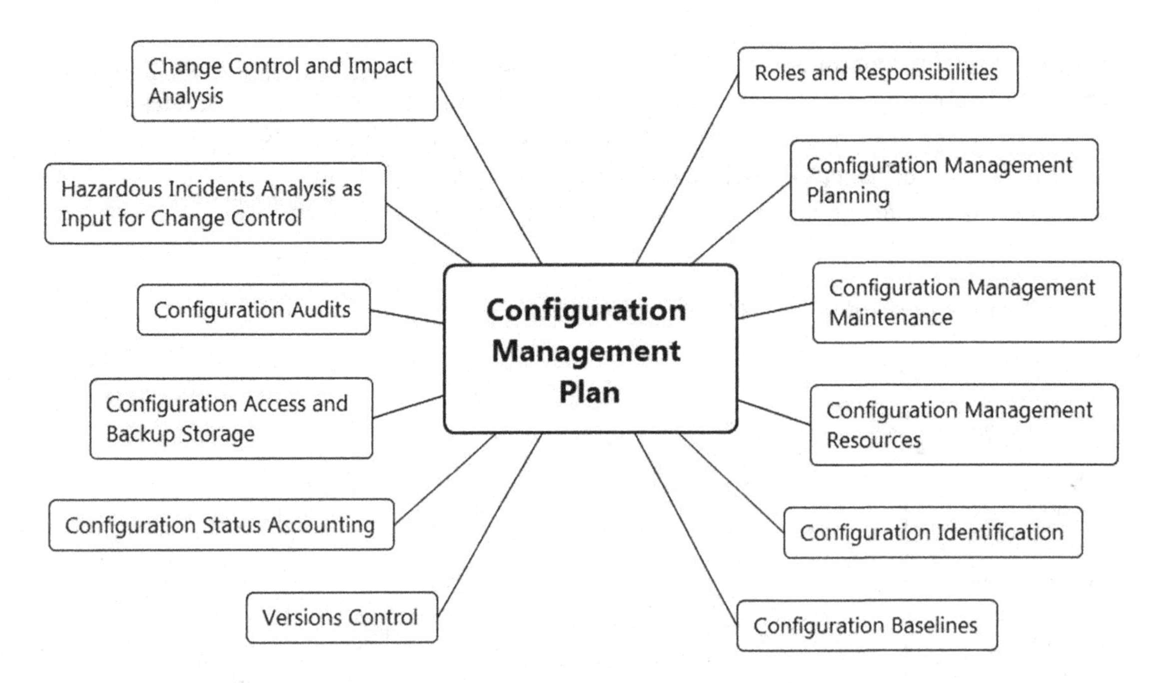

- The roles and responsibilities of project participants in the configuration management process; the Configuration Management & Change Control Board of the key project participants should be organized with all those, whose opinions are important to consider when making changes;
- An approach to planning and maintaining the configuration management process;
- Resources of the configuration management process, first of all, the applied tools of electronic document management (SVN, Git, etc.);
- The procedure for the identification of the configuration items and the formation of baselines (basic versions);
- The procedure for applying tools to control the versions of software and hardware components of the product and to account for their status;
- The procedure for accessing configuration components and backup storage;
- The procedure and periodicity for configuration audits;
- The procedure for analyzing and eliminating the detected defects and bugs including those found during operation;
- The procedure for change control, including impact analysis and validation of changes.

Tools Selection and Evaluation

The IEC 61508 "Functional safety of electrical/ electronic/ programmable electronic safety-related systems" states the following tools classification depending on the degree of influence on the final product, system, or software (see Figure 5):

- Class T1 tools do not generate any outputs that directly affect the executable code; it includes text and image editors, configuration management tools (those do not directly generate code), action & bug trackers;
- Class T2 tools support testing and other types of verification and validation (for example, static code analysis or test coverage analysis); there is no direct impact on the executable code, however, a problem in the test tools may lead to errors in the software that may not be detected; this class should include not only software, but also software / hardware simulators of input / output signals; it should be noted that design tools for mechanical, electrical and electronic components (for example, printed circuit boards design tool) can also be assigned to class T2;
- Class T3 tools generate outputs that directly affect the executable code, such as translators and compilers that are components of Integrated Development Environments (IDE) & Software Development Kits (SDK), scripts to support builds and controller logic configuration.

Figure 5. Tools classification

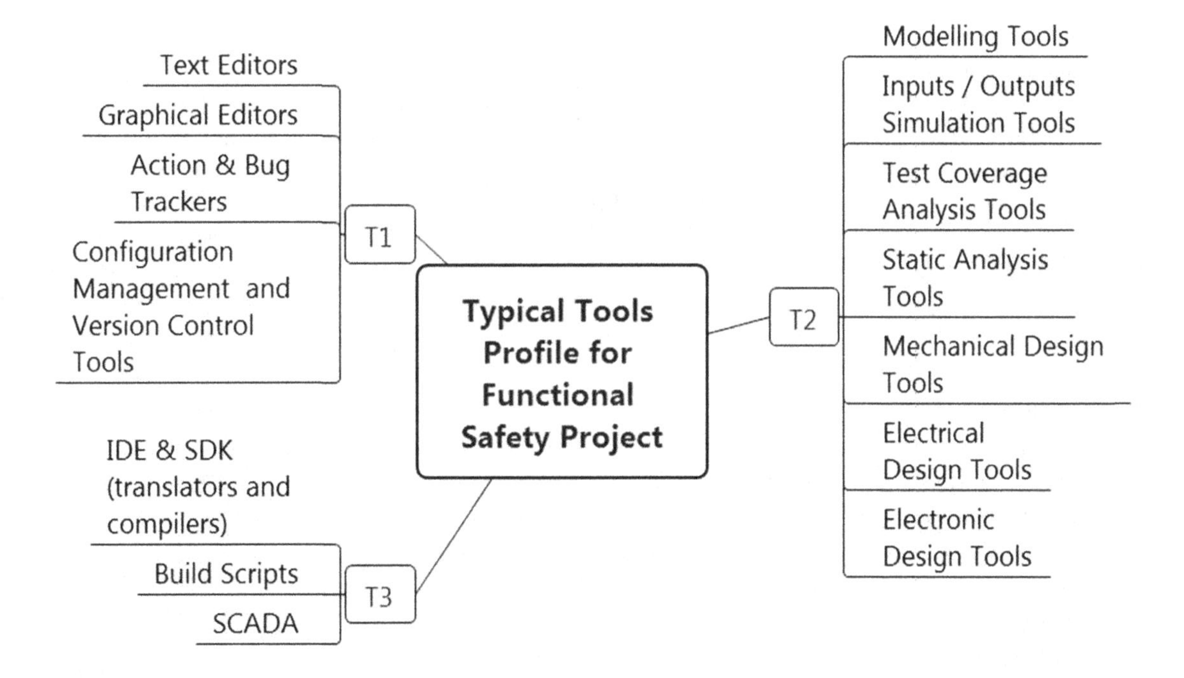

To ensure compliance with safety and security requirements, it is advisable to develop a special report on the selection and evaluation of tools that shall cover the following issues (see Figure 6):

Figure 6. Structure of the tools selection and evaluation report

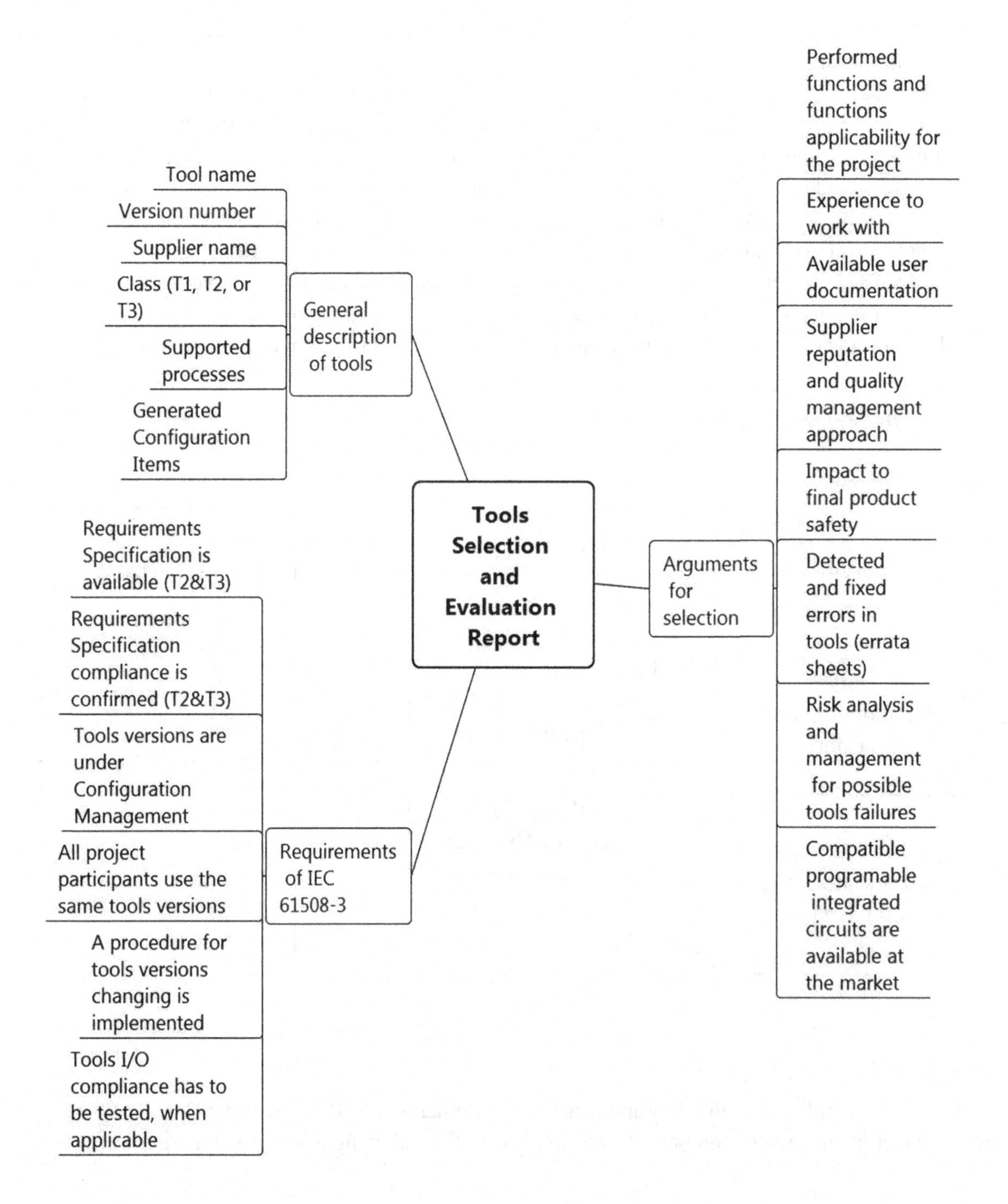

- A description of the used stack of tools (both software and hardware, both commercially available and in-house) used for product development, testing, and supporting processes (configuration management, documents processing, project management, etc.) for each of the tools you should specify: type (to support which process is used), name, version number, supplier name, class (T1, T2 or T3), as well as generated outputs in terms of configuration items;

- Results of evaluation (analysis) of tools according to a set of predetermined criteria, such as, for example: the functions performed and their applicability in this project, experience of use, available documentation, information about the supplier (market reputation, quality management system, approach to configuration management and etc.), the impact on the safety of the product, the errors found and eliminated, the possible risks of use in terms of failures and the strategy for managing these risks, the availability of compatible products on the market programmable chips (for software development and electronic projects);
- The results of the analysis for compliance with the requirements for the tools specified in IEC 61508-3, such as:
- for tools of classes T2 and T3 requirements specifications or user documentation should be available that uniquely describe how the operation takes place;
- for tools of classes T2 and T3 their compliance with the requirements specification or user documentation has to be documented (for example, in the form of a certificate);
- the versions of the tools used should be monitored, since not all versions can meet the specified conditions; all project participants must use the same version; for transitions between versions the appropriate procedure should be applied;
- if the tools are used as a single technological complex (for example, code and tests are generated based on the specification), their compatibility with each other should be tested.

Figure 7. Structure of the documentation plan

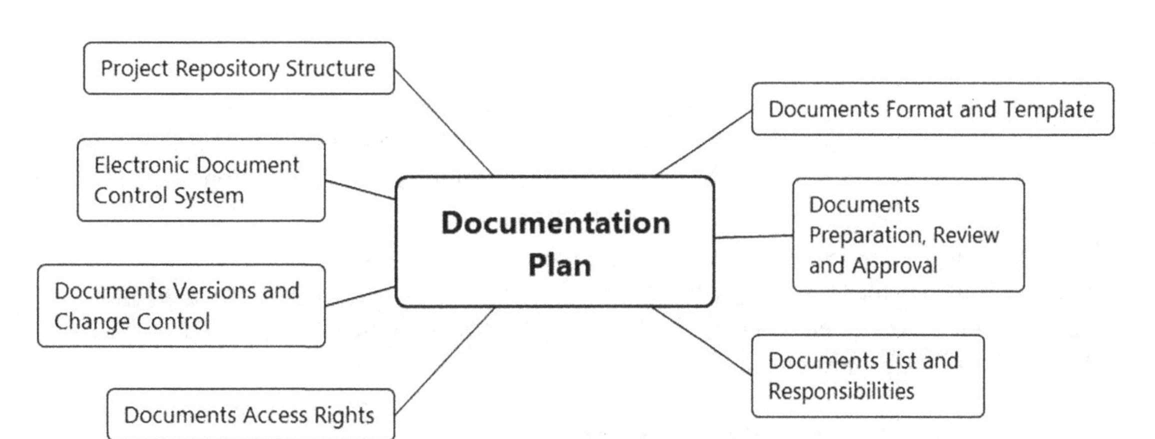

Documentation Management

For detailed documentation management a related documentation plan has to be developed. That plan does not apply to the organization as a whole, but only to the participants of a considered project for developing a product important to safety and security. The documentation plan has to cover the following issues (see Figure 7):

Figure 8. Structure of functional safety audit plan

- Requirements to identification, development, execution, coordination and approval of documents;
- Review procedures and criteria for evaluating documents (for example, in the form of checklists);
- A list of project documents and allocation of responsibility for the development, review and approval;
- The procedure for access to documents and access rights of project participants;
- The procedure for making changes to documents, accounting policy and version changes;
- Requirements to use of electronic document management system;
- A structure of the project repository.

Management of Safety and Security Assessment

Periodic safety and security audits are conducted for assessing safety and security during the project. These audits can be either internal (conducted by the project team) or external (performed by the third party). Another kind of audits is the certification audit, which is conducted upon completion of the project work by the certification authority. According to the results of the certification audit, a certificate of compliance with the standards requirements is issued. In addition, the certification authority may also participate in periodic audits. Audits should be conducted according to pre-developed plans. In the audit plan, the following issues have to be defined (see Figure 8 as an example for functional safety audit):

- Periodicity of audits (for example, at the completion of each of the development stages);
- Areas of assessment in terms of the structure of products and processes;
- Involved participants, organizations and other required resources (temporary, financial, required tools, etc.);
- The level of independence of auditors; as noted above, audits can be internal and external; in general, the issue of independence in evaluating safety has its traditions in various industries and countries;
- Competencies of the employers performing the audit;
- Expected results;
- Corrective actions performance;
- An approach to document audit results and requirements for the content of audit report, which shall be issued based on the results of audits;
- Checklists, including a specific set of requirements (issues), compliance with which should be evaluated during the audit; the initial data for compiling an audit checklist are the requirements of SSMP and other plans related to ensuring of safety and security.

SOLUTIONS AND RECOMMENDATIONS

In this section, the abstract enough requirements of the IEC 61508 for safety and security management are interpreted for implementation in working processes. This approach has been tested in practice by successful certification projects. Safety requirements are integrated with security a requirement that allows implementing the common safety and security management framework.

To comply with the managerial requirements it is recommended to develop a special document named the safety and security management plan. The main areas of management requirements are personnel management, configuration management, tools selection and evaluation, documentation management, and safety and security assessment.

For detailed planning of personnel management, an appropriate plan (human resource management plan) is being developed, which does not apply to the organization as a whole, but only to participants of the certification project.

Configuration management depends on the used electronic document management tools. To implement the process, it is necessary to develop the configuration management plan. Configuration items include not only program code components, but also development and testing tools, a full set of design, user and any other relevant documentation, including design documentation, according to which all mechanical, electrical and electronic components of I&C systems are produced.

Software and hardware tools are classified and evaluated depending on the degree of influence on the final product (system and software). Most attention is paid to tools that directly generate program code, like translators and compilers. According to the requirements of IEC 61508, when developing software, only those compilers and translators that have passed full validation should be used. Many developers of compilers and translators themselves carry out their certification for compliance with the requirements of IEC 61508. It is advisable to document the results of the tools evaluation in the appropriate report on the selection and evaluation of tools.

For detailed documentation management planning, an appropriate documentation plan has to be developed. The types of documents developed in the certification project include: documents for planning, requirements specification, system architecture, hardware design, software project, documents for verification and validation, documents related to tools (user manuals, specifications requirements, certificates of conformity, information on suppliers), change management documents, user manuals, procedures and instructions used in the project for the organization of work, as well as documents for safety and security assessment.

To assess the safety and security during the course of the project, periodic audits are carried out. These audits can be either internal (conducted by the certification project team) or external (performed by persons not participating in the project, including a third-party organization). Audits should be conducted in accordance with audit plans. Based on the results of the audits, relevant audit reports are issued.

FUTURE RESEARCH DIRECTIONS

We propose to consider the following future research directions for the safety and security management area:

- Close integration of the processes related to safety and security;
- Formalized and automated implementation of new appeared standards requirements to safety and security management;
- Formalized and automated assessment of safety and security management processes compliance with regulatory requirements.

CONCLUSION

A set of requirements related with safety and security management is considered in this chapter. The management requirements are divided into subsets including human resource management, configuration management, computer tools selection and evaluation, documentation management, and planning of safety and security assessment. The structure for the umbrella safety and security management plan is proposed, as well as the structures of the additional lower level planning documents. All the above allows to create a regulatory framework, which is applicable for Nuclear Power Plants I&C systems.

REFERENCES

Medoff, M., & Faller, R. (2010). *Functional Safety – An IEC 61508 SIL 3 Compatible Development Process. Exida L.L.C.*

Sklyar, V. (2016). Safety-critical Certification of FPGA-based Platform against Requirements of U.S. Nuclear Regulatory Commission (NRC): Industrial Case Study. *Proceedings of the 12th International Conference on ICT in Education, Research and Industrial Applications*, 129-136.

Sklyar, V., & Kharchenko, V. (2017). Assurance Case Driven Design based on the Harmonized Framework of Safety and Security Requirements. *Proceedings of the 13th International Conference on ICT in Education, Research and Industrial Applications*, 670-685.

Smith, D., & Simpson, K. (2004). *Functional Safety. A Straightforward Guide to applying IEC 61508 and Related Standards*. Oxford, UK: Elsevier Butterworth–Heinemann.

ADDITIONAL READING

EPRI 1015313. (2010). *Computerized Procedure Systems: Guidance on the Design, Implementation, and Use of Computerized Procedure Systems, Associated Automation, and Soft Controls*, Electric Power Research Institute, Palo Alto, CA, USA.

IAEA. (2007). Terminology used in Nuclear Safety and Radiation Protection: IAEA Safety Glossary.

IAEA. (2011). IAEA NP-T-3.12. Core knowledge on instrumentation and control systems in nuclear power plants.

IAEA. (2016). IAEA SSG-39. Design of instrumentation and control systems for nuclear power plants.

IAEA (2018). IAEA NP-T-2.11. Approaches for Overall Instrumentation and Control Architectures of Nuclear Power Plants.

IEC. (2010). IEC 61508. Ed.2. Functional safety of electrical/electronic/programmable electronic safety related systems.

IEC. (2011). IEC 61513. Ed.2. Nuclear power plants – Instrumentation and control important to safety – General requirements for systems.

U.S. Nuclear Regulatory Commission. (2018). NUREG-0800. Standard Review Plan for the Review of Safety Analysis Reports for Nuclear Power Plants. Chapter 7. Instrumentation and control.

Chapter 12
Assurance Case for I&C Systems Safety and Security Assessment

Vladimir Sklyar
National Aerospace University KhAI, Ukraine

Vyacheslav S. Kharchenko
National Aerospace University KhAI, Ukraine

ABSTRACT

Two existing notations for the assurance case (claim, argument and evidence [CAE] and goal structuring notation [GSN]) are considered. Supporting software tools for development of the assurance case are considered. Some ways for improvement and modification are proposed for both assurance case notations (CAE and GSN). For CAE, the authors obtained annex with acceptance and coverage criteria as well as an algorithm of the assurance case update through life cycle stages. For GSN, they improve structured argumentation with support of structured text using. Recommendations for using the assurance case notations and tools for I&C systems are formulated.

INTRODUCTION

Final safety and security assessment is running after completion of all development, verification and validation stages. In this section we discuss how can all project artifacts be represented for safety and security assessment, and what is the way to most effectively confirm compliance with the safety and security requirements? The answer to these questions is provided by the Assurance Case methodology, which is widely used in the practice of safety and security assessment.

The Assurance Case is a structured set of arguments and documentary evidence that justify the compliance of a system or service with specified requirements (GSN, 2011).

Licensing and certification authorities check the Assurance Case, as an integral document proving compliance with the entire set of requirements to safety and security. The Assurance Case can be either compiled by the project team or outsourced.

DOI: 10.4018/978-1-7998-3277-5.ch012

The Assurance Case is used to assess the safety and security of control systems in such fields as energy, automotive, railway, weapons, aviation and space technology, medicine, petrochemical, and others (Evidence, 2011). Researchers identify the following four goals to be achieved for systems (Rushby, 2015):

1. Requirements meet the needs of the customer;
2. Assumptions about the use of systems correctly describe the environment of use;
3. The system design meets the requirements and assumptions regarding the application;
4. The developed system corresponds to the documented design.

There is an international community in the area of the Assurance Case that is engaged in theoretical research and practical application of this methodology. Important acquisitions include the development of documents such as the GSN Community Standard (GSN, 2011) and the Structured Assurance Case Metamodel (SACM, 2016).

Thus, the Assurance Case is a proven methodology with a 20-year history of application, which is constantly evolving and getting support from leading safety and security experts. It seems appropriate to use the Assurance Case in the practice of certification and licensing of Instrumentation and Control (I&C) systems.

BACKGROUND

The historical and theoretical origins of the Assurance Case lie in the field of logical reasoning, such as operations with logical predicates, including the implication In 1958, the British philosopher Stephen Tulmin published the book "The Uses of Argument", Toulmin, 1958, in which he expanded the operation of logical inference with the degree of confidence and additional arguments and counter-arguments. In addition, Toulmin proposed to present the argument in graphical form, and this approach has since become widespread.

The predecessor of the Assurance Case is historically the Safety Case. The concept of the Safety Case originated in the 1950s, although the term itself appeared later. The first regulatory document requiring the development of a Safety Case for hazardous industrial facilities is the European Union's "CIMAH (Control of Major Accidents Hazards) Regulations". The widespread introduction of the Safety Case into practice began to occur after an unprecedented accident on the Piper Alpha oil platform in the North Sea, which claimed the lives of 167 people in 1988, Cullen, 1990.

All of the above has led to new approaches in safety assessment and assurance. In the 1990s. Tulmin's argument was used as the basis for the development of semi-formal notations to justify safety, Kelly, 1998. The work was done in the UK, at the University of York, where Goal Structuring Notation (GSN) was developed. Adelard developed the Claim, Argument and Evidence (CAE) notation in parallel. These notations are used in the present, and then we consider them in more detail, Maksimov et al., 2018.

Initially, the focus was on functional safety issues (Safety Case), then with the advent of the information security problem, a similar approach was extended to the Security Case, and with it came the understanding that it was necessary to work simultaneously on providing both safety and security features. Currently, the term Assurance Case means the justification of both safety and security.

We found the following areas of the Assurance Case applications:

- The Assurance Case for attributes assessment allows to develop Dependability Case, Quality Case, Risk Case etc. combining different kinds of critical attributes of software, systems and infrastructures. Firesmith et al., 2006;
- The Assurance Case based certification is going to integrate the Assurance Case regime with existing practices in certification as well as to extract requirements from standards. Hawkins et al., 2013;
- Assurance Based Development considers simultaneous design of systems and their assurance argumentation, which finally shall be represented in a view of the Assurance Case, Graydon et al., 2009;
- Assurance Case for knowledge management supports associated activities such as business management strategy, change and maintenance management, documents management, software test management, etc., Kobayashi et al., 2018;
- Improvement of argumentation is directed to improve confidence and to eliminate uncertainty in the Assurance Case argumentation with qualitative or quantitative approaches, Rushby, 2015, Goodenough et al., 2015.

Among the numerous applications of the Assurance Case, it should be noted a notable class of works aimed at improving the argumentation in the construction of the Assurance Case structure, GSN, 2011. The roots of this question lie in the fundamental problems of knowing a person around the world. On the one hand, there are theoretically insufficient arguments that would be supported by completely satisfactory evidence. On the other hand, one of the expectations of using the Assurance Case is that, ideally, it will help correct safety and security problems, even if they do not explicitly follow from the available documentation.

To address this problem, researchers have turned to the foundations of epistemology, which takes into account the study of the nature of knowledge, the rationale and rationality of trust ("what makes reasoned trust really valid?").

Among the works devoted to the problem of argumentation for the purpose of constructing speakers, the following directions should be identified, Rushby, 2015:

- The application of the "eliminative induction" paradigm (which originates from the work of Francis Bacon) to obtain a qualitative assessment of confidence in a goal, which, from this point of view, is a hypothesis; doubts about the validity of a hypothesis are identified and eliminated, and this can be represented, for example, as a graphical map with specific notation;
- Assurance Case converting into different notations to increase confidence in the correctness of arguments;
- Using of quantitative probabilistic approaches (for example, Bayesian trust networks or Dempster-Schaefer theory) to determine the likelihood of confidence in the correctness of arguments, and the use of combinations of logical and probabilistic methods;
- Structuring arguments and applying different types of arguments.

However, there are some shortcomings in the existing works, which are due to the lack of satisfactory practical argumentation techniques. Thus, in order to apply the Assurance Case methodology, in order to ensure and evaluate the safety and security of I&C systems, Sklyar & Kharchenko, 2017, it is necessary to select and improve the appropriate mathematical and methodological apparatus for structuring the system of argumentation, and this is the objectives of the chapter.

ASSURANCE CASE NOTATIONS AND TOOLS

Claims, Arguments and Evidence (CAE) Notation Description

The CAE (Claim, Argument and Evidence) notation operates with three specified entities: claim indicates the achievement of the required system properties, evidence provides a documented basis for argumentation, demonstrating the achievement or non-achievement of goals, and arguments are built using inference rules and link evidence with objectives. Arguments such as deterministic (or logical), probabilistic, and qualitative are commonly used. To designate claims, arguments and evidence, graphic primitives are introduced that have different shapes (see Figure 1).

Figure 1. Claim, Argument and Evidence (CAE) notation: main components

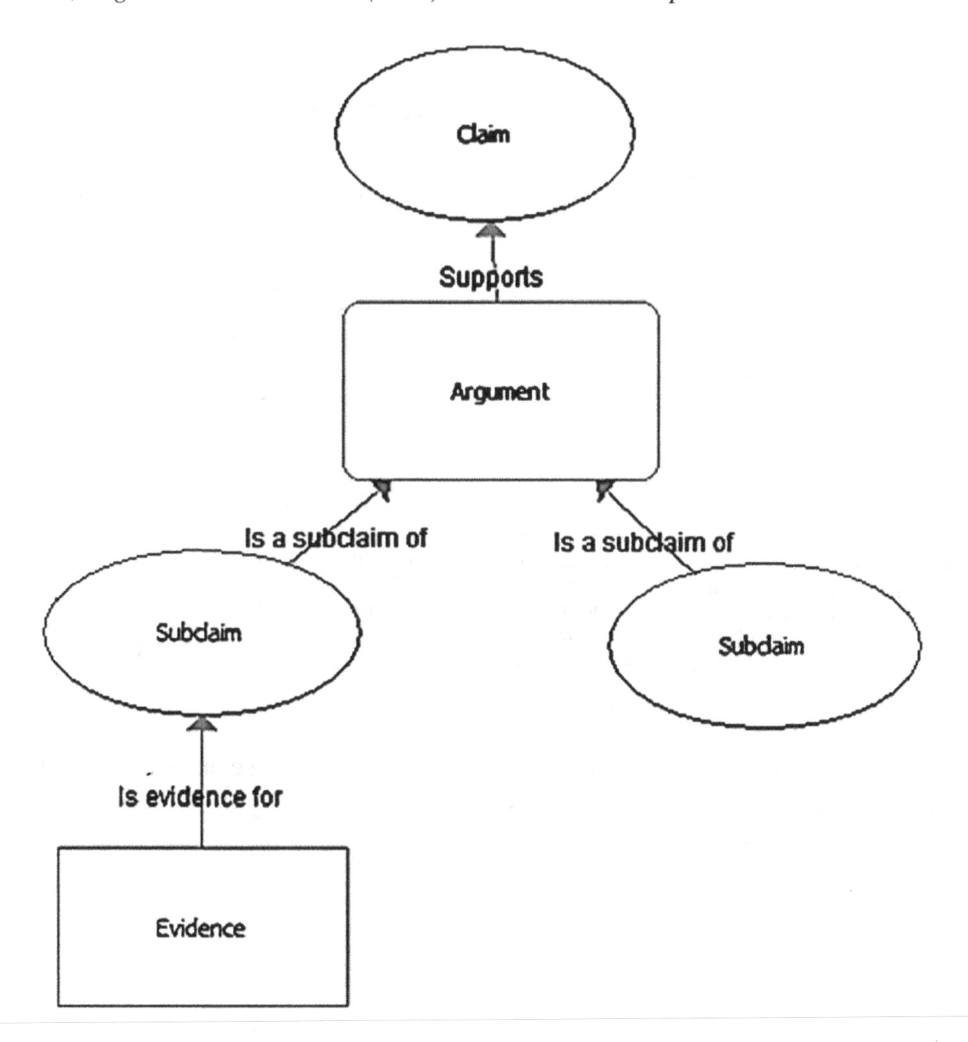

Building a hierarchy of goals and sub-goals is the first step in the development of the Assurance Case. As shown in the diagram (see Figure 1), the structure of goals, arguments and evidence is not necessarily three-level, for example, additional sub-goals can be used to support the argument.

As an example of using CAE notation, consider the general case of the formation of requirements for system functional safety (Ye & Cleland, 2012). The main goal is adequate, accurate and complete wording of the requirements. For this, the following subgoals must be achieved (see Figure 2):

- Requirements for the management of functional safety have to be defined;
- Regulatory requirements established in standards, laws and other regulatory documents have to be defined;
- Safety criteria have to be defined;
- Integration requirements have to be defined.

Figure 2. CAE notation: an example for functional safety

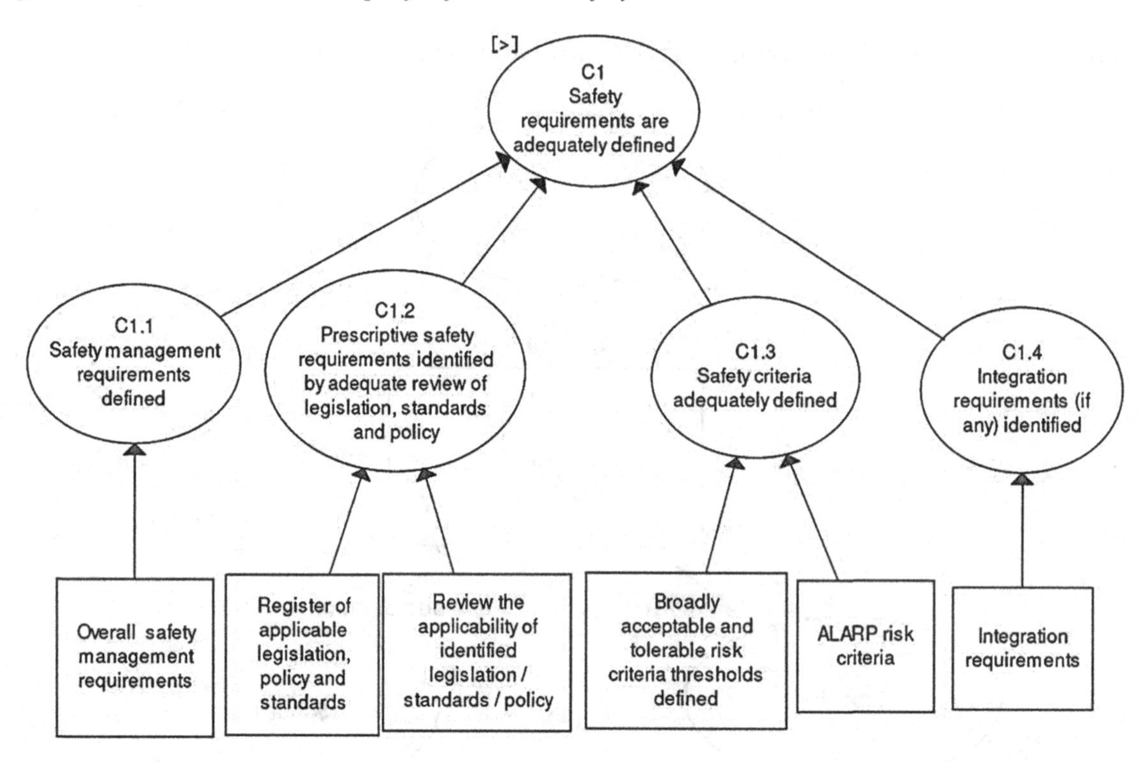

This diagram does not show the argumentation system, since this is a general case, and the argumentation strategy may be different. The requirements stated in regulatory documents, the results of risk analysis, etc. are used as evidence.

Update and Application of Claims, Arguments and Evidence (CAE) Notation

Usually CAE notation is applied in graphical view, but tabular view can also be used. Claim, Argument and Evidence should be located respectively in the fields of the table (see Table 1). Let's consider an example from the standard IEC 61508 "Functional safety of electrical/ electronic/ programmable electronic safety-related systems" relating to personnel management, Sklyar & Kharchenko, 2017.

Table 1. A table view of CAE notation

IEC 61508	Claim	Argument	Evidence
1/6.2.1	Responsibilities of the project participants	**HR1**: Organizational Chart. **HR2**: Project Roles Description	–
1/6.2.3	Understanding by the project participants of their roles and responsibilities	**HR6**: Participants and Signature List	–
1/6.2.4	Communications of the project participants	**HR5**: Participants Communications Plan	–
1/6.2.13	Evaluation and assurance of the project participants competencies	**HR3**: Competency Matrix. **HR4**: Training Plans and Training Records Reference	–
1/6.2.14	Issues affected to the project participants competencies	**HR3** **HR4**	–
1/6.2.15	Documentation of the project participants competencies	**HR3** **HR4**	–
1/6.2.16	Monitoring of safety management processes	**HR1** **HR2**	–

Table 1 describing CAE contains fields used according to the following purpose:

- IEC 61508 – a reference to part (before the slash "/") and clause of IEC 61508;
- Claim – a brief statement of the requirement; note that for convenience, the entire requirement can be placed in a table according to the text of the standard; in the table under consideration, only those requirements related to personnel management are selected;
- Argument – an approach to represent compliance with the requirement; several approaches can be applied to ensure compliance with the same requirement (one-to-many relationship), and the same approaches can be used for different requirements (many-to-one relationship or many-to-many relationship); if we consider the human resource management plan, it becomes clear that its structure is determined by the arguments derived from the requirements of IEC 61508; a graphical representation of the structure of the human resource management plan confirms the effectiveness of using the graphic Mind Map notation for a simplified description of the Assurance Case; the arguments are assigned the numbered identifiers from HR1 to HR6, also according to the order of their entry into the structure of the human resource management plan;
- Evidence – in this example, that field of the table is not populated, since the assessment of compliance with the requirements is determined for each specific project based on an audit of the developed documents and the implemented processes.

The table describing the Assurance Case may also include fields for independent evaluation by a third party and description of corrective actions. Consider, by the example of the human resource management plan, the application of the process approach to managing and evaluating the safety at all stages of the life cycle. To fulfill this task, we modify the CAE notation. A reasoning strategy may be supported by compliance criterion and coverage criterion. Compliance criterion clarifies how compliance with requirement and claim can be achieved. Coverage criteria apply to multiple hierarchical requirements (for example, when all requirements must be verified during the testing process). Thus, CAE notation is transformed into CAEC notation (Claim, Argument, Evidence and Criteria) (see Figure 3).

Figure 3. Claim, Argument, Evidence and Criteria (CAEC) notation: main components

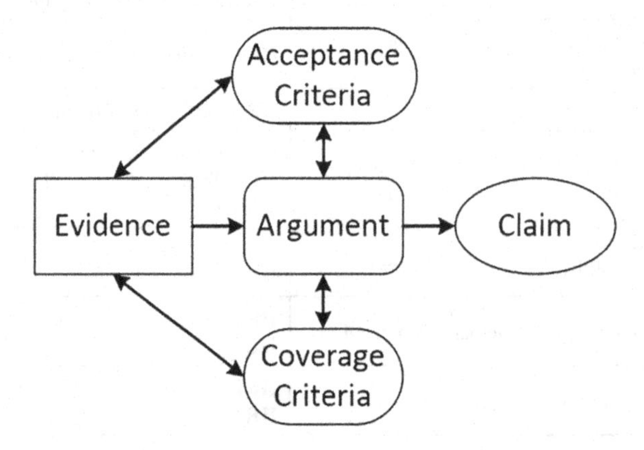

The second component of the amended methodology is the notation describing the promotion of the Assurance Case through the stages of the life cycle. V-shaped life cycle is implemented for I&C system, which includes phased development and phased verification and validation. Thus, the Assurance Case must be supplemented after each of the stages of development, verification and validation (see Figure 4).

Figure 4. The relationship between the components of the life cycle (development, verification and validation, assurance case)

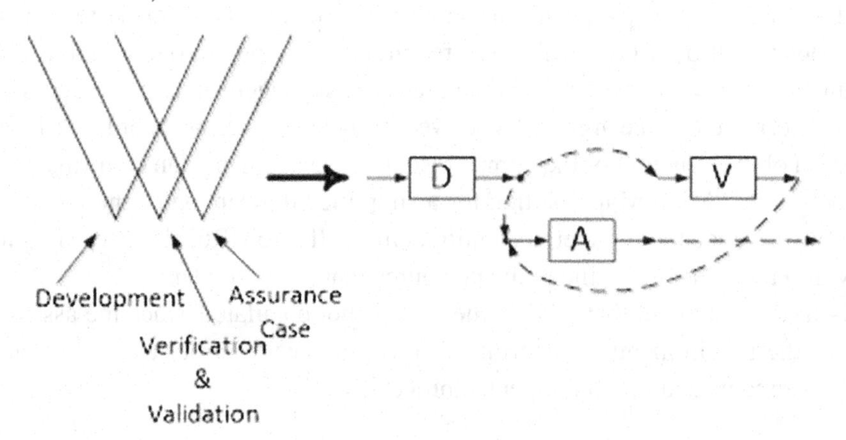

This approach is described in the form of DVA notation (see Figure 5), what means Development, Verification & Validation, and Assurance Case.

Figure 5. The relationship between the components of the life cycle (development, verification and validation, assurance case)

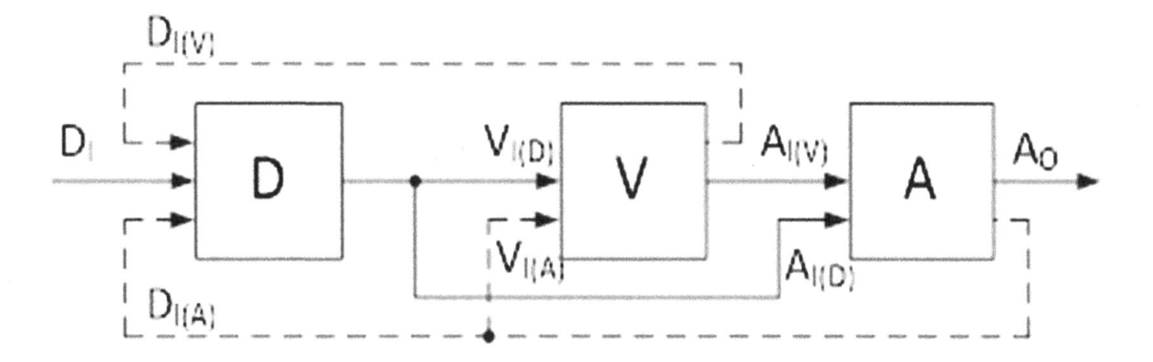

The DVA notation includes the following data sets transmitted between components:

- $D_I = \{d_{i1}, d_{i2}, ..., d_{iK}\}$ – input development process data transmitted from the previous stage of the life cycle;
- $V_I(D) = \{v_{id1}, v_{id2}, ..., v_{idL}\}$ – the input data of the verification and validation process transmitted from the development process;
- $A_{I(D)} = \{a_{id1}, a_{id2}, ..., a_{idM}\}$ – input data of the Assurance Case process, transmitted from the development process;
- $A_{I(V)} = \{a_{iv1}, a_{iv2}, ..., a_{ivN}\}$ – input data of the Assurance Case process, transmitted from the verification and validation process;
- $D_{I(V)} = \{d_{iv1}, d_{iv2}, ..., d_{ivP}\}$ – input development process data transmitted from the verification and validation process (feedback);
- $D_{I(A)} = \{d_{ia1}, d_{ia2}, ..., d_{iaQ}\}$ – input development process data transmitted from the Assurance Case process (feedback);
- $V_{I(A)} = \{v_{ia1}, v_{ia2}, ..., v_{iaR}\}$ – input data of the verification and validation process transmitted from the Assurance Case process (feedback);
- $A_O = \{a_{o1}, a_{o2}, ..., a_{oS}\}$ – output data of the Assurance Case process (this is also output data of the life cycle stage), transmitted to the input of the next life cycle stage after resolution of all findings and anomalies.

The application of the considered CAEC and DVA notations constitutes an approach called Assurance Case Driven Design, Sklyar & Kharchenko, 2017. The goal of the approach is to reduce certification costs by consistently preparing and correcting the Assurance Case, starting from the very first stages of the life cycle. Thus, the Assurance Case supports and guides the development, verification and validation process.

From the point of view of life cycle organization, the application of the Assurance Case methodology should be coordinated during development, quality assurance, safety and safety assurance, as well as during assessment and certification, like DevOps (development and operation) methodology (see Figure 6).

Figure 6. The diagram of components interaction for development of the assurance case

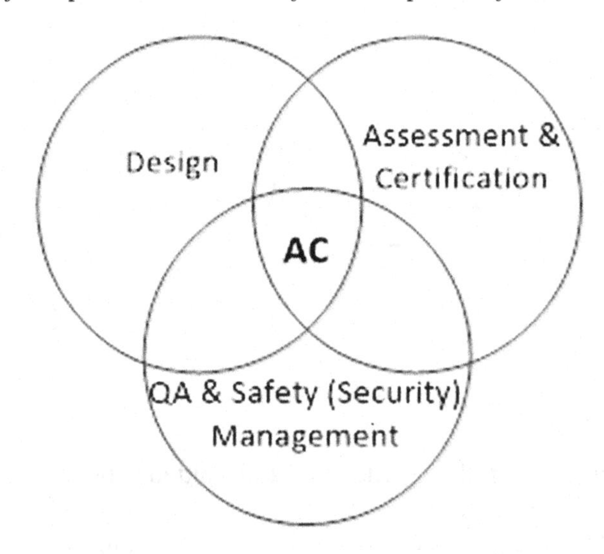

Table 2. The assurance case driven design application through safety and security life cycle

SSLC stage	ID	HR1	HR2	...	HR6
Concept	D1	A(D1,HR1)	A(D1,HR2)	...	A(D1,HR6)
SRS	D2	A(D2,HR1)	A(D2,HR2)	...	A(D2,HR6)
SRS Review	V2	A(V2,HR1)	A(V2,HR2)	...	A(V2,HR6)
SAD	D3	A(D3,HR1)	A(D3,HR2)	...	A(D3,HR6)
SAD Review	V3	A(V3,HR1)	A(V3,HR2)	...	A(V3,HR6)
HW Design	D4	A(D4,HR1)	A(D4,HR2)	...	A(D4,HR6)
HW Design Review	V4	A(V4,HR1)	A(V4,HR2)	...	A(V4,HR6)
FMECA	V5	A(V5,HR1)	A(V5,HR2)	...	A(V5,HR6)
SW Design	D5	A(D5,HR1)	A(D5,HR2)	...	A(D5,HR6)
SW Design Review	V6	A(V6,HR1)	A(V6,HR2)	...	A(V6,HR6)
SW Coding	D6	A(D6,HR1)	A(D6,HR2)	...	A(D6,HR6)
Code Analysis and Review	V7	A(V7,HR1)	A(V7,HR2)	...	A(V7,HR6)
SW Testing	V8	A(V8,HR1)	A(V8,HR2)	...	A(V8,HR6)
Fault Insertion Testing	V9	A(V9,HR1)	A(V9,HR2)	...	A(V9,HR6)
Integration Testing	V10	A(V10,HR1)	A(V10,HR2)	...	A(V10,HR6)
Validation Testing	V11	A(V11,HR1)	A(V11,HR2)	...	A(V11,HR6)

Let's consider applying the Assurance Case methodology throughout the life cycle stages. To do this, we use the example of assessing the compliance of the human resource management plan with the requirements of IEC 61508 (see Table 2). Below is a list of the stages of the safety and security life cycle, including development, verification and validation (Table 2). At each stage, the compliance of the human resource management process with each of the requirements of the human resource management plan should be verified. The use of the Assurance Case methodology allows determination of compliance with the requirements at the argument level {HR1, ..., HR6}. Records of compliance checking and the associated results are phased into the documented Assurance Case.

Goal Structuring Notation (GSN) Description

GSN (Goal Structuring Notation), like CAE, operates with entities such as goal (indicated by a rectangle and is analogous to a claim), argumentation strategy (indicated by a parallelogram and is analogous to argument), and a solution (indicated by a circle and is analogous to evidence) (see Figure 7).

Figure 7. Goal structuring notation (GSN) notation: main components

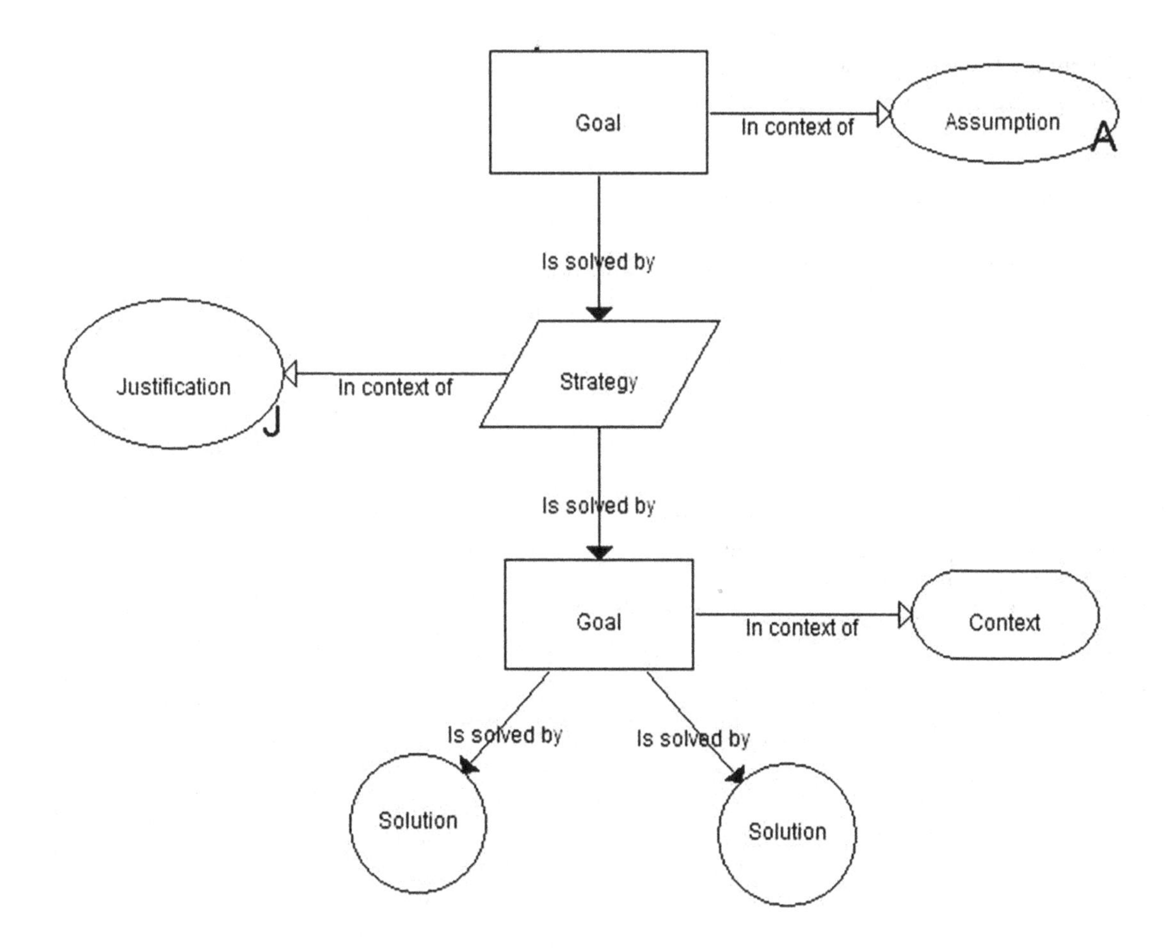

The context is used for informational support of goal setting. Assumptions and justifications can be used to support argumentation. The goal structure is also hierarchical.

When comparing CAE and GSN, it should be noted that CAE pays more attention to justifying individual arguments. To do this, a detailed construction of the argumentation steps is performed. GSN focuses more on typical argument structures (so-called patterns). The use of both notations can be sufficiently subjective, that is, the approach to constructing arguments depends on the person performing the task. A detailed analysis of the notations also reveals some gaps in the multiple components, which are due to the fact that not all elements of semantics are identified in the notations, Rushby, 2015).

Given that none of the notations is perfect, it is advisable to use the associated availability of the software tools as selection criteria. From this point of view, the clear leader is GSN. The GSN format is anchored in the Goal Structuring Notation (GSN) Community Standard, GSN, 2011, as well as in the Structured Assurance Case Metamodel (SACM) data model, SACM, 2016, by Object Management Group (OMG). In terms of tools support, CAEs can only be developed using Adelard ASCE, while there are several CAEs for GSN development. Therefore, it is more appropriate to use GSN for I&C systems assessment.

Tools for Development of Assurance Case

Today, there are three of the most functional software tools that are used to create and maintain the Assurance Case. All of them have a paid license.

Figure 8. Adelard ASCE program interface

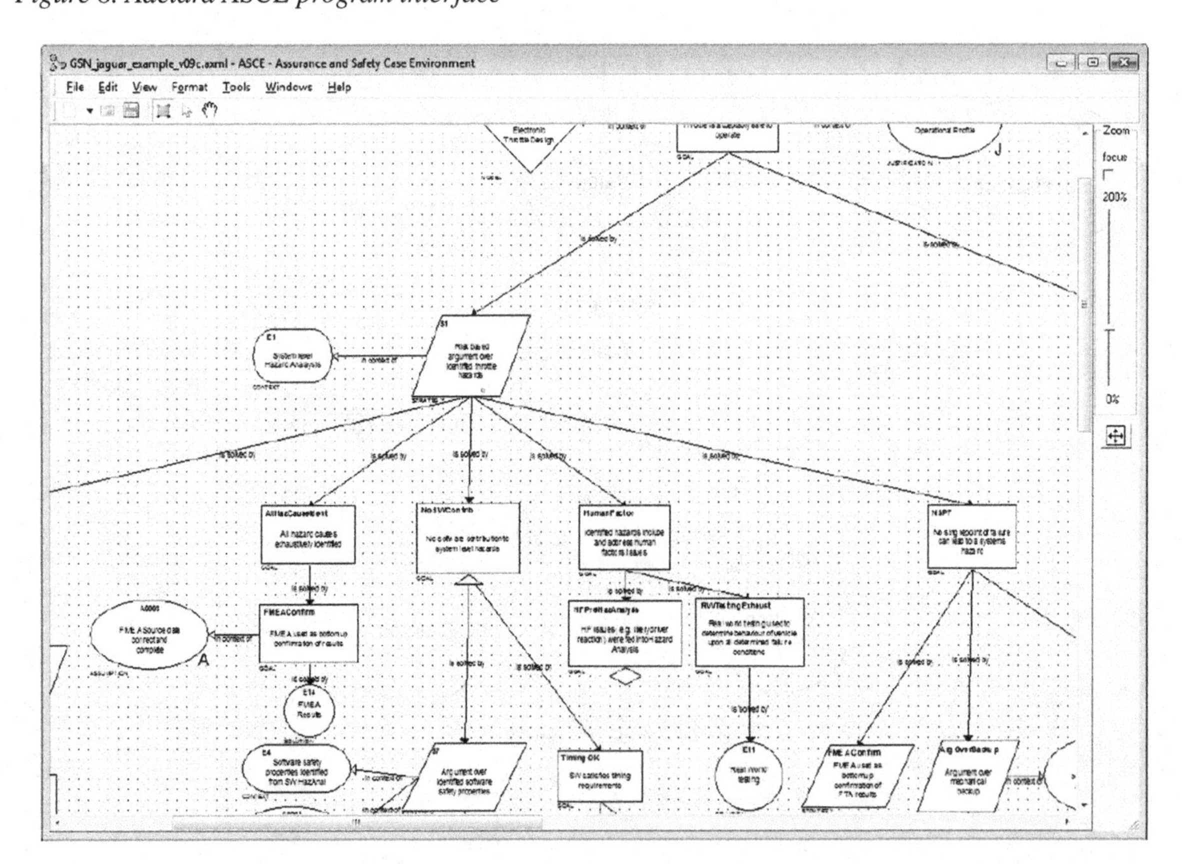

The first and the most widely used tool is the ASCE (Assurance and Safety Case Environment), which has been developed and maintained by the British company Adelard since the 1990s. In the UK, the development of the Assurance Case is required by laws and standards in many areas related to security, so ASCE has a fairly large market here (see Figure 8).

Adelard ASCE supports both CAE and GSN. The main part of the tool is a graphic editor, in which additional text or hyperlink information may be attached to graphic blocks. The program supports the export of charts in HTML and MS Word formats. It is impossible to download the ASCE software from the Adelard website on your own; you must fill out a request for either a 30-day trial version or an academic license, after which the request will be reviewed by the company.

The next software tool is Astah GSN (see Figure 9) developed by Change Vision company from Japan. The company was created in 2006. Astah GSN was developed as a part of the Astah Professional toolkit, which is a media for complex systems modeling. As the name suggests, this program supports only GSN. In addition, it can create Mind Map diagrams. In the graphical editor, you can attach text and hyperlinks to graphic symbols. Charts are saved in the internal format of the program (*.agml). It supports the export of diagrams in the form of figures, as well as in the XMI format (XML Metadata Interchange).

Figure 9. Astah GSN program interface

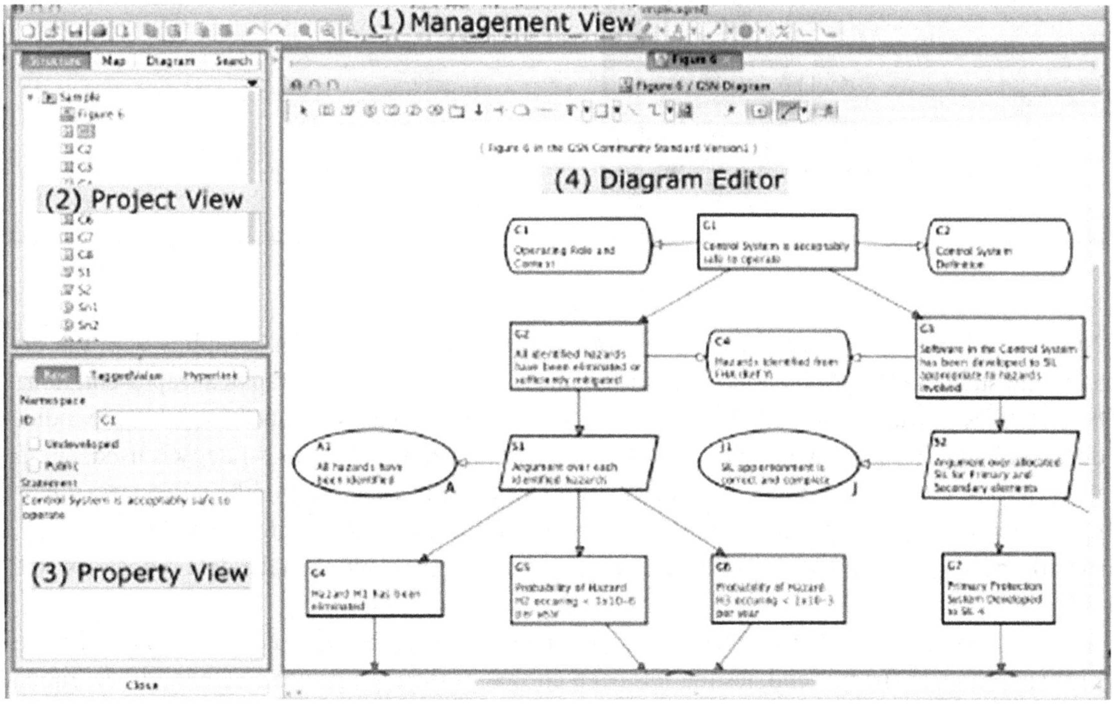

You can download a trial version of the software from the Astah GSN website. Supported operating systems are Windows, MacOS, and Linux. The trial version will work 50 days. User manuals and video demonstrations are also available on the site.

The software tool NOR-STA (see Figure 10) was developed by the Polish company Argevide, which was founded by the staff of the University of Gdansk.

Figure 10. NOR-STA program interface

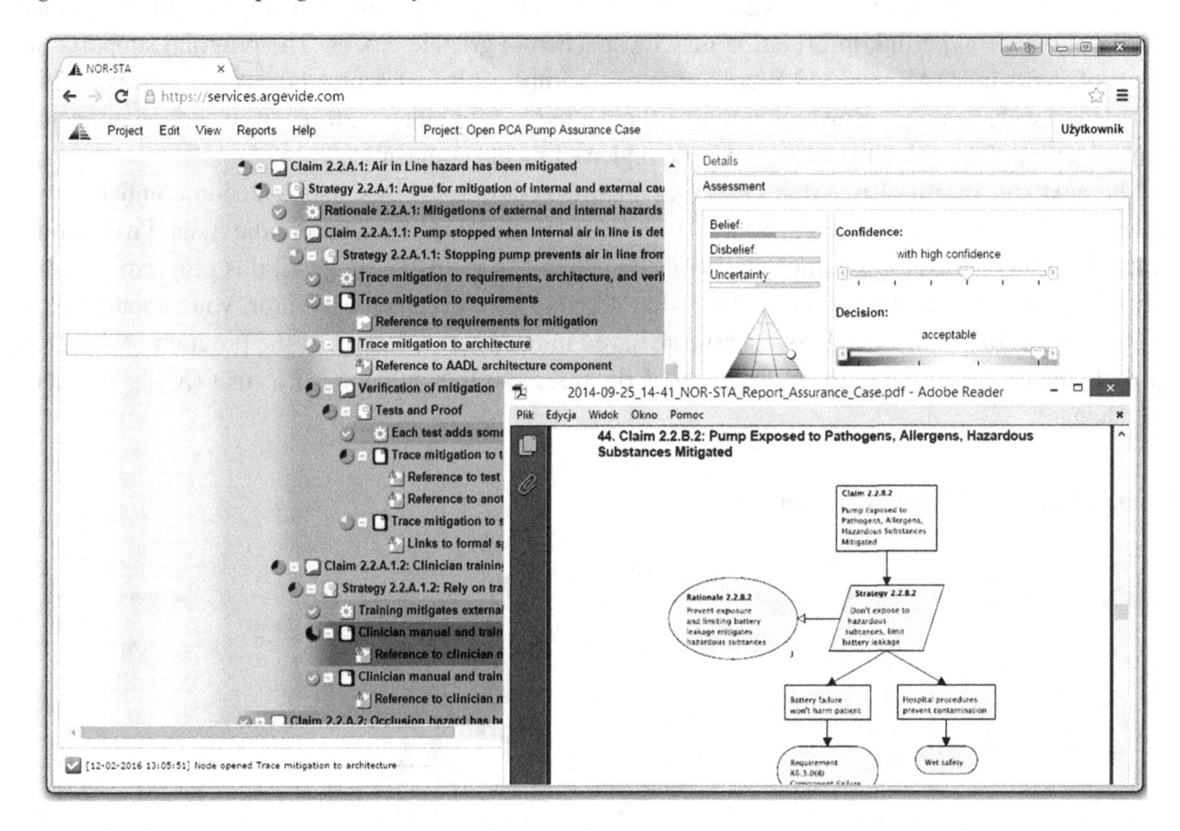

NOR-STA supports its own TRUST-IT notation (see Figure 11), which complies with the provisions of the standard ISO/IEC 15026. The difference is that, instead of a graphical representation, the NOR-STA uses a structured hierarchical list. Entities in hierarchical Assurance Case list are indicated by different icons. To confirm compliance with the claim, the argumentation strategy is used, and facts or observations, rationale, assumptions and sub-claims are used as analogue of the evidence. Unlike the two previous desktop applications, NOR-STA is used online and supports distributed team work. For privacy purposes, you can install NOR-STA on a dedicated server, and then the data repository will be stored on it.

In the considered example (see Figure 11), the main goal is to demonstrate the absence of errors in the software module. To this end, testing has been chosen as the argumentation strategy. The rationale for the strategy is the development and execution of reliable tests. The actual confirmation of compliance is that the test report does not contain unresolved errors. An additional sub-goal is to cover all the requirements for the software module with tests. An own argumentation strategy can be developed for this purpose. As an assumption we assume that the testing tools used are reliable. Data can be presented as a GSN diagram, and you can also convert to Word, Excel, PDF, and XML formats. At the request of the user, a 30-day trial access can be provided on the NOR-STA website for using this software.

Figure 11. Trust-IT notation and an example of its application

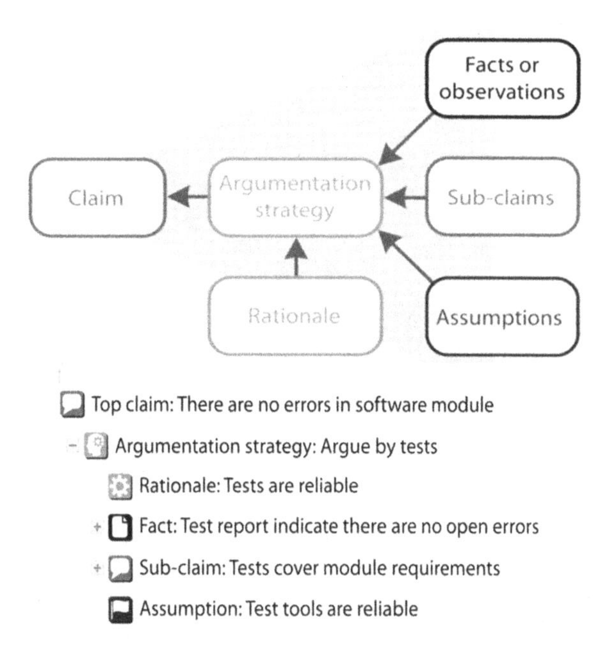

Our recommendation for tools using in I&C systems area is Astah GSN, since it support GSN and can be loaded free from the website to support a trial time of operation.

ARGUMENTATION IMPROVEMENT WITH STRUCTURED TEXT TEMPLATES

The argumentation in the Assurance Case corresponds to the implication in logic, when the truth of the conclusion depends on the truth of the conditions. A logical rule involves a logical multiplication in the form of:

SC1 AND SC2… AND SCn IMPLIES C,

where SCi are subgoals, which also can be complex expressions.

As noted above, in the STATE-OF-THE-ART there are some drawbacks in the existing papers that are related to the lack of argumentation techniques. One of the few authors who have attempted to address this gap is John Rushby, who in his technical report, Rushby, 2015, offers an approach to developing structured arguments based on a modified GSN. In this section we use and develop this approach.

Classical application of GSN (Figure 7) is characterized by support for argumentation steps (AS) of any claim (C) at the same time as subclaims (SC) and evidence (E). This approach has some drawbacks, which are due to the inability to always have a regular and typical argument structure (Figure 12).

Therefore, in Rushby, 2015, modification of argumentation steps is proposed to reduce them to a typical two-step structure (Figure 12). The first step, called the reasoning step (RS), is an analysis of subgoals that are aimed at achieving the primary goal, but there is no recourse to the evidence at that step. In the second step, called the evidential step (ES), the evidence for supporting the subgoals that

Figure 12. Transformation of a simple argument form to a structured argument form

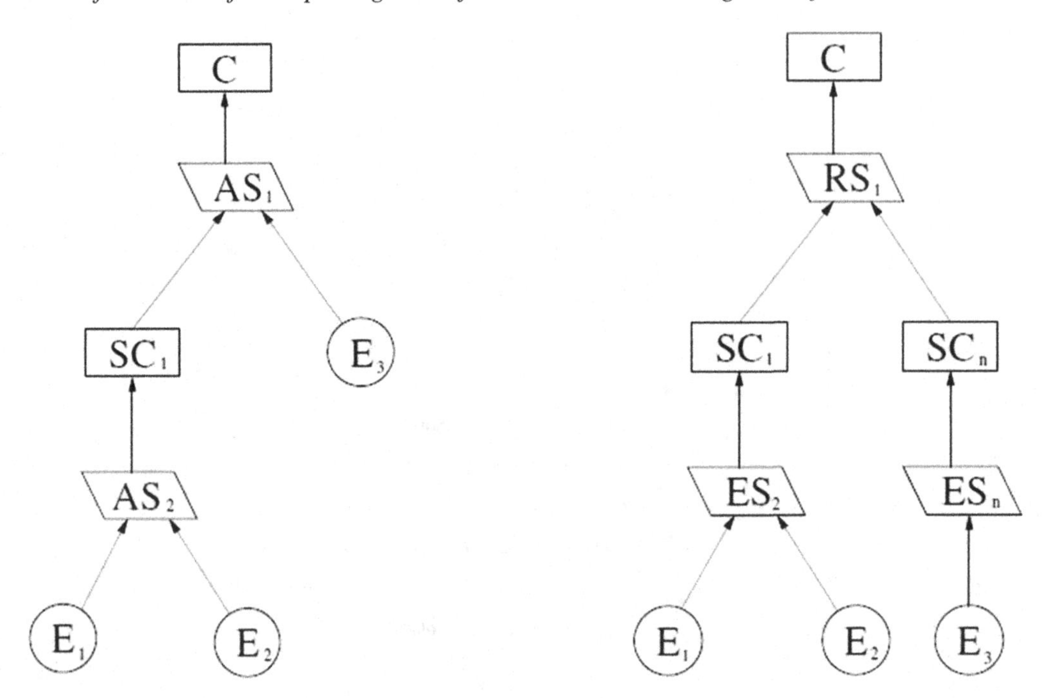

was formulated in the previous step is formulated. Thus, the graph of the argumentation structure is transformed as shown in Figure 12. This allows us to make a connection between the concept of safety and security (goal) and our knowledge of the physical world (evidence).

To further formalize the steps of RS and ES in Rushby, 2015, it is suggested to use structured text. This approach is appropriate, but in our opinion, it has a number of opportunities for improvement, such as the following:

- There is not a general algorithm for the development of the Assurance Case;
- Relations of graph with snippets of structured text are not explicitly explained;
- Structured text does not have a clear template.

In addition, the development of the Assurance Case is in many ways a creative process, which many depends on the human factor. The following is an improvement of the approach described in Rushby, 2015, which, in our view, will allow us to move further in structuring the arguments of the Assurance Case and eliminate the above shortcomings. It should also be noted that our studies use the principles of modularity and hierarchy of the Assurance Case, as suggested in (Rushby, 2015). A further development of the object-oriented approach for the Assurance Case in this work is the application of the encapsulation principle, which allows localization of the content of arguments in the overall graph of the Assurance Case structure.

Let us demonstrate the possibility of explicitly combining the Assurance Case with structured text components. Let's present a hierarchy of requirements that create the structure of the Assurance Case in the form of a pyramid. In most regulatory requirements for I&C systems, the structure of requirements is not too multilevel, as a rule, it is three or four levels (Figure 13).

Figure 13. Hierarchy of requirements to I&C systems and a relation of requirements with argumentation steps

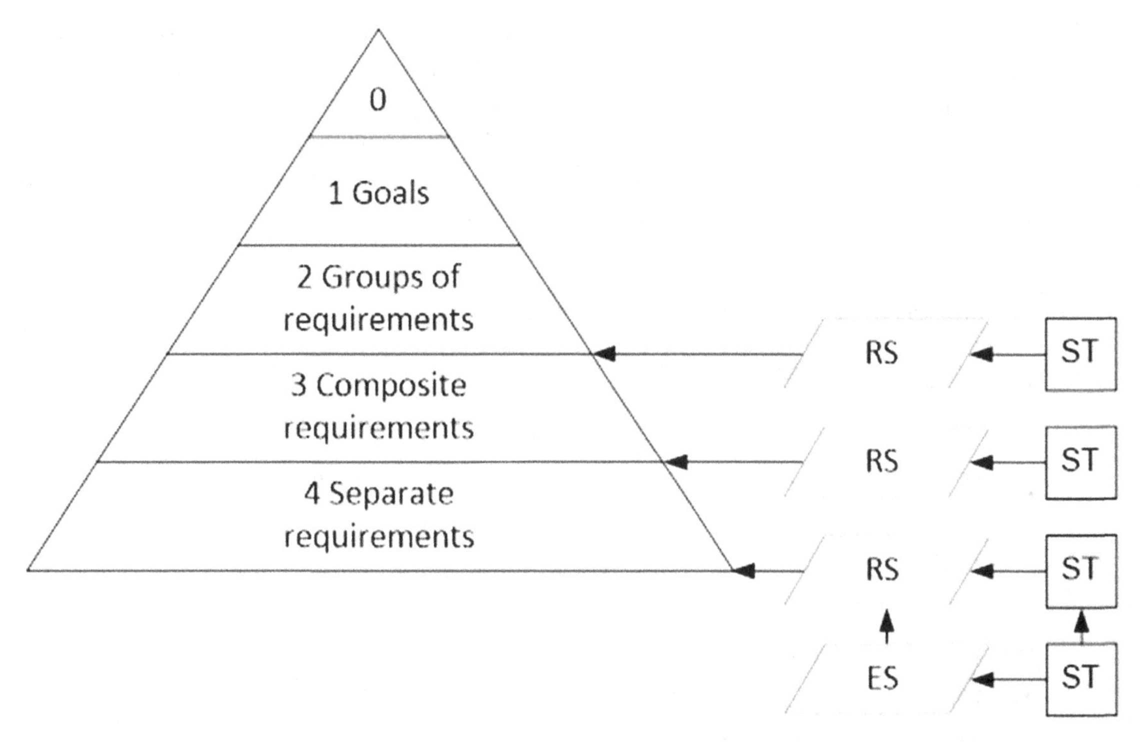

Zero level is a meta goal according to which the I&C system must meet all safety requirements. At the first level, global security goals are achieved, for example, according to functional safety requirements (IEC 61608):

- The safety and security management system shall achieve all safety objectives, including safety and security assessments;
- Safety and security life cycle should be implemented during system development;
- A sufficient set of measures against random failure must be applied to the system;
- A sufficient set of measures against systematic and software failures, including cyberattack defense, must be applied to the system.

The requirements groups contain thematically related requirements and support one or other of the global safety and security goals. For example, the requirements for safety and security management in IEC 61508 include requirements to human resource management, configuration management, documentation management, and others, Sklyar & Kharchenko, 2017.

The structure of the links between the zero, first and second levels is transparent enough and does not require detailed elaboration of the arguments, since these arguments are typical and well tested. However, structured arguments are required when moving from the second level to the lower levels. The requirements of the lower levels may be either composite (such as include a number of separate requirements) or separate. If all requirements are separate, this level becomes third, and then it is directly related to the subgroups of requirements.

Figure 13 combines the overall structure of the Assurance Case and the algorithm for constructing structured arguments. Such arguments should be developed for the second, third and fourth (if any) levels. An approach to argument structure is introduced in Figure 12. For the lowest level, besides the reasoning steps (RS), evidential steps (ES) should also be applied. Since it is not appropriate to add detailed information about the content of the arguments on the graph structure, each of the nodes of the Assurance Case graph, starting with the second level, is marked with an argument description using so-called structured text (ST). It should be noted that the Assurance Case graph is not a strict tree because the same evidence can support different arguments or subgoals.

Figure 14. A temlate of structured text for reasoning step

Reasoning Step

Context

Connection with the Assurance Case graph in relation with high and low levels

Docs

Technical documents related with arguments and evidences

Claim

Goal related with argument

Subclaims

Subgoals demonstrated the goal (Claim) achievment

Justification

Structure and content of subgoals (Subclaims)

END Reasoning Step

Figure 15. A temlate of structured text for evidential step

Evidential Step

Context

Connection with the Assurance Case graph in relation with high and low levels

Docs

Technical documents related with arguments and evidences

Claim

Subclaims from Reasoning Step become Claim

Evidence

Proofs, which support Claim achiement

Justification

Structure and content of Evidence

END Evidential Step

Let's develop a typical structured text configuration for the reasoning and evidential steps using the GSN components (Figure 7). The structured text has a template with a set of fields that are denoted by service words that correspond to the GSN components. We need to provide two templates, for the reasoning step (RS) and the evidential step (ES) (Figure 14, 15). In these templates, the names of the service words are given in bold, and italics provide a brief description of the content that should fill the template fields.

SOLUTIONS AND RECOMMENDATIONS

The Assurance Case is a proven in use methodology with a 20-year history of application, which is constantly evolving and getting support from leading safety and security experts. It seems appropriate to use the Assurance Case in the practice of certification and licensing of Nuclear Power Plants I&C systems.

In this chapter we analyzed semiformal notations which are used for the Assurance Case development as well as the appropriate computer tools for the Assurance Case. There two the main notations widely used in the Assurance Case, including Claim, Argument and Evidence (CAE) notation and Goal Structuring Notation (GSN). When comparing CAE and GSN, it should be noted that CAE pays more attention to justifying individual arguments. To do this, a detailed construction of the argumentation steps is performed. GSN focuses more on typical argument structures (so-called patterns). The use of both notations can be sufficiently subjective, that is, the approach to constructing arguments depends on the person performing the task. A detailed analysis of the notations also reveals some gaps in the multiple components, which are due to the fact that not all elements of semantics are identified in the notations, Rushby, 2015.

Given that none of the notations is perfect, it is advisable to use the associated availability of the software tools as selection criteria. From this point of view, the clear leader is GSN. The GSN format is anchored in the Goal Structuring Notation (GSN) Community Standard, GSN, 2011, as well as in the Structured Assurance Case Metamodel (SACM) data model, SACM, 2016, by Object Management Group (OMG). In terms of tools support, CAEs can only be developed using Adelard ASCE, while

Figure 16. The top level of the IEC 61508 goals in Goal Structuring Notation

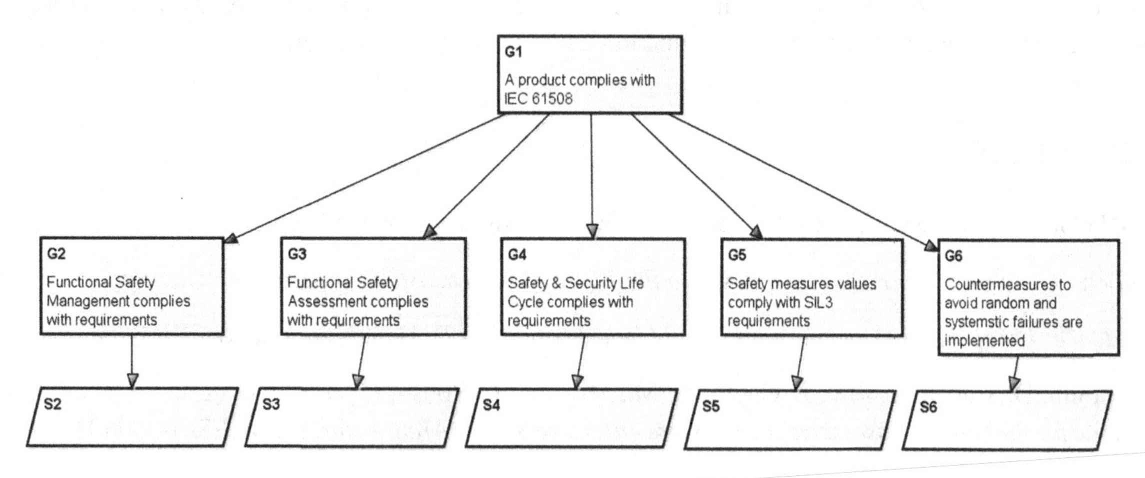

there are several CAEs for GSN development. Therefore, it is more appropriate to use GSN for I&C systems assessment. Concerning available software tools, it is possible using in Astah GSN, developed by the company Change Vision (Japan), since it support GSN and can be loaded free from the website to support a trial time of operation.

Application of the Assurance Case in area of Nuclear Power Plants I&C systems can improve level of argumentation in safety and security assessment as well as improve visualization of the obtained results. For example, Figure 16 represents the top level of requirements of the standard IEC 61508 "Functional safety of electrical/ electronic/ programmable electronic safety-related systems".

FUTURE RESEARCH DIRECTIONS

We propose to consider the following future research directions for the safety and security management area:

- Update of the overall algorithm and method for the development of the Assurance Case;
- Improvement structured text templates for describing arguments;
- Design templates for typical Assurance Case applicatiobs.

CONCLUSION

Notations and software tools for support of the Assurance Case methodology were analyzed. The analysis of existing approaches to the development of the Assurance Case is conducted. Existing works have some drawbacks due to the lack of satisfactory practical argumentation techniques. One of the few authors who has attempted to address this gap is John Rushby, who in his technical report, Rushby, 2015 offers an approach to developing structured arguments based on modified GSN and structured text. In this chapter we use and update this approach.

Thus, in order to apply the methodology of the Assurance Case for safety and security evaluation of Nuclear Power Plants I&C systems, a mathematical and methodological apparatus for structuring the argumentation was selected and improved. In this chapter we propose the general algorithm of the Assurance Case development, the structure of the Assurance Case graph, which is based on the structure of the arguments and is developed in conjunction with the structured text templates.

REFERENCES

GSN Community Standard. (2011). Version 1. Origin Consulting (York) Limited.

Cullen, W. (1990). *The Public Enquiry into the Piper Alpha Disaster*. Department of Energy.

Evidence: Using safety cases in industry and healthcare. (2012). Health Foundation.

Firesmith, D., Capell, P., Elm, J., Gagliardi, M., Morrow, T., Roush, L., & Sha, L. (2006). *QUASAR: A Method for the Quality Assessment of Software-Intensive System Architectures*. CMU/SEI-2006-HB-002.

Goodenough, J., Weinstock, C., & Klein, A. (2015). *Eliminative Argumentation: A Basis for Arguing Confidence in System Properties, Technical Report, CMU/SEI-2015-TR-005*. Pittsburgh, PA: CMU/SEI.

Graydon, P., & Knight, J. (2009). *Assurance Based Development.* Technical Report CS-2009-10. University of Virginia.

Hawkins, R., Habli, I., Kelly, T., & McDermid, J. (2013). Assurance Cases and Prescriptive Software Safety Certification: A Comparative Study. *Safety Science*, *59*, 55–71. doi:10.1016/j.ssci.2013.04.007

Kelly, T. (1998). *Arguing Safety: A Systematic Approach to Managing Safety Cases* (PhD thesis). University of York.

Kobayashi, N., Nakamoto, A., Kawase, N., Sussan, F., & Shirasaka, S. (2018). What Model(s) of Assurance Cases Will Increase the Feasibility of Accomplishing Both Vision and Strategy? *Review of Integrative Business and Economics Research*, *7*(3), 1–17.

Maksimov, M., Fung, N., Kokaly, S., & Chechik, M. (2018). Two decades of assurance case tools: A survey. In B. Gallina, A. Skavhaug, E. Schoitsch, & F. Bitsch (Eds.), *Computer Safety, Reliability, and Security*. Springer.

Rushby, J. (2015). *The Interpretation and Evaluation of Assurance Cases.* Technical Report SRI-CSL-15-01. SRI International.

Sklyar, V., & Kharchenko, V. (2017). Assurance Case Driven Design based on the Harmonized Framework of Safety and Security Requirements. *Proceedings of the 13th International Conference on ICT in Education, Research and Industrial Applications*, 670-685.

Structured Assurance Case Metamodel. (2016). v2.0. Object Management Group.

Toulmin, S. (1958). *The Uses of Argument*. Cambridge University Press.

Ye, F., & Cleland, G. (2012). *Weapons Operating Centre Approved Code of Practice for Electronic Safety Cases*. Adelard LLP.

ADDITIONAL READING

Bishop, P., & Bloomfield, R. (2000). A Methodology for Safety Case Development. *Safety and Reliability*, *20*(1), 34–42. doi:10.1080/09617353.2000.11690698

Calinescu, R., Gerasimou, S., Habli, I., Iftikhar, M., Kelly, T., & Weyns, D. (2018). Engineering Trustworthy Self-Adaptive Software with Dynamic Assurance Cases. *IEEE Transactions on Software Engineering*, *44*(11), 1–30. doi:10.1109/TSE.2017.2738640

Duan, L., Rayadurgam, S., Heimdahl, M., Ayoub, A., Sokolsky, O., & Lee, I. (2017). Reasoning About Confidence and Uncertainty in Assurance Cases: A Survey. In M. Huhn & L. Williams (Eds.), *Software Engineering in Health Care* (pp. 64–80). Springer. doi:10.1007/978-3-319-63194-3_5

Firesmith, D., Capell, P., Elm, J., Gagliardi, M., Morrow, T., Roush, L., & Sha, L. (2006). *QUASAR: A Method for the Quality Assessment of Software-Intensive System Architectures, CMU/SEI-2006-HB-002*. Pittsburgh, PA: CMU/SEI.

Haddon-Cave, C. (2009). *The Nimrod Review. An independent review into the broader issues surrounding the loss of the RAF Nimrod MR2 Aircraft XV230 in Afghanistan in 2006*. London, UK: Crown Copyright.

Hawkins, R., Habli, I., Kelly, T., & McDermid, J. (2013). Assurance Cases and Prescriptive Software Safety Certification: A Comparative Study. *Safety Science*, *59*, 55–71. doi:10.1016/j.ssci.2013.04.007

Hitchcock, D. (2005). Good Reasoning on the Toulmin Model. *Argumentation*, *19*(3), 373–391. doi:10.100710503-005-4422-y

ISO (2011). ISO 26262. Road vehicles – Functional safety (in 10 parts).

ISO, (2019). ISO/IEC/IEEE 15026:2019. Systems and software engineering – Systems and software assurance (in 4 parts).

Jee, E., Lee, I., & Sokolsky, O. (2010). Assurance Cases in Model-Driven Development of the Pacemaker Software. In T. Margaria & B. Steffen (Eds.), *Leveraging Applications of Formal Methods, Verification, and Validation* (pp. 343–356). Springer. doi:10.1007/978-3-642-16561-0_33

Littlewood, B., & Wrigh, D. (2007). The Use of Multilegged Arguments to Increase Confidence in Safety Claims for Software-based Systems: A Study Based on a BBN Analysis of an Idealized Example. *IEEE Transactions on Software Engineering*, *33*(5), 347–365. doi:10.1109/TSE.2007.1002

Rhodes, T., Boland, F., Fong, E., & Kass, M. (2010). Software assurance using structured assurance case models. *Journal of Research of the NIST*, *115*(3), 209–216. doi:10.6028/jres.115.013 PMID:27134787

Sklyar, V., & Kharchenko, V. (2019). Green Assurance Case: Applications for Internet of Things. In V. Kharchenko, Y. Kondratenko, & J. Kacprzyk (Eds.), *Green IT Engineering: Social, Business and Industrial Applications* (pp. 351–371). Springer. doi:10.1007/978-3-030-00253-4_15

Sun, L., Zhang, W., & Kelly, T. (2011). Do safety cases have a role in aircraft certification? *Procedia Engineering*, *17*, 358–368. doi:10.1016/j.proeng.2011.10.041

Wei, R., Kelly, T., Dai, X., Zhao, S., & Hawkins, R. (2019). Model Based System Assurance Using the Structured Assurance Case Metamodel. *Journal of Systems and Software*, *154*, 211–233. doi:10.1016/j.jss.2019.05.013

Weinstock, C., Goodenough, J., & Hudak, J. (2004). *Dependability Cases. CMU/SEI-2004-TN-016*. Pittsburgh, PA: SEI/CMU.

Zhao, X., Zhang, D., Lu, M., & Zeng, F. (2012). A New Approach to Assessment of Confidence in Assurance Cases. In F. Ortmeier & P. Daniel (Eds.), *Computer Safety, Reliability and Security* (pp. 79–91). Springer. doi:10.1007/978-3-642-33675-1_7

Chapter 13
Cyber Security Assurance in the Design, Implementation, and Operation of NPP I&C Systems

Oleksandr Klevtsov
https://orcid.org/0000-0001-5665-5039
State Scientific and Technical Center for Nuclear and Radiation Safety, Ukraine

Artem Symonov
https://orcid.org/0000-0001-6971-523X
State Scientific and Technical Center for Nuclear and Radiation Safety, Ukraine

Serhii Trubchaninov
State Scientific and Technical Center for Nuclear and Radiation Safety, Ukraine

ABSTRACT

The chapter is devoted to the consideration of the issues concerning the cyber security assurance of NPP instrumentation and control systems. A brief overview of the international regulatory framework in the field of cyber security for nuclear facilities is given. The different approaches to the categorization of NPP instrumentation and control systems by cyber security are expressed. The basic principles of cyber security assurance of NPP instrumentation and control systems are considered. The specific measures of cyber security assurance (i.e., graded according to the cyber security levels) on the stages of development, implementation, and operation of NPP instrumentation and control systems are presented.

INTRODUCTION

The problem of cyber security assurance for nuclear facilities (especially, NPPs) is real for the modern "digital" world because there were several cyber-attacks against NPPs and other nuclear facilities during the last years. The I&C systems important for NPP safety were the objects of such cyber-attacks and would likely be targets for cyber-attacks in the future.

DOI: 10.4018/978-1-7998-3277-5.ch013

International organizations, as well as national governments, strive for the implementation of practical protective measures against cyber threats at nuclear facilities and the development of the appropriate regulatory framework.

The goal of the chapter is the consideration of main principles and measures of cyber security assurance for NPP I&C systems described in international and national rules, guides and standards.

BACKGROUND

International Atomic Energy Agency (IAEA), International electrotechnical commission (IEC) and national nuclear and radiation regulatory authorities (USA, Germany, etc.) pay special attention to the issues of nuclear power plants (NPP) cyber security assessment and assurance. This is due to three main factors: the wide-ranging transfer from analog to digital NPP instrumentation and control (I&C) systems, the vulnerabilities of digital I&C systems to cyber threats and the increase of the number of cyber-attacks against digital I&C systems the potential for serious consequences for safety-critical infrastructures.

The cyber security assurance is an important task because the cyber-attacks may adversely affect nuclear security as well as the nuclear and radiation safety of NPPs and other nuclear facilities. The majority of countries realize the appropriate measures for protection against cyber threats and develop the regulatory framework in the field of cyber security considering the recommendations from IAEA documents (e.g. IAEA, 2011). Ukraine has also started the development of the legislative and regulatory framework and the realization of practical measures of protection against the cyber threats at NPPs and enterprises-developers of I&C systems.

Cyber security is a specific aspect of information security referring to computer systems, networks, and digital systems. Also, cyber security is an important part of the general (nuclear) security of the nuclear facility. The aim of cyber security is the protection of characteristics of confidentiality, integrity and availability of electronic data, computer systems and processes.

It is necessary to note that different international and national standards use three terms which are the synonyms: "cyber security" (widely used in documents of US Nuclear Regulatory Commission (NRC), particularly, in RG, 2010), "computer security" (used in documents of IAEA and IEC), "IT security" (rarely used in some publications). Authors use the term "cyber security" in this chapter.

The term "safety" according to IAEA, 2006 means the protection of people and the environment against radiation risks. Safety concerns the facilities and activities related to radiation risks Safety includes the safety of nuclear facilities, radiation safety, safety for the management of radioactive waste, the safety of transportation of radioactive materials.

This chapter is limited only to consideration of I&C systems which are directly involved in monitoring and control of technological processes at NPP. The authors exclude from consideration the other computerized systems operating at NPPs (e.g., like physical protection systems, business-processing systems, etc.).

According to IAEA, 2011 all I&C systems are graded by security levels for the identification of required cyber security measures. US NRC uses similar term «cyber security level» (e.g., in RG, 2010). IEC uses the equivalent term «computer security degree» (e.g., in IEC, 2019). All these terms are synonyms.

REGULATORY FRAMEWORK

The international regulatory framework in the field of cyber security of NPP I&C systems is formed based on:

- general requirements to NPP I&C systems;
- industrial standards on cyber security of networks and computerized systems;
- standards and guides on nuclear security (because cyber security is a part of nuclear security of nuclear facilities); and
- standards on information security (because cyber security is a specific aspect of information security).

This subchapter is focused on a short overview of international standards that contain the requirements to cyber security of the nuclear facility and I&C systems.

IAEA Documents

IAEA considers cyber security is a part of the nuclear security of nuclear facilities. IAEA Nuclear Security Series includes more than 35 documents (by 2019) highlighting different aspects of nuclear security. Some new documents of this series are under development.

Let us consider IAEA documents directly related to cyber security.

Document IAEA, 2011 is one of the most important technical guides in which the general principles of cyber security assurance of nuclear facilities are presented. The special attention is given to the importance of the description of envisaged measures in the general security plan of a nuclear facility.

Document IAEA, 2015 considers the issues of information security in the nuclear industry. The main theme of this document is the protection of information that is critically important for safety assurance. The document describes the aspects of identification, detection, sharing and protection of confidential and critically important nuclear information with limited access.

Document IAEA, 2018 highlights the issues of cyber security of digital I&C systems that perform safety functions or auxiliary functions at the nuclear facilities. It considers the role of cyber security in the protection of digital I&C systems, measures of cyber security assurance, I&C systems classification by the cyber security levels and zones, coordination between the cyber security and safety, cyber security assurance on all life-cycle stages of I&C systems. In addition, the recommendations for the development of the main documents as cyber security policy and cyber security plan are given.

It is necessary to note, that document IAEA, 2016 sets the requirements for NPP I&C systems design but also contains some requirements for the cyber security of NPP I&C systems. These are the requirements for coordination between cyber security and safety, access control to the digital I&C systems important for NPP safety, protection of lines for connection with the emergency centers, functions of cyber security assurance during the operation of I&C systems, etc.

IEC Standards

The main IEC standards on cyber security are IEC, 2019 and IEC, 2016.

Standard IEC, 2019 is directly devoted to the issues of cyber security of NPP I&C systems. It sets requirements for the development and implementation of cyber security programs for NPP I&C systems, defines the life-cycle and describes the main measures of NPP I&C systems cyber security assurance. Standard is oriented for use during the whole life-cycle of I&C systems in case of modernization of existing NPPs or design of new NPPs.

An important aspect considered in IEC, 2019 is the definition of three security degrees for I&C systems, depending on the categories of functions (according to IEC, 2009) performed by the I&C system. Security degrees define the severity of the requirements for cyber security for the specific I&C system.

Standard IEC, 2016 contains the principles of coordination between cyber security and safety on the level of overall NPP I&C as well as on the level of individual I&C systems.

US NRC Documents

It should be noted that some US NRC documents on cyber security are not available for public access. Therefore, the authors consider only US NRC documents, which are open to public access.

The Code of Federal Regulations (CFR) is on the highest level in the hierarchy of US NRC documents. The requirements of CFR, 2015 are obligatory to meet by all licensees.

Paragraph 73.55 CFR, 2015 contains the requirements concerning the implementation of a security program and the cyber security program (as a part of the general security program) for a nuclear facility.

Paragraph 73.56 CFR, 2015 addresses the issues of authorization of personnel access to a nuclear facility. It is one of the protection measures for physical and cyber security assurance.

Paragraph 73.58 CFR, 2015 sets the requirements for the connection between safety and security. It considers the measures for solving potential conflicts between safety and security that should be considered during the development and implementation of cyber security measures.

Paragraph 73.54 CFR, 2015 contains the requirements that address the cyber security of digital assets (computers, communication systems, and networks), which are the parts of systems used for performing safety, security and emergency preparedness functions and auxiliary systems.

Paragraph 73.77 CFR, 2015 defines the requirements for reporting cyber security events by licensees to US NRC.

Methods to implement the requirements are detailed in regulatory guides.

RG, 2011 is the general regulatory guide that concerns the issues of use of computers in NPP safety systems. Cyber security is considered one of the above-mentioned issues. RG, 2011 addresses both the secure development environment and secure operating environment.

RG, 2010 sets the requirements to cyber security program for nuclear facilities and includes the development, implementation (with the description of appropriate stages) and realization of cyber security plan. This essential US NRC document on cyber security of nuclear facilities describes the life cycle of cyber security assurance process: development of cyber security program, its implementation at nuclear facilities, ongoing monitoring of the program, periodical review of the program, changes management and documentation maintenance.

The development, implementation, and realization of cyber security program require such actions as analysis of digital systems and networks at nuclear facility, identification and assessment of safety-critical resources, implementation of safe architecture according to appropriate rules, analysis of potential risks of breach of cyber security and activity for maintenance of cyber security program.

RG, 2015 is devoted to consideration of cyber security issues. This regulatory guide provides the acceptable method for reporting cyber security events. These requirements enhance the analysis of the reliability and effectiveness of the licensee's cyber security program. The reporting requirements have an important impact on the assurance of protection of digital computer networks and systems from cyber-attacks.

GENERAL PRINCIPLES OF CYBER SECURITY ASSURANCE

Each organization that takes part in activity connected with the development, implementation, and operation of NPP I&C systems should elaborate a set of specific measures for protection against cyber threats.

Cyber security encompasses the administrative (e.g., policy, procedures, permissions, etc.), technical (e.g., doors, locks, seals, etc.) and software (e.g., passwords, users authentication, etc.) measures that ensure:

- Prevention of malicious actions through the counteractions and protection of I&C systems, hardware, network equipment, software, data, and spare parts;
- Malicious actions management through the use of measures for the identification, delay, and response; and
- Mitigation of consequences of malicious actions including measures for a resumption of normal operation of I&C systems, hardware, and network equipment.

Cyber security measures should ensure that any single inadvertent actions or single personnel cannot breach all layers of defense for a given threat vector. However, the basis for cyber security assurance is compliance with the fundamental principles that will be considered below.

Legislative Acts and Regulatory Requirements

The primary task for protection against cyber threats is the development and implementation of national legislative acts and regulatory requirements.

Because of the nature of cyber security, its assurance requires the legislative acts that allow consideration of the specific character of cybercrimes and operation modes of computer systems. Due to the fast evolution of IT-technologies, it is important that the national legislative acts continuously be reviewed and renewed with the consideration of the new types of potential threats for cyber security.

The regulatory requirements to physical protection as well as to nuclear and radiation safety should be considered for the elaboration of regulatory requirements to cyber security. According to IAEA, 2011 the regulatory requirements to cyber security should include the provisions concerning:

- Responsibility of administration for cyber security assurance;
- Segregation of duties between the operational personnel and cyber security team;
- Cyber security policy;
- Cyber security program;
- Realization of cyber security plan;
- Audit – internal, external and carried out by the regulatory authorities; and

- Current assessment of threats, the results of which should be continuously brought to the attention of the administration and operational personnel.

Cyber Security Policy

Organizations and companies that take part in the life cycle of NPP I&C systems should develop, implement and support cyber security policy as a part of the general policy of NPP security assurance.

The top management of an organization should approve the cyber security policy. It defines the high-level objectives of cyber security as well as the main tasks and procedures of cyber security assurance in accordance with the regulatory requirements. According to requirements IAEA, 2017 cyber security policy should be documented, cover all information carriers, and be known to all staff.

Cyber security policy is elaborated in compliance with the requirements of guides, rules, and standards on cyber security and should consider the impacts on nuclear and radiation safety.

The periodic review and updates of cyber security policy and procedures is carried out for ensuring the confidence that it continues to address (in appropriate way) the risks for I&C systems which are protected by this policy. Also, new cyber security threats, the evolution of technologies, modifications of I&C systems, changes in the structure of organizational and other aspects should be considered.

Cyber security policy is used for any I&C system or any group of I&C systems and covers the following important aspects:

- Physical and logical access management;
- Configuration management;
- Check of I&C system and its components;
- Delivery of I&C system;
- Risk and threat management;
- Incidents response and recovery; and
- Assessment of compliance with the requirements for cyber security.

The requirements of cyber security policy are expressed in detail in a cyber security program and a cyber security plan that is used for the realization of this policy and for monitoring of its implementation. The policy should be viable, achievable and auditable.

Cyber Security Culture

The implementation of cyber security culture is an essential factor for cyber security assurance. The top management of the operating organization provides the complete integration of cyber security culture into general security and safety culture at a nuclear facility.

The basis of cyber security culture is the understanding of existing of real threats and the importance of cyber security by all persons who are involved in activity connected with the regulation, management, and operation of a nuclear facility.

Cyber security culture is ensured by the performing of activity directed to the informing of personnel and the improvement of understanding of the cyber security issues (for instance, by means of posters, reminders, training, instructions, testing, etc.).

Compliance with the following requirements should be confirmed during the assessment of cyber security culture in the organization:

- The cyber security requirements are strictly documented and personnel understand them well;
- The processes of I&C systems operation are documented;
- The employees understand and recognize the importance of fulfillment of monitoring measures in frame of cyber security program; and
- The maintenance of the I&C system ensures their protection and operation according to the fundamental cyber security principles and procedures.

Defense in Depth

Cyber security assurance should include the defense in depth against cyber-attacks directed to I&C systems at a nuclear facility. Defense in depth involves the implementation of conception of multiple sequential defensive layers and measures. It is one of the most important principles for protection against cyber threats (in particular, according to the IAEA, 2011).

The defense in depth is ensured by the combination of multiple sequential defensive layers and only the failure of all these layers can lead to an adverse effect from a cyber attack on I&C systems. The defensive layers and connected protective measures provide the prevention or delay of cyber attack progress. The failure of one layer does not lead to a failure of functions or to a worsening of I&C system characteristics, because of existing of additional layers or barriers that ensure sufficient protection. In case of adequate organization of defense in depth any single technical, human or organizational failure cannot lead to an adverse effect of a cyber attack on an I&C system, but the combinations of failures that can lead to cyber incidents are unlikely. The necessary element of the defense in depth is the independence and the effectiveness of different defensive layers and their protection against common cause failures at all life-cycle stages of an I&C system. Also, each layer realizes the protection against cyber threats that may occur at adjacent layers.

The defense in depth should be implemented not only at nuclear facilities, but also at companies that develop and manufacture I&C systems (in particular, the RG, 2010 requires this). I&C systems developers and operational organization should implement, use and support the defense-in-depth strategy (considering defensive architecture and physical protection measures) for the ensuring of possibilities for the identification, prevention, response, mitigation of consequences and recovery after cyber-attacks.

The defense-in-depth strategy should be described in a cyber security program.

Graded Approach for the Cyber Security Assurance

One of the fundamental principles of cyber security assurance is the graded approach that consists in the forming of requirements for cyber security according to the cyber security classification defined for I&C systems. Protective measures are used proportionally to potentially possible consequences of cyber-attacks. The nature of protective measures depends on the cyber security level of a specific I&C system (i.e., higher cyber security level requires more stringent protective measures).

Practical realization of the graded approach usually requires the logical grouping of I&C systems with the same cyber security levels into cyber security zones for administrating and implementation identical protective measures. Criteria for the definition of cyber security zones are the architecture, physical placement of I&C systems, nature of interconnections interfaces, the topology of local networks.

The cyber security level is assigned for each cyber security zone and it defines the protective measures for every I&C system in this zone. The same cyber security level is assigned for every I&C system and their components within the same zone.

Cyber security zones reflect the logical and physical grouping of I&C systems, but cyber security levels define the degree of necessary protection. The same cyber security level may be assigned for several cyber security zones if they require a similar levels of protection.

The following principles should be used for the implementation of the zone model:

- Each zone covers I&C systems that have similar cyber security levels;
- Similar protective measures are used for I&C systems within one zone;
- I&C systems within one zone form the area of reliable connection that does not require the use of additional protective measures;
- The mechanisms of data flows separation are realized on the zone borders for prevention of the unauthorized access and spread of errors from the zone with the less cyber security level to the zone with the higher cyber security level;
- The direct connection channel that passes through more than two zones is not allowed;
- Network equipment (communicators, cables, etc.) is placed in the same cyber security zone as I&C systems that use this equipment;
- Network equipment that is used for connecting I&C systems from two different cyber security zones should belong to the zone with the higher cyber security level and the appropriate cyber security measures should be applied for this equipment;
- Data transfer should be fulfilled through the protected communication channels and should be unidirectional from the zone with the higher cyber security level to the zone with the lower cyber security level (reverse data transfer should be justified with the use of cyber security risk analysis and should be initiated from the zone with the higher cyber security level by the query to the zone with the lower cyber security level);
- Borders between two cyber security zones should be equipped by the hardware and software for the separation of data flows in compliance with the requirements related to the zone with the higher cyber security level;
- Data flows between different cyber security zones should be monitored for ensuring the effectiveness of the defensive architecture; and
- mobile devices or temporary equipment for access to I&C systems should be used only within the one cyber security zone or within the defined set of zones with a similar cyber security level.

Coordination Between Cyber Security and Safety

Cyber security tools (including antivirus software) should be evaluated to ensure that, their failures and their maintenance should not have an adverse impact (e.g., prevention, delay, or distortion) on the human-machine interface, technical characteristics and performing of I&C systems functions, important for safety, in conditions of normal or abnormal operation.

The negative impact from the cyber security measures implemented in the specific I&C system on the adjacent I&C systems should be prevented. The use of internal cyber security tools should be minimized in the safety systems (i.e., preference should be given to external cyber security tools).

The developers of I&C systems should analyze the potential risks of the negative impact of cyber security tools on operation of I&C systems and should minimize risks.

An additional risk assessment is carried out in case of any modifications of an I&C system, upgrade of software or change of its configuration. The absence of the negative impact of cyber security tools of operation and technical characteristics of an I&C system confirms during the verification of software, testing, and validation of the I&C system.

CATEGORIZATION

The requirements established in the IAEA, IEC and US NRC documents depend on the accepted categorization of I&C systems on cyber security, but in all cases a graded approach to cyber security is used (i.e. protection measures are applied proportionally to the potential consequences of cyber-attacks). In particular, a graded approach consists in the separation of I&C systems into cyber security levels and zones, depending on their relevance to security. Protection measures of various degrees are applied for different cyber security levels and zones (the more stringent protection should be provided for the I&C systems more important for safety).

Categorization of I&C Systems on Cyber Security by IAEA

The structure of the possible cyber security levels and the relationship of these levels with the corresponding cyber security measures are presented in IAEA, 2011. In accordance with this document, the cyber security level is understood as an abstraction that determines the degree of protection required for various computer systems (i.e., not just I&C systems directly involved in the control of technological processes, but also physical security computer systems, office work systems, etc.) at nuclear facilities (including NPPs).

The main purpose of introducing cyber security levels is to provide a framework for forming of a graded set of protective measures for various computer systems based on their cyber security levels.

For each cyber security level, it is necessary to implement a different set of protective measures that meet the security requirements of this level. At the same time, part of the protective measures is applied for all systems at all levels, and some measures are specific for a certain cyber security level.

The relationship between the critical importance of systems and the cyber security measures applied to them is shown in Figure 1. It shows that protective measures of a basic level should be provided for all computer systems. In addition, special protection measures should be applied for each cyber security level: minimum protection is required for level 5, and maximum protection I necessary for level 1. However, some protective measures may be repeated for several levels.

Figure 1. Security level and the severity of the measures

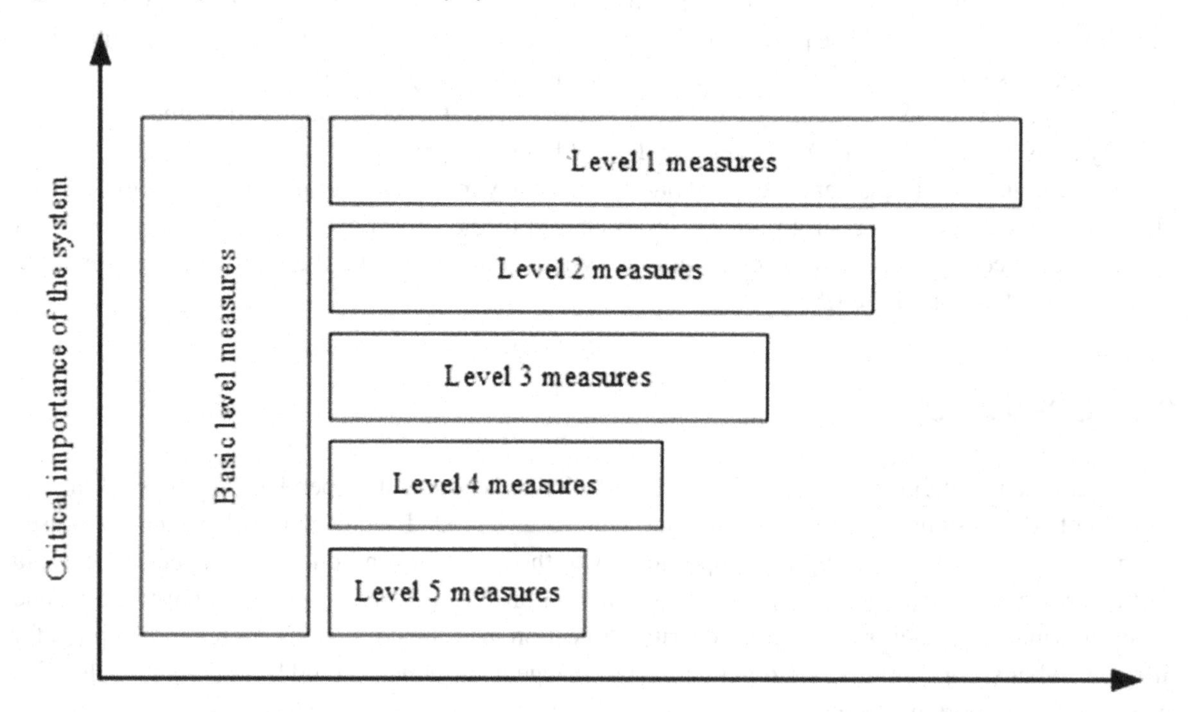

There are no criteria for the assignment of systems to the particular cyber security levels in IAEA, 2011. Note that IAEA, 2011 provides sample lists of basic and graded (for different cyber security levels) protective measures; an accurate choice of levels and related protection measures should be made in accordance with the specific environment, the specifics of the nuclear facility and the results of a cyber security risk analysis.

IAEA, 2011 introduces the concept of "computer security zone", which is a logical and physical concept that allows the grouping of computer systems that are equally important for safety, in order to facilitate the application of the graded administrative management, communication and the application of protective measures inside the zone.

According to IAEA, 2011, the following recommendations should be observed in case of applying the zonal model:

- Each zone should include systems that have equal or comparable importance for physical security, as well as for nuclear and radiation safety of the nuclear facility;
- Similar protective measures are applied concerning the systems belonging to one zone;
- Computer systems within one zone form an area of reliable communication that does not require the use of additional cyber security measures; and
- Mechanisms for separation of data flows are implemented at the borders of zones for the prevention of unauthorized access and spread of errors from a zone with lower security requirements to a zone with higher security requirements.

Technical and administrative measures for the separation of cyber security zones should consider the requirements of appropriate cyber security levels. A direct connection channel passing through more than two zones is not allowed.

Each cyber security zone consists of systems of equal or comparable importance. Therefore, each zone can be assigned to a cyber security level that defines protective measures for computer systems in this zone.

The relationship between cyber security zones and levels is not one-to-one. If several zones require the same degree of protection, they can be assigned to the same cyber security level. Zones reflect the logical and physical grouping of computer systems, while levels determine the required degree of protection.

IAEA 2018 notes that cyber security levels and safety classes are different but related concepts. Safety classification is based on the effect of the functions (and their failures) of the I&C system and its components on the safety of a nuclear installation. Cyber security levels are determined based on the vulnerability assessment of threats and consequences of failures or malfunctioning of the I&C system and its components due to cyber-attacks.

According to IAEA, 2018, nuclear and radiation safety measures, on the one hand, and cyber security measures, on the other hand, should be designed and implemented in such a way as to ensure the relationship between these two areas and exclude the possibility of their mutual negative influence on each other.

IAEA, 2018 indicates the necessity of taking into account the possibility that many components may be at risk due to a single cyber attack (e.g., different channels of the same system) or the attack may be aimed at different targets and combine different types of threats. This requires a special analysis to assess the potential consequences of cyber-attacks on the safety of a nuclear facility.

According to IAEA, 2018, it is necessary to identify the I&C system components for which malfunctioning may adversely affect the safety or performance of emergency monitoring functions. Cyber security levels are set depending on the importance of such components for the safety of a nuclear facility.

It is emphasized that the implementation of cyber security zones for systems can lead to the necessity of assigning a higher level of cyber security than originally intended for some components. For example, the communicator is not a component important for safety, however, if the communicator provides communication between two sets of reactor trip system, it is included into the same cyber security zone as this system. Accordingly, the communicator should also be assigned to the same level of cyber security as the reactor trip system, because there is a potential danger of its use for disrupting the operation of one of the safety systems. In this case, the inclusion of the communicator and the reactor trip system in the same cyber security zone eliminates the necessity of implementation of additional security measures in the system for protection against potential cyber security threats from the communicator.

Categorization of I&C Systems on Cyber Security by IEC

IEC, 2019 introduces the categorization by cyber security degree. The term “cyber security degree” is similar to the term “cyber security level” adopted by the IAEA. Below authors will use the term “cyber security level” instead of “cyber security degree” for the unification of the terminology in the frame of this chapter.

The cyber security level for NPP I&C systems is determined based on an analysis of the possible maximum consequences of a successful cyber attack against a system in terms of its impact on the safety or power generating by NPP. The more dangerous such consequences, the higher cyber security level is set for the system.

The categorization scheme in IEC, 2019 is based on the following principles.

1. The consequences of cyber-attacks that affect the safety of NPP should be considered as more serious than those that affect power generation.
2. NPP systems should be considered from a functional point of view. It is necessary to evaluate the impact of a possible cyber attack on the safety and power generation by NPP regarding the most sensitive and significant function of the system, the harmful impact on which can lead to the most serious consequences.
3. The analysis of a specific system should consider the possibility that other I&C systems of NPP may be the object of the same cyber attack and it can significantly complicate the overall situation.

IEC, 2019 defines three cyber security levels: S1, S2, and S3. Note that the categorization of cyber security, according to IEC, 2019, applies only to the NPP I&C systems directly involved in the control and monitoring of technological processes, and does not apply to other computer systems of NPP. IEC, 2019, establishes a relationship between cyber security levels and categories of functions, according to IEC, 2009, performed by the NPP I&C systems.

The cyber security levels in IEC, 2019 are determined for the I&C systems in accordance with the categories of functions, performed by the NPP I&C systems, and maximum consequences of their failures.

An unambiguous correspondence between the cyber security level and the category of functions performed by the NPP I&C system is not required. A higher cyber security level can be assigned to a system if the maximum consequences of malicious influences on any of its functions require the adoption of more stringent measures to ensure cyber security (e.g., asset or power generation protection business needs).

In addition, IEC, 2019 states that, if necessary, the number of cyber security levels can be increased to establish requirements for cyber security in relation to other computer systems that are not involved in the control of technological processes.

Similarly, with the approach adopted by the IAEA, the concept of cyber security zones is also introduced in IEC, 2019 in order to the practical implementation of a graded approach by logical combining of systems with the same cyber security levels into groups for administering and implementing identical protective measures. The criteria for determining of cyber security zones can be the architecture and physical placement of systems, the organization of intersystem interfaces, or the topology of local networks.

According to IEC, 2019, the application of the zonal model must comply with the following principles:

1. Each zone includes systems that have the same cyber security level. If for architectural or other reasons, a certain NPP I&C system has a lower cyber security level than other systems in a particular zone, its level is increased and these systems should comply with the requirements for the I&C system of other systems in this zone.
2. Additional protective barriers between systems belonging to the same cyber security zone are not required. However, for interzonal interfaces, barriers can be an effective protection measure.
3. Network equipment (communicators, cables, etc.) is located in the same cyber security zone as the associated NPP I&C systems. If network equipment is used for connecting systems belonging to different zones, appropriate separation of this network equipment into zones is implemented and the requirements of the same cyber security level are established for it as for systems included in the corresponding zone.

4. The exchange of data is initiated from the side of the zone containing systems of a higher cyber security level by requesting to the zone containing systems of a lower cyber security level.
5. The boundaries of the zones are equipped with technical means for separating data flows in accordance with the requirements for the cyber security level of the NPP I&C systems.

Figure 2. Simplified cyber security architecture (example)

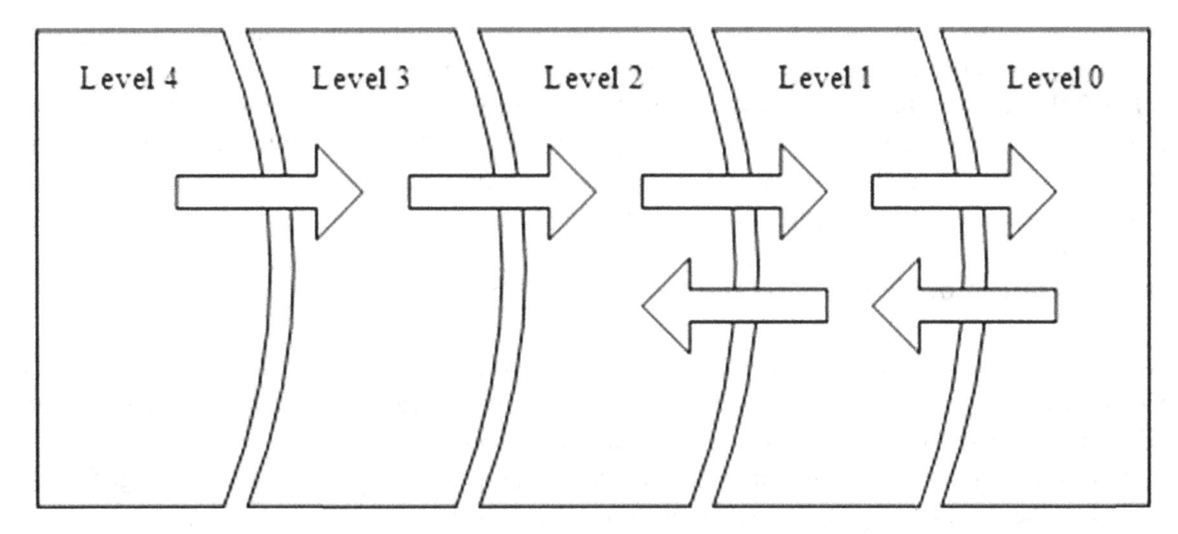

The relationship between zones and cyber security level is not straightforward. For example, if necessary, several zones can have the same cyber security level.

Similarly, with the IAEA approach the IEC, 2019 describes possible protection measures (i.e., common for all systems and graded by different cyber security levels).

Categorization of I&C Systems on Cyber Security by US NRC

RG, 2010 defines requirements for defensive cyber security architecture.

According to RG, 2010, a defense in depth against cyber-attacks aimed at critical digital resources (systems, computers, and hardware significant from a cyber security point of view) should be used in the general cyber security strategy for a nuclear facility. The defense-in-depth strategy is described in a cyber security plan. One of the acceptable methods for implementation of this strategy is the use of a defensive architecture framework, which establishes formal communication boundaries (or cyber security levels) where protective measures are used for detection, prevention, delay, mitigation and recovery in case of cyber-attacks. A defensive architecture framework includes a series of concentric cyber security levels that correspond to physical protection zones existing at a nuclear facility (e.g., internal zone, protected zone, controlled zone, corporate zone, public zone).

An example of an acceptable cyber security architecture framework, according to RG, 2010, is shown in Figure 2. Cyber security levels are separated by defensive boundaries, for example, using tools such as diodes and firewalls for control and protection of communications between levels. Systems requiring a higher degree of cyber security are placed at a higher level with a large number of protective barriers.

The logical model shown in Figure 2 does not always have to correspond directly with the physical location of the system in a particular physical protection zone. For example, a system assigned to level 3 may be located in a physical security zone corresponding to level 2, however, more stringent protective measures (in terms of cyber security) corresponding to the level 3 established for it should be provided for this system.

Note that, according to RG, 2010, the systems and hardware involved in the performing of functions important for the safety or physical protection of a nuclear facility, as well as auxiliary systems and equipment that could adversely affect the performing of these functions, should be assigned to the highest cyber security level and should be reliably protected from all lower levels.

RG, 2010 describes the basic principles of interaction between systems belonging to different cyber security levels (for example, hardware-based one-way transfer of information from level 4 to level 3 and from level 3 to level 2), and it gives recommendations for configuring of systems and protective barriers to ensure appropriate degree of protection against cyber threats.

Categorization of NPP I&C Systems on Cyber Security in Ukraine

SSTC NRS is now developing a regulatory guide on cyber security for the Ukrainian NPP I&C systems. One of the most important tasks is the determination of cyber security levels of the NPP I&C systems and grading of requirements depending on the established level.

According to the results of the analysis performed, it is proposed, to use as a basis the categorization adopted by the IEC as the most acceptable for the needs of Ukraine:

- The IEC, 2019 considers the same object for ensuring cyber security as in the developed regulatory guide (i.e., the NPP I&C systems, but not all computer systems of a nuclear facility); and
- The cyber security levels in the IEC, 2019 document are tied to the categories of functions of the NPP I&C systems, which gives the possibility to harmonize them with the current NP, 2015, in which the similar categories of functions for the NPP I&C systems are established.

The categories of functions performed by the I&C systems important for the NPP safety are defined in the NP, 2015. Functions performed by the I&C systems important for safety are referred to categories A, B, C, depending on their role in safety ensuring as well as on the possible consequences caused by the failure or incorrect performing of the function. Functions performed by the I&C systems that are not related to safety are not classified by categories.

The regulatory guide on cyber security developed for Ukrainian NPP I&C systems defines the cyber security levels of I&C systems which are determined according to the categories of functions performed by these systems (see Figure 3).

1. Cyber security level K1 is set for I&C systems that perform category A functions.
2. Cyber security level K2 is set for I&C systems that perform:
    a. Category B functions;
    b. Category C functions required for real-time operation; and
    c. Unclassified functions required for real-time operation.
3. Cyber security level K3 is set for I&C systems that perform:
    a. Category C functions not related to cyber security level K2; and

   b. Auxiliary unclassified functions required during operation and maintenance.
4. Cyber security level K4 is set for computer systems that have links with I&C systems that are assigned to cyber security levels K1, K2 or K3 but are not involved in the control of technological processes.

Figure 3. Cyber security levels proposed for Ukraine

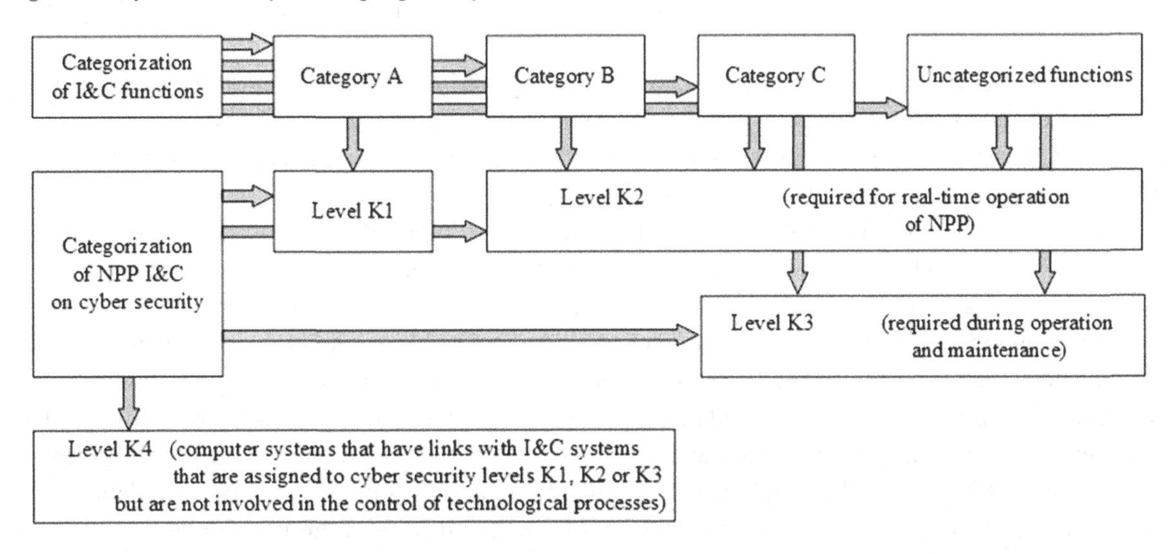

All components of I&C systems belong to the same cyber security level as whole I&C systems. As it can be seen from the analysis presented, despite some differences, the IAEA, IEC, US NRC, and Ukraine use the similar ideology regarding the application of a graded approach for cyber security ensuring, based on the appropriate categorization of systems by cyber security levels (or degrees).

CYBER SECURITY ASSURANCE ON THE STAGE OF I&C SYSTEMS DEVELOPMENT

General Design Cyber Security Measures for NPP I&C Systems

I&C system design should define measures for all its life cycle stages that ensure compliance with cyber security requirements.

The development of an I&C system is intended the minimization of potential vulnerabilities and the implementation of general and additional (i.e., depending on the cyber security level of a specific I&C system) cyber security measures in the system.

The analysis of own developed software code (including automatically generated code) is performed as one of the preventive measures against the presence of hidden functions in the software of I&C systems and their components. If possible, the decompiling of the executable code and the comparison of the obtained results with the source code is fulfilled for confirmation of their identity and prevention of the entering of malicious fragments into software code by the compiler.

The software of I&C systems, important for NPP safety, should be verified in accordance with the NP, 2015. During the software verification the check of the implementation of cyber security measures and the absence of their negative impact on the fulfillment of functions of the I&C system, important for safety, is performed.

Verification of tools used for the development of software is carried out in accordance with the requirements for software of the I&C systems and their components. Controls should be established that ensure that adverse effects of the developed I&C systems on the cyber security of other I&C systems are prevented.

During the design of I&C systems, it should be considered that the connection between the I&C systems of different cyber security levels is initiated by the I&C systems of a higher cyber security level. The design measures in the I&C systems should ensure the prevention of the negative impact of other I&C systems.

Means of monitoring and alarming about physical access to the I&C systems, hardware, software, system configuration changes or abnormal I&C systems behavior should be used. The number of access points to computer networks should be minimized in the I&C systems. Means for prevention of negative impact on the I&C systems from the side of special testing and maintenance equipment should be implemented. User access to the I&C systems is implemented using authentication tools. Only authorized personnel are allowed to access and modify of existing configuration, software, or hardware. The human-machine interface (used during the operation or maintenance) provides access to the software only for authorized personnel and only for the minimum required amount of functions and data.

At the same time, the interception or distortion of the data displayed through the human-machine interface, aimed at preventing or delaying the operator's actions in case of performing security functions should be prevented. Means for limiting of access to programmable elements of the I&C systems and preventing creation of new unauthorized paths for access to these elements should be implemented.

Design measures are realized for ensuring sufficient confidence that the protection of the I&C system assigned to a certain cyber security level will not degraded as a result of the impact of the I&C system assigned to the lower cyber security level. Any pre-developed component (including one provided by third parties) is selected, configured and adjusted to minimize the vulnerability of the I&C systems.

Effective security measures should be implemented during the design, configuration, and settup of the parameters for programmable hardware for:

- Management of selective user access to software functions and memory;
- Data transfer lines to the I&C systems with the lower levels of cyber security; and
- Monitoring of modifications of software and parameters.

The validation of the I&C systems (or hardware) in their final configuration should demonstrate the effectiveness of cyber security measures and the absence of any adverse effect of these measures on the functions related to safety.

During validation, tests are performed for confirmation of the adequacy and correctness of the cyber security measures implemented in the I&C systems and for the identification of potential vulnerabilities of the cyber security of the I&C systems.

Testing shall be conducted in a secure environment. Used test equipment should be subjected to the cyber security measures similar to those applicable to the tested I&C systems, considering the appropriate cyber security level of the I&C systems.

Cyber security measures that cannot be integrated directly into the I&C systems should be implemented separately. Additional administrative controls are used for the application and maintenance of such individual devices.

Any information relating to the design, manufacturing, implementation, and operation of the I&C systems shall be identified and, where appropriate, designated as restricted information for which appropriate protection measures against unauthorized disclosure, theft, distortion or destruction should be used.

Means for protection against unauthorized access to the components of the I&C systems and network equipment shall be provided. Measures should be realized against the presence of hidden functions in the software of the I&C systems.

Any new software is designed in such a way as to minimize the vulnerabilities of the I&C systems. During the designing of software, the results of the cyber security assessment of software should be considered. The settings and configuration of any earlier developed software should minimize the vulnerability of software (for example, by minimizing functions to the required limit or by using existing cyber security measures for software).

The possibility of changing the software stored on data carriers or as spare parts should be excluded. The design documentation defines the functions critical for cyber security and the elements of cyber security applied in the software.

If user access to the I&C systems is critical for cyber security, then an authentication procedure should be applied with use of technical means based on a combination of knowledge information (such as a password), personal property (such as a key or a card with a built-in microprocessor) and personal characteristics (such as fingerprint).

User access rights should be limited considering the possibilities and consequences of potential threats to cyber security. One of the possible ways for proper implementation of this requirement is the cryptographic method (data encryption).

Remote access from any location outside the technical environment of the I&C systems is prohibited. Antivirus protection and regular updating of antivirus databases should be provided in the I&C systems.

Cyber Security Assurance on the Stage of Development of NPP I&C Systems and Software

The developers of I&C systems, their components, and software provide reliable and monitored processes for cyber security assurance. Cyber security should be periodically evaluated during the development of I&C systems, their components and software.

The development stages of I&C systems, their components and software are implemented in a secure environment, with the realization of appropriate cyber security measures for prevention of intrusion of malicious software code or data, and for protection of sensitive information related to the developed I&C systems, their components, and software. The security of the development environment and the adequacy of the applied cyber security measures should be periodically assessed.

The secure development environment uses administrative cyber security measures, such as configuration management, limitation, and monitoring of the use of mobile devices and removable media, etc.

Unauthorized access to I&C systems, their components and software should be prevented during the development stage. Software is developed using licensed or verified tools. Isolation of software development tools from external computer networks is provided. The potential threats to cyber security should be considered during the software development.

Measures are provided against the implementation of hidden functions in application or system software (particularly, code verification). Potential software modification tools that can cause incorrect performing of functions of the I&C systems are identified. Verification confirms the ability to detect such a modification.

Any technical and software development tools, as well as test equipment, must be checked for confirmation of prevention of adversely impact from them and creation of potential paths for intrusion of malicious code or data into a secure development environment or developed I&C systems, their components or software. Cyber security measures used for development tools and test equipment should be similar to those implemented in a secure development environment with respect to the developed I&C systems, their components or software.

The final stage of the development of an I&C system is validation, which is conducted in particular for confirmation of the compliance with the cyber security requirements and the absence of a negative impact of the implemented cyber security measures on the functions, technical characteristics and safety features of the system.

Cyber security measures are used for the prevention of any harmful intrusion during the transportation of I&C systems, hardware or software from the developer to the NPP.

Cyber Security Assurance for NPP I&C Systems and Software Developed by Third-Party Organizations

The input control of all purchased products used by the developer for the manufacturing of the I&C systems and their components is performed.

Purchased software used by the developer for the manufacturing of the I&C systems and their components should be the subject of mandatory testing, which should confirm that the I&C system behaves in a deterministic way without adverse impacts to the overall system performance from the cyber security measures.

The use of purchased software for performing safety functions should be minimized. Purchased hardware and software are verified for compliance with cyber security requirements through appropriate testing.

The I&C systems should use only licensed purchase software. The functionality of the purchased software (including operating systems) is limited only by the range of capabilities required for proper performing of the functions of the I&C systems.

CYBER SECURITY ASSURANCE ON THE STAGE OF I&C SYSTEMS IMPLEMENTATION AND OPERATION

Cyber Security Assurance During the Installation, Commissioning and Testing of I&C Systems at NPP

The I&C systems installation, commissioning and testing at NPP shall be carried out in a secure environment with the use of equipment to which cyber security measures are applied similar to those applied to the implemented I&C systems taking into account cyber security level of this systems.

The equipment used during installation, commissioning and testing should be checked to verify that it does not create new paths for the intrusion of malicious software code or data into a secure environment or into components of the I&C systems. Cyber security measures are realized for managing and monitoring the transfer of data, software, and hardware inside or outside the secure environment.

The access of specialists from other organizations (for installation, repair, testing, etc.) to the I&C systems should be limited according to their tasks, both in terms of the duration of access and the specific list of the I&C systems (or their components) to which access is allowed.

The staff of other organizations should be instructed about cyber security assurance during the performing of procedures with the I&C systems at NPP. The operator provides monitoring over the actions of personnel of other organizations during any operations with access to the I&C systems at NPP (in particular, the "two-person rule" is applied).

General Measures of Cyber Security Assurance During the Operation of NPP I&C Systems

The operator implements demilitarized zones firewalls that separate the external networks from the general-purpose networks, as well as the general-purpose networks from the local networks, which provide information exchange between the I&C systems or their components, which are directly involved in the monitoring and control of the technological processes at the NPP.

Security tools are implemented at the NPP, which separate the I&C systems of different cyber security levels. Cyber security assessment of I&C systems is required in the cyber security program and the cyber security plan. If the assessment shows that the planned measures are insufficient, the additional requirements for protection measures are determined.

The on-site assessment of the cyber security of the I&C systems configuration and parameters is performed for confirmation of the implementation of appropriate protective measures against potential cyber threats.

The I&C systems modification activities are regularly planned and carried out with the considering of potential cyber threats. The I&C systems operation logs are periodically checked from the cyber security point of view. The number of access points to the local networks is minimized as far as possible.

Measures for detection and appropriate response on attempts of unauthorized access or connection to I&C systems or local networks should be implemented and the negative impact of these measures on I&C systems operation shall be excluded.

Access to I&C systems or local networks should be strictly monitored to prevent unauthorized persons from intrusion into I&C systems operation. This is achieved through the implementation of technical protective measures (e.g., locks on the cabinets, control of physical access to the rooms, etc.), software restriction and detection of unauthorized access, and the introduction of appropriate organizational measures that are set according to the cyber security levels of specific I&C systems.

In the context of configuration management, any permanent or temporary changes related to the construction and communication of I&C systems, access or connection of additional communication lines for test or maintenance devices, software updates shall be identified and recorded to detect changes that could adversely affect the cyber security.

Software configuration and hardware changes are prohibited if they are not planned, approved, or documented. The procedure for the prompt recovery of an acceptable operability level of I&C systems after a cyber attack should be determined. Measures are realized for minimization of the probability that the specified recovery procedures will be vulnerable to the same cyber threat.

Independent inspections are carried out periodically for the check of the adequacy of the implemented cyber security measures and their compliance with the requirements of cyber security regulations and standards and the cyber security program. Testing of cyber security means realized in the I&C systems is performed periodically at the NPP.

The operator monitors the cyber security of I&C systems continuously for the detection of cyber threats, system malfunctions, unauthorized access, or changes. The results of monitoring should be archived and protected against deletion or modification. The appropriate human-machine interface is implemented for the support of staff during the monitoring of cyber security, identification, reporting and alarm of cyber threats in all operating modes of NPP.

Measures are realized for the prevention of the creation of additional passes for the data transferring between I&C systems assigned to the different levels of cyber security through the equipment and communication lines used for monitoring, maintenance, and recovery.

Maintenance of the I&C systems should cover cyber security and include:

- Periodic testing, review of I&C systems operation logs and on-line monitoring of I&C systems operation;
- Activities for identification, prevention, and mitigation of the effects of degradation of components of I&C systems; and
- Actions for diagnosis, recovery, or replacement of failed components.

Cyber security measures are implemented during maintenance for the prevention of the introduction of malicious software code or data into I&C systems or their components.

For the equipment used during the maintenance cyber security measures are implemented similar to those applicable to the relevant I&C systems, considering their cyber security level. Any connections of the maintenance equipment to the I&C systems is prohibited if it is not necessary or if appropriate maintenance activities are not carried out.

If certain cyber security means are turned off temporarily for maintenance than appropriate compensatory cyber security measures should be realized during the period of these activities. Remote access to the I&C systems from outside the NPP is prohibited during maintenance.

After the maintenance or testing of the I&C systems, the software configuration and settings should be checked for the prevention of their unauthorized change. The activities of personnel during the operation, maintenance, and testing of the I&C systems are controlled in accordance with the procedures prescribed by the NPP (in particular, the two-person rule applies).

If any removable storage media are used during operation, maintenance and testing than their contents should be monitored before connecting to I&C systems for prevention of intrusion of malicious software or data to I&C systems and after disconnecting from I&C systems for prevention of unauthorized copying of system data.

In case of replacing certain elements of I&C systems during modification, maintenance, or repair, any data and software should be removed from the replaced elements for the prevention of the use of this information for the preparation and conduction of cyber-attacks. If data and software cannot be removed from replaced elements of I&C systems, then such elements shall be destroyed or stored in accordance with appropriate physical and cyber security measures and protection against unauthorized access.

Measures of Cyber Security Assurance During the Operation of NPP I&C Systems (Level K1)

The connections of the I&C systems assigned to cyber security level K1 are limited by other I&C systems assigned to cyber security levels K1 and K2 as well as related tools.

Communication is provided from the I&C systems assigned to cyber security level K1 to the I&C systems assigned to cyber security level K2. The reverse data transmission should be initiated by the I&C systems assigned to cyber security level K1.

Data transfer from the I&C systems assigned to cyber security level K2 to the I&C systems assigned to cyber security level K1 is limited to the mandatory data without which the performing of functions is impossible and is only allowed on condition of detailed analysis and justification of the cyber security risks. Any data transferred from the I&C systems assigned to cyber security level K2 to the I&C systems assigned to cyber security level K1 is reliably protected by appropriate restrictions (e.g. data format, control of transfer time, checksums, etc.).

Software updates and configuration changes to the I&C systems assigned to cyber security level K1 are performed only with the use of local hardware locks (e.g., keys) and only in one system channel at a time.

Bidirectional data transfer between the I&C systems assigned to cyber security level K1 and the maintenance equipment is performed using an individual communication line that is separated from other networks. This communication line is protected by technical, software and administrative means. Allowed access to the I&C systems assigned to cyber security level K1 as well as access to software or changes of configuration should be monitored by alarms in the main and supplementary control rooms or local control panels.

Unauthorized persons and non-designated I&C systems cannot read or modify data and software of the I&C systems assigned to cyber security level K1. At the same time, the necessary access should be provided for authorized persons and designated I&C systems.

The main security measures (including technical protection, the connection of blocking devices, etc.) are implemented inside the I&C systems. Basic software requirements for the I&C systems can supplement the system-level security measures.

The software is configured and configured to collect important information for periodic check of cyber security assurance for I&C systems and for preparing an appropriate report.

Measures of Cyber Security Assurance During the Operation of NPP I&C Systems (Level K2)

Communication is realized from the I&C systems assigned to cyber security level K2 to the I&C systems assigned to cyber security level K3. At the same time, the I&C systems assigned to cyber security level K2 acts as the initiator of communication. These requirements (i.e., orientation and initiation of communication) are implemented by the use of appropriate technical means (such as special equipment for filtering of data flows).

The transfer of data from the I&C systems assigned to cyber security level K3 to the I&C systems assigned to cyber security level K2 is restricted and used only in properly justified cases. Changes of the software and configuration of the I&C systems assigned to cyber security level K2 by the I&C systems assigned to cyber security level K3 should be prevented.

Software updates and configuration changes of the I&C systems assigned to cyber security level K2 are performed in one channel of the system at a time, during the predetermined time intervals, and are protected by appropriate blocking. Bidirectional data transfer between the I&C systems assigned to cyber security level K2 and the maintenance equipment is performed using an individual communication line that is separated from other networks. This communication should be protected by technical, software and administrative means.

I&C systems assigned to cyber security level K2 do not use input communication lines from outside the NPP or from the I&C systems assigned to cyber security level K4.

Measures of Cyber Security Assurance During the Operation of NPP I&C Systems (Level K3)

Access from computer systems assigned to cyber security level K4 that could affect functions of the I&C systems assigned to cyber security level K3 should be justified and applied only if it does not jeopardize the compliance with safety requirements established for this I&C system.

The connection between the I&C systems assigned to cyber security level K3 and the computer systems assigned to cyber security level K4 is initiated by the I&C systems assigned to cyber security level K3. Exceptions are properly justified, and communication is monitored.

Measures of Cyber Security Assurance During the Operation of NPP I&C Systems (Level K4)

Special cyber security measures do not apply for computer systems assigned to cyber security level K4. Negative impact from the computer systems of the cyber security level K4 on the I&C systems, their components and software of the cyber security levels K1, K2, K3 should be prevented.

Monitoring of Changes During the NPP I&C Modification

The necessary software modifications are fulfilled considering the potential cyber security threats. An on-site assessment of modification is fulfilled for checking that appropriate protection measures against potential cyber security threats are used.

Specific modes determine during the commissioning and modification of the I&C systems. These modes should cover interfaces and special features of the I&C systems, which are blocked during power generation, the function of the activation of an alarm system that is blocked during the modification, the use of service stations and tools, and the necessity of operator's actions from a fixed working place.

Any special modes should be compensated by additional measures during or after modification that is fulfilled for ensuring the functioning and cyber security of I&C systems. The tools and equipment used during the on-site modification of software are selected based on the level of their potential threat for the cyber security of I&C systems. New data sets or new versions of software that make changes related to cyber security are subject to verification for confirmation that cyber security requirements are properly addressed. The procedure of on-site installation of software or data includes the checks of the software functionality, which are carried out before the complete commissioning of I&C systems.

SOLUTIONS AND RECOMMENDATIONS

All NPP I&C systems should be sufficiently protected against potential cyber threats because these systems are safety important. Therefore, it is recommended:

- To develop the regulatory framework for cyber security assurance of NPP I&C systems;
- To categorize all NPP I&C systems on cyber security levels; and
- To implement the set of cyber security measures for all NPP I&C systems based on their cyber security levels.

FUTURE RESEARCH DIRECTIONS

The future research directions can be as follows:

- Development of the regulatory framework as well as detailed methods of cyber security assurance;
- Realization of general technical decisions for cyber security assurance at all NPPs;
- Creation of cyber security teams and the center of operative response on cyber threats and cyber-attacks; and
- Use of diversity with the application of analog hardware for performing the most important functions (e.g., reactor trip).

CONCLUSION

Cyber security assurance of the NPP I&C systems as well as the development of an appropriate regulatory framework are relevant in many countries, including Ukraine. A comprehensive analysis of the various aspects of cyber security should be conducted for achieving this aim.

Adequate protective measures should cover all stages of the NPP I&C systems life cycle and provide a set of organizational, technical and software measures.

According to the results of an analysis of the existing regulatory framework for cyber security of nuclear facilities and the NPP I&C systems, it should be noted that its development is extremely relevant for both international organizations and national regulators (the most actively it carried out in the USA).

It should be noted that in addition to regulatory documents with regulatory requirements for cyber security, it is also advisable to develop documents where technical aspects will be considered, potential cyber threats will be analyzed, and specific methodological recommendations will be given for cyber security measures of NPP I&C systems at different life cycle stages taking into account the specific features of used technologies.

During the development of regulatory documents, it is advisable to consider detailed cyber security requirements for NPP I&C systems in combination with the requirements for nuclear safety, as well as considering the requirements of international standards.

The requirements for cyber security for NPP I&C systems should be based on the categorization of these systems by cyber security levels and cyber security zones. Cyber security measures are graded according to the appropriate cyber security levels of NPP I&C systems.

The set of cyber security measures described in the chapter is based on the requirements of international and national rules, guides and standards.

REFERENCES

CFR. (2015). *Title 10, Code of Federal Regulations*. Washington, DC: U.S. NRC.

IAEA. (2006). *Safety Standards Series No. SF-1. Fundamental safety principles: safety fundamentals*. Vienna, Austria: IAEA.

IAEA. (2011). *Nuclear Security Series No. 17. Computer security at nuclear facilities: reference manual: technical guidance*. Vienna, Austria: IAEA.

IAEA. (2015). *Nuclear Security Series No. 23-G. Security of nuclear information*. Vienna, Austria: IAEA.

IAEA. (2016). *Safety Standards Series No. SSG-39. Design of Instrumentation and Control Systems for Nuclear Power Plants*. Vienna, Austria: IAEA.

IAEA. (2017). *Nuclear Security Series No. 28-T, Self-assessment of nuclear security culture in facilities and activities: technical guidance*. Vienna, Austria: IAEA.

IAEA. (2018). *Nuclear Security Series No. 33-T. Computer security of instrumentation and control systems at nuclear facilities*. Vienna, Austria: IAEA.

IEC. (2009). *61226. Nuclear power plants – Instrumentation and control important to safety — Classification of instrumentation and control functions*. Geneva, Switzerland: IEC.

IEC. (2016). *68859 + AMD1:2019 CSV. Nuclear power plants — Instrumentation and control systems — Requirements for coordinating safety and cybersecurity*. Geneva, Switzerland: IEC.

IEC. (2019). *62645. Nuclear power plants — Instrumentation, control and electrical power systems — Cybersecurity requirements*. Geneva, Switzerland: IEC.

NP. (2015). *306.2.202. Requirements on nuclear and radiation safety to instrumentation and control systems, important for safety of nuclear power plants*. Kyiv, Ukraine: SNRIU.

RG. (2010). *5.71. Cyber security programs for nuclear facilities*. Washington, USA: U.S. NRC.

RG. (2011). *1.152, Revision 3. Criteria for use of computers in safety systems of nuclear power plants*. Washington, USA: U.S. NRC.

RG. (2015). *5.83. Cyber security event notifications*. Washington, DC: U.S. NRC.

ADDITIONAL READING

Bozkus Kahyaoglu, S., & Caliyurt, K. (2018). Cyber security assurance process from the internal audit perspective. *Managerial Auditing Journal, 33*(4), 360–376. doi:10.1108/MAJ-02-2018-1804

Chumak, D., & Klevtsov, O. (2015). Computer security at nuclear facilities in Ukraine: Areas of interactions between nuclear safety and security [in Ukrainian]. *Nuclear and Radiation Safety, 3*(67), 60–64.

EPRI. (2019). *3002015794. Cyber Security Baseline Configuration Topical Guide*. Palo Alto, CA, USA: Electric Power Research Institute.

IAEA. (2008). *Nuclear Security Series No. 7. Nuclear security culture: implementing guide*. Vienna, Austria: IAEA.

IAEA. (2018). *Nuclear Security Series No. NP-T-2.11. Approaches for overall instrumentation and control architectures of nuclear power plants*. Vienna, Austria: IAEA.

Klevtsov, O., Symonov, A., & Trubchaninov, S. (2016). Computer security of NPP instrumentation and control systems: Categorization [in Russian]. *Nuclear and Radiation Safety, 4*(72), 65–70. doi:10.32918/nrs.2016.4(72).10

Klevtsov, O., & Trubchaninov, S. (2015). Computer security of NPP instrumentation and control systems: Cyber threats [in Russian]. *Nuclear and Radiation Safety, 1*(65), 54–58.

Klevtsov, O., Yastrebenetsky, M., & Trubchaninov, S. (2015). Computer security of NPP instrumentation and control systems: Normative base [in Russian]. *Nuclear and Radiation Safety, 4*(68), 51–57.

Lia, Y.-F., & Liu, S.-Z. (2018). On the Recent Research Advancements of Cyber Security of Nuclear Power Plants. Probabilistic Safety Assessment and Management PSAM 14. Los Angeles, CA, USA.

NEI. (2010). 08-09 [Revision 6]. Cyber Security Plan for Nuclear Power Reactors. Nuclear Energy Institute.

NRC. (2016). 2015-0179. Rulemaking for Cyber Security at Fuel Cycle Facilities. Regulatory Basis Document. U.S. Nuclear Regulatory Commission.

Rowland, M. T., Dudenhoeffer, D. D., & Purvis, J. S. (2017). Computer Security for I&C Systems at Nuclear Facilities. NPIC&HMIT 2017, San Francisco, CA, USA.

Symonov, A., & Klevtsov, O. (2018). About the Problem of Regulatory Activity for Computer Security of NPP Instrumentation and Control Systems in Ukraine, In *Conference Proceedings of 2018 IEEE 9th International Conference on Dependable Systems, Services and Technologies DESSERT 2018* (pp. 7-12). Ukraine, Kyiv. 10.1109/DESSERT.2018.8409089

Symonov, A., Klevtsov, O., & Trubchaninov, S. (2017). Computer security of NPP instrumentation and control systems: Protection measures from computer threats [in Russian]. *Nuclear and Radiation Safety, 2*(74), 46–50. doi:10.32918/nrs.2017.2(74).09

Symonov, A., Klevtsov, O., Trubchaninov, S., & Lazurenko, O. (2019). Computer security of NPP instrumentation and control systems: Computer security justification documents. [in Russian]. *Nuclear and Radiation Safety, 4*(84), 73–81. doi:10.32918/nrs.2019.4(84).09

Tehan, R. (2017). *Cybersecurity: Critical Infrastructure. Authoritative Reports and Resources*. Congressional Research Service.

The White House (2018). *National Cyber Strategy of the United States of America.*

WG. (2017). *D2.38. Framework for EPU Operators to Manage the Response to a Cyber-Initiated Threat to Their Critical Infrastructure*. CIGRE.

Wheeler, T., Denman, M., Williams, R. A., Martin, N., & Jankovsky, Z. (2017). *Nuclear Power Plant Cyber Security Discrete Dynamic Event Tree Analysis (LDRD 17-0958) FY17 Report*. Sandia National Laboratories. doi:10.2172/1395751

Yastrebenetsky, M., Klevtsov, O., Rozen, Yu., & Trubchaninov, S. (2018). Elaboration of the System of the Standards, Related to Safety and Security of Instrumentation and Control Systems of Ukrainian Nuclear Power Plants. In *Conference Proceedings of 2018 IEEE 9th International Conference on Dependable Systems, Services and Technologies DESSERT 2018* (pp. 13-17). Ukraine, Kyiv. 10.1109/DESSERT.2018.8409090

KEY TERMS AND DEFINITIONS

Cyber Security: A particular aspect of information security that is concerned with computer-based systems, networks, and digital systems.

Cyber Security Incident: An occurrence that actually or potentially jeopardizes the confidentiality, integrity, or availability of a computer-based, networked or digital information system or the information that the system processes, stores, or transmits or that constitutes a violation or imminent risk of violation of security policies, security procedures, or acceptable use policies.

Cyber Security Policy: Aggregate of directives, regulations, rules, and practices that prescribes how an organization manages and protects computers and computer systems.

Defense in Depth: The combination of successive layers of systems and measures for the protection of targets from nuclear security threats.

Information Security: The preservation of the confidentiality, integrity, and availability of information.

NPP I&C Safety: A part of nuclear and radiation safety, relating to jointly operating I&C systems and manufacturing equipment of NPP power units and depending on proper operation of the I&C systems.

NPP Nuclear and Radiation Safety: A feature of non-exceeding determined limits of radiation effects on personnel, the population and the environment under NPP normal operation, operational events and design basis accidents and also of restricting radiation effects in beyond design basis accidents.

Chapter 14
NPP-Smart Grid Mutual Safety and Cyber Security Assurance

Eugene Brezhniev
National Aerospace University KhAI, Ukraine

Oleg Ivanchenko
https://orcid.org/0000-0002-5921-5757
University of Customs and Finance, Ukraine

ABSTRACT

The smart grid (SG) is a movement to bring the electrical power grid up to date so it can meet current and future requirements to fit customer needs. Disturbances in SG operation can originate from natural disasters, failures, human factors, terrorism, and so on. Outages and faults will cause serious problems and failures in the interconnected power systems, propagating into critical infrastructures such as nuclear industries, telecommunication systems, etc. Nuclear power plants (NPP) are an intrinsic part of the future smart grid. Therefore, it is of high priority to consider SG safety, mutual influence between NPP and SG, forecast possible accidents and failures of this interaction, and consider the strategies to avoid them.

INTRODUCTION

There are following development trends in SG (Y. Saleem, et.al. 2019) such as:

- **Implementation of Open Protocols**: Open industry standard protocols are replacing vendor-specific proprietary communication protocols.
- **Interconnected to Other Systems**: Connections to business and administrative networks to obtain productivity improvements and mandated open access information sharing.
- **Reliance on Public Information Systems**: Increasing use of internet and public telecommunication systems the for portions of the control system, etc.

DOI: 10.4018/978-1-7998-3277-5.ch014

The SG always needs to be available, and locking the system during an emergency could cause safety issues and security issues.

The SG security objectives are confidentiality, integrity and availability. In most industries confidentiality and integrity have higher precedence over availability. In the electrical power system, electricity must always be available, so this is the most important security objective. Integrity is the next important security objective followed by confidentiality. Availability is the most important security objective.

Integrity is the next important security objective in the SG. The SG uses data collected by various sensors and agents. This data is used to monitor the current state of the electrical power system. Unauthorized medication of the data, or insertion of data from unknown sources can cause failures or damage in the electrical power system. The electricity in the power grid not only needs to always be available, but it also has to have quality. The quality of the electrical power will be dependent on the quality of the current state estimation in the power system (A. Basit, G. A. S. Sidhu, 2016).

One of the main concerns for SG is the connectivity, automation and tracking of large number of devices, which requires distributing monitoring, analysis and control through high speed, ubiquitous and two-way digital communications. It requires distributing automation of SG for such devices or "things".

There are a lot of different types of influences between NPPs and SG, which stipulate NPP's safety levels. These influences cause the change of state of each PG subsystem during its life cycle. The balance of influences is considered as a basis for the SG stability and NPP safety. The change of these influences could lead to a balance violation. In its turn these violations lead to the NPP and SG subsystem's states changing. This chapter represents an approach for formalization of different types of influences between the NPP and SG. It helps to analyze the complex behavior of NPP and SG as a system of systems (SoS) and predict their safety levels considering the change of states.

The application of hybrid methods makes operator less dependent on information from instrumentation and control system (I&C). The illustrative example for the NPP reactor safety assessment is considered in this chapter.

BACKGROUND

A SG is comprised of four main subsystems (see Figure 1), power generation, power transmission, power distribution and power utilization.

IoT technologies (W. Shu-wen, 2016) can be applied to all these subsystems and appears as a promising solution for enhancing them, making IoT a key element for SG. In the power generation area, IoT can be used for the monitoring and controlling of energy consumption, units, equipment, gas emissions and pollutants discharge, power use/production prediction, energy storage and power connection, as well as for managing distributed power plans, pumped storage, wind power, biomass power and photo-voltaic power plants. In the power transmission area, IoT can be used for the monitoring and control of transmission lines and substations, as well as for transmission tower protection. In the power distribution area, IoT can be used for distributed automation, as well as in the operations and equipment management. In the power utilization area, IoT can be used for smart homes, automatic meter reading, electric vehicle charging and discharging, for collecting information about home appliances energy consumption, power load controlling, energy efficiency monitoring and management, etc.

Figure 1. Four main subsystems of smart grid

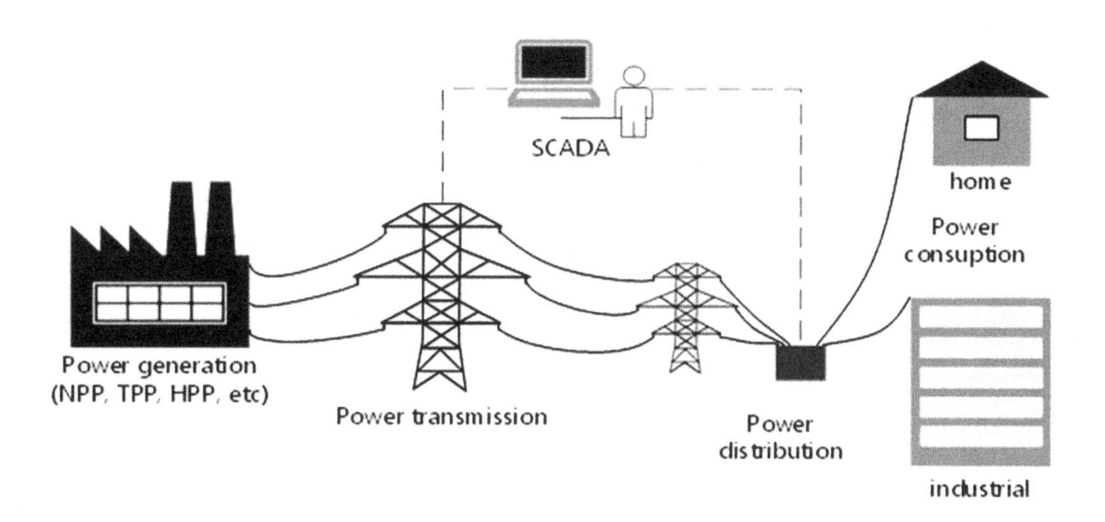

IoT supports information and material flows (see Figire 2) in SG making it cost-efficient for SG's clients. But SG also faces new, complex challenges (Y. Wang, 2012). There are several challenges that the SG might face, including:

- **Security Challenges:** Security is the most important challenge for the SG present and future (TLP, 2016). As more connected devices get added to the grid, it will become increasingly important to make sure they are secure, as each and every device will offer a new avenue for hackers to maliciously exploit (W. Wang & Z. Lum, 2013);
- **Interoperability:** Ensuring interoperability for connected devices is a must. Effective communication between devices ensures SG resiliency, reliability, and power management and provides greater visibility into grid operations; As embedded devices, IoT technologies are being introduced into SG architecture and this process might lead to new risks to safety of SG;
- **Logistics and Scalability:** To increase the efficiency of a SG, the utilities industry needs scalability at the forefront.

Cyber threats to the SG can also emerge from attacks directed via IoT devices connected to networks. IoT devices have been increasingly targeted by botnet malware (whereby the hacker takes over the operation of a large number of infected devices) to launch denial-of-service or other cyberattacks.

If such IoT cyberattacks were able to access electric utility operational or industrial control systems, they could potentially impair these systems or cause electric power networks to operate based on manipulated conditions or false information.

Figure 2. Information and material flows in SG

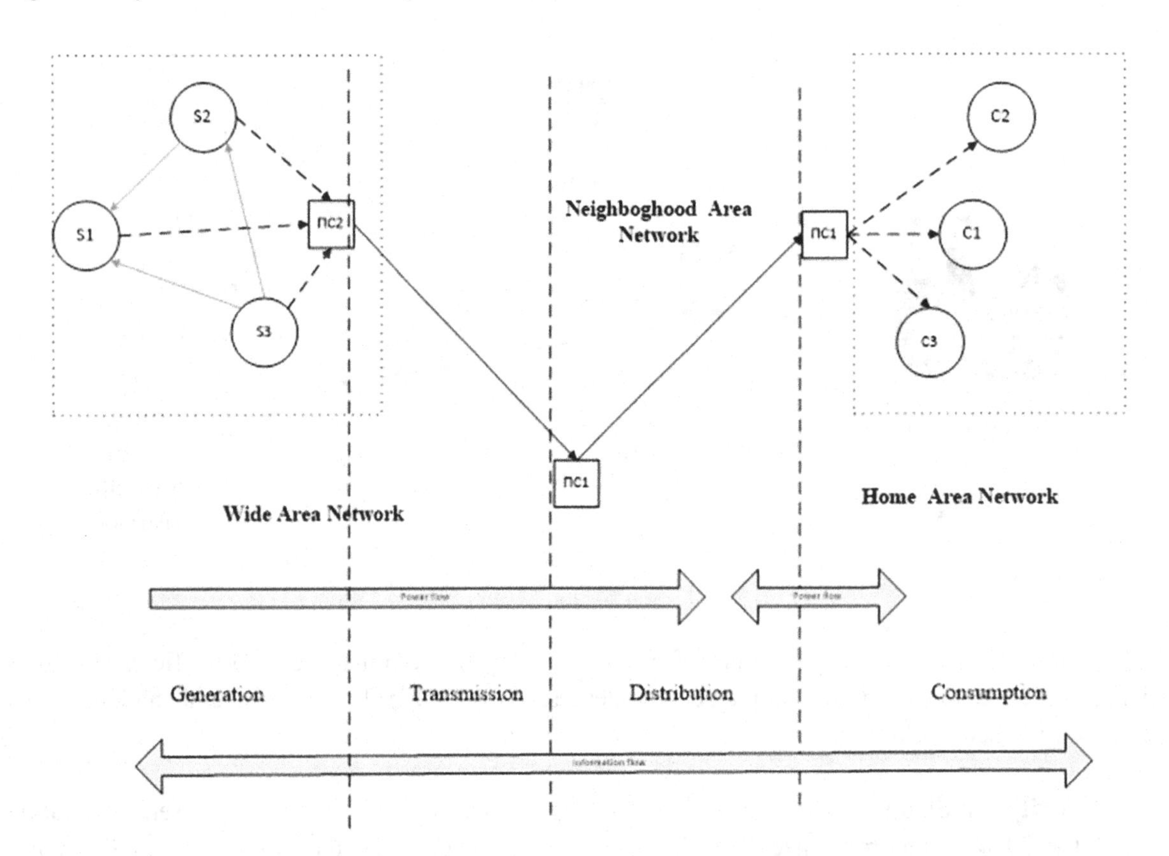

ANALYSIS OF IOT BASED SMART GRID AND I&C SAFETY FACTORS

SG safety is affected by many factors regarding its design, manufacturing, installation, commissioning, operation and maintenance. Consequently, it may be extremely difficult to construct a complete mathematical model in order to assess the safety because of inadequate knowledge about the basic failure events. This leads inevitably to problems of uncertainty in SG safety assessment.

The SG is a very complex system. It is characterized by huge number of nodes and links between nodes with increasing structural complexity; links between nodes could change over time, have different weights, directions, etc.

There are a lot of risks as the inherited essences of SG life cycle. Due to high complexity, its dynamic nature these risks are not static. Moreover, SG life cycle is characterized by complicated risk flow when safety and reliability issues might endanger the cyber security and vise verse. The risk associated with SG weakest link could compromise the safety and reliability of SG as a whole.

Main safety factors of IoT based Smart Grid and their relationship are shown in Figure 3.

Figure 3. Main safety factors of SG and their relationship

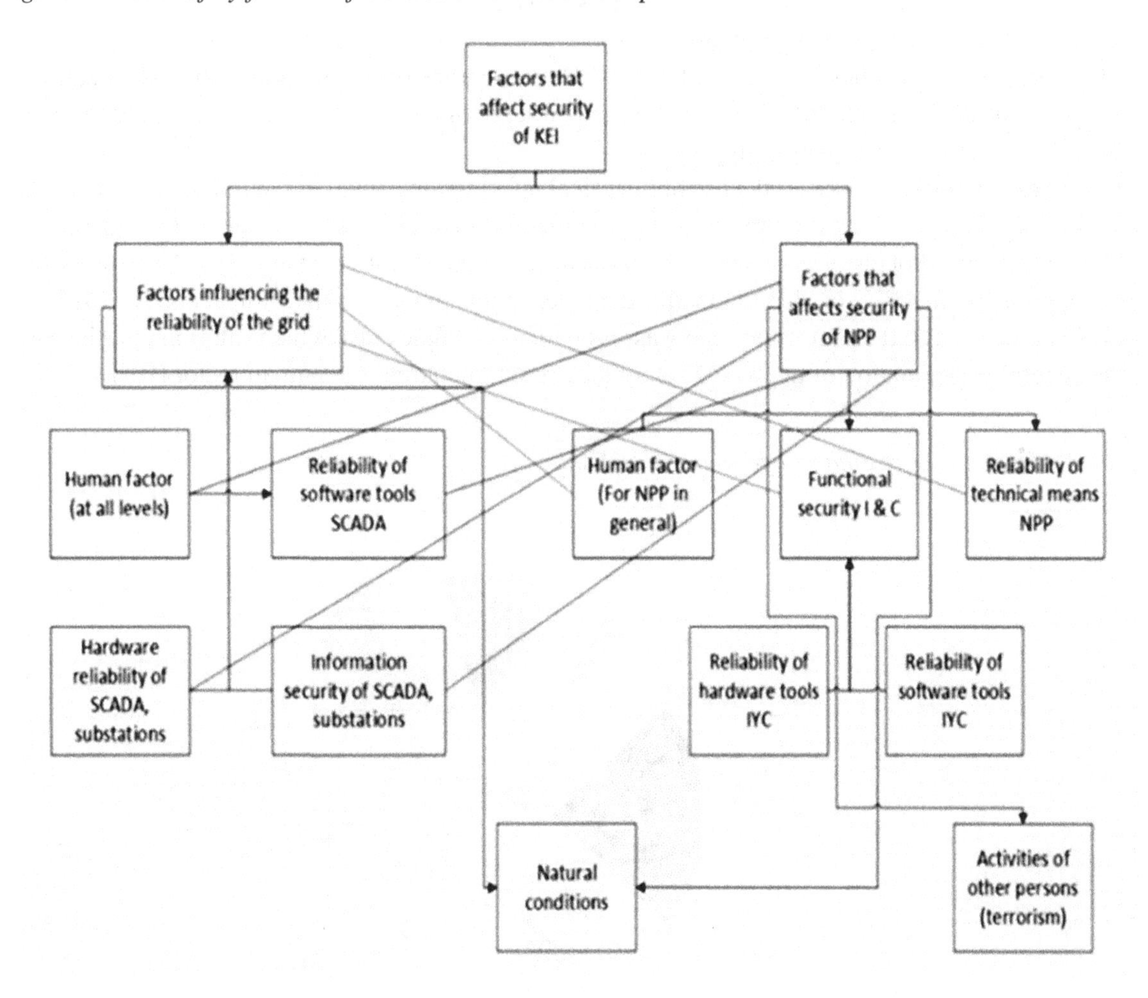

Natural Factors Influence Analysis

Analysis of the SG's accidents and failures causes confirms the vulnerability of its assets to the effects of natural factors (earthquakes, volcanic eruptions, landslides, etc.) due to centralized architecture, geographic distribution, balancing between generation and demand, and the complexity of integrating alternative sources, etc.

Figure 4 shows main causes of accidents and failures in the SG work that occurred over the past 10 years.

The main methods that take into account the natural factors influence include: risks indexation method, environmental impact assessment, geospatial analysis, hazard mapping, historical analysis, etc.

Risks indexation method is a qualitative approach that allows monitoring the risks for the public at the national level. The disadvantage of the method is simplification and usage of averaged data, which prevents a specific assessment for a specific region.

Environmental impact assessment is a mean of supporting decisions to assess the impact of companies on the infrastructure vulnerability to natural disasters. The disadvantage of the method is the focus on post-event analysis of the natural disasters consequences. The approach is not integrated into the disaster management planning process.

There are also approaches to mapping hazards. The advantage of the method is the visualization of information for decision-making. A disadvantage is a large amount of necessary information, which leads to the need to use computing support.

Historical analysis is based on the use of historical information about the natural disasters area. The method is based on weighing different aspects of vulnerabilities. The disadvantage is the high requirements for the database of historical events, which also needs support (filling, evaluation of reliability, etc.)

All approaches make it possible to identify the risks nature and assess their impact on the IoT based Smart Grid security. At the same time, safety assessment is a complex task which solution must be based on the general consideration of the whole safety factors set, including the human factor (HF).

Figure 4. Main causes of Smart Grid major accidents

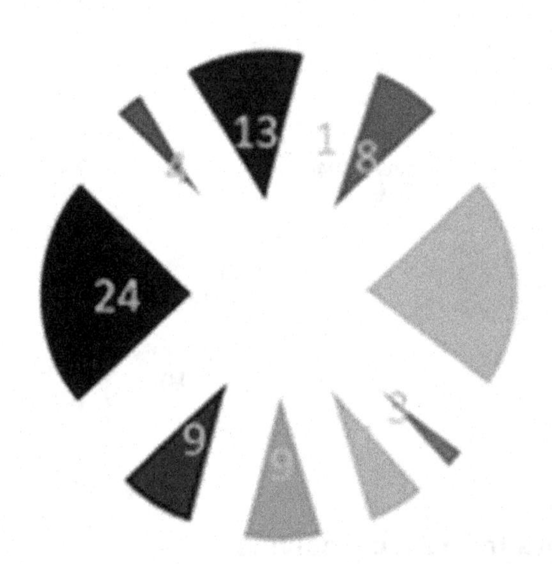

Human Factor Influence Analysis

IoT based SG consists of sophisticated human-machine systems, in which the impact of HR on reliability (P. Pekka, 2000) and security is determinational at any level of its hierarchy.

There are many approaches to human-operator reliability assessment in many areas of their activities (M. Madonna, et.al., 2009) primarily related to security. The main approaches to incorporating HF into IoT based Smart Grid security analysis include: Human Error Probability (HEP) quantitative methods; holistic and decomposition methods. Holistic models are used to assess the human operator's overall performance without breaking them into small sub-actions, as suggested by using the Technique for Human Error Rate and Prediction (THERP) (A. Swain, 1964) and Techique for Human Event Analysis (ATHENA). Decomposition methods consider the human functions quality as a combination of sub-actions.

NPP's Safety and Cyber Security Risks

Nuclear energy occupies a unique position in the debate over global climate change as the only carbon-free energy source. Nowadays it is already contributing to world energy supplies on a large scale, and has potential to be expanded if the challenges of safety, nonproliferation, waste management and economic competitiveness are addressed and technologically fully mature. So it might be concluded that Nuclear Power Plants (NPP) are an intrinsic part of future IoT based SG. I&C NPP will be communicating with other I&Cs of SG power generation facilities.

Risk is an inevitable factor throughout the life cycle of the smart grid and I&C NPP (E. Brezhnev, 2010). Risk R(t) is estimated as follows

$$R(t) = P(t) \times S(t),$$

where P(t) – adverse event (failure, accident) occurrence probability caused by negative interactions between SG systems (subsystems) of the I&C NPP; and S(t)– adverse event (accident, failure) consequences severity in the I&C NPP measured in terms of damage (economic, health and people life, etc.).

Within the IoT based Smart Grid, risk decomposition can be done by the levels of its hierarchy (Figure 5).

Let's highlight the main risk groups in the IoT based Smart Grid (table 14.1).

The risks on the IoT based Smart Grid are the sum of systems local risks (LR) that were not reduced during the design process, as well as emergent risks (ER) $R_{CI}^{emerg}(t)$, due to negative interactive systems when they are combined within Smart Grid:

$$R_{CI}(t) = \sum_{h=1}^{H} \beta_h R_{CI}^{h}(t) + \sum_{i=1}^{N} \alpha_i R_i(t),$$

where $R_{CI}^{h}(t)$ – ER h-th type,

$$R_{CI}^{emerg}(t) = \sum_{h=1}^{H} \beta_h R_{CI}^{h}(t);$$

$R_i(t)$ – LR i-th asset SG system;

$$R_{S_1}^{Closed}(t) = \sum_{i=1}^{N} \alpha_i R_i(t);$$

α_i, β_h - risk priorities $\alpha_i, \beta_h \in [0,1]$.

The closed system risks set and its risks within SG set differ from each other. This is due to the presence of ER in the SG, that is:

$$R_{!\,I}^{emerg}(t) = \sum (R_{CI}^{physic}(t) + R_{CI}^{inf\,orm}(t) + R_{CI}^{organiz}(t) + ... + R_{CI}^{logic}(t)).$$

Figure 5. Hierarchical risk structure in Smart Grid

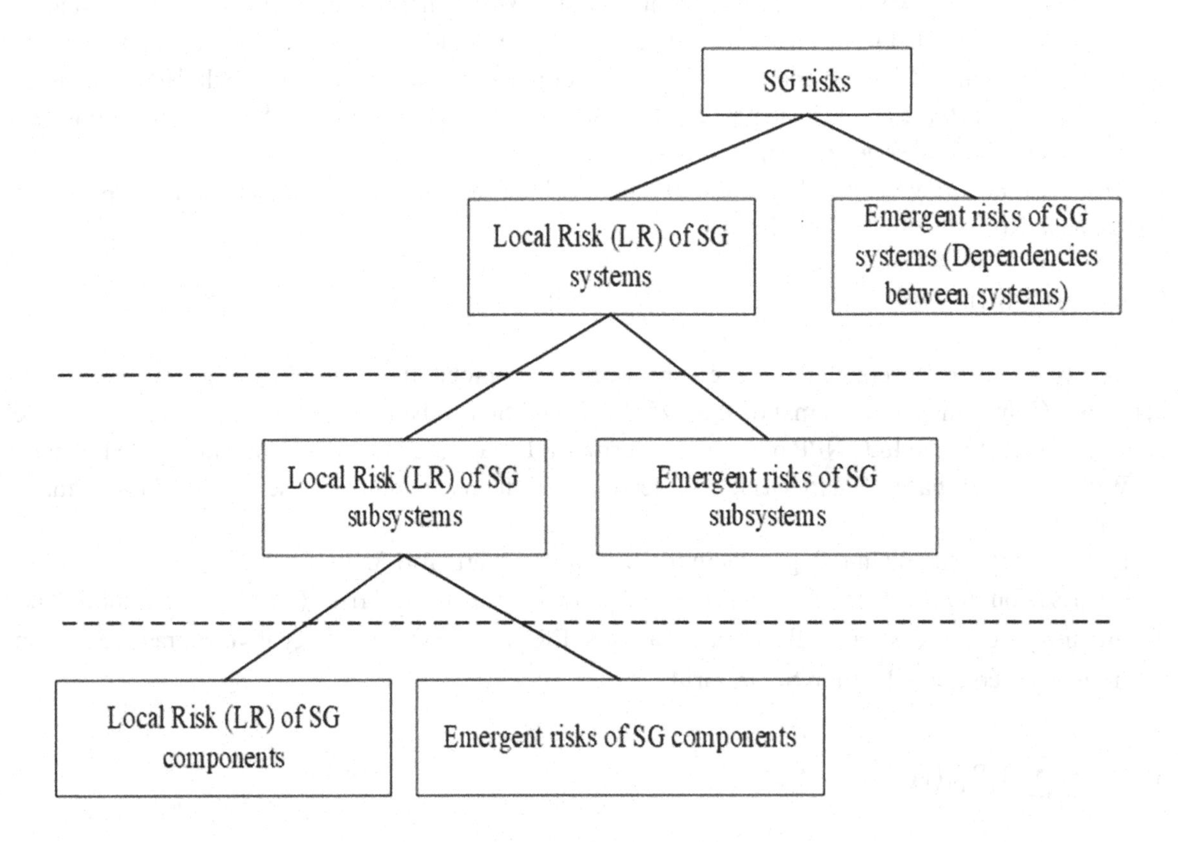

The dependencies between the safety and reliability levels on the IoT based Smart Grid is shown on Figure 6. IoT based Smart Grid security is determined by the security of the NPP and the SG reliability. The NPP's safety is determined by the functional safety and I&C cyber security. SG reliability is determined by its equipment reliability, as well as its smart devices and components information security.

If risk management strategies do not ensure their reduction, it leads to their transfer to the level of the systems, in particular, for the NPP, $R_{S_1}^{emerg} * (t) \rightarrow R_{NPP}(t)$. Then ER, due to interconnected power supply and NPP, can be transferred to NPP: $R_{grid}(t) \rightarrow R_{NPP}(t)$, which causes a decrease in safety (for example, due to a loss of power supply).

Table 1. Major risk groups in the smart grid hierarchy

Types of risks on IoT based Smart Grid	Notation
EP systems	
Risks due to the negative geographic impact between IoT based Smart Grid systems	$R_{CI}^{geogr}(t)$
Risks due to negative informational influence between IoT based Smart Grid systems	$R_{CI}^{inform}(t)$
Risks due to negative physical effects between IoT based Smart Grid systems	$R_{CI}^{physic}(t)$
Risks due to the negative organizational influence between IoT based Smart Grid systems	$R_{CI}^{organiz}(t)$
Risks due to negative logic influence between IoT based Smart Grid systems	$R_{CI}^{logic}(t)$
Risks of using inaccurate (inadequate) models	$R_{model}(CI)$
The risks of not accounting for uncertainties	$R_{uncertain}(CI)$
Type of Local Risks	
Risks related to physical assets (hardware): – hardware;	$R_{hard}(S_i)$
– electric cables; – peripheral measuring equipment (sensors);	$R_{ecabl}(S_i)$ $R_{sensor}(S_i)$
– substations and transformers, etc.	$R_{substr}(S_i)$
Risks associated with organizational assets of systems (Management, Organizational Risks): – culture of safety; – imperfect security policies.	$R_{organ}(S_i)$
Risks of multiple failures of IoT based Smart Grid assets: – Common Cause Failure (CCF) risks;	$R_{MF}(S_i) = R_{CFF}(S_i) \cup R_{casc}(S_i)$
– risks of cascading accidents.	$R_{cascad}(S_i)$
Risks of using non-accurate security models	$R_{model}(S_i)$
Risks not taking into account uncertainties	$R_{uncertain}(S_i)$
Emergent risks for systems due to the interplay between their subsystems	$R_{\bar{S}_i}^{emerg}(t)$
Local risks for subsystem (I&C level)	
Risks associated with information assets (I&C software and hardware). Functional I&C safety risks	$R_{I\&C_q^i}(t) = R_{I\&C_q^i}^{hard}(S_i) \bigcup R_{I\&C_q^i}^{soft}(S_i)$
Risks caused by interfacing I&C subsystems (by Emergent risks)	$R_{I\&C_i}^{emerg} * (t) = R_{I\&C_q^i}^{inf\,orm}(\bar{S}_i) \bigcup R_{I\&C_q^i}^{geogra}(\bar{S}_i) \bigcup R_{I\&C_q^i}^{log}(\bar{S}_i) \ldots$
Risks associated with organizational assets (management, organizational risks)	$R_{organ}\left(I\&C_q^i\right)$
The risks of multiple I&C failures: - CCF risks, - risks of cascade failures of assets.	$R_{MF}\left(I\&C_q^i\right) = R_{CCF}\left(I\&C_q^i\right)$ $R_{casc}\left(I\&C_q^i\right)$
Risks of using inaccurate I&C security models	$R_{model}\left(I\&C_q^i\right)$
The risks of not accounting for uncertainties	$R_{uncertain}\left(I\&C_q^i\right)$

Figure 6. Dependencies between safety and reliability levels on IoT based Smart Grid

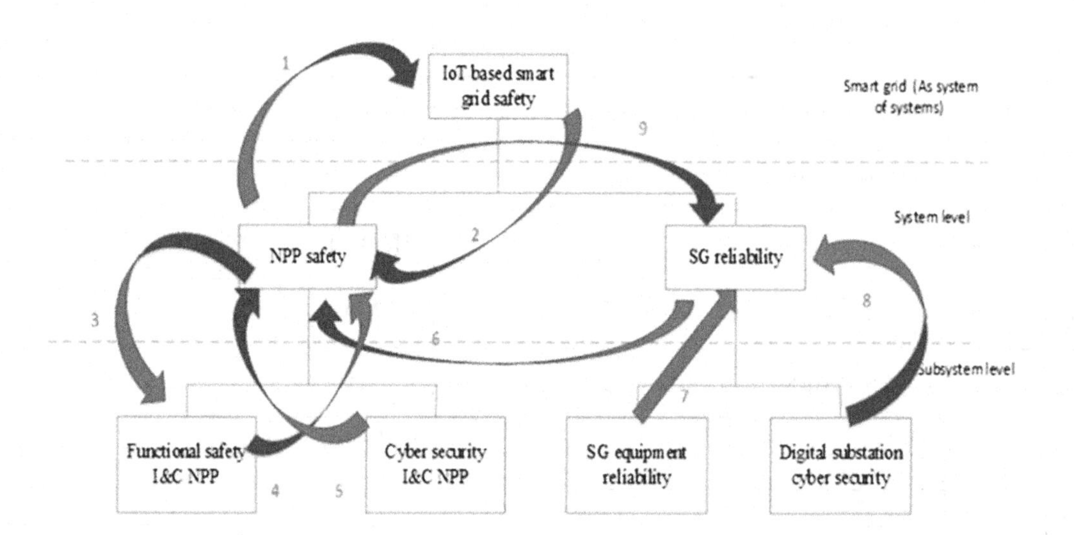

I&C risks $R_{I\&C_q^i}(t)$ are formed by combining sets of NPP risks $R_{NPP}(t)$ risks of functional I&C safety $R_{I\&C_q^i}(t)$, risks $R_{CI}^{emerg}*(t)$, which have not been reduced at the level of IoT based Smart Grid systems, as well as the risks inherent in I&C subsystems interacting $R_{I\&C_i}^{emerg*}(t)$:

$$R_{I\&C_i}(t) = R_{NPP}(t) \bigcup R^*_{I\&C_i}(t) \bigcup R^{emerg}_{I\&C_i} * (t) \bigcup R^{emerg}_{CI} * (t).$$

Table 2. Interoperability description

Nº	Marking	Comment
1	$SI_{NPP}(t) \rightarrow SI_{CI}(t)$	Security of the NPP (System) determines the security of the IoT based Smart Grid
2	$SI_{CI}(t) \rightarrow SI_{NPP}(t)$	Security IoT based Smart Grid ensures the safety of the NPP
3	$SI_{NPP}(t) \rightarrow FSI_{I\&C^i}(t)$	Safety of NPPs determines functional safety of I&C NPP
4	$FSI_{I\&C^i}(t) \rightarrow SI_{NPP}(t)$	Functional safety I&C NPP determines safety NPP
5	$CSI_{I\&C^i}(t) \rightarrow SI_{NPP}(t)$	Information security I&C NPP determines safety NPP
6	$RI_{grid}(t) \rightarrow SI_{NPP}(t)$	The reliability of the grid (system) determines safety NPP (system)
7	$RI_{subst}(t) \rightarrow RI_{grid}(t)$	The reliability of the substation (subsystem) determines the reliability of the grid (system)
8	$CSI_{substr}(t) \rightarrow RI_{grid}(t)$	Information security of the substation (subsystem) determines the reliability of the grid (system)
9	$SI_{NPP}(t) \rightarrow RI_{grid}(t)$	NPP safety determines the reliability of grid (systems)

I&C risks can be transferred to other NPP systems, so $R_{I\&C}R\rightarrow_{NPP}$

This results in an increase in all local risk (LR) NPPs associated with its equipment failure.

Thus, there are dependencies at all levels of the IoT based Smart Grid: the negative interaction of safety states causes the ER appearance; IoT based Smart Grid asset hierarchy allows you to depict a hierarchy of risks: ER IoT based Smart Grid – system risks – I&C risks; risk reduction is possible due to ER reduction.

SELECTION OF SMART GRID MODEL

Formalized IoT based SG can be represented as a hierarchy. The main levels include:

- **SG level** (first level), which contains different systems: systems generating (NPP, TPP, alternative genereation sources, etc.) (S_1), services delivery (S_2), services distribution (S_3), etc.;
- **Subsystem level** (second level), for example, for NPP (as systems), subsystems are normal operation systems, subsystems of distribution and transmission, etc.

SG hierarchical representation is shown in Figure 7. If necessary, the levels of components and elements can be considered.

The nodes of each level of SG are interconnected by $I_{geo}^{NPP}(t)$ connections (influences) $I(t)_h^{CI}(S_i \rightarrow S_j)$ different types, namely:

$$I(t)_h^{CI}(S_i \rightarrow S_j) \in \{I(t)_{geo}^{CI}(S_i \rightarrow S_j), I_{geo}^{CI}(t)(S_i \rightarrow S_j),$$
$$I_{phys}^{CI}(t)(S_i \rightarrow S_j), I_{org}^{CI}(t)(S_i \rightarrow S_j), I_{inf\,orm}^{CI}(t)(S_i \rightarrow S_j), I_{log}^{CI}(t)(S_i \rightarrow S_j)\},$$

where

$I_{geo}^{CI}(t)(S_i \rightarrow S_j)$ - geographic impact due to the spatial closeness SG systems, which leads to an increase in ER associated with the spread of the effects of accidents between systems;

$I_{phys}^{CI}(t)(S_i \rightarrow S_j)$ – physical impact due to flows of material resources (for example, electricity) between systems; loss of which leads to an increase in ER associated with sources of supplies of resources, resources, energy;

$I_{org}^{CI}(t)(S_i \rightarrow S_j)$ – organizational impact due to the impact of staff errors of one system on the safety of another system on the IoT based Smart Grid;

$I_{inform}^{CI}(t)(S_i \rightarrow S_j)$ – information interactions associated with the flows of information (data) between information assets IoT based Smart Grid;

$I_{lcg}^{CI}(t)(S_i \rightarrow S_j)$ – logical influence associated with the logic of the functioning of systems.

SG model has a graph like:

Figure 7. Smart grid hierarchical representation

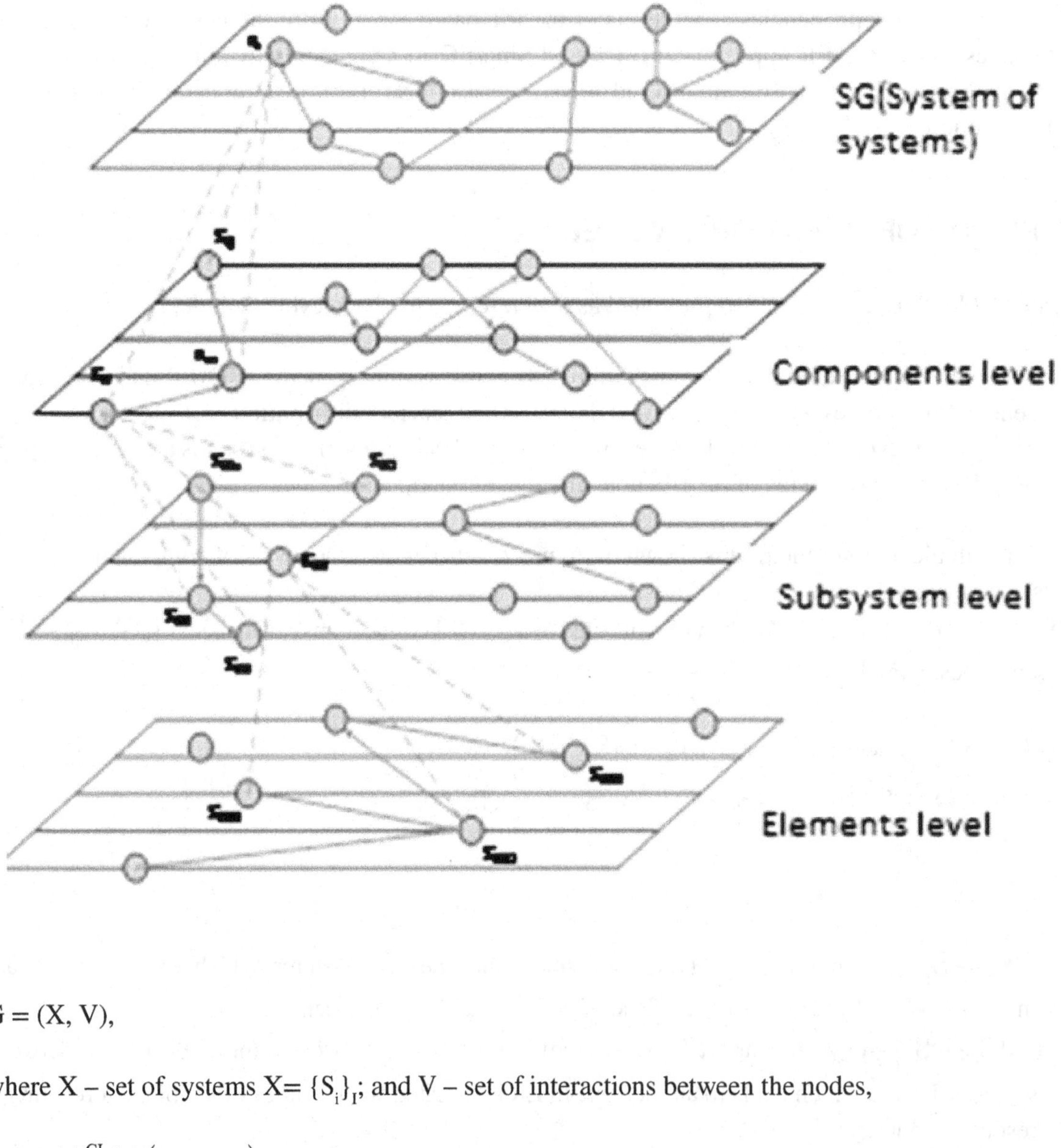

SG(System of systems)

Components level

Subsystem level

Elements level

$$G = (X, V),$$

where X – set of systems $X= \{S_i\}_I$; and V – set of interactions between the nodes,

$$V= \{I(t)_h^{CI}(t)(S_i \rightarrow S_j)\}_H .$$

Thus, Smart Grid can be presented as a hierarchical graph, where each i-th node of the j-th level of the hierarchy can also be represented as a subgraph of the form:

$$G_{ij} = \{(X_{ijs})_{IJS}, (V_{kij})_{KIJ}\},$$

where $(X_{ijs})_{IJS}$ – s-th subsystem of the i-th system (node) of the j-th level of the hierarchy; $j=\overline{1,J}$, $s=\overline{1,S}$; and $(V_{ikj})_{IKJ}$ – k-th rib of the i-th system of the j-th level of the hierarchy, $k=\overline{1,K}$, which is a kind of mutual influence.

Among the whole set of subsystems a subset is allocated I & C $\left\{I\,\&\,C_q^i\right\}_Q$, connected with other subsystems by information links $I(t)_{inform}^{CI}\left(S_i \rightarrow S_j\right)$.

The NPP as a part of SG interacts with other elements of SG. All influences (or relationships) existed in SG could be divided into several hierarchy's levels.

The influences between different systems of IoT based Smart Grid could be described (or formalized) by means of the Influence vector. The Influence vector is characterized by the value and direction. The direction points the initial influence source and systems being under influence. The value characterizes the strength of influence.

The influences between NPP and IoT based Smart Grid systems could be represented by a matrix of influence shown in the table 3

Table 3. Matrix of influence

	NPP	TPP	HPP	Digital substation
NPP	-	M	H	L
TPP	L	-	H	L
HPP	M	L	-	
Digital substation	M	L	H	-

The influence matrix shows how elements of the system influence each other and strength of their influence. As an example, NPP influences TPP with a strength – medium and HPP with high level of influence. Generally, influence is an ability of one system to determine the state, characteristics and behavior of other systems.

To evaluate the influences between NPP and SG systems we need to have the metrics by which this influence could be measured and compared. Two types of metrics are suggested: linguistic and numerical. The linguistic metric operates with the linguistic values used to evaluate the strength of influence. The different values as high, medium and low are applied to consider and predict the smart grid's system state changing provided the accident in other SG systems occurred. Numerical values, as ranks, are used in the similar way, and the different ranks stand for the different strength of influence. Expert judgments are considered as the basis for taking the influence values. The influence database is completed for each NPP. These values are regularly updated.

NPP could influence the SG in the different ways, such as physical, geographical, organizational, by means of information, logical, societal, etc. Thus, we introduce the space of influence. Physical, geographical, organizational, informational, logical, societal is a particular influence.

Total influence might be represented as:

$$I_t^{NPP}\left(I_{geo}^{-NPP}(t), I_{phys}^{-NPP}, I_{org}^{-NPP}, I_{soc}^{-NPP}, I_{log}^{-NPP}(t)\right)$$

where $\overline{I}_{phys}^{NPP}(t)$ – a physical reliance on materials flow between NPP and other SG systems;

$\overline{I}_{inf}^{NPP}(t)$ – a reliance on information transfer between NPP and other elements of SG (via I&C systems);

$\overline{I}_{geo}^{NPP}(t)$ – a local environmental event affects components of NPP-SG (usually the transmission lines) due to physical proximity;

$\overline{I}_{log}^{NPP}(t)$ – an influence that exists between NPP - SG that does not fall into one of the about categories;

$\overline{I}_{org}^{NPP}(t)$ – organizational influences through policy, regulation, markets;

$\overline{I}_{soc}^{NPP}(t)$ – influences that SG components may have on public opinion, fear and confidence.

FEATURES OF SMART GRID SAFETY AND SECURITY ASSESSMENT AND ASSURANCE

As it was mentioned above Smart Grid is considered as a set of systems $\{S_i\}_I$, their I&C system (subsystems) $\{I\&C_q^i\}_Q$ and the h-type bonds (mutual influences) $I(t)_h^{CI}(S_i \rightarrow S_j)$ between the systems.

It is important to note that interactions occur at all levels of the IoT based Smart Grid, at the subsystem level, as well as at the system-subsystem level, that is

$$I(t)_h^{Si}\left(\overline{S}_{ik} \rightarrow \overline{S}_{lm}\right), I(t)_h^{Si}\left(S_i \rightarrow \overline{S}_{ik}\right)$$

The main means of monitoring system security is I&C $\left\{I\&C_q^i\right\}_Q$. Between security states I&C systems subsystems are also mutually influential $I(t)_h^{I\&C_q}\left(\overline{S}_{ik} \rightarrow \overline{S}_{jl}\right)$.

IoT based Smart Grid is characterized by security (safety index, SI), $SI_{CI}(t)$, the cost of systems (subsystems) (C) (M, resources to boost SI and risk reduction), as well as reliability indicators $RI_{S_i}(t)$. I&C system is characterized by indicators functional safety $FSI_{I\&C}(t)$. There is an interconnection between the IoT based Smart Grid security features and the functional I&C system safety. Systems Si (subsystems $\overline{S}_{ij}$) are characterized by a state of safety $\{St_l^{S_i}(t)\}_L(\{St_g^{S_{ij}}(t)\}_G)$.

The value of the current SI of the IoT based Smart Grid does not match the required value $SI_{CI}^{req}(t)$, $SI_{CI}(t) \notin \Omega_{SI_{CI}^{accept}(t)}$. The current risks of IoT based Smart Grid do not match their acceptable value, $R_{CI}(t) \notin \gamma_{R_{CI}^{accept}}$ where Υ- set of acceptable risks. IoT based Smart Grid SI meets the required value only when the current IoT based Smart Grid risks $R_C(t)$ are acceptable, i.e.

$$SI_{CI}(t) = SI_{CI}^{Accept}(t) \Leftrightarrow R_{CI}(t) = R_{CI}^{accept}(t).$$

The risks of Smart Grid are determined by the local (own) system risks $R_{S_i}^{Closed}(t)$ and emergent risks $R_{CI}^{emerg}(t)$, caused by the negative impact between systems on the IoT based SG. The individual target function of the i-th system in the IoT based SG has the form:

$$f_{S_i}(x) \to \max, \varphi_i(x) \le 0, i = \overline{1,m}, x \in X,$$

where X - set of alternatives to achieve the goal;

f:X→R` and φ:X→R`- specified functions.

Target independence between IoT based Smart Grid systems is recorded as:

$$\left\{f_{s_i}(X), f_{s_j}(Y), \text{¼}\left(f_{s_i}(X) f_{x_j}(Y) = 0\right) \wedge \left\{X, Y, \text{¼}(X, Y) = 0\right\}\right\}.$$

Safety dependence looks like:

$$\left\{St_g^{S_i}(t), St_g^{S_j}(t), \text{¼}\left(St_g^{S_i}(t), St_g^{S_j}(t)\right) \neq 0\right\}.$$

The I&C system is characterized by risks that may be greater (less) than acceptable I&C system risks $R_{I\&C_q^i}(t) <> R_{I\&C_q^i}^{accept}(t)$.

Safety dependency between IoT based Smart Grid and I&C system looks like:

$$\left\{St_g^{S_i}(t), St_k^{I\&C_q^i}(t), \text{¼}\left(St_g^{S_i}(t), St_k^{I\&C_q^i}(t)\right) \neq 0\right\}.$$

Between the risks $R_{I\&C_q^i}(t)$ and $R_{CI}(t)$ there is an interconnection, i.e.:

$$R_{I\&C_q^i}(t) \subseteq R_{CI}(t), R_{CI}(t) \subseteq R_{I\&C_q^i}(t).$$

So,

1)
$$\begin{aligned}
&CI = \{\{S_i\}_I, \{I\,\&\,C_q\}_Q, \{S_{ij}\}_{IJ}, I(t)_h^{CI}(S_i \to S_j), I(t)_h^{Sij}(S_{ik} \to S_{lm}), I(t)_h^{Si}(S_i \to S_{ik})\},\\
&\exists SI_{CI}(t) \notin \Omega_{\underline{SI}_{CI}^{accept}(t)}, \exists R_{CI}(t) \notin \Upsilon_{\underline{R}_{CI}^{accept}(t)}, \exists!\ _{CI}(M_{CI}) \in \Omega_{accept};\\
&I\,\&\,C_q = \{\{\bar{S}_{ij}\}, I(t)_h^{I\&C_q}(\bar{S}_{ik} \to \bar{S}_{jl})\}, \exists SI_{I\&Cq}(t) \notin \Omega_{SI_{I\&C_q}^{accept}(t)}, \exists R_{I\&C_q}(t) \notin \Upsilon_{R_{I\&C_q}^{accept}(t)}.
\end{aligned}$$

It is necessary to ensure an acceptable level of Safety Index of Smart Grid by identifying, evaluating and reducing the ER by diversifying the systems (subsystems) on Smart Grid and redistributing resources, which ensures that the ER is reduced to an acceptable level:

2)
$$CI^* = \{\{S_i^*\}_I, \{\overline{S}_{ij}^*\}_{IJ}, I^*(t)_h^{CI}(S_i^* \rightarrow S_j^*), I^*(t)_h^{Sij^*}(\overline{S}_{ik}^* \rightarrow \overline{S}_{lm}^*), I^*(t)_h^{Si^*}(S_i^* \rightarrow \overline{S}_{ik}^*)\},$$
$$\exists SI_{CI^*}(t) \in \Omega_{SI_{CI^*}^{accept}(t)}, \exists R_{CI}(t) \in \Upsilon_{R_{CI}^{accept}(t)}; \exists!\ _{CI^*}(M_{CI^*}) \in \Omega_{accept}$$
$$I\&C_q^* = \{\{\overline{S}_{ij}^*\}, I^*(t)_h^{I\&C_q}(\overline{S}_{ik}^* \rightarrow \overline{S}_{il}^*)\}, \exists SI_{I\&Cq^*}(t) \in \Omega_{SI_{I\&C_q^*}^{accept}(t)}, \exists R_{I\&C_q^*}(t) \in \Upsilon_{R_{I\&C_{q^*}}^{accept}(t)}.$$

At the same time, the reduction of infrastructure risks should not increase the risk for I&C and reduce its functional safety.

The complex of risk reduction measures (diversification and redistribution) in Smart Grid is limited by resources $C_{CI}(M_{CI}) \in \Omega_{accept}$.

ANALYSIS OF SMART GRID SAFETY AND SECURITY

The generic vulnerabilities of smart grid (C. Lopez, 2015; F.Aloul, 2012) which might lead to emergency situation specific areas (safety and security) are the following:

- **Increasing Dependence and Interdependence** between SG systems.
- **Natural events and accidents.**
- **Blunders, errors, and omissions.** By most accounts, incompetent, inquisitive, or unintentional human actions (or omissions) cause a large fraction of the system incidents that are not explained by natural events and accidents.
- **Insiders.** Normal operation demands that a large number of people have authorized access to the facilities or to the associated information and communications systems. If motivated by a perception of unfair treatment by management, or if suborned by an outsider, an "Insider" could use authorized access for unauthorized disruptive purposes.
- **Recreational hackers.** For an unknown number of people, gaining unauthorized electronic access to information and communication systems is the most fascinating and challenging game.
- **Criminal activity.** Some people are interested in personal financial gain through manipulation of financial or credit accounts or stealing services. In contrast to some hackers, these criminals typically hope their activities will never be noticed, much less attributed to them.
- **Industrial espionage.** Some firms can find reasons to discover the proprietary activities of their competitors, by open means if possible or by criminal means if necessary. Often these are international activities conducted on a global scale.
- **Terrorism.**
- **National intelligence.** Most, if not all, nations have at least some interest in discovering what would otherwise be secrets of other nations for a variety of economic, political, or military purposes.
- **Information warfare.** Both physical and cyber attacks on our infrastructures could be part of a broad, orchestrated attempt to disrupt a major US military operation or a significant economic activity.

Thus, IoT application stipulates an appearance of **specific smart grid vulnerabilities**, such as:

- **Customer security**: Smart meters autonomously collect massive amounts of data and transport into the utility company, consumer, and service providers. This data includes private consumer information that might be used to infer consumer's activities, devices being used, and times when the home is vacant.
- **Greater number of intelligent devices**: A smart grid has several intelligent devices that are involved in managing both the electricity supply and network demand. These intelligent devices may act as attack entry points into the network.
- **Physical security**: Unlike the traditional power system, smart grid network includes many components and most of them are out of the utility's premises. This fact increases the number of insecure physical locations and makes them vulnerable to physical access.
- **The lifetime of power systems**: Since power systems coexist with the relatively short-lived IT systems, it is inevitable that outdated equipments are still in service. This equipment might act as weak security points and might very well be incompatible with the current power system devices.
- **Implicit trust between traditional power devices**: Device-to-device communication in control systems is vulnerable to data spoofing at the points where the state of one device affects the actions of another. For instance, a device sending a false state makes other devices behave in an unwanted way.
- **Using Internet Protocol (IP) and commercial off-the-shelf hardware and software**: Using IP standards in smart grids offer a big advantage as it provides compatibility between the various components. However, devices using IP are inherently vulnerable to many IP-based network attacks such as IP spoofing, Tear Drop, Denial of Service, and others.

It is woth noting that the new challenge which is stipulated by IoT application is **cyber common cause events (CCCE)**. CCCEs might be determined as events when cyber assets' availability, confidentiality and integrity (of one system or different systems with the same functionalities) are compromised within a specified (short) time interval.

The reasons are the common vulnerabilities, coupling within networks between equipments which might lead to security violation due to human errors, shared input data equipment, environmental events (flooding, storm and cyber attacks).

NPP: SMART GRID SAFETY AND SECURITY STRATEGIES

The interaction (informational, physical, geographical, etc.) between systems in SG leads to new (emergent, hereinafter ER) risks that cannot be identified in the early stages of LC. If the local risk (LR) is measured and reduced at the design stage of the system, the uncertainty in the estimation of ER remains one of the main danger sources to SG and its safety. Thus, increasing the SG safety can be achieved by identifying and reducing ER.

Since each of the SG systems has some resources (ability to load-sharing), then one of the possible strategies is to reduce ER is the redistribution aimed (see Figure 8) at the one system excess resources use to reduce another system ER, due to the interaction.

Figure 8. General view of SG (S_0) and the interaction between its systems

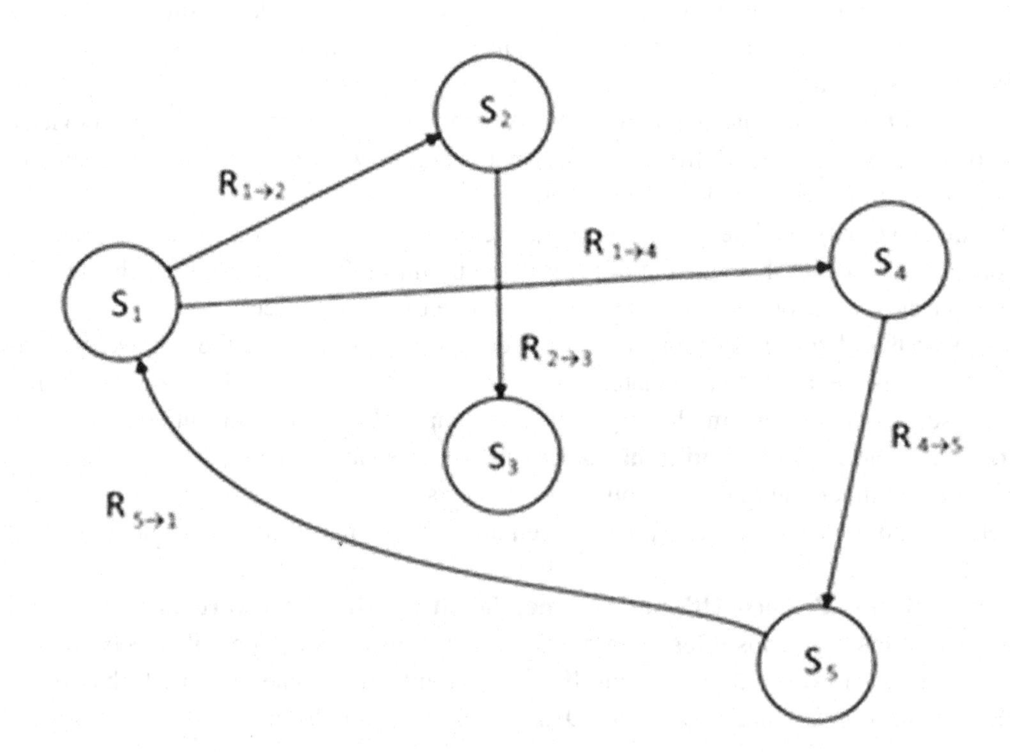

SAFETY STRATEGIES: REDEPLOYMENT OF RESOURCES IN SMART GRID

All systems in SG are open, i.e. systems which exchange resources (power loadable reserve, etc.) and information with other systems during their LC. The degree of openness (DO) is characterized by many parameters: number of ties with other systems type, period of interaction. DO of any SG system changes as all the parameters of the interaction change. Open systems are subject to the negative influence of other systems in SG. The more the system is, the greater its vulnerability to negative influence is.

The results of SG risk analysis depend on the adopted (under studies) system DO. It is quite difficult to reliably determine DO, especially at the project analysis stage. The underestimation of interaction makes the system conservatively "closed" in some sense. This leads to the fact that risk analysis becomes inaccurate, deterministic, which reduces the reliability of the results and effectiveness of countermeasures aimed at reducing ER. Please do not include section counters in the numbering.

SG can be represented as a set of interacting systems, linked by bonds of different nature (electricity flow, information, logical, etc.).

In the general case SG (S_0) can be represented as a set of the interacting systems.

Risks in open and closed (without interaction) systems are different. In the general case, the total amount of LR for a closed system S1 (without taking into account the interference from other systems in SG) at time t can be represented by the additive convolution of the form:

$$R_{S_1}^{Closed}(t) = \sum_{i=1}^{N} \alpha_i R_i(t),$$

where $R_{S_1}^{Closed}(t)$ – cumulative LR for the closed-loop S_1 at time t; $R_i(t)$– specific LR for S_1 at time t; and α_i– coefficient describing the priority of LR for the system.

We can assume that for system S1, some SG component part of S0 in the presence of interference between S1 and S2, the magnitude of the resulting risk to system S1 is related to the S_2 system will be greater than the magnitude of the risk for S1 closed the effect of the system S_2.

In other words, the cumulative LR for the closed-loop system S1 is less than its total risk as part of a system S0, containing systems S1 and S2.

The difference between the total risk of a closed and open system can be represented as:

$$R_{S_1}^{emerg}(t) = R_{S_1}^{S_0}(t) - R_{S_1}^{Closed}(t),$$

where $R_{S_1}^{emerg}(t)$ – the value of ER that occurs in the system S_1 as a result of interaction with S_2 in the system S_0; and $R_{S_1}^{S_0}(t)$ – the risk of system S_1 to the system SG (S_0).

Thus, at any point in time, the system Sj is characterized by the value of LR and ER due to the negative influence of h-type (informational, physical, etc.) from another system Si. Assume that each system Si has a resource (reactive power) which can be used to reduce LR.

In the general case, the task of resources redistribution within the SG to reduce ER caused by interference between the systems can be formed. A shared SG resource is an additive amount of system resources, which is a limitation in redistribution.

Condition. The system may not transfer the resources if the current safety index (SI) and the ER value do not match the required values.

Assumption. These resources are only used to reduce ER.

In general, the task of reallocating resources in SG can be formulated as:

- there is some system Si in SG, with resources MSi. Safety Index and ER do not meet the required values, i.e.

$$SG = \left\{ \{S_i\}_I, \{M_{S_i}\}, \left\{ SI_{S_i}(t) \notin ©_{SI_{Si}^{accept}(t)} \right\}, \left\{ R_{Si}^{emerg}(t) \notin ¥_{R_{Si}^{accept}(t)} \right\} \right\}.$$

- it is necessary to provide an acceptable level of LS and ER system by redistributing resources within SG, i.e.

$$SG = \left\{ \{S_i\}_I, \{M_{Si}^*\}, \left\{ SI_{S_i}(t) \in ©_{SI_{Si}^{accept}(t)} \right\}, \left\{ R_{Si}^{emerg}(t) \in ¥_{R_{Si}^{accept}(t)} \right\} \right\}$$

subject to the restrictions that $M_{CEI} = \sum_{i}^{I} M_{Si}$.

Within SG there are various strategies of resources redistribution to reduce ER.

Strategy of Redistribution With the Mandatory Allocation of Sufficient Resources

The system (a subject of influence, donor system) transfers to another system (object of influence) the amount of resources necessary to reduce ER that it creates. In this case, the donor system must allocate the number of resources sufficient to maintain SI of another system (As Safe As Reasonably Practical, ASARP).

This approach may not be rational for the donor system, because there are risks of situations in which it will not be able to reduce its ER due to the influence of other systems.

Condition. The donor system Si can allocate resources to reduce ER just in case when the current SI and the level of ER are acceptable,

$$SI_{S_i}(t) \notin \Omega_{SI_{Si}^{accept}(t)}, R_{Si}^{emerg}(t) \in \Upsilon_{R_{Si}^{accept}(t)}.$$

The critical system condition Crt (Si) is considered as SI.

Assumption. The current resource system provides a reduction of its LR, i.e. the system must have sufficient resources to reduce the LR.

We introduce an indicator which characterizes the vulnerability of the resource (resource vulnerability index, RVI) systems in SG:

$$RVI_{Si} = \frac{N_{Si \to Sj}}{N_{Sj \to Si}},$$

where $N_{Si \to Sj}$ – the number of outgoing ties (the system Si is the subject of influence); $N_{Sj \to Si}$ – the number of incoming ties (the Si system is the object of influence). The higher the value of RVI is, the more resource insecurity system in SG risks there are.

We introduce an additional indicator – the ratio of the current LS Crt (Si) to the value of RVI – overall vulnerability index (OVI) of the system. The smaller OVI is, the more vulnerable the system is (low safety and high risks of resource insecurity).

The resources reallocation within the framework of the first strategy includes the following steps:

- Evaluation of ER $R_{S_i}^{emerg}(t)$, current SI Crt (Si) of systems and resource vulnerability index Li, OVI.
- Ranking of SG systems to a minimum of OVI with the purpose to select the system for the redistribution initialization. Note that the resources distribution between systems in SG begins with systems with a minimum OVI value. This means that the system has low safety and high risks of resource insecurity. Thus, from the ranked series we take the vulnerable system, from which the improved safety is begun.

- Systems-subjects possibilities determination (associated with the affected system) for the resources transfer to reduce ER system with a minimum OVI. If you have multiple subjects of influence, the resources transfer for the object of influence begins from the "strong" system (the highest of OVI). Resources transfer from the donor resources system (with a minimum of OVI) is performed only if the current ER and the donor safety indicator are acceptable.
- The resource transfer by the donor system for its objects of influence with the aim of reducing ER (subject to the requirements of LS and ER). Because the donor system transfers as much as you need, ER system of object of influence is reduced after the resources transfer.
- Further, we considered the following system in ranked-set specified in step 2. Step 3 is repeated.
- In the case when the donor system is not satisfying the conditions of LS and ER admissibility, the resources transmission is not performed. Next, we study the following donor system (including ties) for the object of influence.
- Resources transfer is over when: it is impossible to transfer one of the systems in SG resources or the conditions of ER and SI for all systems in SG admissibility are satisfied.

Thus, each donor system must provide the resources to its object of influence, sufficient to reduce its ER.

The algorithm of the first strategy implementation is shown in Figure 9.

Strategy of Redistribution With the Possible (but Insufficient) Resource Allocation

The donor system transfers to the object of influence as many resources as possible with a view to ensuring the required SI level (with the existing ER relating to the negative influence of other systems), i.e., it allocates a certain surplus of resources, leaving itself just exactly what it needed.

This strategy is acceptable for the donor system because the remaining resource is sufficient to reduce its ER. For the system-object of influence, this approach may not be rational, as the allocated resources may be insufficient to reduce the ER created by the donor system.

Condition. The donor system Si may allocate resources only if its current SI and ER are acceptable, i.e.

$$SI_{S_i}(t) \notin \Omega_{SI_{si}^{accept}(t)},\ R_{Si}^{emerg}(t) \in \Upsilon_{R_{Si}^{accept}(t)}.$$

Assumption. The current resource of system provides the donor system LR reduction.

It should be noted that in deployment (for both strategies), there are risks in which the donor system will give more resources than you get from the other systems.

For example, the system S4 should give part of its resources to the system S1 and S5. System S1 receives resources from the systems S4 and S5, transferring the portion of the resources of the system S2. The larger the index of RVI system is, the greater the risk of its insecurity resource to reduce LR and ER. The second strategy implementation algorithm is shown in Figure 10.

Figure 9. The first strategy algorithm

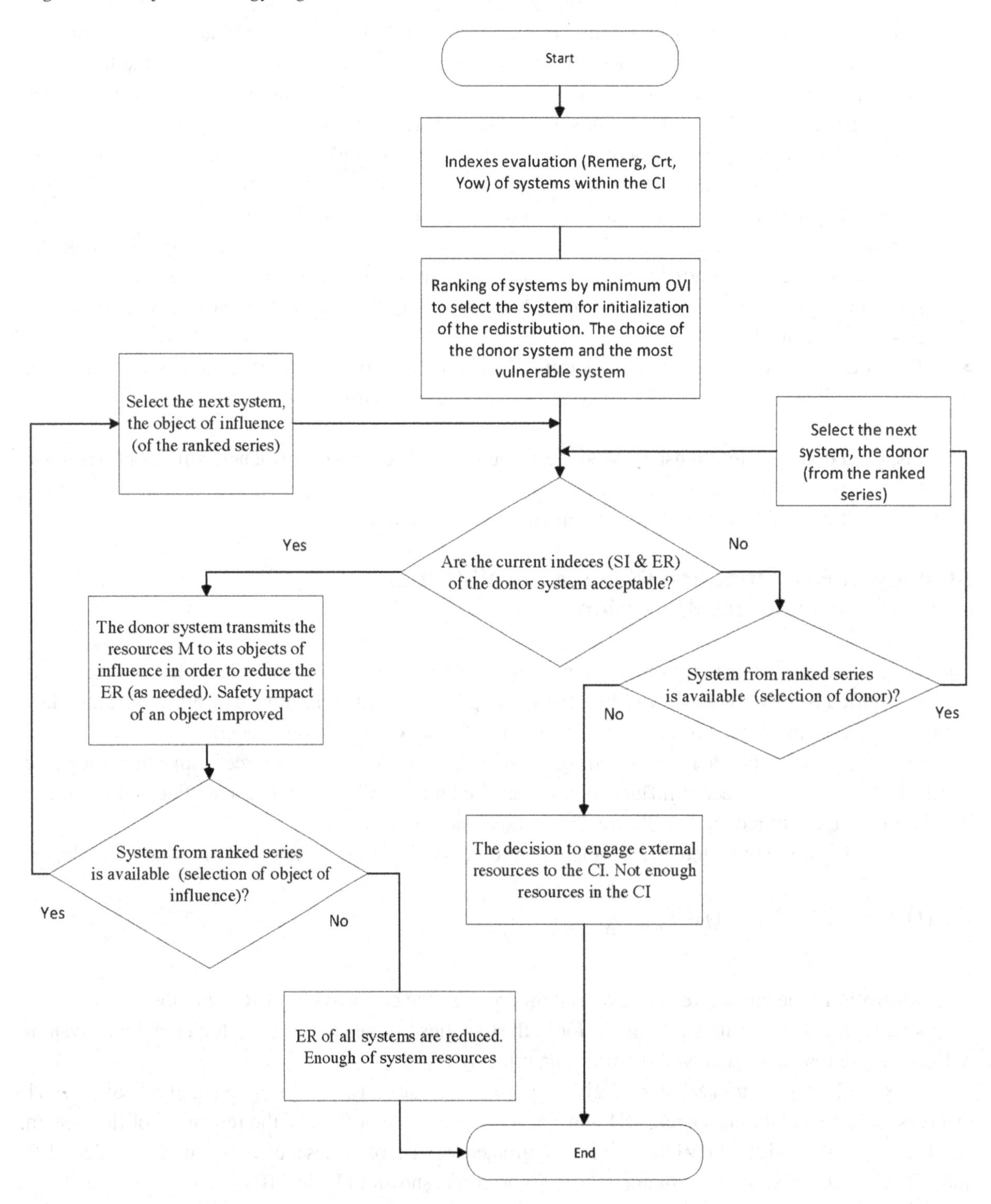

Figure 10. The second strategy algorithm

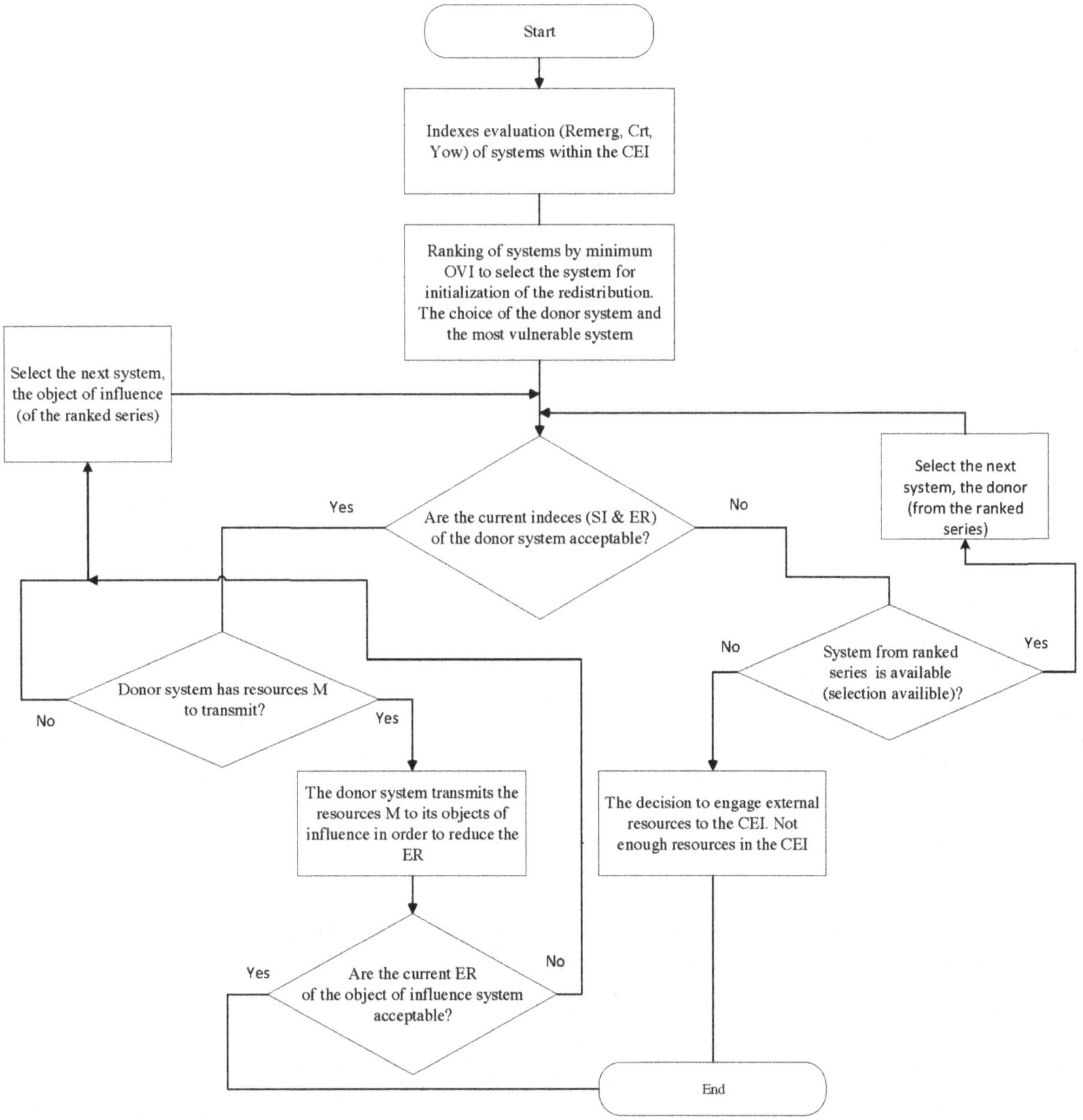

Start

Indexes evaluation (Remerg, Crt, Yow) of systems within the CEI

Ranking of systems by minimum OVI to select the system for initialization of the redistribution. The choice of the donor system and the most vulnerable system

Select the next system, the object of influence (of the ranked series)

Are the current indeces (SI & ER) of the donor system acceptable?

Yes

No

Select the next system, the donor (from the ranked series)

System from ranked series is available (selection availible)?

Yes

No

Donor system has resources M to transmit?

No

Yes

The donor system transmits the resources M to its objects of influence in order to reduce the ER

The decision to engage external resources to the CEI. Not enough resources in the CEI

Are the current ER of the object of influence system acceptable?

Yes

No

End

There are 2 special cases in the second strategy framework:

- The amount resources M^{**}_{Si}, transferred by the donor system, are not enough for full ER compensation, i.e. only some of the same resources, the size of which does not compensate for ER, are transferred. This means that you must use the following donor system (if any);

Table 4. Initial data for the strategies' implementation

System	Local risk	Emergent risk	System resource	Safety index, SI	RVI	SI/RVI, OVI
S_1	$R_{S_1}^{Closed}(t)$	$R_{S_1}^{emerg}(t)$	M_{s1}	Crt (S1)	RVI1	L1
S_2	$R_{S_2}^{Closed}(t)$	$R_{S_2}^{emerg}(t)$	M_{s2}	Crt (S2)	RVI2	L2
S_3	$R_{S_3}^{Closed}(t)$	$R_{S_3}^{emerg}(t)$	M_{s3}	Crt (S3)	RVI3	L3
S_4	$R_{S_4}^{Closed}(t)$	$R_{S_4}^{emerg}(t)$	M_{s4}	Crt (S4)	RVI4	L4
S_5	$R_{S_5}^{Closed}(t)$	$R_{S_5}^{emerg}(t)$	M_{s5}	Crt (S5)	RVI5	L5

Table 5. The final risks and resources reallocation for the first strategy

#	Resource state, T_0	Resource state due to its redistribution, T_1	The ability to fend off the ER	The ability to fend off the LR
S_1	M_{S1}	$M_{S1} + M^*_{S5} - M^*_{S1} - M^{**}_{S1}$	+	-
S_2	M_{S2}	$M_{S2} + M^*_{S1} - M^*_{S2}$	+	-
S_3	M_{S3}	$M_{S3} + M^*_{S2}$	+	+
S_4	M_{S4}	$M_{S4} - M^*_{S4} + M^{**}_{S1}$	+	-
S_5	M_{S5}	$M_{S5} + M^*_{S4} - M^*_{S5}$	+	-

- Resources transferred to another system fully ensure the ER reduction. This case leads to the first approach, therefore only the first case is considered further.

Table 6. The final reallocation of risks and resources for the second strategy

#	Resource state, T0	Resource state due to its redistribution, T1	The ability to fend off the ER	The ability to fend off the LR
S_1	M_{S1}	$M_{S1} + M^{**}_{S5} - M^{**}_{S1} - M^{***}_{S1}$	-	-
S_2	M_{S2}	$M_{S2} + M^*_{S1} - M^*_{S2}$	-	-
S_3	M_{S3}	$M_{S3} + M^*_{S2}$	-	+
S_4	M_{S4}	$M_{S4} - M^*_{S4} + M^{**}_{S1}$	-	-
S_5	M_{S5}	$M_{S5} + M^*_{S4} - M^*_{S5}$	-	+

Figure 11. The initial data for the first strategy

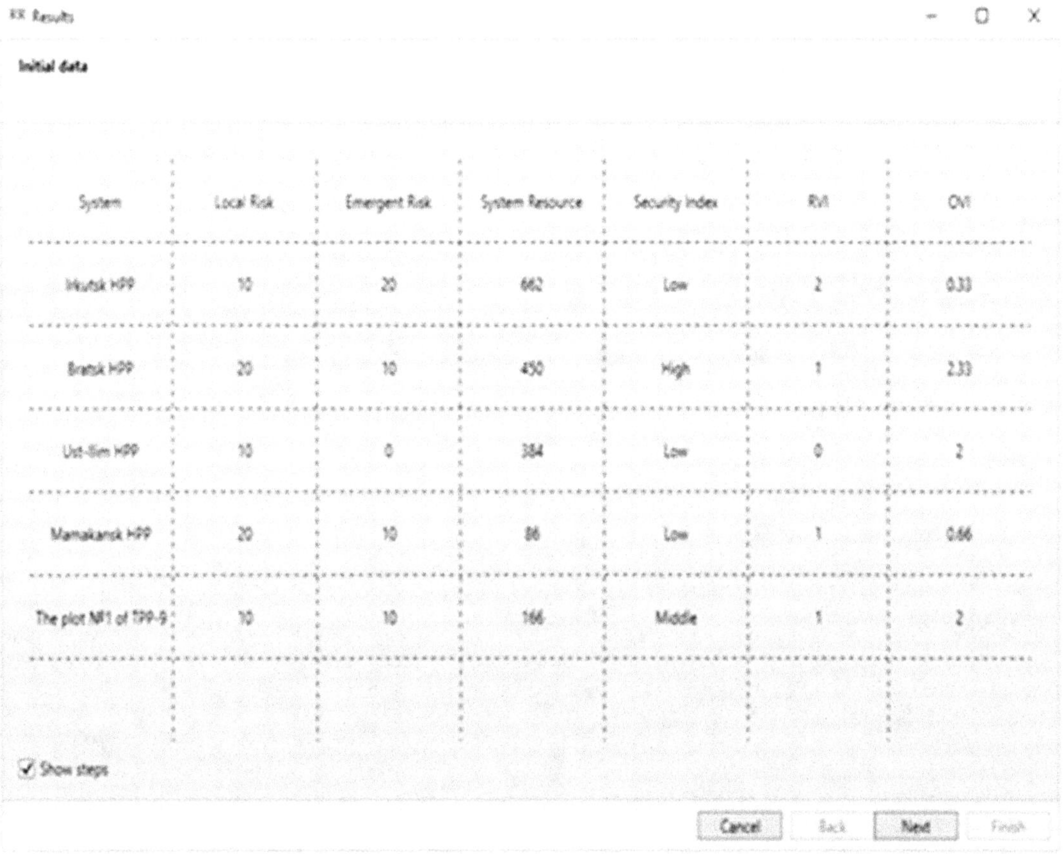

ER Results

Initial data

System	Local Risk	Emergent Risk	System Resource	Security Index	RVI	OVI
Irkutsk HPP	10	20	662	Low	2	0.33
Bratsk HPP	20	10	450	High	1	2.33
Ust-Ilim HPP	10	0	384	Low	0	2
Mamakansk HPP	20	10	86	Low	1	0.66
The plot №1 of TPP-9	10	10	166	Middle	1	2

☑ Show steps

Cancel Back Next Finish

The second strategy is more complicated for ER object system of influence reducing, all of the donor systems resources (including ties) can be used, which could lead to a situation where the donor systems can reduce their own ER (because they are subject to influence from other systems). For the first strategy, the number of possible donors will be less because more strict requirements are defined on the resources transfer to another system. Let's consider the illustrative example of the two strategies implementation.

The source data for the strategies implementation are shown in Table 4.

In the present example we obtained: final risks and resources redistribution (first strategy, see table 14.5); final risks and resources redistribution (second strategy, see table 14.6). A specific SG feature is to conduct risk management for all SG in general. The decrease in ER for a particular system is due to the allocation of resources across SG. For example, the first strategy is implemented in two stages: 1) a system with low-level LS and a high resource failure risk level revealed; 2) this system provides resources to other systems (subject to the restrictions on relations), provided that its ER can be parried with the available resources of these systems.

Comments. When using the first strategy the common ER of all SG are balanced, because the systems transmit their resources to reduce them. In this case, if the system originally allocated the excess resources to reduce LR (with some margin), then LR are balanced in SG. It is obvious that the second strategy does not only allow you to reduce ER but also LR for individual systems (provided that the transferred resources are not sufficient to mitigate ER). In this regard, SG cannot provide the desired LS level. For SG, it is advisable to allocate resources.

Figure 12. The final data for the first strategy

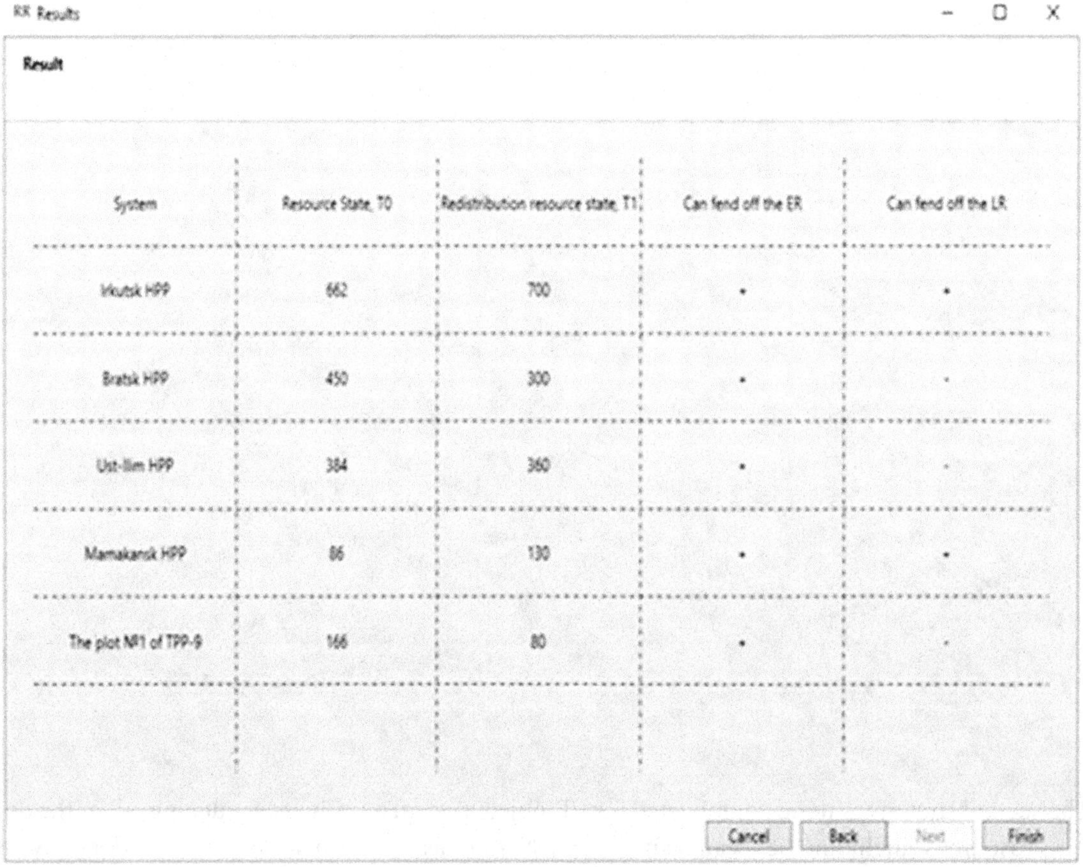

RR Results

Result

System	Resource State, T0	Redistribution resource state, T1	Can fend off the ER	Can fend off the LR
Irkutsk HPP	662	700	•	•
Bratsk HPP	450	300	•	·
Ust-Ilim HPP	384	360	•	·
Mamakansk HPP	86	130	•	•
The plot №1 of TPP-9	166	80	•	·

Cancel | Back | Next | Finish

A software tool has been developed to support this method. This program gives an ability to define system structure, describe elements state and predict a better resource redistribution strategy as a result of calculations. The software tool was developed on top of Microsoft .NET platform using WPF framework to achieve a fast result and a high user experience. It gives an ability to easily create extensions and improvements to an application as an applied realization of the method.

The safety modeling of Siberia power grid was performed as a practical case of strategy application (see Figure 11). All generation systems (S1-S5) pull electricity into grid according to their capacities (resources). The systems' resources are their power (MW) capacity. ER for each station is threat of taking additional power load in case when other systems are lost due to accidents. The system might be unable to take this additional load and the whole grid will be lost (black out). OVI is calculated after the defuzzification of SI.

All systems have a reserve capacity that can be allocated to support the power grid balance (production and consumption balance). ER is associated with the possible exclusion of one of the stations from the process of power generation.

The remaining systems will have to take the burden if it happens. If the station reserve has been decreased, it means that the system took over a part of someone's load.

Physically, it started additional energy generation (for example, a generator for a power plant). If the reserve was increased, it should turn off these facilities, because the other system has taken over the part of the total load.

Analysis of final data (see Figure 12) shows that after resource distribution only two generation stations (Irkusk and Mamakansk HPPs) have improved their capacity to cope with ER, but decreased their capacity to cope with LR. If such strategy had been considered by Syberia operator during decision-making, it is likely that Sayano–Shushenskaya HPP accident would have been avoided as it would have been based on the station abilities to decrease ER.

Thus, the strategies implemented by the resources reallocation between systems in SG allow you to provide the required level of SI and ER with constraints on resources. Each of the offered strategies takes into account individual preferences in SG systems: to provide as many resources as needed or as possible, taking into account the assessment of individual risks. The first strategy is preferable with the increasing of accidents risks, the second involves the functioning of SG in the moderate risk. Thus, the strategy is chosen due to the current SI systems, interaction dynamics, SG resource safety. LR and ER monitoring will allow you to flexibly choose a particular strategy.

It is advised to support decision-making and implementation approaches.

The implementation of the proposed approaches to the resources reallocation is supported by the tool. This tool allows simulating the redistribution taking into account the characteristics of the systems included in SG (strategy, initial resources amount, communications, LR and ER).

SDOE-Based Development of Resilient Smart Grid Substation

The cyber security aspect of the digital substations' development is becoming relevant. This is due to the increasing number of cyber incidents related to attacks on SCADA type systems.

Trends analysis indicates a steady increase in the number of attacks on industrial systems since 2010 (after Stuxnet). For example, in February 2011, a massive Night Dragon attack was conducted on five oil refining companies. In 2012, in a number of large companies operating in the banking sector of Syria, Lebanon, Sudan, malware ("Flame") was detected, designed to perform spyware, record talks, etc. Thus, there is an obvious increase in the number of attacks on industrial system that will be saved in the future.

Industrial systems developers seek to increase their security against external attacks, reduce the risks to cyber security and potential vulnerabilities. Implementing the IS requirements for systems, developers are striving to develop a secure product. Many requirements for information security can be divided into: requirements of the regulator, regulatory framework, customer, etc. The general requirements group is inconsistent and not harmonized very often. For application development companies, one of the initial tasks is to analyze the requirements, systematize them and determine their priorities.

Taking the results of this analysis into account, the company-developer seeks to develop practical approaches aimed at meeting the entire requirements set.

It should be noted that the requirements analysis and the approaches development of are not separated tasks. We have to deal with a set of agreed tasks, the outputs and inputs of which are interconnected, very often. Thus, we can talk about the need to design (develop) the process of ensuring the information product security in the company. Such process is, in fact, one of the important business company processes, aimed at achieving customer satisfaction with the final product properties, at its cyber security. An effective process design for ensuring the information security of a product can be based on the use of business process engineering approaches.

When developing processes in a company, it is necessary to take the company's resources, the personnel training level, product requirements, the maturity of the technologies used, etc, into account. The QMS is an important basis for the process design, which is an essential business management system component. It contains a description of all of the company business processes aimed at the quality products obtaining.

Since information security is also an important business process for a company, its design should be carried out taking into account other processes of the company within the framework of the quality management system.

The following process approach to the information security for a product providing can be applied to the development of digital substations or SCADA systems.

Data hubs are an example of industrial control systems that are subject to high demands on functional and information security. So, in particular, in the USA, the Nuclear Regulatory Commission (US NRC) actively uses a number of documents in its practice, the most important of which, from the information security of NPP DH point of view, include the following: 10 CFR Part 50, Appendix B, "Quality Assurance Criteria for Nuclear Power Plants and Fuel Reprocessing Plants"; Regulatory Guide (RG) 1.152-2011, Revision 3, "Criteria for Use of Computers in Safety Systems of Nuclear Power Plants, etc.

Such documents regulate many aspects, covering stages from the safe development and operation environment to the critical systems security properties, including information security. However, the activities related to information security into the DH life cycle model integration issues remain relevant and unresolved.

It is noted that the computerized systems software security refers to the ability to survive due to unauthorized, unwanted, and unsafe interventions throughout the life cycle of such systems.

Thus, there are many requirements for an information security product. Their implementation can be achieved in the implementation framework of one of the company's important business processes – ensuring information security. The ensuring information security process design should be based on the relationships with other processes in the company. The ensuring information security process should be an integral part of the company's QMS.

A secure development environment is defined as a condition for the availability of appropriate physical, logical, and software measures during the system design stages to ensure that non-unwanted, unnecessary, and undocumented functionality (for example, redundant code) is not incorporated into security-critical digital systems. A safe functioning environment is defined as a condition for the availability of appropriate physical, logical and administrative measures at an enterprise to ensure reliable operation of I&C systems through the absence of their degradation due to incorrect behavior of connected systems, as well as events generated by unintended access to such I&C systems.

In general, for both environments the main component types are:

- hardware (including technical means of functioning and ensuring the development and operation infrastructures security, equipment for development and operation);

- software (including what is related to development and operation processes, as well as directly to the infrastructures under consideration);
- data network (related to both the development and operation environment);
- staff;
- product (DH, which is developed and then operated).

In addition to the above types of components, an important aspect subject to separate detailed analysis is their configuration, which includes many factors that encompass the physical, logical and behavioral relationships. The configuration largely depends on the quality and completeness of the QMS implementation in the development and operation environments, personnel qualifications, and the quality of the software and hardware components used.

The implementation of SDOE, in the context of RG 1.152-2011, relates to the following aspects:

- Measures taken to implement a safe environment for the development of safety-critical DHs that prevent undocumented, redundant and unwanted changes;
- Protective measures that prevent a predictable set of undesirable actions that may affect the integrity, reliability or functionality of a DH during its operation.

Cascade life-cycle model phases make it possible to form a structure of special guides on the critical from a security point of view digital systems protection as well as on the SDOE implementation by identifying and mitigating potential weaknesses or vulnerabilities in each of the phases that may lead to the SDOE or the reliability of such systems degradation.

Security audits are an integral part of the security process during the development, implementation and maintenance of SDOE. Such audits include:

- Audit of the development environment: after the establishment of the infrastructure in which the development will be carried out, and before the start of the development project for the I&C system;
- Periodic audits: prior to each development phase.

The development environment audit is designed to assess the "basic" level of development environment security; mainly, the relevant measures are implemented by technical means and relate to the such infrastructure networks involved in the development process, physical means of protection, etc.

In turn, periodic audits are intended to assess additional specific measures required for the implementation of each development phase in the life cycle. Each of these audits includes the following steps:

- Additional measures for the development phase implementation verification (i.e., what is implemented and how);
- Verification of compliance with the guidelines and instructions of the organization related to a safe development environment by personnel (i.e., a practical assessment of whether employees follow the relevant QMS requirements).

The input for each of the audits is the corresponding plan (and, possibly, documents relating to certain aspects of the organization of the development environment or the implementation of processes), and the output is the SDOE audit report, in which the results are documented.

Thus, the information security provision is an important business process of the company. Its design (development) must be carried out taking into account the specifics of the company, its resources and the technologies used. The output of this process is a product that satisfies the customer in terms of quality. The ensuring information security process should be implemented within the company's QMS. When designing the ensuring information security process it is necessary to create a secure application development environment. The basis for developing a secure application can be the creation of a secure development environment, which can be based on a nuclear QMS and its principles.

CONCLUSION

No doubts the problem of the safe interaction between SG and a PG is topical. In the future they will be connected together forming a very complicated and dynamical system of systems (SoS). To make this highly interconnected SoS safe, cyber secure and reliable the development of safety assessment technique and tool are required. Beside this we have to take into account the future trend of development of both of them. Smart grid is the future power grid which combines a traditional PG with an "intelligent" information and communications technology (IoT technologies) infrastructure to create a smart power system. To assure the NPP safety, it is necessary to consider and thoroughly analyze the risks between SG and NPP due to their influences.

The complex nature of NPP and SG mutual interaction calls for the need of development of new approaches to NPP and smart grid safety/cyber assessment. The chapter considers an approach to influences formalization and series integration of safety assessment methods as well. This Chapter presents a comprehensive description of SG safety and security modeling considering influences between SG and NPP. In addition, this study gives an overview of challenges in smart grid safety and security in context of NPP risks (local and emergent).

This chapter study also gives the short overview of safety and security assessment and assurance's approaches, presents SG safety strategies. The role of IoT devices (such as sensors) in implementation of safety/security management is described.

The development process of digital substation in SG should be cyber protected while being implemented within the company's QMS. When designing secure digital substation it is necessary to create a secure development environment. The basis for developing a secure digital substation is a secure development environment, which can be based on a nuclear QMS and its principles. These practices could be adapted by desing companies which produce digital substations.

FUTURE RESEARCH DIRECTIONS

Future R&D are the following:

- development of software tool which supports the SDOE vulnarabilities assessment of digital substation;

- risk analysis of IoT technologies for NPP;
- smart grid and NPP safety/cyber security joint assessment. Security issues of smart grid which can undermine the safety of NPP. There is also need to develop software supporting this assessment. Next important steps of research and development activities, related to assurance of safety of NPP is to develop an framework for NPP safety/security management considering all possible risks from smart grid and IoT technologies.

REFERENCES

Aloul, F., Al-Ali, A. R., Al-Dalky, R., Al-Mardini, M., & El-Hajj, W. (2012). Smart Grid Security: Threats, Vulnerabilities and Solutions. *International Journal of Smart Grid and Clean Energy*, *1*(1), 1–6. doi:10.12720gce.1.1.1-6

Basit, A., Sidhu, G. A. S., Mahmood, A., & Gao, F. (2017). Efficient and autonomous energy management techniques for the future smart homes. *IEEE Transactions on Smart Grid, Volume*, *8*(2), 917–926.

Brezhnev, E. (2010). Risk-analysis in critical information control system based on computing with words' model. *Proceeding of 7th International Workshop on Digital Technologies, Circuit Systems and Signal Processing*, 67-72.

10 . CFR Part 50, Appendix B, Quality Assurance Criteria for Nuclear Power Plants and Fuel Reprocessing Plants.

Lopez, C., Sargolzaei, A., Santana, H., & Huerta, C. (2015). Smart Grid Cyber Security: An Overview of Threats and Countermeasures. *Journal of Energy and Power Engineering*, *9*, 632–647.

Madonna, M., Martella, G., Monica, L., Pichini Maini, E., & Tomassini, L. (2009). The human factor in risk assessment: Methodological comparison between human reliability analysis techniques. *Prevention Today*, *5*(1/2), 67–83.

Pekka, P. (2000). Human reliability analysis methods for probabilistic safety assessment. Technical Research Centre of Finland.

Regulatory Guide (RG) 1.152-2011, Revision 3, Criteria for Use of Computers in Safety Systems of Nuclear Power Plants.

Saleem, Y., Crespi, N., Rehmani, M., & Copeland, R. (2019). Internet of Things-aided Smart Grid: Technologies, Architectures, Applications, Prototypes, and Future Research Directions. *IEEE Access: Practical Innovations, Open Solutions*, *7*, 62962–63003. doi:10.1109/ACCESS.2019.2913984

Shu-wen, W. (2011). Research on the Key Technologies of IOT Applied on Smart Grid. *International Conference on Electronics, Communications and Control (ICECC)*, 2809–2812. 10.1109/ICECC.2011.6066418

Swain, A. (1964). *THERP.* Sandia National Laboratories, SC-R-64-1338.

Wang, W., & Lum, Z. (2013). Cyber security in the Smart Grid: Survey and challenges. *Computer Networks*, *57*(5), 1344–1371. doi:10.1016/j.comnet.2012.12.017

Wang, Y., Lin, W., Zhang, T., & Ma, Y. (2012). Research on Application and Security Protection of Internet of Things in Smart Grid. In *International Conference on Information Science and Control Engineering (ICISCE)*. TLP: White Analysis of the Cyber Attack on the Ukrainian Power Grid. https://www.nerc.com/pa/CI/ESISAC/Documents/E-ISAC_SANS_Ukraine_DUC_18Mar2016.pdf

ADDITIONAL READING

Brezhnev, E., & Kharchenko, V. (2013). BBN-based approach for assessment of Smart Grid and nuclear power plant interaction, *East-West Design & Test Symposium (EWDTS): proceedings of XI IEEE symposium*.

Brezhnev, E., Kharchenko, V., Fesenko, G., V. Levashenko, V., Zaitseva, E. (2015). Smart grid substation availability assessment: Recovered MSS-based approach, *Safety and Reliability of Complex Engineered Systems (ESREL),* 1477-1484.

Kharchenko, V., Brezhnev, E., & Boyarchuk, A. (2011). *The Smart Grid's Safety: Dynamical Hierarchical Criticality Matrices-Based Analysis* (p. 65). Innovative Smart Grid Technologies.

Vuković, O., & Dan, G. (2014). Security of Fully Distributed Power System State Estimation: Detection and Mitigation of Data Integrity Attacks. *IEEE Journal on Selected Areas in Communications, 32*(7), 1500–1508. doi:10.1109/JSAC.2014.2332106

Chapter 15
I&C Platforms:
Equipment Families

Yuri Rozen
https://orcid.org/0000-0002-9366-5794
State Scientific and Technical Center for Nuclear and Radiation Safety, Ukraine

Serhii Trubchaninov
State Scientific and Technical Center for Nuclear and Radiation Safety, Ukraine

Kostyantyn Gerasymenko
Severodonetsk Research and Production Association "Impulse", Ukraine

ABSTRACT

This chapter contains information on the platforms (equipment families) that increasingly serve as basic tools in the development of instrumentation and control (I&C) systems for different applications, including nuclear power plants (NPPs). The advantages of the platforms are indicated, the stages of their design and development are shown, and major differences between the platforms belonging to different generations are addressed. The main characteristics and features of the modern platforms used in NPP automation are provided. The procedure for development and qualification of the platforms delivered to NPPs is discussed.

INTRODUCTION

Almost 50 years ago, the suppliers of hardware for I&C systems began to show interest in developing and using a special product that later became known as the "equipment family", "aggregate complex", or "platform". Currently, such platforms are most often used to develop automatic process control and equipment for facilities in various industries, including NPPs. The properties of the platforms thus play a significant role in ensuring the safety and efficiency of the design, manufacture, testing, commissioning, and maintenance of such systems.

DOI: 10.4018/978-1-7998-3277-5.ch015

The purpose, characteristics, and approaches adopted for the development, qualification, and application of platforms for the automation of NPPs are described below.

BACKGROUND

An I&C platform (equipment family) is a set of hardware and software components that may work cooperatively in one or more defined configurations (structures), designed to implement a series of specific instrumentation and/or control systems for various purposes whose number is not determined beforehand.

The design and use of a platform allows the hardware and software developed within the platform to be used instead of segmented products from different suppliers that are often not compatible with each other. This ensures high efficiency and required quality of activities through:

- Functional, structural, and constructive completeness for the main application. The completeness of a platform is determined by the capability to create various I&C systems on its basis, whose functions, structure, design features, technical parameters, and characteristics can satisfy the requirements of a sufficiently wide number of specific customers (end users).
- Informational, power, constructive, operational, and other types of compatibility. The compatibility of I&C platform components is determined by the coherence of their input and output signals, interfaces, power supply parameters, shapes and sizes (including basic, overall, and mounting dimensions with their maximum deviations), allowable working and limiting operating conditions, and other properties, allowing smooth exchange of information between the components, supply of electricity with required quality, mechanical combination, and protection against the adverse effects of temperature, humidity, radiation, and mechanical and other factors that may arise in normal conditions of use, operational occurrences, and accidents.
- Software, metrological, guidance, informational, and other types of support.

The above peculiarities make it possible to design various I&C systems differing by the composition, structure, and technical characteristics on the basis of one and the same platform tailored to the customer (end user) requirements and, at the same time, ensure the well-targeted use of tools and effective activities in the design, programming, manufacturing, and operation of these systems. This approach allows the development, implementation, and safety justification of each particular system to be focused mainly on its applied functions, relying the implementation of basic functions on technical solutions adopted in developing and tested in applying the platform (nevertheless, the adoption of individual products or programs that are not part of a specific platform may be deemed to be reasonable in technically and economically feasible cases).

In addition to hardware and software directly oriented at the implementation of basic and applied functions, an I&C platform also includes service equipment, software, and documents that allow the user to minimize efforts in the design, integration, inspection, commissioning, maintenance and to avoid potential errors.

The importance of problems connected with platforms for NPP I&C systems is confirmed by the elaboration of a new special IEC document (IEC, 2019) devoted to these platforms.

Fundamental differences between an I&C platform and a software and hardware complex (SHC) should be mentioned. An I&C platform is a conceptual object and exists only as engineering, software, and process documents, methodologies, instructions, standards, etc. Therefore, a I&C platform generates a series of various SHCs, each representing a real object (produced commercially). An I&C platform is not bound to any specific system: it is intended for typical functions, structures, information environment, and conditions of use and operation conditions that are common for a sufficiently wide, though restricted, range of different systems whose number is not defined beforehand. Each specific SHC that can be implemented on a specific platform is commonly designed for the only one particular system. An individual SHC uses only a part of the capabilities peculiar to a specific I&C platform (restricted listing of hardware and software, support structures, signals, interfaces, etc., required and sufficient for a specific application).

An I&C platform is usually intended to perform a series of standard functions, which can be selected, variously combined, and customized according to the tasks of a specific I&C system. The hardware and application software of SHCs are configured using relevant tools and service software recommended or supplied by the I&C platform designer.

The design of an I&C platform is commonly preceded by research efforts to identify actual problems and needs for a certain range of potential users, and then the research findings are used to develop technical solutions that can effectively meet these needs. An I&C platform is more valuable to the end user than a simple combination of its components because it features a well-thought-out ideology, optimal equipment composition, all the necessary types of software. An I&C platform also uses adopted components: standards, techniques, design automation tools, maintenance and service infrastructure, etc. For an I&C platform to acquire actual value, all its components must be compatible and well combined with each other.

An I&C platform can represent a product of a specific manufacturer or combine a set of items gathered and adapted by the supplier. The marketing policy of I&C platform designers can provide for one or several options of its use for developing I&C systems:

1. Manufacture and delivery of individual parts of an I&C platform. In this case, the customer (end user or organization that assembles and delivers the complete I&C system to the end user) formulates requirements for the I&C system, selects and orders the necessary hardware and software (including, if necessary, products from other manufacturers), integrates and adapts the parts to perform the required functions, analyzes compliance with the requirements and performs necessary experimental verifications, and produces operational documentation.
2. Development, manufacture, and delivery of the central part of the system being developed (software and hardware complex (SHC)) to the customer. In this case, the developer of the I&C platform:
    - Based on the customer's input data, formulates requirements for supply of the SHC and determines its composition.
    - Manufactures the necessary hardware, integrates it with the basic software and adapts to perform the assigned functions.
    - Analyzes compliance with the established requirements, including necessary experimental tests of the SHC.
    - Issues operational documentation and conducts training for end-user operating personnel.
    - Delivers the SHC to the customer together with service equipment, documents, operation recovery kit (spare parts).
    - Takes part in the installation, adjustment, and testing of the SHC at the place of operation.

The customer (end user):

- Determines the composition of peripheral components (components of the I&C system), implementing the required (assigned) applied system functions together with the SHC.
- Selects and orders or purchases such products from their manufacturers and/or suppliers.
- Integrates the peripheral components with the SHC hardware and software and adapts them to perform the applied system functions.
- Issues operational documents for the I&C system and conducts the necessary experimental tests to confirm that the system meets the requirements.

FIRST GENERATIONS

In the early stages of implementing the conceptual approach to the design and use of I&C platforms, it was assumed that the equipment developed for the assembly of new and the upgrade of existing I&C will be manufactured and supplied according to the first option described above. In this regard, a unified hardware complex (UHC) was developed in Russia and Ukraine back in the 1980s. The UHC represented an I&C platform (equipment family) intended for the layout of various I&C systems important to NPP safety. The UHC included a limited, though rather wide, range of electronic units made on sliding mounting boards with unified sizes and universal support structures (cabinets) for their placement, mechanical protection, power supply, cooling, electrical connection, connection of mated equipment, and external energy sources. The composition of blocks in each UHC cabinet depended on the function performed. Individual blocks were intended for input, galvanic separation, and generation of continuous and discrete signals, implementation of logic functions ("Memory", "AND", "OR", "NOT"), generation of signals with adjustable delay, gain and output of signals, etc. The UHC was used to design (without engineering surveys and factory tests) various I&C systems to receive and process signals from temperature, pressure, flow, and neutron flux sensors and tools for manual entry and storage of instructions issued by operating personnel, generation and issue of commands to control valves, actuators, indicators, and alarms.

In 1998-2005, Ukraine developed upgraded UHC blocks and cabinets with digital signal processing and of noise-resistant design with built-in monitoring and diagnostic tools designed to replace the outdated products that were in operation at Ukrainian NPPs. At the time of upgrade, they met the requirements of regulations, rules, and standards for nuclear and radiation safety that were in force in Ukraine, including those pertaining to operating conditions, immunity to electromagnetic interference, seismic resistance, and service life (Grabenko, Bakhmach, et al., 2005). Instrument-making enterprises of Ukraine manufactured and supplied UHC blocks and cabinets to all Ukrainian NPPs that served as the basis for implementing local control devices for process equipment, safety control systems, emergency and preventive protection systems, systems of normal operation for the reactor and turbine departments, etc. The Russian UHC manufacturer implemented similar upgrades at that time.

Although the upgrades significantly improved the properties of individual UHC blocks and cabinets, the operating experience of the systems implemented on their basis revealed a number of fundamental flaws. These flaws are due to the concept adopted in the development and upgrade of the platform that envisaged that individual parts would be manufactured and supplied for the I&C platform, the original composition of the cabinets and blocks would remain unchanged, and physical (at the level of connectors) compatibility of the new products with the UHC equipment to be replaced would be ensured.

The following root causes of the flaws revealed can be defined:

- Many properties of the I&C systems, such as noise immunity, seismic resistance, environmental resistance, insulation quality, etc., were determined not only by the rated characteristics of typical blocks and cabinets because the specific composition, location, and electrical connections of blocks in UHC cabinets could have a significant impact. Compliance of said I&C properties with the requirements could only be verified by tests of each assembled cabinet. However, since the UHC blocks were delivered individually and the cabinets were configured in situ (NPP), the necessary tests were difficult or impossible to perform.
- The requirement to retain the functional composition and physical compatibility with the UHC cabinets and blocks developed originally (based on outdated chips) made further use of more advanced components totally ineffective.
- All communications between the UHC blocks and cabinets were made using continuous and discrete signals and electrical connecting lines, which significantly impaired the noise immunity of the I&C systems in the actual electromagnetic environment in NPP rooms. The use of modern noise-resistant data transmission methods was impossible within the UHC configuration ideology.
- Diagnostics covers only UHC blocks and cabinets, while the lack of data exchange between redundant channels did not allow one of the most effective diagnostic criteria - data inconsistency in these channels – to be used.

The approach adopted for the manufacture and delivery of components for these platforms was partially due to the fact that they used exclusive hardware methods and microcircuits with a low degree of integration. Therefore, to implement the instrumentation and control functions, it was required to manufacture a large number of blocks and cabinets, develop complex electrical wiring and connection diagrams, and create unique test methods and tools. In these conditions, the SHC manufacturing, assembling, configuring, and testing processes were so complex that did not allow the platform developer to undertake a significant number of orders related to the SHC layout and implementation at the NPP site, especially considering the tight schedule for the work (within refueling outage).

These difficulties could be resolved after a new generation of platforms was developed. These platforms were characterized by software-based implementation of control and instrumentation functions, greater integration of the components used, and minimized composition of hardware and included service programs, tools, and techniques to ensure the unification of technical solutions, improve the efficiency and quality of efforts focusing on the integration of components and the development of SHC application programs.

This made it possible to implement the second of the above options, according to which the platform developer and/or its official distributors used their design, software, guidance, and regulatory documents for the assembly, manufacture, factory testing, and delivery of ready-for-service SHCs and participated in their installation on the customers' (end users') sites, integration with other components, and checks when specific systems were commissioned.

The aggregate hardware complex for local information and control systems, developed by the scientific and production association for automated control systems (NPO SAU, Kharkiv, Ukraine), was one of the first platforms of this generation. This platform had been produced at NPO SAU until the mid-1990s under the MicroDAT trademark (Didenko & Rozen, 1985) and was planned for widespread use in the Soviet Union as a technical basis for the automation of facilities in thermal power, ferrous and nonferrous metallurgy, petroleum, petroleum refining, coal, gas, food, microbiological, and other industries and in nonindustrial spheres (utilities, transport, communications, etc.). The potential use of MicroDAT in nuclear energy was considered and highly evaluated (for control and protection systems of fast neutron reactors, control systems for electric heating of communications and equipment filled with liquid metal coolant, and nuclear fuel reloading control systems).

Many of the projects were implemented by 1991, but the collapse of the USSR and subsequent destruction of the unified economic mechanism did not allow full implementation of the capabilities achieved to proceed with the intended development and support the commercial production of this platform in Ukraine. Nevertheless, the characteristic features of MicroDAT that were further elaborated when modern platforms were developed should be mentioned:

- Aggregation at several subdivision levels (see Figure 1): level 1 – components of I&C systems (software and hardware complexes and periphery equipment); level 2 - operationally stand-alone parts of software and hardware complexes (devices in floor-based and hinged cabinets, tables, drawer unit); level 3 - built-in parts of devices (functional blocks in unit and complete plug-in frames); level 4 - built-in parts of functional blocks (functional elements on pullout mounting boards).

- Standardization of the basic and connecting sizes of functional blocks and functional elements, unification of connectors for their electrical linkage and connection to external circuits.
- Standardization of input and output signals, interfaces, and data exchange protocols between functional elements (intrablock interface), functional blocks (interblock interface), and operationally stand-alone parts of the SHC (intrasystem interfaces).
- The presence of carrier structures with unified wiring for placement, mechanical protection, electrical integration, and connection of external information and power circuits to functional blocks (in floor-based and hinged cabinets, tables, drawer units) and elements (in functional blocks).
- Optimized composition of functional elements, necessary and sufficient for the implementation of information and control functions specific to the main proposed application of the platform.
- Consistency of requirements for operationally stand-alone devices, built-in blocks, and elements in terms of resistance to external factors typical of the expected conditions of their operation.
- Formalization and automation of the design layout of functional blocks, devices, and software and hardware complexes using the developed informational, regulatory, and guidance support and associated hardware and software tools.
- Implementation of the concept of distributed control, relying on the use of intelligent peripheral devices, standard signals, networks, and data exchange interfaces between geographically separated I&C systems, their components, as well as SHC parts.

Similar I&C platforms were developed also in Germany, France, USA, and other countries at that time.

Figure 1. Platform aggregation principle

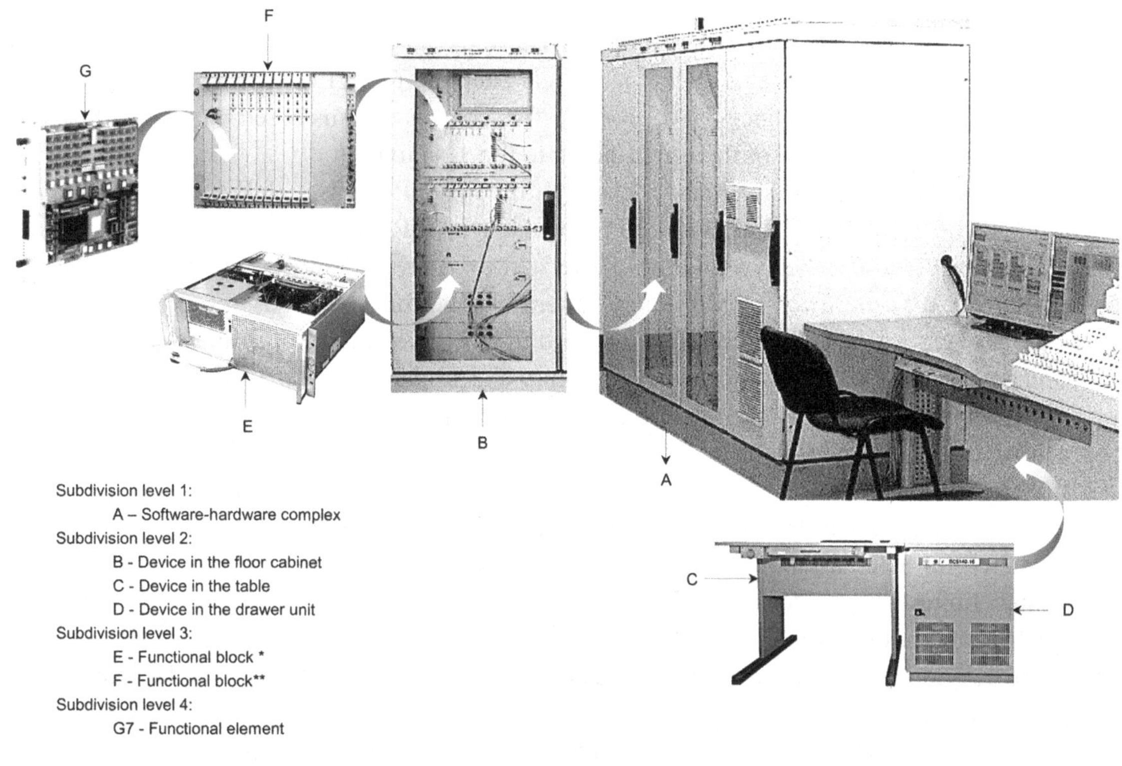

* Further subdivision is impossible or inappropriate

** Further subdivision is foreseen

MODERN I&C PLATFORMS (OVERVIEW)

The I&C platforms that have been widely used at present at NPPs are as follows:

- TELEPERM XS (Siemens Energy, Germany / AREVA NP, France).
- ALS (Westinghouse Electric Company, USA).
- Spinline & Hardline (Rolls-Royce Civil Nuclear, UK).
- FirmSys & FitRel (China Techenergy Co., Ltd., China).
- POSAFE-Q (POSCO ICT Co., Republic of Korea).
- MSKU-4 / PS 5140 (SRPA “Impulse”, Ukraine).
- VULCAN/VULCAN-M (LLC “Westron”, Ukraine).

The RadICS platform developed by the research and production corporation (RPC) “Radiy” (Ukraine) will be described in the next chapter.

TELEPERM XS I&C Platform

TELEPERM is the general name of the I&C platforms developed by Siemens for implementation of the central part of industrial I&C systems including nuclear energy. The TELEPERM-M platform became the first one in the early 1980s. Since 1980, about 15,000 control systems based on this platform have been installed in the world, many of them continuing to function today.

Two platforms were developed in 2006 specially for the first NPP units with French EPR-1600 reactors (Olkiluoto NPP, Finland): TELEPERM XP (new name is SPPA-T2000) designed for systems that perform category B and C functions and are not related to safety and TELEPERM XS designed for implementation of the central part of NPP safety systems that perform functions of the highest category A (there are also cases when TELEPERM XS is used in systems not related to safety, such as turbine control and protection systems, for which high reliability and productivity of this platform is important). The developers transferred the documentation and the right to produce the TELEPERM XS platform to the French AREVA NP company, while SPPA T2000 remained their property.

TELEPERM XS includes all the necessary hardware and software (operating system, library of functional blocks, etc.) and design tools for specific applications (graphical editor, code generators, etc.) that are required for SHC layout, factory testing, commissioning testing, maintenance, and in-service detection and elimination of defects.

Typical SHCs based on TELEPERM XS consist of stand-alone functional blocks that receive, generate, filter, and process input signals, control executive elements, and notify on the state of the technological process and defects revealed.

Blocks consist of functional input/output elements (modules) and communication modules made on sliding circuit boards (Figure 2), whose composition and number are determined by the types and number of mated peripheral devices and data transmission channels.

The core of each block includes a 32-bit processing module to control the execution of applied functions, performance of input/output modules, interblock data exchange, and self-testing procedures (module program is stored in flash memory and executed cyclically). TELEPERM XS input/output modules provide direct connection of sources and receivers with standard signals. The modules use microcomputers whose programs provide cyclic implementation of the main function and check of input and output channels and communication with the processing module.

The function blocks are connected by TXS Profibus data channels with the Field Data Link protocol (EN, 1996). TXS Ethernet channels with Logical Link Control protocol (IEEE, 2012) are used to connect top-level computers, service devices, and displays.

The concept of the TELEPERM XS platform was evaluated by the German Society for the Safety of Plants and Reactors (GRS)/Institute for Safety Technology (ISTec) and the United States Nuclear Regulatory Commission (U.S. NRC). The platform has passed qualification to confirm that it can be applied in safety systems of NPPs that perform category A functions according to the requirements of IEC, ISO, and IEEE standards. The qualification process involved a detailed independent expert analysis of the results obtained at all stages in the development of hardware, software, and design tools and included standard tests of a representative HSC based on TELEPERM XS in the configuration typical of applications.

The standard tests included equipment testing (in accordance with German national standard KTA, 2015) and software verification taking into account the requirements of IEC (2006). Experts evaluated the tests for completeness and correctness and the interpretation of their results for validity.

Figure 2. Aggregation of TELEPERM XS

Modules (functional elements) of TELEPERM XS

Computer module

Input / output module

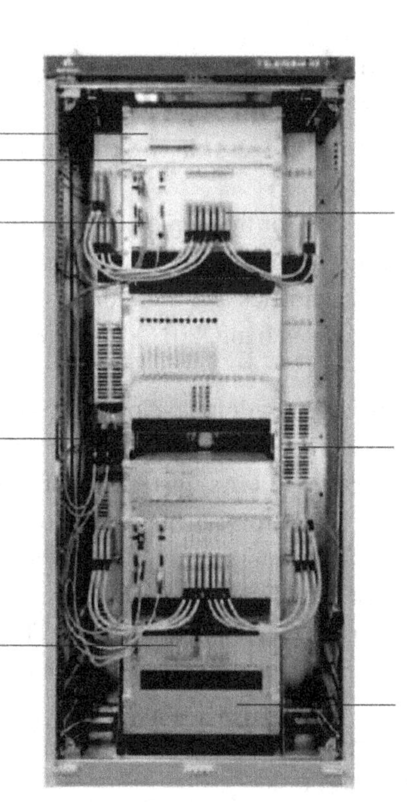

Device in the floor cabinet

The first applications of the TELEPERM XS platform began more than ten years ago. Since then, new applications are being introduced on its basis annually. TELEPERM XS equipment is used in the I&C systems of Mochovce NPP Units 3 and 4 (Slovakia), Beznau NPP Units 1 and 2 (Switzerland), Paks NPP Units 1-4 (Hungary), and Kola NPP Units 3 and 4 (Russia). In China, seven NPP units are equipped with similar systems. By 2015, TELEPERM XS equipment was installed or ordered for 81 NPP units in 16 countries. AREVA NP received U.S. NRC approval in 2010, which issued permission to use TELEPERM XS to upgrade the I&C system at the Oconee NPP (South Carolina, USA).

Standard applications of TELEPERM XS are safety systems designed for emergency protection of nuclear reactors, engineered safety feature actuation system (ESFAS), etc., as well as safety-related systems that measure parameters, control and limit the reactor power, monitor neutron flux, and control the movement of neutron-absorbing rods, automatic startup, and standby diesel generators.

ALS I&C Platform

The Westinghouse Electric Company has been designing instrumentation and control systems for NPPs for many years using various platforms: earlier its own WDPF II and later more sophisticated Ovation from Emerson Process Management and, in safety systems, Eagle from Hirschman Automation and Control (see Yastrebenetsky & Rozen, 2014). Currently, the Westinghouse Electric Company uses the following platforms:

- Ovation (Panfilo, 2011) - for the implementation of information and control functions of categories B, C, and not related to safety in normal operation systems of the reactor and turbine compartments, display of safety parameters, process alarms, control of feedwater pumps, management and control of the turbine, etc.
- Common Q (Westinghouse, 2018) - for the implementation of safety and safety-related functions of categories A, B, and C in reactor protection systems, actuation of engineered safety features, neutron flux monitoring, core diagnostics and monitoring, post-accident monitoring, etc.
- Advanced logic system (ALS) (Xu & Jiang, 2016) - for the implementation of diverse safety and safety-related systems.
- ALS is an I&C platform of new generation developed in 2004 by the CS Innovations Company (currently a subsidiary of Westinghouse Electric), whose technology is based on the use of programmable logic integrated circuits (FPGAs) instead of software methods for implementing information and control functions. The platform is highly reliable due to its simple hardware architecture, advanced diagnostic functions, and progressive system design processes. Built-in technical diagnostic means are capable of detecting failures in the operation of the system and initiating the necessary corrective actions aimed at maintaining safety. The diverse safety systems may be implemented based on ALS to reduce the probability of common-cause failures of critical functions.

The ALS is based on a set of functional elements (boards) that can be combined in various combinations to perform monitoring and control tasks:

- Input boards (IPB) perform filtering, conversion, input of continuous direct current and voltage signals, and input of discrete signals.

- Output boards (OPB) ensure transformation of digital codes, output of continuous direct current and voltage signals and output of discrete signals.
- Communication boards (COM) provide data exchange with other boards via independent channels with standard serial interfaces.
- Basic boards (CLB) perform control data exchange between boards and with other function blocks, implementing the assigned applied functions of the system.

Each input channel of the IPB board is galvanically isolated and contains filtering and overvoltage suppression circuit. Each output channel of the OPB board is isolated and protected against short circuit and overvoltage. All boards have self-testing functions to monitor and diagnose most of their elements and circuits without affecting the implementation of basic functions. All boards have a LED panel that displays the status of each board.

The HSC is a combination of general-purpose boards (IPB, OPB, and COM) and an CLB board focused on a specific system task. The boards are placed in standard block plug-in frames (subracks) up to 10 boards in one frame to form functional blocks, one of them being the main block and the other (up to five) representing expansion blocks. Data are transferred from IPB and/or COM to CLB and from CLB to COM and/or OPB of the main block in fixed format via bi-directional duplicated RAB1 and RAB2 buses using the EIA-485 interface and the UART protocol. Self-test results for the blocks are displayed on their front panels and transmitted to CLB on a separate TAB bus. The RAB1, RAB2, and TAB buses are implemented on connectors on the rear panel of the block plug-in frame, to which all the boards installed in it are connected (Figure 3).

Figure 3. One of the possible configurations of the function block, implemented on the ALS platform

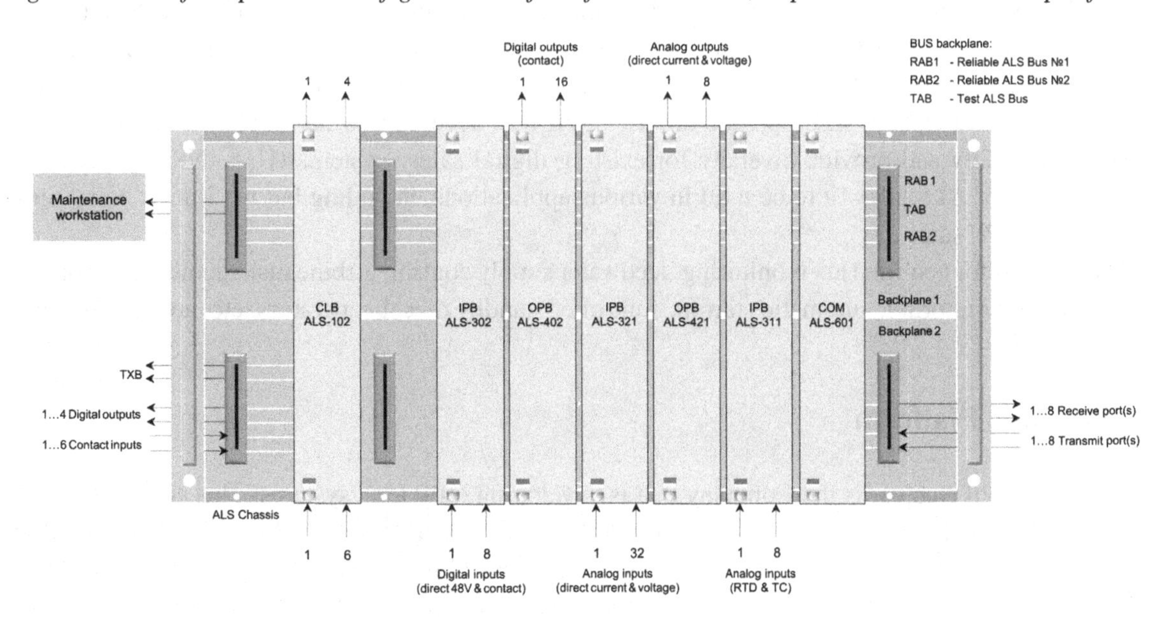

Each transaction is initiated by a base CLB board ("master") and terminated by a "slave" (IPB, OPB, or COM) board. The base board reads all the input data and derives all the output data in each fixed time interval, whose duration is set upon completion of the system. If any malfunctions occur on the signal path or the transmission is not completed within a fixed time interval, error signals are be sent via TxB to notify operators on the need for corrective action.

Data exchange between functional blocks is done through the ports of communication boards using a point-to-point structure, EIA-422 serial interface, and UART protocol. The COM port tuned for reception sends the received data to the buffer, checks their integrity, and only then makes them accessible via RAB buses. The COM port tuned for transmission receives data on the RAB buses, checks their integrity and converts and transfers them to the conjugated port of the communication board of another function block. In the event of an error, retransmission of data is not performed: the port goes into a predetermined state and an error signal is issued to the TxB output of the CLB board to attract the operators' attention.

Functional blocks and power supply units are installed in earthquake-resistant floor cabinets with mounting panels and elements for external connections. Control panels and a service station can be integrated into the cabinets. Remote I/O cabinets can be installed in close proximity to peripherals.

The "deterministic behavior" of each system is ensured by the ALS organization of data exchange between the boards of the function block and between the blocks, rigid duration of the exchange interval on the RAB buses for each application, as well as zero need to use the operating system, multitasking, interruptions, and dynamic scheduling of tasks.

The system performance is uniquely determined by the connection structures of the FPGA logical elements, which are fixed in their electronic designs and physically implemented in the programming process by the SHC manufacturer. Thus, potential unintentional or deliberate unauthorized intervention in operation of the system through external wired and/or nonwired circuits, local networks, or removable storage media is completely excluded. There is only one way to change the algorithm – use of the service station – which is possible only after the system is decommissioned.

The ALS platform was first applied in 2009 on the steam supply line of the Wolf Creek NPP (USA). This enabled the CS Innovations Company to submit the documents necessary to qualify ALS as an I&C platform for safety systems to the U.S. NRC. The U.S. NRC reviewed the submitted documents, as well as the standard test results, and approved the use of ALS to develop new and replace existing systems important to safety and provide diversity for existing digital safety systems (U. S. NRC, 2013). The characteristics of ALS allow it to be used in various applications, including but not limited to systems important to NPP safety.

The systems for neutron flux monitoring, feedwater supply control, turbine management and control, post-accident monitoring, automatic startup, control of standby diesel generators, etc. are implemented based on ALS.

Spinline I&C Platform

Rolls-Royce Civil Nuclear is the Company that is developing NPP I&C systems. This company developed the first digital reactor protection system, installed at Paluel NPP Unit 2 (France) in 1984. Spinline is a new platform developed by Rolls-Royce Civil Nuclear taking into account experience in creating and using the previous generations of platforms, from which Spinline differs by an advanced network architecture, fiber optic data lines, 32-bit microprocessor, Field-Programmable Gate Arrays (FPGA), and a new development environment for systems and application software.

Spinline includes embedded functional elements (boards) that implement:

- Input of 32 discrete or 16 continuous signals (Input boards).
- Output of 32 discrete or 12 continuous signals (Output boards).
- Input of signals from radiation detectors.
- Digital data processing (Processor boards).
- Data exchange (Communication boards).

FPGAs with a capacity of 30,000-40,000 gates that implement simple standard or more complex application functions (a higher level of complexity is used for equipment not related to safety) can be used in input, output, and communication boards.

Functional elements made on sliding mounting plates of standard design (Figure 4) are installed in a standard subrack (chassis) and united with the connectors of one or two printed crossboards on its back wall, forming a functional block. Functional elements can be inserted, removed, and replaced during system operation. A mechanical key determines only one predefined type of elements that can be installed at the specific position of the subrack.

Spinline elements as part of a function block exchange data on a simplified and protected version of the standard VME bus (IEC, 1991), which does not use the capabilities of the Multimaster and the bus arbiter. The deterministic behavior ensures that the transaction time limit is not exceeded and avoids critical overload. For this purpose, if an error is detected in the received data (or if new data have not been received within a predetermined time interval)m the output boards automatically transfer their outputs to predefined safe states.

Functional blocks (from one to five) installed in a standard (IEC, 2008) floor cabinet form am operating stand-alone device with protection IP 32 according to IEC, 2013. To support the human-machine interface, Spinline provides devices controlled by industrial personal computers:

- Local display unit - checks the presence of monitored parameters in the specified (allowable) ranges.
- Monitoring and maintenance unit - constantly diagnoses the correct performance of system components, signals events that require attention of personnel (for example, equipment failures, opening of the cabinet door, etc.), supports event register with printing and archiving functions.
- Automatic testing unit - displays current data, analyzes and archives the system checks in power operation and shutdown of a power unit.

Data exchange between functional blocks and/or stand-alone devices (stations) is carried out through a NERVIA local network using the broadcasting protocol when any message sent by one station is received simultaneously by all other stations in the network. The time of each transaction is limited by a predetermined maximum value to avoid the hang-up of several devices if one of them fails.

In order that functional elements that are not related to safety could not interfere with the performance of functions important to safety, the separation of data streams across different NERVIA networks, physical separation (geographical dispersion) of stations, electrical separation (mutual galvanic isolation) of their electrical circuits, and the use of fiber optic data lines are provided.

The procedure in which the functional blocks use the data received over the network is strictly defined and fixed: the blocks can process the data intended for them sequentially (one after another) and/or in parallel (simultaneously). Parallel mode is used, for example, in redundant systems.

Figure 4. Spinline functional modules and blocks

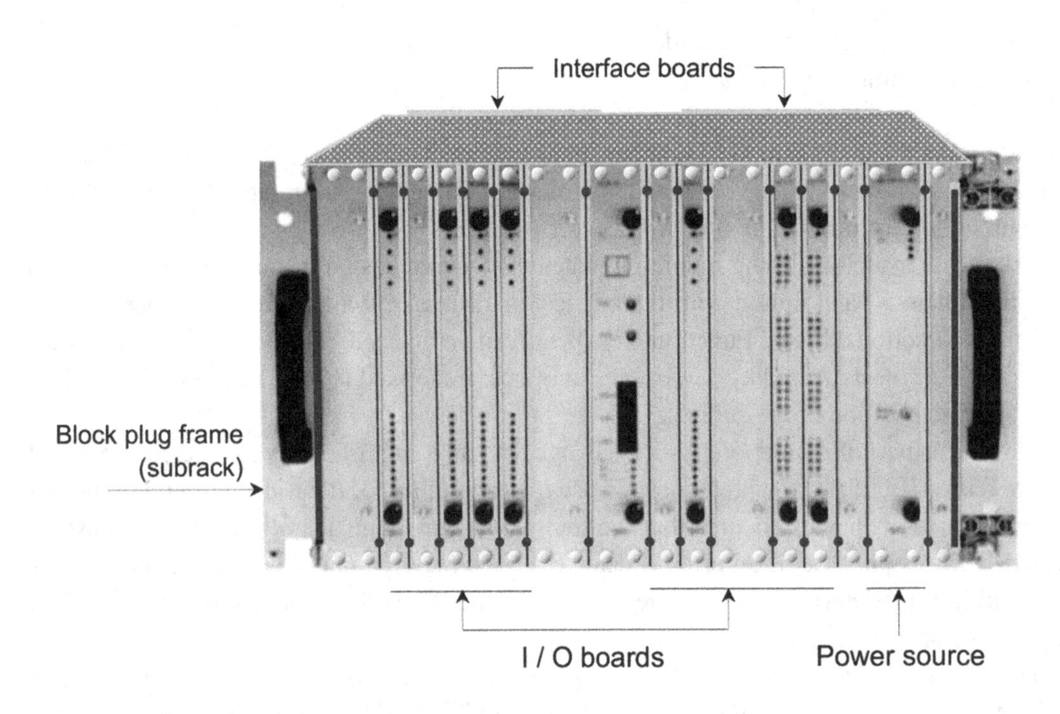

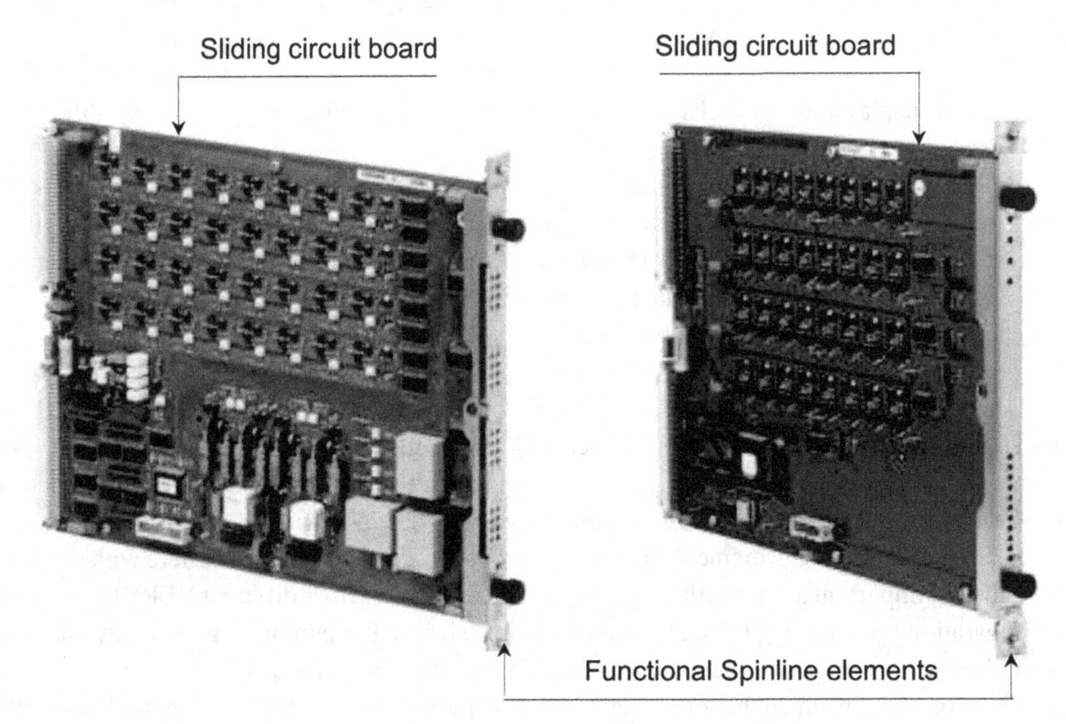

Spinline provides the capability to create I&C systems of practically any complexity. Spinline equipment is delivered to the customer as functional blocks (for integration into cabinets of the existing system) or as stand-alone devices that are connected at the site of operation to form a hardware-software complex – the central part of the new system. The composition of the supplied functional blocks and devices is determined by the requirements of the system for which they are intended.

Spinline software includes the operating system, application programs and system and software development environment.

The operating system is supplied as a standard software component for use on processor boards. It provides control over the collection and transfer of data and services that are necessary for the execution of application programs; in particular, it checks the control and data flows automatically stopping the processor when any inconsistencies are detected to will shift the board outputs to a predetermined safe state.

Application programs are developed for each project; they support the performance of functions that are appropriate for the purpose of the system being developed and may also include means for verifying the configuration for protection against possible errors that occur in the design or operation. The periodic check of functions for correct execution can be automated, in particular, through software substitution of test values instead of current values of the input signals.

The CLARISSE system and software development environment provides software tools and libraries needed to configure Spinline processor boards and Nervia networks, adapt the operating system to application needs, and develop application software for specific I&C systems.

The Spinline platform allows developing systems with redundancy; a group of redundant channels (usually three or four) can be located in one or different functional blocks. The identity of the channels and data recovery in accordance with the set logical condition ("two out of three", "two out of four", etc.) is verified by a special functional element that receives data from all channels of this group or a program that simulates the operation of this element. To protect systems based on the Spinline platform against common cause failures, functional diversity can be used to provide distinction between the algorithms implemented by the channels.

The Spinline platform provides a safe environment for the development, testing, implementation, and protection of systems against unauthorized, inadvertent, and unsafe changes in their operation and maintenance (Bach, Duthou, et al., 2017).

In 2011, Rolls-Royce Civil Nuclear submitted an application package for licensing its Spinline platform in the USA to the U.S. NRC. The positive conclusions on the application made by U.S. NRC in 2014 allowed the company to use Spinline in the sphere of nuclear energy in the United States and several other countries. Électricité de France, the largest state-owned energy generating company in France, and AREVA NP chose the Spinline platform to upgrade the I&C systems at twenty power units with a capacity of 1300 MW, and Finnish state energy company Fortum decided to use Spinline to upgrade the I&C systems at Units 1 and 2 of the Loviisa NPP.

The Spinline platform, initially intended only for I&C systems used at NPPs and combining software methods for implementing functions and hardware solutions based on FPGA, allows the development and effective operation of systems important to safety (including safety systems that perform category A functions according to IEC, 2006). Standard applications include reactor emergency protection systems, neutron flux monitoring systems, engineered safety feature actuation systems, etc. They are supplemented by rod control systems based on another platform, Rodline (Rolls-Royce, 2019), developed specially for the implementation of such systems.

Hardline I&C Platform

The Hardline platform (Rolls-Royce, 2018) is a new generation of nonprogrammable (hardware) platforms designed to implement category A, B, or C functions (IEC, 2009) in systems of safety classes 1, 2, 3 (IEC, 2011) and 1E (IEEE, 2003). Hardline is the evolution of previous Rolls-Royce hardware platforms and is based on modern components, architecture, and system design processes critical to nuclear safety.

All required system functions are performed by Hardline hardware. The absence of software-controlled functions makes such systems immune to common cause failures and prevents unintentional or deliberate unauthorized intervention in their operation.

The Hardline platform includes functional elements (modules) that perform the simplest operations: receive signals from thermoelectric and thermoresistant sensors, provide galvanic separation (isolation) of signal circuits, filter electromagnetic interference, set and change threshold values, generate and issue output signals and actuator control commands, activate signaling devices, implement standard logical functions, ensure redundancy with restoration in accordance with the accepted logical conditions (one out of two, two out of three, two out of four), provide priority drive control, etc. A set of functional modules as part of the Hardline platform is necessary and sufficient to implement basic functions of software and hardware complexes (central part of the I&C systems for which this platform is intended).

The functional modules are implemented on standard-size sliding circuit boards that are mounted in typical 19" block plug-in frames and connected to a communication bus installed on the backplane of the frame.

The modules installed in the frame form a functional block (Figure 5). The communication module (the last one in the block) is the gateway through which information is transmitted to external systems: it receives serial data on the status of inputs, outputs, and settings of the functional modules via the communication bus of the frame and transmits them over a local area network with a standard Modbus TCP interface to the upper-level system that monitors, diagnoses, and maintains up to 22 functional blocks (stations).

In the operation process, the Hardline modules carry out continuous self-monitoring for quick detection of most hardware failures. Data on the status of each module are transmitted via the communication bus and local network to the monitoring, diagnostic, and maintenance system.

The circuits connecting each module to the communication bus are electrically isolated and separated from the rest of this module. The isolation and separation of circuits are certified in accordance with the requirements of IEC (2015) so that failure of the communication bus cannot affect the operation of the modules connected to it. Data are transmitted via the communication bus and local area network in only one direction, from the periphery to the center. Therefore the transmission means can in no circumstances prevent the functional modules from correct performance of their operations.

Hardline can be supplied either as individual functional units for installation in standard 19" cabinets of the existing system or as one or more devices consisting of functional units installed in Rolls Royce floor cabinets (up to six units in one cabinet). The composition of the functional blocks and devices is adapted to the requirements of the new and/or upgraded system. The unification of the internal electrical wiring of block frames and cabinets makes it possible to "delay differentiation", which can be carried out at the final stage of the factory production of functional blocks and devices.

Hardline allows the SHC to be implemented for I&C systems of any size: from simple with several functional modules without redundancy to the most complex ones, such as an integrated system with more than 200 functional modules, fourfold redundancy (with failure recovery in accordance with accepted logical conditions), priority control of drives, as well as remote monitoring, diagnostic, and maintenance devices (based on personal computers for which necessary application software was developed and delivered within the Hardline platform). At the same time, it should be noted that the use of simplest operations performed by Hardline functional modules to implement such rather complex system functions, such as control of process parameters, priority drive control, generation of transmitted data, etc., leads to the need for nontrivial physical connections of inputs and outputs of these modules and settings for their characteristics. For this purpose, a special built-in configuration board connected to the frame backplane is used.

Figure 5. Functional block realized on the base of Hardline platform

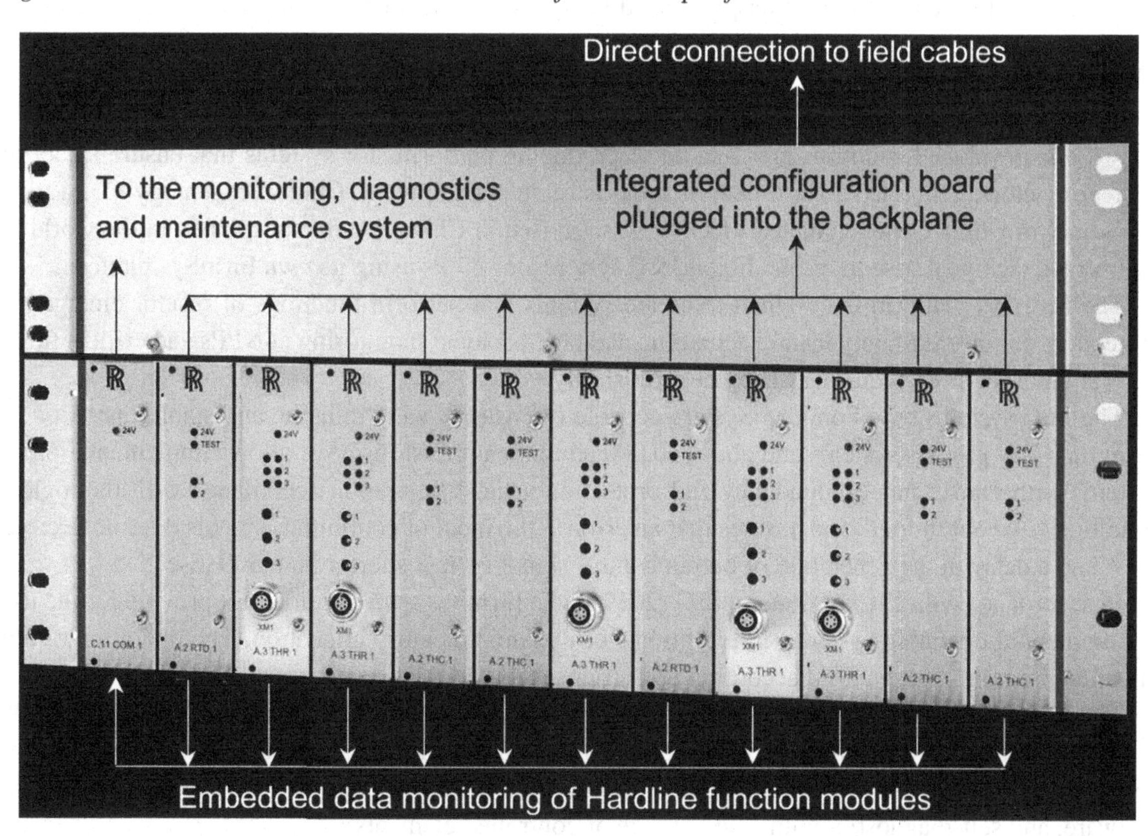

Safety systems based on the Hardline platform allow periodic offline and online testing. In the process, conditions are simulated that would lead to the actuation of protection for early detection of failures that can prevent the safety system from initiating the necessary protective actions.

Qualification requirements were determined so as to cover all possible conditions for the use of HSC in the most critical configurations. Qualification was carried out by analysis and testing (qualification tests) in certified laboratories, which covered the first manufactured samples of modules of each type and functional units and devices in the most representative selected configurations. The mechanical, electrical, and functional characteristics of the hardware, quality of the products manufactured, resistance to environmental impacts (dry and wet heat, cold, rapid temperature fluctuations), resistance to mechanical impacts (sinusoidal vibrations, shock, imitation of earthquakes), and electromagnetic interference were verified during the tests.

The HSC integrated into a specific system is qualified at the place of operation by comparing the configuration and settings of the devices included in it with previously selected and qualified representative configurations. The lack of software simplifies the licensing process for the Hardline platform and the associated HSC by the regulator.

As part of the upgrade project for a 1,300 MW French reactor, Rolls-Royce upgraded existing nonprogrammable I&C systems developed in the 1980s and introduced new features using the Hardline platform. Two nonprogrammable safety systems were delivered to the Finnish Loviisa NPP: one sets the priority of operational personnel instructions aimed at ensuring safety over the commands of other systems that are not related to safety and the other provides diversified implementation of critical safety functions.

FirmSys Platform

Only some developed countries are capable of producing platforms for systems that ensure safety of nuclear reactors. China has long relied on imports of such equipment. China Techenergy Co., Ltd, a subsidiary of China General Nuclear (CGN), was the first in China and one of the few in the world to use its own technical base to create digital I&C systems for NPPs using its own FirmSys platform.

The FirmSys platform is designed to create systems that perform functions of reactor emergency protection, engineered safety feature actuation, and post-accident monitoring at NPPs and critical functions at other facilities requiring the highest reliability.

The I&C systems based on FirmSys are capable of ensuring safe, reliable, and stable operation of more than 260 process systems and about 10,000 equipment units. FirmSys allows implementation of systems with four-channel redundancy and protection against failures in accordance with the logical condition "two out of four" and ensures that an error in the input of continuous signals does not exceed 0.1% and a delay in the generation of output control signals is less shorter than 300 msec.

In accordance with IEC (2006) and IEC (2012a), the FirmSys software does not provide for the use of commercial operating systems, interruptions, or dynamic memory allocation. The operation of application programs is organized according to a strict schedule.

The FirmSys-based systems meet extremely high requirements for functional safety and effectively prevent unintentional or deliberate unauthorized intervention in their operation due to the use of a "point-to-point" structure with unidirectional data transfer, their own compiler and specially developed software, and self-diagnostics with a wide range of controlled elements.

The FirmSys platform ensures that control systems important to safety be completely in the "home hands" since China Techenergy Co has independent intellectual property rights. The FirmSys platform was certified by China's National Nuclear Safety Administration (NNSA).

The first safety system based on the FirmSys platform was implemented at Yangjiang NPP Unit 5 built by CGN. The unit was connected to the power system in May 2018, which was noted in China as an important milestone in the development of domestic nuclear energy. The FirmSys platform has already been used or is planned to be used for upgrade of existing and building of new NPP units: Yangjiang NPP Unit 6, Hongyanhe NPP Units 5 and 6, Fangchenggang NPP Units 3 and 4, Tianwan NPP Units 5 and 6, a high-temperature gas-cooled reactor in Shidao Bay, etc. FirmSys is also expected to be used in aviation and shipbuilding.

FitRel I&C Platform

The FitRel platform (Chen, 2017) was developed by China Techenergy Co., Ltd, and became China's first digital hardware platform based of the FPGA technology.

There is no software in the FitRel platform: all functions are implemented through internal connections between elements of the corresponding FPGAs. If the same functions important to safety are performed simultaneously and independently by two different systems – software-controlled (based on the FirmSys or similar platform) and nonprogrammed ("hardware", for example, based on the FitRel platform) – the maximum achievable diversity, practically eliminating common cause failures, will be provided. Sufficient diversity can also be provided in one system if, for example, two redundant channels are implemented on a platform with software-controlled functions (FirmSys) and the other two channels perform the same functions using hardware components (FitRel).

Other FitRel advantages are that all necessary data are not processed sequentially as in the programmable processor systems but simultaneously ("in parallel") into several independent FPGAs to provide rapid response of the system to external and internal events.

In addition to the main functions (appropriate to the purpose of the system), the FitRel platform performs additional functions during operation, such as condition monitoring and continuous diagnostics of equipment to reduce the frequency of periodic inspections, minimize the risk of potential human errors, and reduces system recovery time in case of failures. The features that perform basic system functions important to safety and additional ones that are not related to safety are implemented by different FPGAs, so that the basic functions do not depend on errors or failures of additional functions.

The main system functions (including functions of redundant channels) can also be divided between different FPGAs to ensure their independence, simplify checks, and mitigate the consequences of errors or failures. Like other modern platforms, FitRel allows each created to be configured based on the functional requirements.

FitRel ensures that an error of input and output of continuous signals does not exceed 0.1% and a delay in the generation of output control signals is less than 150 msec.

As part of the FitRel platform, the following were developed (Figure 6):

- Hardware (functional elements) that perform the functions of input/output of continuous and discrete signals, data processing, communication. and additional functions.
- Support structures (block plug-in frames and floor cabinets) for the placement of functional elements, their electrical combination, and connection to external circuits.
- Software tools needed for users to design specific systems based on the FitRel platform and end users at different stages of the life cycle of these systems.

Figure 6. Aggregation of FitRel platform

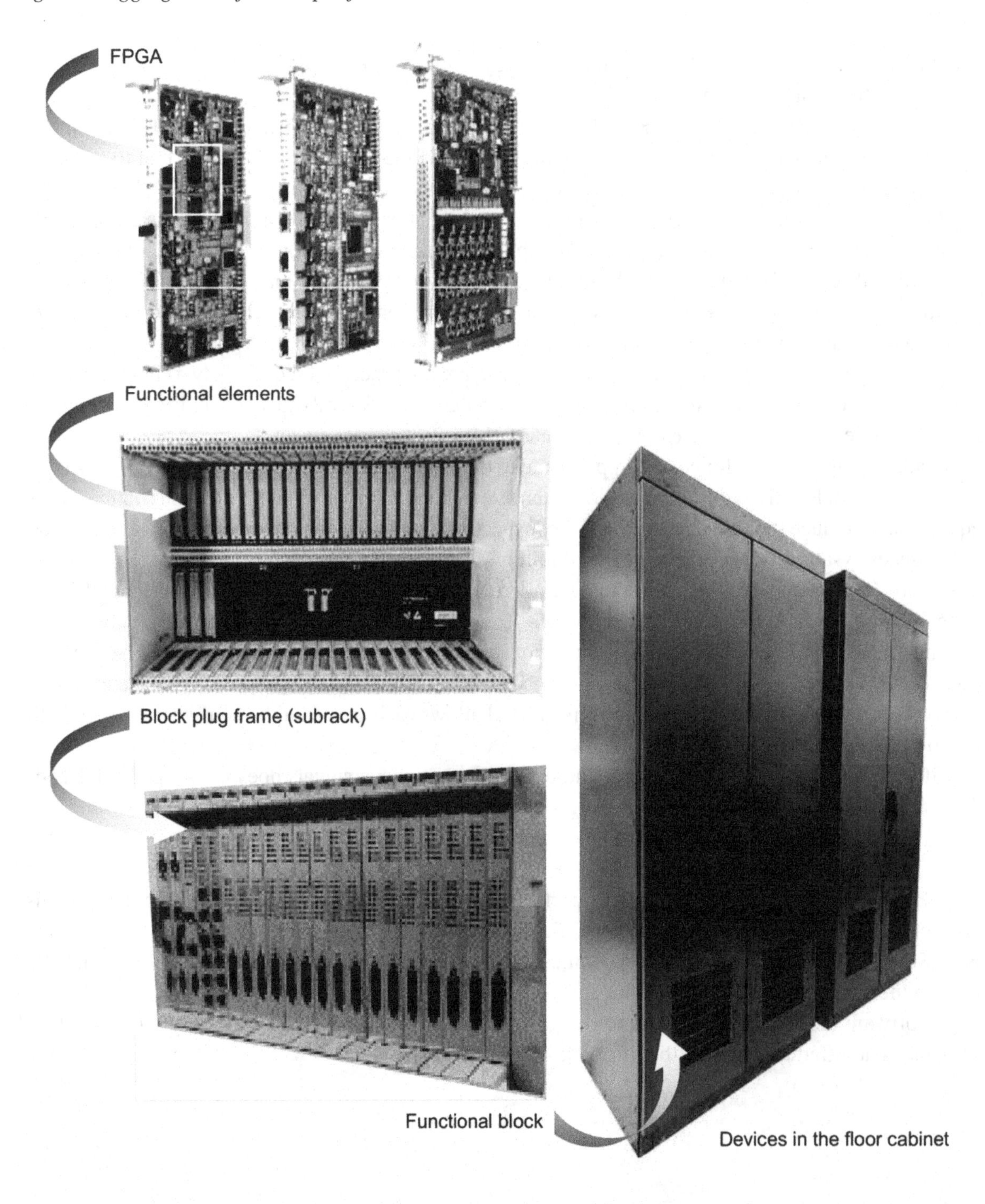

Compared to programmable data processing means, the FitRel functional element for data processing has simple hardware structure: it requires only FPGA to perform the same functions instead of a microprocessor and one or more integrated circuits of operative and programmable read-only memory connected by multi-bit interface buses on a printed circuit board. This results in lower failure rate, lower power consumption, and easier replacement during upgrades (no development, verification, or validation of application software required). FPGAs are also used to implement physical and link layers of the local Ethernet area network that connects each FitRel function block with top-level equipment that configures and manages the entire system.

In operation of the system, it is possible to change the set and/or maximum permissible values of the monitored parameters, display the results of continuous diagnostics, and start automatic testing of the system.

The FitRel platform has been in development since 2013, and the first hardware samples were released in 2014. Based on them, projects of diverse actuation systems were developed and implemented in 2015 - 2016 at Yangjiang NPP Units 5 and 6, Hongyanhe NPP Units 5 and 6, Tianwan NPP Units 5 and 6, and Fangchenggang NPP Units 3 and 4.

Standard testing of representative SHCs based on the FitRel platform was performed in accordance with the procedures governed by:

- A series of IEC 60068-2 standards "Environmental Testing Package", establishing methods for testing resistance to vibration, shock, temperature, dry heat, and other external factors (DES, 2019);
- United Kingdom national standard GB, 2018, establishing methods for seismic qualification of electrical equipment important to safety of NPPs;
- international IEC standard, 2009b, establishing requirements for testing the electromagnetic compatibility of I&C systems important to safety of NPPs.

In June 2018, Rolls-Royce Civil Nuclear and China Techenergy Co signed an agreement on the representation of integrated I&C solutions for the global nuclear market. In accordance with this agreement, the companies also undertook to release a new I&C platform combining technical solutions adopted for Rolls-Royce (Spinline and Hardline) and China Techenergy (FirmSys and FitRel) platforms. This new I&C platform will provide both particles with the opportunity to better adapt to the needs and requirements of consumers.

POSAFE-Q I&C Platform

The POSAFE-Q platform is owned by POSCO ICT Co. (Republic of Korea) and the Korean Atomic Energy Institute as part of a national research program on management and control systems. POSAFE-Q is focused on creating programmable logic controllers (PLCs) designed for use in NPP safety systems (Lee, Song & Yun, 2009). The development of the platform began in 2001 and ended with the receipt in 2009 of a license safety certificate in the Republic of Korea.

As part of the POSAFE-Q hardware, 22 types of modules (functional elements) were developed, such as power supply, input/output modules of continuous, discrete, and pulsed signals, a processor module, communication modules, etc. There is a continuous (on-line) self-diagnosis of modules, whose results are displayed on their front panels and issued through communication modules on the local network to devices of a higher hierarchical level.

The digital input/output modules are oriented at the input signals of direct current (24 V and 48 V) and alternating current (120 V and 230 V) and output signals of direct current (24 V, 48 V, and 125 V) of electromechanical and solid state relays. Continuous signal input/output modules are oriented at input and output direct current and voltage signals and input signals from thermoelectric and thermoresistor converters. The processor module uses a Texas Instrument processor. Communication modules include Data Link and Data Network modules, using the Fieldbus protocol (PROFIBUS-FDL) with electrical or optical communication lines.

Modules made on slide-in mounting plates of a standard constructive system are installed in block plug-in frames to form functional blocks (Figure 7). Two independent power supply modules and up to 15 POSAFE-Q modules of other types, including one or two processor module(s), can be installed in the frame. The processor module(s) are connected by the local bus with the I/O and communication modules installed in the frames. In use of local bus extenders, the number of I/O modules as part of the PLC can be increased. Each power supply module is capable of individually providing power to all elements that can be installed in the functional block.

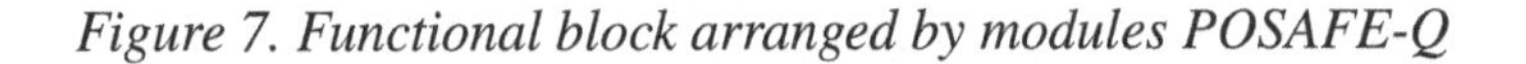

Figure 7. Functional block arranged by modules POSAFE-Q

For the POSAFE-Q platform, the Real Time Operating System (RTOS) and the Software Engineering Tool (an integrated development and debugging environment for application programs implemented on a personal computer that connects to the RS-32 port of the processor module for loading) were developed application program).

In development of the POSAFE-Q platform, requirements for NPP equipment set forth in international and national standards were considered and, in accordance with the accepted criteria, hardware and software were planned, designed (developed), evaluated, and tested. The RTOS software was developed as part of the life cycle including the phases of planning, formulation of requirements, design, implementation, integration with hardware, installation, and support for operation and maintenance. The software was verified and validated at all stages of its life cycle in accordance with IEEE (2003).

The POSAFE-Q platform was verified in the development process through tests of its elements, integration test, system test, interoperability test, and functionality and productivity tests. The normal operation mode and effects caused by potential failures of the PLC components were analyzed in detail and the expected failure rate was assessed.

The POSAFE-Q platform was qualified to demonstrate the efficiency of functional elements and blocks in generic operating conditions possible in the intended locations of equipment and in specific conditions regulated in the specifications of each supplied PLC. The qualification program included experimental tests in an independent testing laboratory:

- Resistance (taking into account aging factors) to the effects of calculated temperature, pressure, and humidity in accordance with IEEE (2003) and EPRI (1996).
- Electromagnetic compatibility (immunity to electromagnetic interference and power surges regulated in a series of IEC 61000 standards) in accordance with U.S. NRC (2013) and EPRI (1996).
- Resistance to seismic effects that may occur during and after operating basis earthquake (OBE) and safe shutdown earthquake (SSE) at the nuclear power plant site (tests with resonance search, five tests for OBE simulation and one test for SSE simulation, with exposure simultaneously in three mutually perpendicular directions).

It should be noted that the design solutions used in POSAFE-Q are more characteristic of general industrial PLCs (installation of functional blocks directly in end-user cabinets, connection of external circuits directly to switching elements on the front panels of modules, etc.). This distinguishes POSAFE-Q from other platforms (which are most often supplied as operationally stand-alone devices, usually in floor cabinets) and can cause problems in ensuring the functional safety of I&C systems.

In addition to the experimental tests in the platform qualification process, the documents submitted by the developer for verification and validation of POSAFE-Q software, reflecting the results of offline testing of software components, their potential integration with hardware, testing of RTOs, and verification of interaction between several operating systems were analyzed (in accordance with IEEE, 2004b) and cyber security was assessed in compliance with the U.S. NRC (2006).

Based on the results of testing and analysis, the POSAFE-Q platform was certified by authorized institutions of the Republic of Korea as a platform for the development of class 1E monitoring and control systems for NPPs. The POSAFE-Q platform was first applied at NPP units with new Korean APR-1400 reactors at Ulchin NPP Units 1 and 2 and Kori NPP Units 5 and 6.

MSKU-4 / PS 5140 I&S Platforms

MSKU-4 and PS 5140 are platforms developed by the Ukrainian Company – SRPA "Impulse", which are used in the development of systems important for safety in nuclear and thermal energy and other industries. The latest versions of the platforms make it possible to compose fault-tolerant hierarchical physically distributed systems for monitoring and control of especially critical facilities, including NPP units, where they perform safety functions (category A) and normal operation functions, important for safety (categories B and C) and high-speed railway transport.

Lower level of platform carry out on the basis of MSKU-4. MSKU-4 collect information from sensors with continuous and discrete electrical signals; perform and processing of data obtained by programs that implement the specified control algorithms (regulation, protections, interlocking, starting and stopping equipment); generation and output of control commands to executive devices; information exchange with top-level management systems that can be implemented on the basis of the PS5140 platform. At this level, data is collected from lower-level systems, archived, displayed to operating personnel.

Additional functions of the lower and upper level systems include continuous monitoring of the technical condition of the equipment, diagnosis of defects, display and output of diagnostic messages.

The main units of aggregation in the layout of the systems are control computer complexes MSKU-4 and PS5140 workstations, which are assembled by design from a set of hardware and software facilities of the corresponding platform.

The MSKU-4 platform includes:

- A set of built-in functional units (modules) made on sliding circuit boards: processor modules, interface modules, control modules, I/O modules (communication with the object);
- Load-bearing structures for mounting, securing, mechanical protection, power supply of modules and connecting external circuits: block plug-in (mounting) frames, floor cabinets (with a degree of protection of at least IP23), cross-section and connecting panels.

The processor module is based on the Intel Atom.

Continuous signal input modules carry out galvanic separation of input circuits, suppress normal and general types of interference, execute analog-to-digital conversion, and output data on the values of informative parameters in a 14-bit format to three independent ports of the Identity Registration Protocol (IRP):

- 16 continuous DC and medium voltage signals;
- 16 continuous low-voltage DC signals;
- 8 signals from thermistor sensors;
- 8 signals from potentiometric sensors.

The input modules of discrete signals carry out channel-by-channel or group galvanic separation of input circuits, conversion to digital format and output of data on logical values to three independent ports of the IRP:

- 16 or 30 (in case of group galvanic separation) of discrete signals in the form of low and high resistance of switching contacts;
- 16 or 48 (in case of group galvanic separation) of discrete signals in the form of low and high levels of DC voltage;
- 16 discrete signals in the form of low and high levels of AC voltage.

Figure 8. General scheme of MSKU-4 (master)

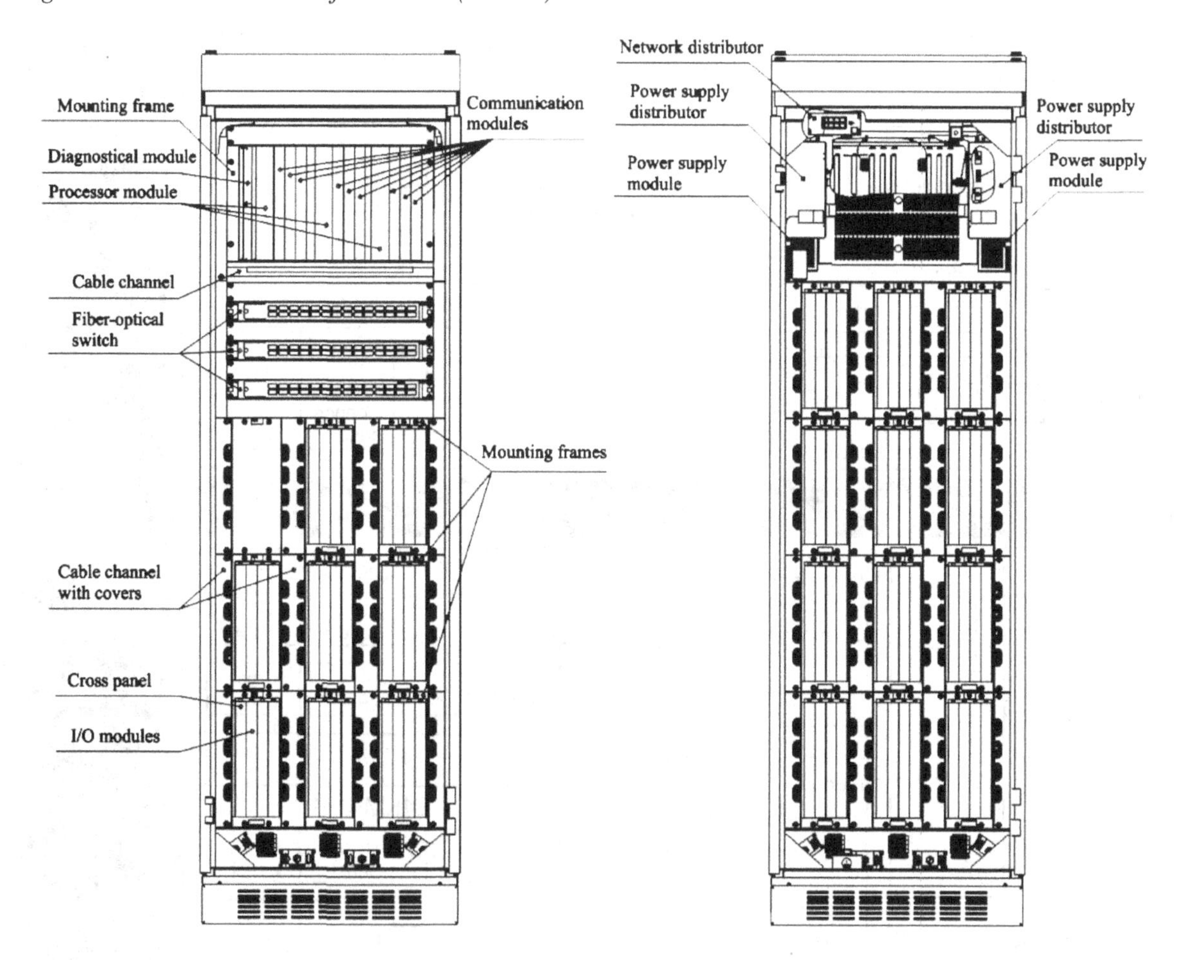

The output modules of continuous signals receive data in 14-bit digital format through three independent ports of the IRP, perform digital-to-analog conversion, galvanic separation of the output circuits and the output of 16 continuous DC signals.

The output modules of discrete signals receive data in digital format via three independent ports of the IRP, perform galvanic separation of the output circuits, generation and output of:

- 16 or 30 discrete signals at the outputs of solid state relays for switching high-level AC or mid-level DC with group separation of circuits;
- 16 discrete signals at the outputs of electromagnetic relays for switching high level AC or DC with channel-by-channel separation of circuits.

The control computer complex (CCC) of MSKU-4 can include one cabinet, called a master, or several cabinets, one of which is the master and the rest are slaves. MSKU-4 can have several variants based on its architecture but the most common one is the «modular» version.

Figure 9. Connection scheme of non-redundant I/O modules to three redundant processor modules

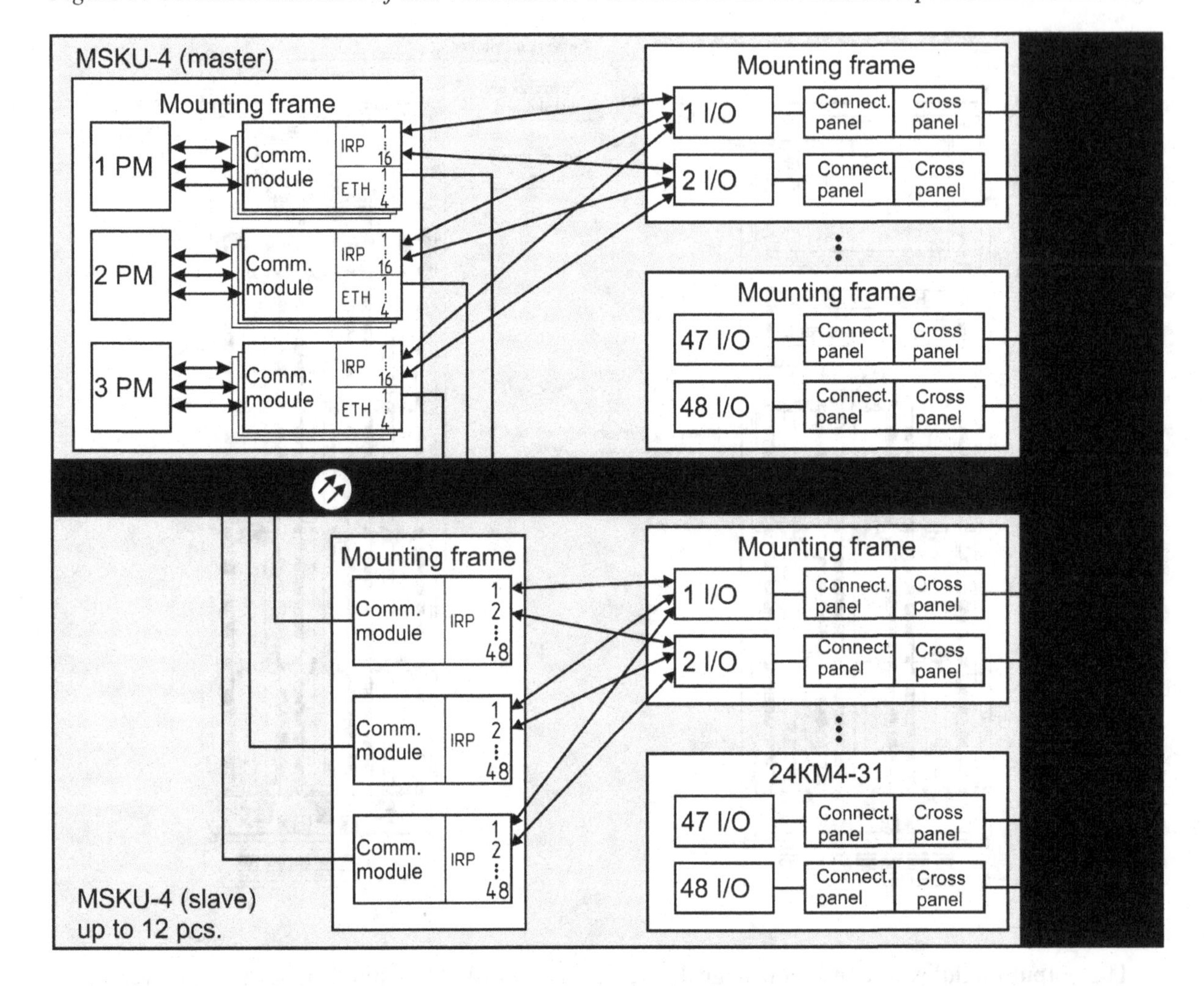

In the "modular" version, the processor unit contains one, two or three processor modules, a control module and up to nine central interface modules installed in the mounting frame. An expansion unit can have one to three I/O modules of any type, located in a small mounting frame combined with connection panels for connecting to external circuit modules. The master cabinet has one processor unit and up to 24 expansion units. I/O modules are connected to the IRP ports of the central interface modules of the processor unit. Each slave cabinet has a connection unit containing one, two or three peripheral interface modules and up to 24 expansion units. I/O modules located in the expansion units are connected to the IRP ports of the peripheral interface modules. Ethernet ports 100BASE-FX of the same modules provide connection of the slave cabinet to the central interface modules of the processor unit located in the master cabinet. Thus, this architecture allows up to 12 slaves to be connected to the master cabinet, with the maximum possible number of I/O modules in each cabinet being 72, and their total number in the CCC MSKU-4 reaches 936.

Architecture and basic configuration schemes of the «modular» version of MSKU-4 are presented in figures 8-10.

Figure 10. Connection scheme of three redundant I/O modules to three redundant processor modules

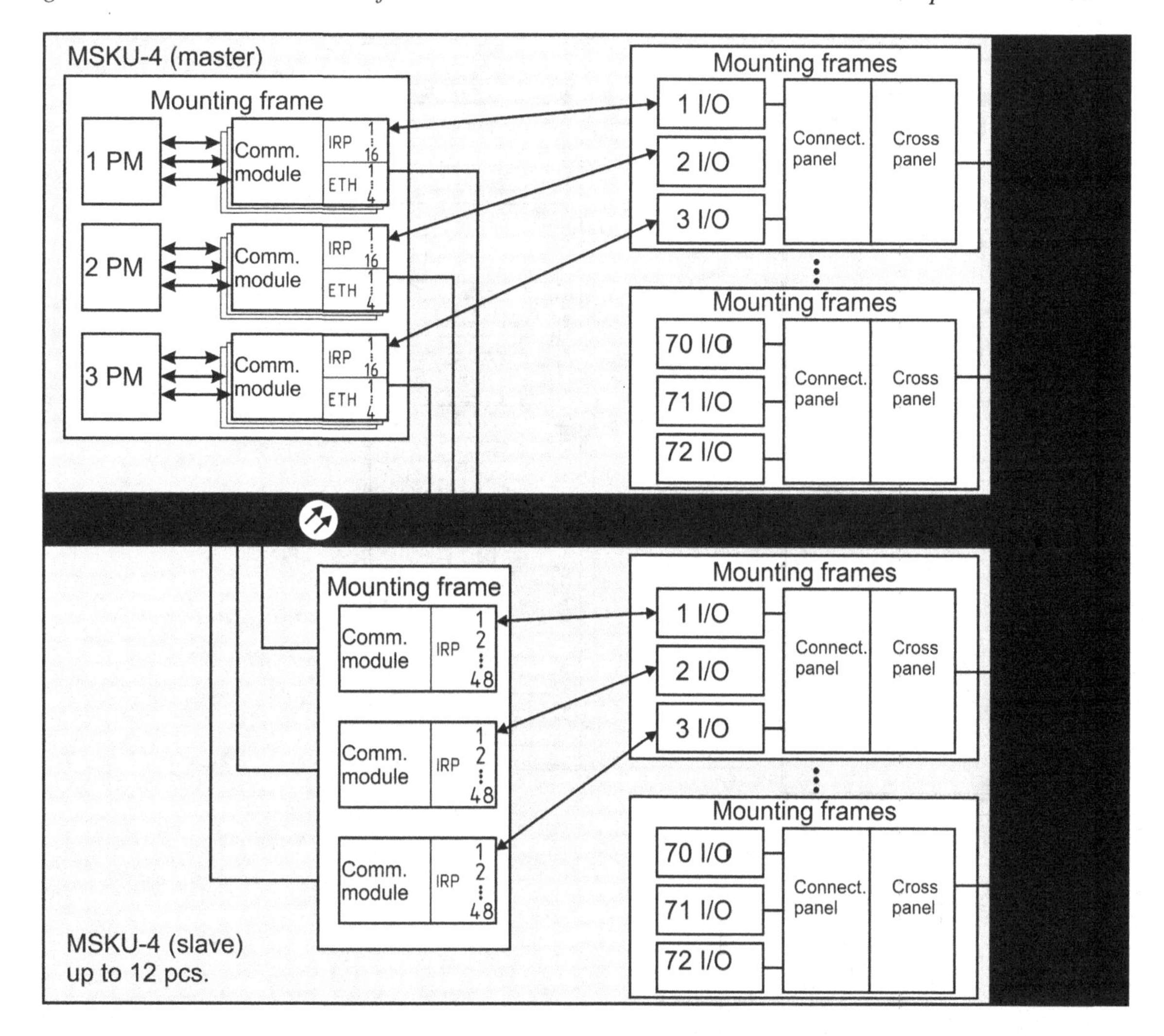

The PS5140 platform includes:

- Processor module;
- Uninterruptible power supply device;
- Display devices (monitors);
- Network hardware;
- Load-bearing structures: tables, pedestals, stands, cabinets.

Processor module is a compact computer based on Intel Core I5/I7 processor.

The uninterruptible power supply device is made as a unit 3U high for rack installation in a standard structural system. Protected against output short circuit. It has redundant power supplies units and batteries, working on the load in parallel and allowing "hot" replacement.

Figure 11. General view of PS5140: a) cabinet version, b) stand version

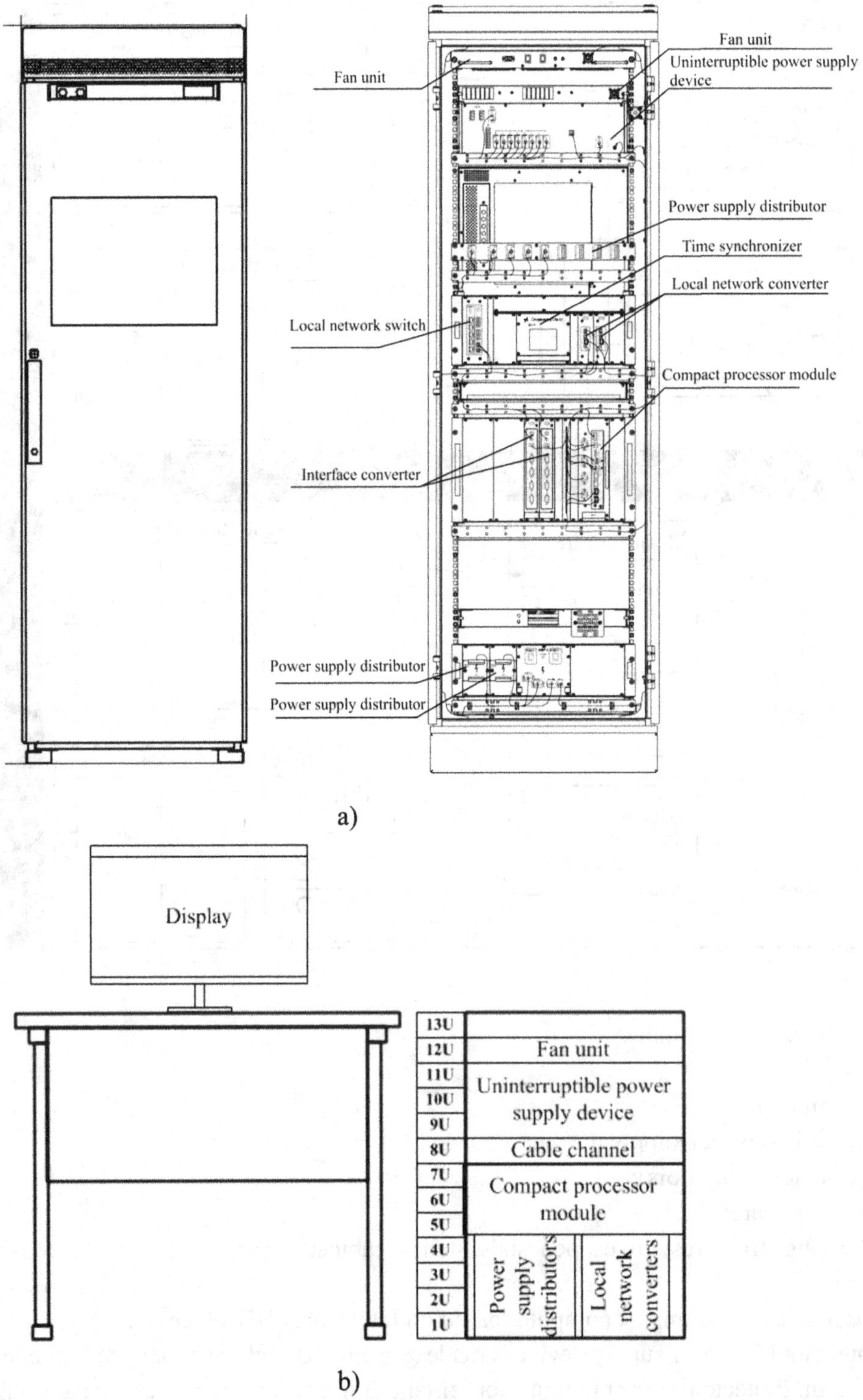

Fan unit
Uninterruptible power supply device
Fan unit
Power supply distributor
Time synchronizer
Local network converter
Local network switch
Compact processor module
Interface converter
Power supply distributor
Power supply distributor

a)

Display

13U
12U Fan unit
11U
10U Uninterruptible power supply device
9U
8U Cable channel
7U
6U Compact processor module
5U
4U
3U
2U Power supply distributors
1U Local network converters

b)

Liquid crystal monitors are used as display devices in the PS5140 – autonomous (desktop) and built-in. The multi-channel operator's console combines functions of a monitor, keyboard, and manipulator for servicing up to 8 processor modules.

Tables and pedestals are used to organize workplaces for operating personnel – placing of autonomous monitors, keyboards, manipulators, and the operator's console. Stands serve for installation of the processor module and uninterruptible power supply. The cabinet allows the installation of one or more processor modules, a built-in monitor, network equipment, uninterruptible power supply devices.

Architecture and basic configuration schemes of PS5140 are presented in figure 11.

The operator's stations use software developed as part of the PS5140 platform, verified and tested in the operation of I&C systems important for the safety of NPPs.

Technical specifications of the MSKU-4 computing complexes and PS5140 workstations are agreed upon by the nuclear regulatory body of Ukraine, the technical means and typical (representative) samples of the MSKU-4 and PS5140 have passed all the tests provided by manufacturer and specialized laboratories according to the methods agreed upon with the regulatory body.

Test results confirmed the ability to create MSKU-4 computing complexes based on the developed platforms with the functions of safety (category A) and normal operation (categories B and C), and PS5140 workstations with the functions of normal operation (categories B and C), in compliance with the requirements of national rules and standards on nuclear safety. The layout, manufacture and factory tests of each MSKU-4 and each PS5140 supplied to the NPP are carried out according to a separate specification agreed with the end user, the operating organization and the nuclear regulatory body. Delivery of equipment to NPP's is permitted based on the results of factory tests confirming its compliance with the requirements of the specification.

Structural solutions, hardware and software, proposed and implemented during the creation of the MSKU-4 and PS5140 platforms, have been tested for their application in safety and normal operation systems (since 2008, more than ten different systems at Ukrainian NPP's).

VULCAN/VULCAN-M I&C Platforms

VULCAN / VULCAN-M – platforms, developed by the Ukrainian Company LLC "Westron" and used to create I&C systems for nuclear enterprises and other industries (WESTRON, 2017). Platforms VULCAN / VULCAN-M:

- Allow you to create software and hardware systems of various architectures and dimensions, to the maximum extent possible taking into account the requirements of users.
- Are based on modern approaches to the construction of distributed hierarchical systems and take into account trends in their development.
- Are based on the long-term experience of LLC "Westron" in the automation of industrial facilities.

SHC has two levels of hierarchy. Means of the lower level carry out the input of electrical continuous and discrete signals from sensors, the primary processing of the received information, the formation and issuance of commands to actuators and alarm elements. Top-level tools produce calculations of enterprise level, archiving, displaying and recording current and archived data, receiving commands and directives from personnel, preparing and issuing information for other (paired) systems.

Additional functions include continuous monitoring of the technical condition of the equipment, diagnosis of defects, display and delivery of diagnostic messages.

The main units of aggregation in the layout of the I&C systems are:

- Subcomplexes of data collection and processing (SCSO) based on the VULCAN-M platform, interfaced with peripheral devices and forming the lower level of the system.
- Workstation subcomplexes (SCRS) based on the VULCAN platform, directly related with operational staff and forming the upper level of the system.

Subcomplexes have a variable composition of equipment and are composed of a set of technical means of the corresponding platform. The composition of the subcomplexes and the layout of each of them are determined by the supplier and / or customer, taking into account the safety class of the system, the functions performed, the number and types of input and output signals, the appointment and placement of workplaces for operational personnel, the organization of data exchange with paired systems, the need for redundancy and other source requirements, and is fixed in the technical specifications for the manufacture and delivery of each subcomplex. Basic software is supplied with subcomplexes, and application software is developed by the supplier, customer, or end user.

The VULCAN-M platform incorporates a set of built-in functional elements (modules) made on sliding Eurocard circuit boards.

The processor module has one or two (controlling each other) microprocessors, operational, permanent and programmable read-only memory chips, 4 independent RS-485 trunk controllers. External computer equipment can be connected to the processor module using duplicated Fast / Gigabit Ethernet and / or RS- 485 channels.

Input modules carry out galvanic separation, indicate the status of input circuits, convert and output values of information parameters via the RS-485 interface line:

- 16 discrete signals in the form of "low" and "high levels of DC voltage.
- 16 discrete signals in the form of switching "dry" contacts.
- 8 continuous DC signals.
- 8 continuous signals of direct current voltage.
- 8 signals from RTD's and thermocouples.

The output modules receive from the RS-485 interface bus, convert, galvanic isolation, indicate the status of the output circuits and issue:

- 16 discrete signals represented by the states of switching proximity keys.
- 6 continuous DC signals.

The conversion module multiplies the input continuous DC signal into 6 directions without changing or with changing the informative parameter of the input signal. Message receiving / issuing modules act as gateways between the RS-485 interface trunk and local networks.

The diagnostic module checks the technical condition and functioning of the SCSO, issues diagnostic messages via the RS-485 interface line and signals the results to the personnel.

Data is exchanged between the processor module and the rest peripheral modules via the main and backup RS-485 lines, which are connected to the ports built into each module. Both highways operate simultaneously; if one of them fails, the other remains operational without any degradation of performance. Two (primary and backup) processor modules and up to 60 peripheral modules located in the same crate, in different crates of the same cabinet and / or different cabinets, including those remote up to 200 m away (using fiber optic, can be connected to one trunk) lines - up to 2000 m). The transmission speed on the RS-485 highway of single elements of a digital data signal is up to 4 Mbit / s.

Data transmission on the RS-485 trunk is done in separate packages, periodically or on request. The broadcast method of transmission is used: a module that has gained access to the backbone, during the allotted (guaranteed) time interval, gives it the whole set of data it generates, while all other modules connected to the same backbone receive this data and select only those which applies to each of them.

For interfacing SCSO with external computer equipment, other SCSO and / or SCRS up to 24 independent galvanically separated RS-485 trunks and / or one or two duplicated pairs of galvanically isolated Fast / Gigabit Ethernet trunks can be organized (transmission distance - up to 100 m via an electric cable and up to 2 km when using an optical fiber line, transmission speed of single elements digital data signal - 100 Mbps or 1000 Mbps).

It is possible to backup the main functions, including input of continuous and discrete signals and digital messages, data processing, output of continuous and discrete signals, primary and secondary power supply. Figure 12 shows an example of the structure of an operational-autonomous device based on the VULCAN-M platform, intended for use in a safety system. SCSO can consist of one such device or have one, two or three operational and autonomous devices made in a floor or wall cabinet (cabinets) in which peripheral modules are installed.

The VULCAN platform allows you to compose SCRS (operator and engineering workstations, computing servers, archiving / documentation servers, etc.) that implement normal operation functions related to categories B and C: data exchange with SCSO and adjacent power unit systems, processing, formation and data output to adjacent systems and / or SCS, archiving, display and registration of current and archive information, support for the human-machine interface. Additional functions include checking the technical condition of equipment and the functioning of SCRS, displaying and issuing diagnostic messages, and signaling malfunctions. LLC “Westron” uses purchased industrial computers and peripheral computer equipment, as well as devices and supporting structures of its own manufacture that are part of the VULCAN platform for linking workstations and SCRS servers:

- embedded industrial computers, switches / routers, electro-optical and optical-electric converters of transmitted digital signals;
- built-in or operational-autonomous video monitors, including those with a touch screen;
- built-in modules for receiving / issuing messages and diagnostic modules made on sliding Eurocard circuit boards;
- operational-autonomous industrial keyboard, mouse (or trackball), laser printer;
- built-in sources of secondary power supply of SCSR components and external devices, power distributors, fan blocks, terminal blocks and / or optical connectors for connecting / crossing external circuits;
- floor cabinets with degree of protection IP20 or IP54, tables and cabinets.

Figure 12. The structure of the VULCAN/VULCAN-M cabinet for use in a safety system

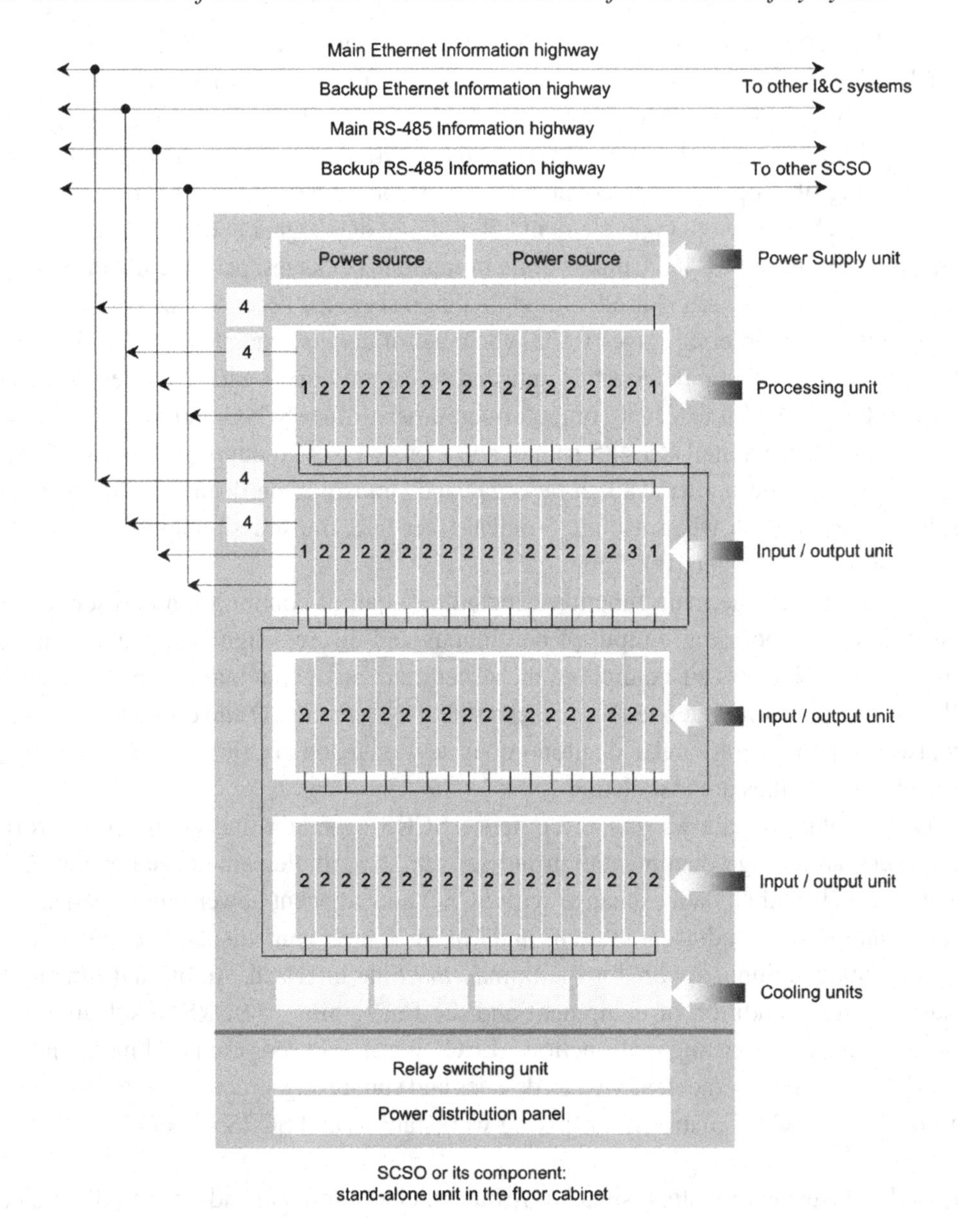

1 - WPB board: receiving, processing, storage and delivery of data, equipment operation management.
2 - WAI board: input of continuous signals of direct current and voltage (low and medium level), or
WDI board: input of discrete signals in the form of DC voltage levels, or
WDO board: output of discrete signals for switching DC or AC circuits, or
WRT board: input of continuous signals from resistance temperature converters, or
WCI board: input of discrete signals in the form of switching "dry" contacts, or
WAO board: output of continuous DC signals.
3 - WCB board: equipment condition monitoring.
4 - Media converter.

Figure 13. The design of operating stand-alone components of SCSO and SCRS

Control system for a backup diesel-electric station of the Armenian NPP-2

Data exchange between SCRS components, adjacent systems and external computer equipment is carried out via one or two duplicated pairs of galvanically isolated Fast Ethernet trunks (the total number of connection points is defined by user, the transmission distance is up to 100 m via an electric cable and up to 2 km when using fiber optic lines, the maximum transmission speed of single elements of a digital data signal is 100 Mbps or 1000 Mbps). Data exchange with SCSO is carried out along the RS 485 trunks (the total number of connection points is up to 62, the transmission distance is up to 100 m). For remote (up to 2 km) connections, electro-optical and optoelectric converters of interface signals are used.

The test results confirmed the ability to create, on the basis of the developed platforms, SCSO subcomplexes that perform safety functions (categories A) and normal operation (categories B and C), and SCRS subcomplexes that perform normal operation functions (categories B and C), in accordance with the requirements of national standards, rules and standards on nuclear safety.

The first systems implemented by LLC "Westron" at Ukrainian NPPs were created on the basis of the WDPF platform, the equipment of which was supplied by the developer of this platform (Westinghouse Electric Company). Then LLC "Westron" developed, mastered its own production and began using

VULCAN / VULCAN-M platforms, and implemented the first system based on them that controls the supply water level in the steam generators of Unit 3 of the South Ukrainian NPP (a similar system was subsequently introduced into operation at Unit 2 of the Armenian NPP). The first application of the VULCAN-M platform in the safety system was ESFAS 3 of Unit 1 of the South Ukrainian NPP. Currently, such systems are operated at units 1 and 2 of the same NPP. The VULCAN platform is used in radiation monitoring systems of Rivne NPP, Zaporizhzhya NPP, South Ukrainian NPP and Dukovany NPP).

On the basis of VULCAN/ VULCAN-M platforms, control systems for standby diesel-electric stations are operated on Ukrainian NPPs and Unit 2 of the Armenian NPP (Figure 13).

SOLUTIONS AND RECOMMENDATIONS

Chapter 15 is devoted to new platforms (equipment families) which have gained wide popularity for NPP I&C (and not only for NPP):

- TELEPERM XS (Siemens Energy, Germany / AREVA NP, France).
- ALS (Westinghouse Electric Company, USA).
- Spinline & Hardline (Rolls-Royce Civil Nuclear, UK).
- FirmSys & FitRel (China Techenergy Co., Ltd., China).
- POSAFE-Q (POSCO ICT Co., Republic of Korea).
- MSKU-4 / PS 5140 (SRPA "Impulse", Ukraine).
- VULCAN/VULCAN-M (LLC "Westron", Ukraine).

A significant part of the main developers of I&C NPP systems from different countries carries out its development on the basis of its own platforms (as for the creation new units, as for modernization (in particular, for the life extension of NPP units.

This trend, apparently, will continue in the future for NPP units.

The question, how much this trend will continue for Small Modular Reactors (SMR) and whether they will use existing platforms or develop new ones - will be addressed in the near future. In any case, we recommend to use big experience of platforms creation for SMR.

FUTURE RESEARCH DIRECTIONS

1. Wider application of complex electronic components based on the Hardware Description Language (HDL). These include Application Specific Integrated Circuits, Complex Programmable Logic Devices, and FPGAs.
2. This will allow the transfer of the basic information and control functions to the level of functional elements ensuring, the visibility of the implementation and full testability, and increase in the noise immunity.
3. Optimal decomposition of functions between different levels of disaggregation (functional elements, blocks, devices), taking into account their importance for safety and the capabilities of used element base.

4. Most platforms provide for the exchange of information between functional blocks through an EN local area network (Industrial ETHERNET). Meanwhile, some developers prefer to use radial structures (point-to-point), optical signals, and fiber-optic message lines, which allows to reach immunity to interference and resistance to failure of elements.
5. Justification the appropriateness of using such structures and signals instead of the common ones proven by many years of experience in using local area networks with backbone and ring structures.
6. Distribution of requirements whose observance should be confirmed in the qualification (acceptance tests) of the platform itself and those that are checked of each sample assembled on the basis of this platform and intended for delivery to the customer (end user).
7. Confidence that the properties that were tested and confirmed in the qualification of the platform would be maintained in the subsequent replication of its components allows one to significantly simplify, speed up, and cheapen the acceptance tests for system.

CONCLUSION

Chapter 15 contains general information on the platforms (equipment families) as tools for developing a predetermined set of I&C systems, including NPP safety important systems. The properties of the platforms play a significant role in ensuring safety of such systems and the effectiveness of their design, manufacture, inspection, commissioning, and maintenance.

The ability to use hardware and software tools developed as part of the corresponding platform instead of disparate products from different suppliers, often incompatible with each other and not always meeting the requirements of customers (users), provides high efficiency and the required quality of development, manufacture, factory testing of components, integration, and verification of the systems at the place of operation. The main requirements for platforms are functional, structural, and design completeness, information, energy, constructive, and operational compatibility, as well as availability of software, metrological, regulatory, guidance, information, and other types of support. In accordance with the customers' requirements, one and the same platform can be used to develop I&C systems that differ in composition, structure, and technical characteristics. The use of service equipment, software, and documents included in the platform allows minimizing the complexity of the design, integration, verification, commissioning, and maintenance of the SHC and avoiding possible errors.

Since their first appearance, several generations of platforms have changed, which took into account the subsequent toughening of the requirements in national and international regulations, rules, and standards for systems and their components and, on the other hand, the opportunities for improving and technical characteristics and improvement of digital technology (increasing the degree of integration of available user-programmable general-purpose microcircuits and electronic components using local area networks, optoelectronic data transfer channels, etc.

ACKNOWLEDGMENT

The authors of this chapter would like to recognize Prof. Michael Yastrebenetsky, State Scientific and Technical Center for Nuclear and Radiation Safety, Ukraine, who was instrumental in the researching and writing of this chapter.

REFERENCES

Bach, J., Duthou, A., Monteil, P., & Burzynski, M. (2017). Computer security approach for ROLLS-ROYCE SPINLINE safety platform Rolls-Royce Civil Nuclear. NPIC&HMIT 2017, San Francisco, CA.

Chen, Y. (2017). *The Application of FPGA-based FitRel Platform in Nuclear Power Plant Diverse Actuation System.* China Techenergy Co., Ltd (CTEC).

Didenko, K., & Rozen, Yu. (1985). MicroDAT. Principles of construction, the main parameters and characteristics. *Instrumentation & Control Systems*, 11.

EPRI. (1996). *TR 107330, Generic Requirements Specification for Qualifying a Commercially Available PLC for Safety Related Applications in Nuclear Power Plants*. Palo Alto, CA: Electric Power Research Institute.

Grabenko, A., Bakhmach, E., Siora, A., & Tokarev, V. (2005). Cabinets modernized in a unified complex of technical equipment UCTM -MD, RT-MD. New developments of control and protection systems for PWR reactors. *Proceedings of the 4th international scientific and technical conference "Ensuring the Safety of NPPs with PWR"*.

IEC. (1991). *60821, VME bus – Microprocessor system bus for 1 byte to 4-byte data.*

IEC. (2006). *60880, Nuclear power plants – Instrumentation and Control Systems important to Safety – Software aspects for computer-based systems performing category A functions.*

IEC. (2008). 60297-3-100, Mechanical structures for electronic equipment – Dimensions of mechanical structures – Basic dimensions of front panels, subracks, chassis, racks and cabinets.

IEC. (2009a). *61226, Nuclear power plants – I&C systems important to safety – Classification of I&C functions*.

IEC. (2009b). *62003, Nuclear power plants – Instrumentation and control important to safety – Requirements for electromagnetic compatibility testing*.

IEC. (2011). *61513, Nuclear power plants – Instrumentation and control for systems important to safety – General requirements for systems*.

IEC. (2012a). *60987, Nuclear power plants – Instrumentation and control important to safety – Hardware design requirements for computer-based systems.*

IEC. (2015). *62808, Nuclear power plants – Instrumentation and control systems important to safety – Design and qualification of isolation devices*.

IEC. (2019). *TR 63084, Nuclear power plants – Instrumentation and control systems important to safety-Qualification of platforms for systems important to safety.*

IEEE. (2003). *Std. 323, Qualifying Class 1E equipment for nuclear power generating stations*. Piscataway, NJ: IEEE.

IEEE. (2004a). *Std. 344, IEEE Standard for Seismic Qualification of Equipment for Nuclear Power Generating Stations*. Piscataway, NJ: IEEE.

IEEE. (2004b). *Std. 1012, Software Verification and Validation*. Piscataway, NJ: IEEE.

IEEE. (2010). *Std. 7-4.3.2, Standard criteria for digital computers in safety systems of nuclear power generating stations*. Piscataway, NJ: IEEE.

Kyun, L. M., Whan, S. S., & Yun, D. H. (2009). *Development and Application of POSAFE Q PLC Platform.* Retrieved from: https://inis.iaea.org/collection/NCLCollectionStore/_Public/43/130/43130436.pdf?r=1&r=1

Panfilo, F. (2011). *Westinghouse Nuclear Automation. Digital Feed Water Upgrades Experience, Validation & Lessons Learned.* Retrieved from: http://fsrug.org/Presentations%202011/Westinghouse%20-%20Ovation%20FW%20Presentation.pdf

RG. (1988). *1.100. Seismic Qualification of Electric and Mechanical Equipment for Nuclear Power Plants*. Washington, DC: U.S. NRC.

Rolls-Royce. (2018). *Technical sheet civil nuclear – systems – instrumentation & control. Hardline. Non-programmed safety I&C platform dedicated to nuclear safety.* Retrieved from: https://www.rolls-royce.com/~/media/Files/R/Rolls-Royce/documents/customers/nuclear/UK_Hardline.pdf

Rolls-Royce. (2019a). *Technical sheet civil nuclear – systems – instrumentation & control. Spinline. Modular I&C digital platform dedicated to nuclear safety.* Retrieved from: https://www.rolls-royce.com/~/media/Files/R/Rolls-Royce/documents/customers/nuclear/UK_Spinline.pdf

Rolls-Royce. (2019b). *Technical sheet civil nuclear – systems – instrumentation & control. Rodline. The most-used digital Rod Control System in the world.* Retrieved from: https://www.rolls-royce.com/~/media/Files/R/Rolls-Royce/documents/customers/nuclear/UK_Rodline.pdf

TR. (2013). *Advanced Logic System Topical Report.* Rockville, MD: U.S. NRC.

Westinghouse. (2018). *Westinghouse COMMON Q Platform.* Retrieved from: http://www.westinghousenuclear.com/Portals/0/operating%20plant%20services/automation/safety%20related%20platforms/NA-0113%20Common%20Q%20Platform.pdf

WESTRON. (2017). *The VULCANVULCAN / VULCANVULCAN-M hardware platform for the control systems of nuclear power plants and thermal power plants.* Retrieved from: http://www.westron.kharkov.ua/files/Vulcan-Vulcan-M_Presentation_Rus.pdf

Xu, Z., & Jiang, D. (2016). A New ALS Based PMS Design and Its Evaluations. *World Journal of Nuclear Science and Technology, 6*(1).

ADDITIONAL READING

AREVA NP. (2012). *Published and copyright. Instrumentation and Control. TELEPERM XS. System Overview.* Retrieved from: http://www.us.areva-np.com/pdf/npc/TELEPERM%C2%AE%20XS%20System%20Overview.pdf

Virag, O., & Pironkov, I. (2009). *Westinghouse New Build Program & Program is I&C Solutions. BULATOM Conference.*

World Nuclear News. (2016). *Chinese I&C system passes IAEA review. World Nuclear Association.* Retrieved from: https://world-nuclear-news.org/C-Chinese-IC-system-passes-IAEA-review-1507164.html

Zhang, C. (2015). *Simplicity and Application simplicity in FitRel R&D.* China Techenergy Co. Ltd.

Chapter 16
FPGA Technology and Platforms for NPP I&C systems

Andriy Kovalenko
https://orcid.org/0000-0002-2817-9036
Centre for Safety Infrastructure-Oriented Research and Analysis, Kharkiv National University of Radio Electronics, Ukraine

Ievgen Babeshko
https://orcid.org/0000-0002-4667-2393
Centre for Safety Infrastructure-Oriented Research and Analysis, National Aerospace University KhAI, Ukraine

Viktor Tokarev
Research and Production Corporation Radiy, Ukraine

Kostiantyn Leontiiev
Research and Production Corporation Radiy, Ukraine

ABSTRACT

This chapter describes an element base of new generation for NPP I&C, namely field programmable gate array (FPGA), and peculiarities of the FPGA application for designing safety critical systems. FPGA chips are modern complex electronic components that have been applied in nuclear power plants (NPPs) instrumentation and control systems (I&CSs) during the last 15-17 years. The advantages and some risks caused by application of the FPGA technology are analyzed. Safety assessment techniques of FPGA-based I&CSs and experience of their creation are described. The FPGA-based platform RadICS and its application for development of NPP I&CS is described.

DOI: 10.4018/978-1-7998-3277-5.ch016

INTRODUCTION

Nuclear power production is a safety-critical process that requires reliable and safe operation. I&CS of NPP play key roles in stable operation ensuring and therefore should be designed in accordance with international requirements on nuclear and operational safety.

One of the contemporary trends is dynamically growing application of novel complex electronic components, particularly, FPGAs in NPP I&CSs and other critical areas. I&CSs based on FPGA technology have been developed and applied in aerospace and process industries since the 1990s. Although the use of FPGAs in NPPSs I&Cs has lagged behind in the past compared to other industries, there is an increasing number of FPGA installations in operating NPPs.

BACKGROUND

The FPGA technology offers an alternative to microprocessor (or computer) technologies and to other types of programmable logic devices. Physically, FPGA is a semiconductor-based complex programmable device which can be configured to perform a required function.

It includes two entities: an FPGA chip, which is a piece of hardware that can be qualified against hardware qualification testing requirements, and the electronic design of the FPGA, implemented into a chip, represented by a set of instructions in hardware description language (HDL) that can be verified against functional requirements.

Problem is development of new approach to assessment of FPGA-based systems for NPP I&CSs taking into account of the FPGA technology features.

FEATURES OF FPGA TECHNOLOGY AND THEIR APPLICATION IN NPP I&CS

This subsection provides results of comprehensive analysis of FPGA technology features, including approaches to implementation of FPGA-based development activities, as well as possible applications of the technology.

Features of FPGA Technology

There are two lines in contemporary programmable logic arrays (Kharchenko, V. S., Sklyar, V. V. (Ed.), 2008; Barkalov, A., Wegrzyn, M. et al., 2006): Complex Programmable Logic Devices (CPLD) and FPGA. CPLDs are a continuation of programmable matrix logic line, whereas FPGAs continue basic matrix crystal line. The desire to combine the advantages of both line led to development of combined architecture VLSIs (see Figure 1). Still, all contemporary FPGAs possess such architecture. We shall discuss FPGAs of APEX II family produced by Intel as representatives of combined architecture.

Figure 1. Architecture types of FPGA

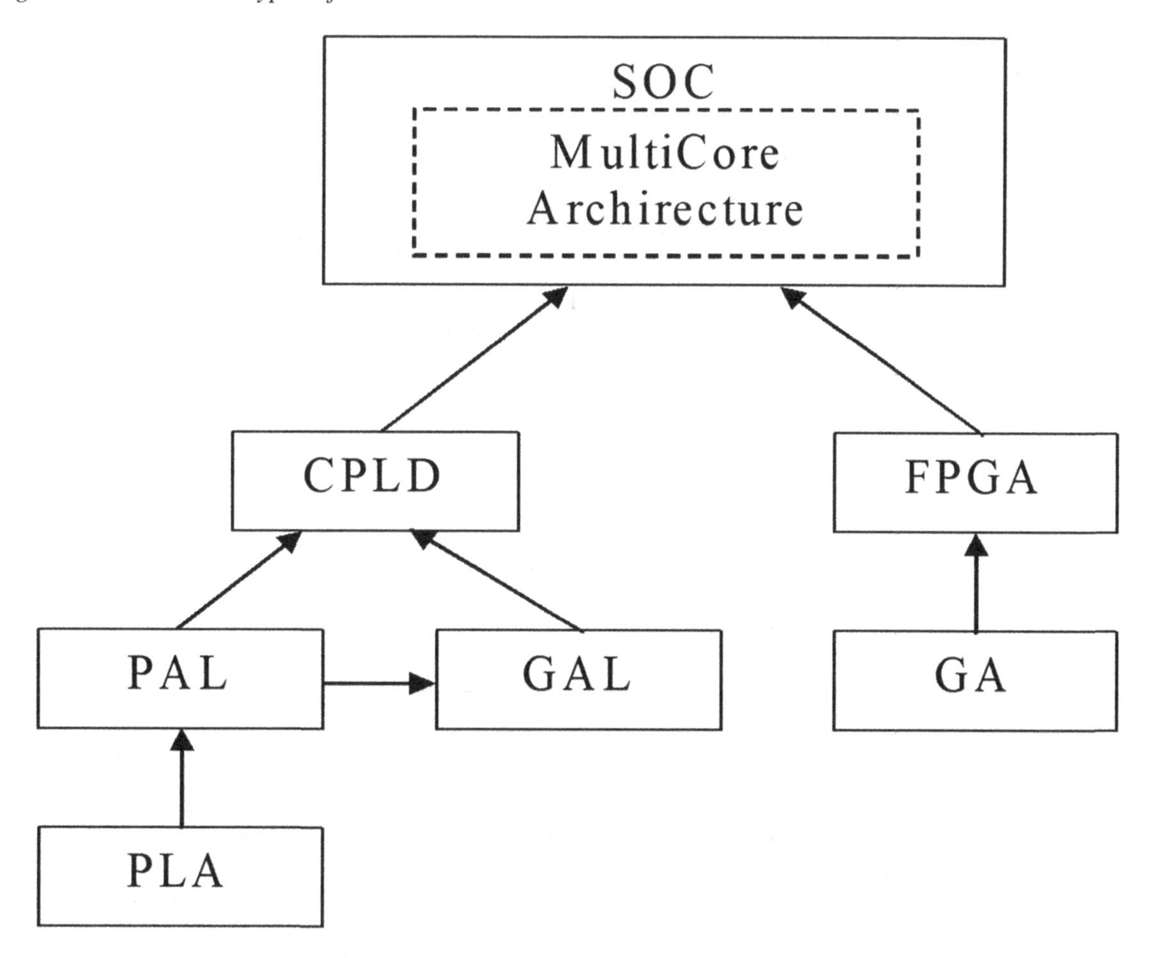

CPLD architecture has its origins in Programmable Arrays Logic (PAL) preceded by Programmable Logic Arrays (PLA) and from Generic Arrays Logic (GAL). Its functional unit consists of microcells, each of them performing some combinatory and/or register functions. Functional logic within the block is a matrix of logic products (terms). A subset of terms may be accessed by each macrocell via term distribution diagram. Switch matrix commutates the signals coming from outputs of the functional unit and I/O unit. As distinct from FPGA (segmented connections), CPLDs have a continuous system of connections (completely commuted connections).

FPGA architecture topologically originates from channeled Gates Arrays (GA). In FPGA internal area a set of configurable logic units is disposed in a regular order with routing channels there between and I/O units at the periphery. Transistor couples, logic gates NAND, NOR (Simple Logic Cell), multiplexer-based logic modules, logic modules based on programmable Look-Up Tables (LUT) are used as configurable logic blocks. All those have segmented architecture of internal connections.

System-On-Chip (SOC) architecture appeared due to two factors: high level of integration permitting to arrange a very complicated circuit on a single crystal, and introduction of specialized hardcores into FPGA. Additional hardcores may be:

- Additional Random Access Memory (RAM) units.
- JTAG interface for testing and configuration.
- Phase-Locked Loop (PLL) – frequency control system to correct timing relations of clock pulses as well as for generation of additional frequencies.
- Processor cores enabling creation of devices with a control processor and a peripheral.

FPGA resources and additional RAM are disposed at the processor address space. The examples of such solutions are the families of Intel Excalibur (Embedded Processor Programmable Solution), Atmel FPSLIC AT94 (Field Programmable System Level Integration Chip).

Tables 1-4 specify the main characteristics of modern FPGA chips from Intel (for more information see intel.com).

Table 1. Maximum resource counts forintel arria 10 GX Devices (GX 160, GX 220, GX 270, GX 320, and GX 480)

	Product line				
	GX 160	**GX 220**	**GX 270**	**GX 320**	**GX 480**
Logic Elements (LE) (K)	160	220	270	320	480
ALM	61,510	80,330	101,620	119,900	183,590
Register	246,040	321,320	406,480	479,600	734,360
M20K memory (Kb)	8,800	11,740	15,000	17,820	28,620
MLAB memory (Kb)	1,050	1,690	2,452	2,727	4,164
Variable-precision DSP Block	156	192	830	985	1,368
18 x 19 Multiplier	312	384	1,660	1,970	2,736
PLL (Fractional Synthesis)	6	6	8	8	12
PLL (I/O)	6	6	8	8	12
17.4 Gbps Transceiver	12	12	24	24	36
GPIO	288	288	384	384	492
LVDS Pair	120	120	168	168	222
PCIe Hard IP Block	1	1	2	2	2
Hard Memory Controller	6	6	8	8	12

Table 2. Maximum resource counts for intel arria 10 GX Devices (GX 570, GX 660, GX 900, and GX 1150)

	Product line			
	GX 570	**GX 660**	**GX 900**	**GX 1150**
Logic Elements (LE) (K)	570	660	900	1,150
ALM	17,080	251,680	339,620	427,200
Register	868,320	1,006,720	1,358,480	1,708,800
M20K memory (Kb)	36,000	42,620	48,460	54,260
MLAB memory (Kb)	5,096	5,788	9,386	12,984
Variable-precision DSP Block	1,523	1,687	1,518	1,518
18 x 19 Multiplier	3,046	3,374	3,036	3,036
PLL (Fractional Synthesis)	16	16	32	32
PLL (I/O)	16	16	16	16
17.4 Gbps Transceiver	48	48	96	96
GPIO	696	696	768	768
LVDS Pair	324	324	384	384
PCIe Hard IP Block	2	2	4	4
Hard Memory Controller	16	16	16	16

Manufacturer guarantees pre-sale testing of 100% FPGAs.

Unlike projects based on Applications Specific Integrated Circuit (ASIC), which have a fixed architecture (fixed IC outlets, IC functionality cannot be altered), FPGAs are reconfigurable.

FPGAs (when the system is fed) are configured by data stored at the configuration device or by those supplied from the system controller. Intel CASE-tools enable programming of devices within the system. They configure FPGA with a consecutive flow of data.

Moreover, FPGAs comprises an optimized interface using microprocessors for serial or parallel, synchronous or asynchronous configuration of those devices. This interface also enables microprocessors to interpret FPGAs as memory and to configure them by recording to memory cell virtual address, thus facilitating the reconfiguration process.

FPGAs consist of several MegaLAB structures. Each MegaLAB comprises 16 LAB – Logic Arrays Blocks, one built-in systemic memory – Embedded System Block (ESB) and a MegaLAB connection which routes signals within the structure between MegaLAB structures and I/O leads via FastTrack connection. Besides, LAB signal fronts may be controlled via local connection using I/O leads.

Each LAB comprises ten Logic Elements (LE), auxiliary transfers of logic elements and stage circuits, LAB control signals and a local interconnection that transmits signals between LEs in the same or an adjacent LAB as well as to IOE or ESB cells.

Table 3. Maximum resource counts for intel arria 10 GT devices

	Product line	
	GT 900	**GT 1150**
Logic Elements (LE) (K)	900	1,150
ALM	339,620	427,200
Register	1,358,480	1,708,800
M20K memory (Kb)	48,460	54,260
MLAB memory (Kb)	9,386	12,984
Variable-precision DSP Block	1,518	1,518
18 x 19 Multiplier	3,036	3,036
PLL (Fractional Synthesis)	32	32
PLL (I/O)	16	16
	72	72
25.8 Gbps Transceiver	6	6
GPIO	624	624
LVDS Pair	312	312
PCIe Hard IP Block	4	4
Hard Memory Controller	16	16

CASE Quartus II compiler places the project-associated logic within LABs or LAB auxiliary blocks, thus permitting usage of quick-acting local interconnections to increase the capacity of system designed. APEX FPGAs use LAB intermittent structure in such a way that each LAB is able to control two areas of local interconnections.

Each LAB structure may control thirty LEs using quick-acting local interconnections. Figure 2 shows LAB structure for APEX II family FPGAs.

Each LE controls left or right area of local interconnections intermittent by LEs, whereas a local interconnection controls LEs within its own LAB or adjacent LABs. This property optimizes the number of available lines and columns of interconnections, thus ensuring their high flexibility.

Each LAB comprises a predestined logic to manage control signals to LEs and ESBs. Control signals may be timing, timing enable, asynchronous reset, pre-installation and loading, synchronous cleanup and synchronous boot. Maximum six signals may be passed simultaneously. Though synchronous boot and cleanup signals are mainly used for counter realization, they may perform other functions as well.

LE is the smallest part of logic in APEX II architecture. Each LE contains a LUT conversion table with four inputs that serves as a functional generator able of quick realization of any four-variables function. Besides, each LE includes a programmable register and transfer and staging circuits.

Table 4. Maximum resource counts for intel arria 10 SX devices

	Product line						
	SX 160	**SX 220**	**SX 270**	**SX 320**	**SX 480**	**SX 570**	**SX 660**
Logic Elements (LE) (K)	160	220	270	320	480	570	660
ALM	61,510	80,330	101,620	119,900	183,590	217,080	251,680
Register	246,040	321,320	406,480	479,600	734,360	868,320	1,006,720
M20K memory (Kb)	8,800	11,740	15,000	17,820	28,620	36,000	42,620
MLAB memory (Kb)	1,050	1,690	2,452	2,727	4,164	5,096	5,788
Variable-precision DSP Block	156	192	830	985	1,368	1,523	1,687
18 x 19 Multiplier	312	384	1,660	1,970	2,736	3,046	3,374
PLL (Fractional Synthesis)	6	6	8	8	12	16	16
PLL (I/O)	6	6	8	8	12	16	16
17.4 Gbps Transceiver	12	12	24	24	36	48	48
GPIO	288	288	384	384	492	696	696
LVDS Pair	120	120	168	168	174	324	324
PCIe Hard IP Block	1	1	2	2	2	2	2
Hard Memory Controller	6	6	8	8	12	16	16
ARM Cortex-A9 MPCore Processor	Yes	Yes	Yes	Yes	Yes	Yes	Yes

Each programmable register in LE may be configured to operate as a D, T, JK or SR trigger. Register timing and cleanup control signals may be accessed using global signals, general purpose I/O leads or any internal logic. To realize combinatory functions registers are omitted (routing performed without them), whereas LUT output controls LE output.

Each LE has two outputs that control such routing structures: local, MegaLAB or FastTrack interconnections. Each output may be operated irrespective from LUT or register output. For instance, LUT may operate one output, while register connects the other one. This property, called "register packing", permits application of register and LUT to realize unconnected functions.

LE may also realize register and non-register options of LUT output.

Figure 2. LAB structure of circuits of APEX II family

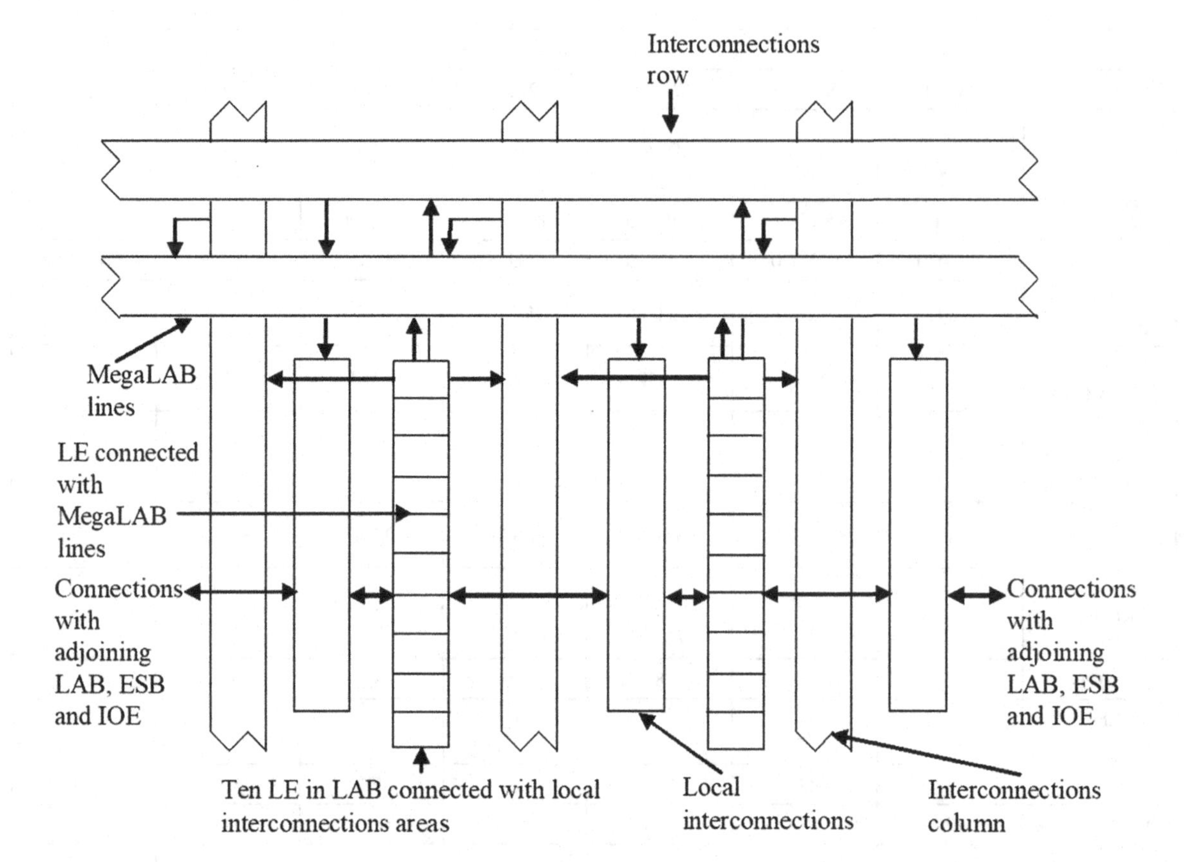

FPGA Technology: Development Tools

During development activities, the use of proven tools is preferred over manual methods. Moreover, for development of safety critical I&CSs, software-based tools shall be purchased only from long-established vendors with a good track record of CM, V&V, problem notification and resolution, including help & training materials. Software tools should be selected and evaluated before their using in lifecycle processes.

One of the tools that can be used in development activities for FPGA-based NPP I&CSs is IDE Quartus II. It supports design and implementation stages, including VHDL coding, RTL synthesis, Netlist synthesis, Placement&Routing, Static timing analysis, and Bitstream generation. It also supports hardware and functional block libraries VHDL design entry, graphical-based design entry methods, and integrated system-level design tools. It integrates design, synthesis, place-and-route, and verification into a development environment.

IDE Quartus II from Intel has a wide range of capabilities such as design entry, simulation, synthesis, verification, and device programming. Generally IDE Quartus II as FPGA design software is widely used in different industries: military, medical equipment manufacturing, automotive electronic manufacturing, financial, bioscience, etc.

I&CS Design Entry is performed according to I&CS Requirements Specifications. The desired circuit is specified either by means of a schematic diagram, or by using a Hardware description language, such as VHDL or Verilog. Inputs documents for design stage typically are: Requirements Specifications,

Electronic Design Architecture description, Electronic Design Detailed description. Results: VHDL files for HPDs.

Correctness of the design (Static Code Analysis) and its compliance with the requirements (Functional Testing) are verified after design phase. Software code verification is performed according to verification plan and testing plans which should be developed and approved before actions on software code verification.

Aldec Advanced Lint (ALINT™) tool can help to detect the design problems early in life cycle, including poor coding styles, improper clock and reset management, simulation, synthesis problems, poor testability and source code issues throughout the design flow. ALINT™ is a programmable design and coding guideline checker that speeds up development of complex system-on-chip designs. Certain rules may be parameterized to fine-tune custom checking policy. Policies combined with various ALINT settings allow development of unique rules checking framework for each design.

VHDL functional testing can be performed with ModelSim Intel tool, which uses either Verilog HDL or VHDL design files, including models for the library of parameterized modules and Intel megafunctions, to generate a functional simulation output of the design based on the set of stimulus applied by the user. Once the design is verified to be functionally correct, the next step is to perform implementation stage (synthesize the design and use the Quartus II software for place-and-route).

ModelSim-Intel software version is compatible with the specific Quartus II tool version. Proper verification of designs at the functional and post place-and-route stages using the ModelSim-Intel software helps ensure design functionality and, ultimately, a quick time-to-market.

Typically, implementation includes the following stages:

- Synthesis (bringing in) of project scheme may be effected using schematic editor (library of elements), hardware description language or automaton state flow graph editor. Functional modules may be developed by various tools, but the data obtained are then united into a single circuit list.
- Translation of data from Electronic Design Interface Format (EDIF) Native Generic Database (NGD) internal format.
- Crystal mapping, i.e. transformation of designed logic elements to their physical counterparts.
- Placement of physical element and routing of their interconnections. On this stage ModelSim tool is used for Verification of Netlist Files & Floor Plan Files generated by Quartus II tool (Logic Simulation, Timing Simulation, Static Timing Analysis).

Application of ModelSim to automate verification environment allows significantly decrease time of verification.

Compliance of developed FPGA-based I&CS with the required functionality can be verified with a testbed, which simulate inputs and allows to test outputs and performance. National Instruments LabView is an automated test software that provides with the tools to create any testing, measurement and control systems. It simplifies system design by offering access to the newest high-performance and low-cost entry points to the reconfigurable I/O platform made by National Instruments Corp., to one of the highest bandwidth vector signal analyzers on digitizers on the market, and to the latest off-the-shelf hardware.

Let us discuss Atera Quartus II tool in details. Quartus II has such basic functional possibilities: usage of hardware description language, project scheme input, compilation, logic synthesis, full timing and functional simulation, analysis of worst timing case, logic analysis, device configuration (Kharchenko, V. S., Sklyar, V. V. (Ed.), 2008).

Figure 3. Stages of tools-based FPGA-projects development

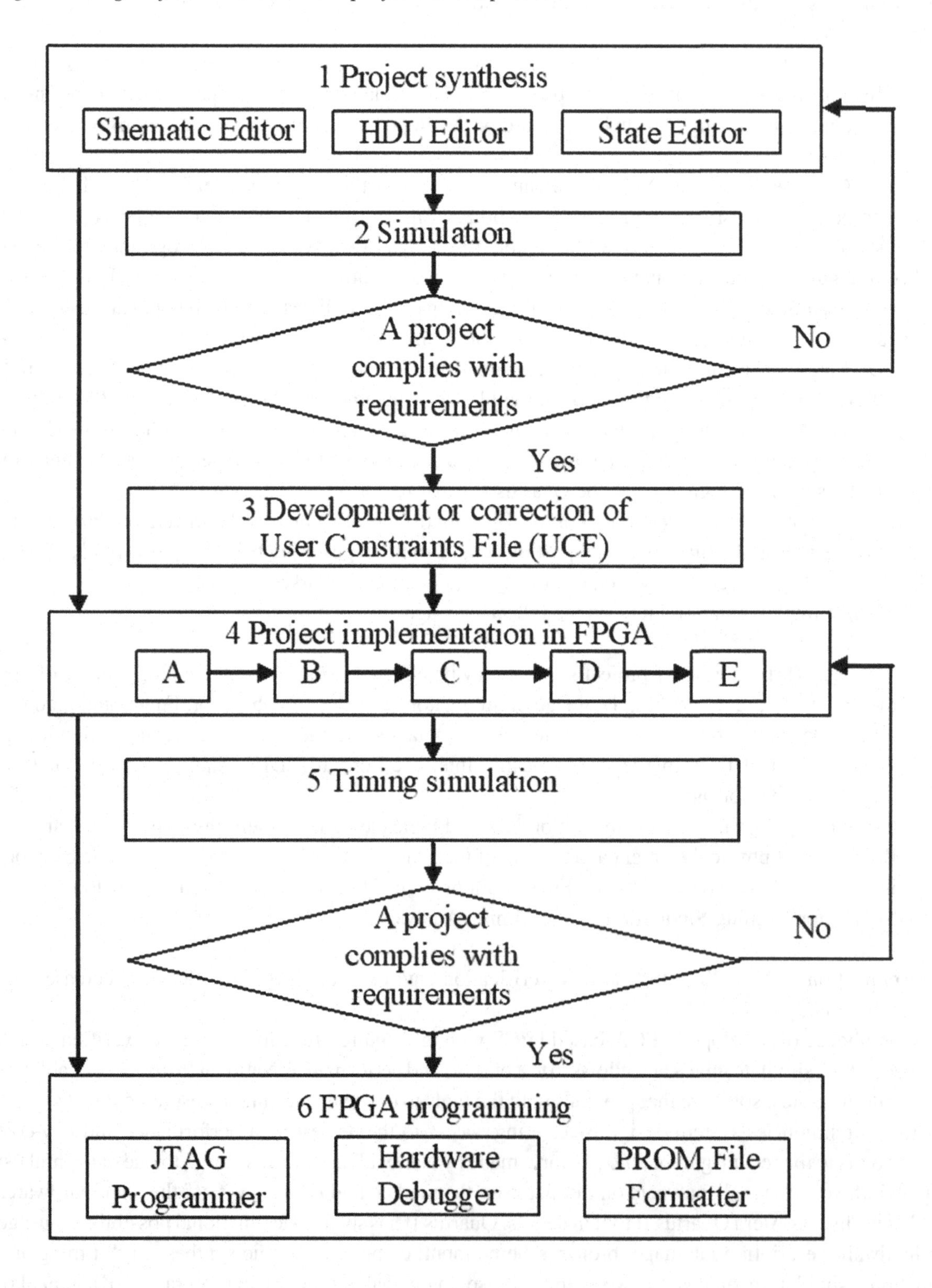

Quartus II includes LogicLock step design package, that permits laying destination of outputs and timing parameters, test functionalities and capacities of designed systems and then establish limitations in order to "lock" (fix) arrangement and characteristics of a specific logic block by applying LogicLock limitations.

The limitations as established by LogicLock functions ensure identical arrangement when logic block is performed within a current project or transferred to another project. LogicLock limitations may "lock" logic at a fixed position within the device. LogicLock may also specify a part of project logic for later optimization of its arrangement in an IC. Addition of logic to a project would not affect the properties of blocks "locked" LogicLock limitations.

The process of design from project synthesis to its realization in a crystal is fully supported by CASE-tools (Melnyk, A. et al., 2007; Tam, S., 2003) (Figure 3). The following is a short description of six design process stages.

1. Synthesis (bringing in) of project scheme may be effected using schematic editor (library of elements), hardware description language or automaton state flow graph editor.

Functional modules may be developed by various tools, but the data obtained are then united into a single circuit list.

2. Simulation is performed in order to test functioning of the project scheme with zero or single delays. Designer forms a diagram of input actions (test vectors).
3. Development or correction of User Constraint File implies description of requirements to arrangement of components and timing relations of signals using an appropriate editor.
4. Project implementation in FPGA includes:
    a. Translation of data from Electronic Design Interface Format (EDIF) Native Generic Database (NGD) internal format.
    b. Crystal mapping, i.e. transformation of designed logic elements to their physical counterparts.
    c. Placement of physical element and routing of their interconnections.
    d. Timing of circuit delays and bit-stream generation.
5. Project is verified by simulation when actual timing values of delays as determined in the crystal are considered instead of zero or single values.
6. Crystal programming permits its JTAG, debugging and PROM file formatting for programmer.

Quartus II presents a large number of library functions for design, including buffers, triggers and latches, I/O registers and logic primitives. A feature of library functions is low integration level.

Besides, Quartus II presents plenty of architecturally optimized macrofunctions that reflect the specifics of FPGA application. The whole set of macrofunctions may be presented by such categories of typical schemes: adders, arithmetic logic devices, buffers, comparators, cipherers, counters, decoders, digital filters, error detectors and correctors, coding-decoding circuits, frequency dividers, latches, multipliers, multiplexors, registers, shift registers, low integration level elements, I/O gates.

A Typical Life Cycle of FPGA-Based I&CS

In development of FPGA-based I&CS LC (Figure 4) it should be taken into consideration that, from the one side, FPGA-based digital devices are complex software-hardware products, thus having much in common with software. Therefore, in analysis and development of FPGA-based I&CS LC the postulates of software engineering standards are reasonably useful (Kharchenko, V. et al., 2004; Kharchenko, V. et al., 2001; Scott, J., Lawrence, J., 1994).

From the other side, FPGA electronic designs as a specific I&CS component have some peculiarities different from software. Therefore the postulates of existing standards regulating software structure cannot be mechanically taken for construction of FPGA-project LC.

Our analysis of software LC showed that for FPGA it should study the section of LC beginning from specification of an I&CS and up to its integration. Such actions may be performed in parallel to software development.

Besides, verification after each stage of development is obligatory for both FPGA electronic designs and software. Software verification is a process aimed to confirm software compliance to defined requirements by way of versatile tests and obtaining verifiable proofs (Lyu, M. R., 1996).

Figure 4. A life cycle of FPGA-based I&CS

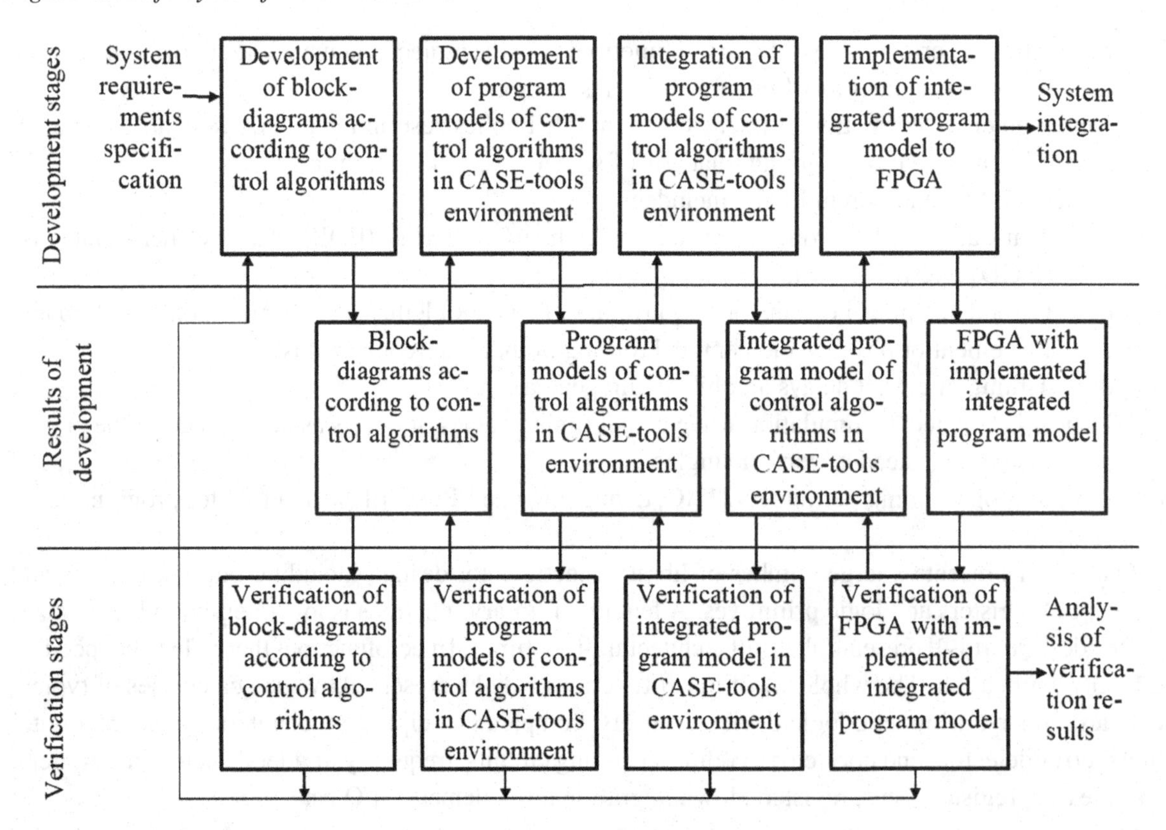

FPGA electronic design is developed on the basis of System Requirements Specification (SRS) document with due consideration of function distribution and non-functional safety requirements between

FPGAs and other hardware. The developed FPGA-based digital device must be integrated within the I&CS, and in future it should be treated as an integral part of I&CS's software and hardware.

Development of FPGA-based digital devices consists of the following stages:

- Development of signal formation algorithm block-diagrams.
- Development of signal formation algorithm program models in design environment which is determined depending on type and/or manufacturer of FPGA realization environment applied.
- Integration of signal formation algorithm program models (development of digital device integrated program model) into design environment.
- Implementation (loading) of integrated digital device model to FPGA.

The key term here is «signal formation algorithm block-diagram» implying a certain functionally finite project module presented in the form of a graphic diagram or a listing in hardware description language (HDL). The result of each step is a new product, the final result being a FPGA with implemented logic structure. At each step the developed product must be verified. The procedures of FPGA-based I&CS development and verification are shown in Figure 4. A description of FPGA-based I&CS LC stages is presented below.

Development of signal formation algorithm block-diagrams. Signal formation algorithm block-diagrams are developed as a direct preparation to development of a digital device block-diagram in CASE-tools (design) environment. Initial data are:

- SRS (functional general and non-functional – general safety requirements) with due consideration of function distribution between software and hardware.
- Process engineering requirements that may be formulated by customer as an addition to SRS requirements.

Initial information in requirements may be both in verbal form and in the form of formalized (problem oriented) languages describing function algorithms.

Signal formation algorithm block-diagrams are developed in the form as close as possible to scheme presentation in FPGA design environment and, as a consequence, should take into consideration the peculiarities of tools applied. For complicated digital devices one of critical issues is structure division into functional modules and formation of series and/or parallel tiers of such modules.

In case the requirements to functioning algorithms are presented in verbal form, at that stage such works may be consecutively performed:

- Development of description of functioning algorithms in a formalized language.
- Development of signal formation algorithm block-diagrams as adapted to current tools.

Development of signal formation algorithm block-diagram program models in CASE-tools environment. The initial data at that stage are signal formation algorithm block-diagrams. Development of signal formation algorithm block-diagram program models in design environment is performed using specialized CASE-tools comprising a library of typical functional elements and blocks.

In the course of development signal formation algorithms and FPGA logic structure are presented in the form of visualized conditional graphic images (block-diagrams). It should be noted that the development of signal formation algorithm block-diagram program models and FPGA program model is similar to the process of software product development in a problem oriented program language using specialized tools.

An alternative to direct diagram drawing may be FPGA structure description in a HDL, such as Verilog or VHDL.

HDL is a formalized record that may be used at all development stages. System function is defined as transformation of input values into output values, operation time in this transform being prescribed in explicit form. General FPGA structure is prescribed by a list of connected components – functional blocks that realize signal formation algorithm block-diagram program models.

At this stage some library modules lacking in standard tools library must be created. Such library modules in FPGA structure are called IP-Cores (Intellectual Property Cores) or IP-functions. Such modules are universal and reliably repetitive, from the one side, and capable of parametric adjustment to a particular project, from the other side. Repeated application of IP-Cores permits to reduce labor costs and design period of digital devices, ensuring their high reliability.

Integration of signal formation algorithm block-diagram program models in CASE-tools environment. At that stage signal formation algorithm block-diagram program models as developed at a previous stage in CASE-tools environment are integrated. An important issue in this is establishment of connections and sequences between developed functional blocks (signal formation algorithm block-diagram program models), including input and output signal formers, in strict conformity with the developed signal formation algorithm block-diagrams.

As well as at the previous stage, integration may be performed both in the form of graphic diagrams and by programming in hardware describing language. The result of this stage is a finite digital device program model ready to be implemented into FPGA chip.

Implementation of integrated program model to FPGA chip. The developed digital device logic structure program model is implemented by adjustment of connections between FPGA logic cells using the appropriate interface equipment (JTAG interface) connected to an instrumental PC. Interface equipment for adjustment of connections between crystal logic cells is selected in accordance with the type and/or manufacturer of components applied.

Thus, this step is a transfer from software implementation of digital device to its final hardware implementation. The product of this step is an FPGA-based digital device that performs certain functions within I&CS.

Verification Approaches for FPGA Electronic Designs

The conformity between verification stages of FPGA-projects, tasks performed at those stages and methods of task performance is explained in Table 5.

Let us give a short characteristic of FPGA-projects verification methods.

Documentation Technical Review Method Applied to Assess Completeness and Correctness of Algorithm Block-Diagrams: Signal formation algorithm block-diagrams are results (products) of a corresponding FPGA-project development stage. The completeness of signal formation algorithm block-diagrams is assessed by comparison between the lists of developed algorithm block-diagrams to signal formation conditions according to SRS agreed with customer. Conformity criterion is coincidence between the list of developed algorithm block-diagrams and signal formation conditions according to SRS.

Correctness of signal formation algorithm block-diagrams is assessed by correctness, unambiguous treatment and preparation quality of the developed block-diagrams. In the course of analysis the following must be confirmed:

- A separate algorithm block-diagram is presented for each signal formation condition.
- Each block-diagrams comprises a strict signal formation condition formulation in accordance with SRS.
- Each block-diagram is performed under established form as a connection of typical structural elements selected from a prescribed standard set.
- Inputs and outputs of each structural element are clearly and unambiguously identified in accordance with established rules.
- Identifiers and names are specified for all input signals, alteration limits being also specified for continuous signals.
- Identifiers, names and destinations are specified for all output signals in the scheme.
- Set-points determining signal formation conditions and return to normal operation are specified in the scheme with necessary accuracy.
- Necessary timing characteristics (back offs under alterations of input signals, signal formation delays, signal issuance time before automatic de-energization, etc.) are specified in the scheme with necessary accuracy.

Conformity criterion is meeting all the above requirements to algorithm block-diagrams preparation.

Table 5. The conformity between verification stages of FPGA-projects, tasks performed and methods of task performance

Verification stage	Verification task	Verification method
Development of signal formation algorithm block-diagrams	Completeness and correctness assessment of signal formation algorithm block-diagrams	Documentation technical review
	Conformity assessment to SRS	Traceability analysis
	Structuredness assessment of algorithm block-diagrams	Complexity assessment
Development of signal formation algorithm block-diagram program models in CASE-tools environment	Testing of algorithm block-diagram program models	Functional and timing simulation in CASE-tools environment
	Completeness of tests assessment	Walk-through of documentation
	Conformity assessment to algorithm block-diagrams	Traceability analysis
Integration of signal formation algorithm block-diagram program models in CASE-tools environment	Testing of digital device program model	Functional and timing simulation in CASE-tools environment
	Completeness of tests assessment	Walk-through of documentation
	Conformity assessment to program models	Traceability analysis
Implementation of digital device program model to FPGA	Testing of FPGA with implemented program model	Blackbox functional testing
	Completeness of tests assessment	Walk-through of documentation
	Conformity assessment to integrated program model	Traceability analysis

Method of Functional and Timing Simulation in CASE-Tools Environment: Functional and timing simulation in CASE-tools environment implies testing of each of signal formation algorithm block-diagram program models in design environment as well as testing of the integrated program model.

Under functional simulation conditions ("input signals") corresponding to normal operation and to each of signal formation conditions are consecutively imitated at test inputs. Altered states of program model outputs and/or of established control points within the program model as caused by such effects are observed at instrumental screen and registered as "hard copies" from the screen. Under timing simulation consecutive state alterations at one of program model inputs are imitated in turns and altered states of test outputs ("timing diagrams") and/or of established control points within the program model are monitored.

Tests performed should imply:

- Testing of new functional blocks (IP-Cores), arranged from typical functional elements in CASE-tools environment.
- Testing of algorithm block-diagram program models arranged from typical functional elements and new functional blocks in design environment.
- Testing of integrated algorithm block-diagram program model arranged from signal formation algorithm block-diagram program models in CASE-tools environment.

New functional blocks are tested directly in CASE-tools environment. As the functional blocks are invariant relative to input data, after verification they may be included into the library of CASE-tools applied and find multiple usages during development of FPGA-projects. Tests for algorithm program models are developed on the basis of signal formation algorithm block-diagrams verified at the previous stage.

Tests for integrated program model are developed on the basis of signal formation algorithm program models verified at the previous stage.

The developed tests and testing results must be presented in FPGA-project verification documents in the form of tables and timing diagrams. In timing diagrams the imitated input states (input signals) of program models in CASE-tools environment, their alterations and altered states of each of outputs should be specified.

Criterion of success is a conclusion that test results correspond to expected results.

Walk-Through Method of Documentation Viewing Used to Assess Testing Completeness: Walk-through of documentation is a kind of inspection of documents correctness, completeness and consistency. We shall mark the peculiarities of one of the key stages – analysis of testing complete coverage of algorithm program models in the process of FPGA-project verification. Such analysis is performed by comparison between the list of qualitatively different combinations of input states and/or of their alterations that cause altered output states and the list of program model input-output states that are imitated and monitored in the course of testing.

Criteria of conformity in this are:

- Presence and completeness of tests for all new functional blocks composed from typical functional elements.
- Presence and completeness of tests for all FPGA-project algorithm program models.
- Presence and completeness of tests for integrated program model.
- Presence and completeness of tests for final FPGA with implemented program model.

Blackbox Functional Testing Method: Functional testing, named also blackbox testing, consists in experimental checking of functions performed by a programmable component with implemented program model to define their conformity to system requirements, signal formation algorithm schemes and user documentation (Scott, J., Lawrence, J., 1994).

Traceability Analysis: This is done to ensure that input requirements of a certain process are exhaustively considered by analysis of their connections to output results as well as all requirements have been defined and brought through the life cycle of development, i.e. from requirement analysis up to final testing.

Traceability analysis includes identification of input requirements and confirmation of the fact that they have been considered by way of inspection of destination documents. For instance, the analysis may inspect translation of system requirements documentation into FPGA-project requirements documentation, or that of FPGA-project requirements documentation into digital device characteristic specification, or that of system requirements documentation into tests, etc. If necessary, traceability analysis may include a requirements confirmation step to ensure that actual requirements, but not simply sections of input documentation, have been traced. The results of analysis must show whether all requirements have been duly considered. For this analysis usually traceability matrices are used comprising comparison between input requirements and the elements of output results.

In the course of FPGA-projects verification traceability analysis is applied to ensure tracing or establishment of connections:

- Between SRS and signal formation algorithm block-diagrams.
- Between signal formation algorithm block-diagrams and their program models in CASE-tools environment.
- Between signal formation algorithm block-diagrams and integrated program model in CASE-tools environment.
- Between FPGA electronic designs integrated program model in CASE-tools environment and FPGA with implemented logic structure.

Conformity of signal formation algorithm block-diagrams to the initial data of SRS is assessed for each block-diagram separately by comparing:

- Logic conditions of signal formation and return to normal operation, the latter being defined by this scheme, to conditions established in specification.
- Numerical values of set points and timing characteristics that define conditions of signal formation and return to normal operation to their values established in specification.
- Identifiers and names of input and output signals and alteration limits of continuous input signals as specified in the scheme to specification data.

Conformity criterion for this verification stage is coincidence of logic conditions, numerical values of set points and timing characteristics, identifiers, names and alteration limits of signals as defined from signal formation algorithm schemes to initial data established in specification.

Conformity of algorithm program models developed in CASE-tools environment to signal formation algorithm block-diagrams is assessed by way of comparison of:

- Identifiers, names and alteration limits of input signals.
- Identifiers, names and formation logic conditions of output signals.
- Connection topologies between structural elements, numerical values of set points and timing characteristics specified in algorithm block-diagrams and "hard copies" from screen that diagrammatically reflect the developed algorithm program models.

Conformity criteria for this stage are:

- Usage in algorithm program models of only those elements that are included into typical functional elements library of CASE-tools environment applied.
- Presence and completeness of tests for all new functional blocks composed from typical functional elements.
- Presence and completeness of tests for all algorithm program models.
- Positive testing results of all new functional blocks and algorithm program models in CASE-tools environment applied.
- Equivalence of developed algorithm program models and protective signal formation algorithm block-diagrams as verified at previous stage.
- Absence of any input, output signals and/or set points in FPGA electronic design model for which inputs and/or outputs exist in none of algorithm program models.

Conformity of FPGA with implemented program model to this program model in CASE-tools environment is assessed by comparison of signal formation conditions and timing characteristics as obtained by testing to logic conditions, numerical values of set points and timing characteristics specified in algorithm block-diagrams and "hard copies" from screen that diagrammatically reflect the developed FPGA logic structure program model.

Conformity criteria for this verification stage are:

- Successful implementation of FPGA electronic design that was verified at previous stage into FPGA-chip.
- Equivalence of output signal formation conditions and timing characteristics as obtained by testing to logic conditions, numerical values of set points and timing characteristics of FPGA electronic design that was verified at previous stage.

Thus, FPGA electronic design traceability analysis method for each verification stage includes such actions:

- Analysis of verification stage input data presentation and separation of component classes (signals, communication lines, nodes, functional blocks, etc.)
- Detailed analysis of components in each class.
- Filling of traceability matrix with input data by systematization of components in each class.
- Analysis of verification stage output data presentation and separation of component classes.
- Conformity analysis between input and output data and filling of traceability matrix with output data by comparison of each of output data component and input data components.
- Analysis of final traceability matrix, formulations of conclusions and recommendations.

- Overpatching and correction of final product in case any discrepancies are found between input data and output result of development stage.

Let us prepare a formal description of FPGA electronic design traceability analysis.

FPGA electronic design is developed and verified in 4 stages, with traceability analysis performed for each stage. Assume FPGA realizing N signal processing algorithms. For each algorithm S classes of omponents exist that belong to FPGA algorithm as well as L input components A_{ij} of FPGA algorithm and M output components B_{ij} of FPGA algorithm that belong to i-th class. Our analysis shows that component classes of FPGA algorithms are equivalent for each development class. Between all input and output algorithm component we must ascertain whether equivalence conformity is met or not. In this context traceability analysis would be successful if each of input elements corresponds to one or more output elements and each of output elements corresponds to one or more input elements:

$$(\forall A_{ij}: \exists \{B_{ij}\}, A_{ij} \Leftrightarrow \{B_{ij}\}) \vee (\forall B_{ij}: \exists \{A_{ij}\}, B_{ij} \Leftrightarrow \{A_{ij}\}). \quad (1)$$

In a case when condition (1) is not met, input and output data of a development stage are not inter-traceable and stage results mist be corrected.

Traceability matrix includes four columns:

- Column of FPGA algorithm input components – its elements are input components A_{ij}, $i = 1,...,S$, $j = 1,...,L$.
- Column of FPGA algorithm output components – its elements are output components B_{ij}, $i = 1,...,S$, $j = 1,...,M$.
- Column of traceability results – elements are conclusions of traceability between FPGA algorithm input and output components; conclusion data take binary values "meeting" ($A_{ij} \Leftrightarrow B_{ij}$ met) or "not meeting" ($A_{ij} \Leftrightarrow B_{ij}$ not met).
- Column of comments – additional data on FPGA algorithm components development and verification.

Complexity Assessment: One of basic FPGA electronic designs characteristics affecting their reliability is complexity. FPGA electronic design complexity metrics may be applied to access the critical scope of signal formation algorithms and integrated program model, above which the probability of bringing errors drastically increases. Complexity assessment includes (McCabe, T. A., 1976):

- Analysis of problem oriented language in which the algorithms have been developed, separation of operator and operand classes.
- Prescription of weights from the point of view of complexity for operator and operand classes.
- Establishment of limit value for integral complexity metric above which the probability of bringing errors into FPGA electronic designs drastically increases.
- Direct complexity measurement including count of the number of operators and operands for each class and determination of integral complexity metric value.
- Analysis of obtained complexity metric values, formulation of conclusions and recommendations.
- Breaking into modules for those algorithms where complexity metric exceeds its limit value.

Key Advantages of FPGA Technology

FPGA is a convenient technology not only for implementation of auxiliary functions (transformation and preliminary processing of data, diagnostics, etc), it is also effective for implementation of safety important NPP I&CS control functions. Application of the FPGA technology is more reasonable than application of software-based technology (microprocessors) in many cases (Kharchenko, V. S., 2008).

The application of FPGA technology has significant advantages that can be utilized both in I&CS modernization projects of existing NPPs and in I&CSs designs for new NPPs. These advantages are the following:

- Design, development, implementation, and operation simplicity and transparency.
- Reduction of vulnerability of the digital I&CS to cyber attacks or malicious acts due to absence of any system software or operating systems.
- Faster and more deterministic performance due to capability of executing logic functions and control algorithms in a parallel mode.
- More reliable and error-free end-product due to reduction in the complexity of the verification and validation (V&V) and implementation processes.
- Relatively easy licensing process of FPGA-based safety systems due to the simplicity and transparency of system architecture and its design process and possibility to provide evidence of meeting licensing requirements, such as independence, separation, redundancy and diversity, in an easier and more convincing way.
- Resilience to obsolescence due to the portability of the HDL code between different versions of FPGA chips produced by the same or different manufacturers.
- Possibility of reverse engineering results implementation via emulation in FPGA of obsolete central processing unit (CPU) without modification of existing software code.
- Specific beneficial properties regarding cyber security compared to microprocessors (no viruses for FPGA).

The following FPGA features are important for safety and dependability assurance:

- Development and verification are simplified due to apparatus parallelism in control algorithms implementation and execution for different functions, absence of cyclical structures in FPGA projects, identity of FPGA project presentation to initial data, advanced testbeds and tools, verified libraries and IP-cores.
- Existing technologies of FPGA projects development (graphical scheme and library blocks in CAD environment; special hardware describing languages VHDL, Verilog, Java HDL, etc; microprocessor emulators which are implemented as IP-cores) allow increasing a number of possible options of different project versions and multi-version I&CS.
- Fault-tolerance, data validation and maintainability are improved due to use of: redundancy for intra- and inter-crystal levels; possibilities of implementation of multi-step degradation with different types of adaptation; diversity and multi-diversity implementation; reconfiguration and recovery in the case of component failures; improved means of diagnostics.

- FPGA reprogramming is possible only with the use of especial equipment (it improves a security); stability and survivability of FPGA projects are ensured due to the tolerance to external electromagnetic, climatic, radiation influences, etc.

FPGA-Based NPP I&CS

This section provides information on FPGA-based NPP I&CSs by the example of systems produced by RPC Radiy.

RPC Radiy developed the RadICS FPGA-based platform, which comprises a set of general-purpose blocks that can be configured and used to implement application-specific functions and systems. The RadICS platform is composed of various standardized modules, each based on the use of FPGA chips as computational engines.

RadICS-based I&CSs provide extensive on-line self-surveillance and diagnostics at various levels, including self-diagnostic and defensive coding of electronic design components, self-monitoring of FPGA circuits, such as control of FPGA power, watchdog timer, cyclical redundancy check (CRC) calculation, state monitoring, and monitoring the performance of FPGA support circuits, I/O modules, communications units, and power supplies.

RadICS FPGA-Based Platform and RadICS-Based Applications

The RadICS platform consists mostly of a set of general-purpose building blocks that can be configured and used to implement application-specific functions and systems. The RadICS platform is composed of various standardized modules, each based on the use of FPGA chips as computational engines.

The basic architecture of the RadICS platform consists of an instrument chassis (Figure 5 below) containing two redundant logic modules, as well as up to 14 other I/O and fiber -optic communication modules. Logic modules gather input data from input modules, execute user-specified logic, and update the value driving the output modules. They are also responsible for gathering diagnostic and general health information from all I/O modules. The I/O modules provide interfaces with field devices (for example, sensors, transmitters, actuators). The functionality of each module is defined by the logic implemented in the FPGA(s) that are part of the above modules.

In addition to the above described general purpose I/O modules, there are also special-purpose I/O modules designed to interface with specific sensors and devices, such as resistance temperature detectors (RTDs), thermocouples, ultra-low voltage analog input boards used for neutronic instrumentation, actuator controller modules, and fiber-optic communication modules that can be used to expand the I&CS to multiple chassis. It is also possible to provide inter- channel communications via fiber-optic based connections between logic modules.

The backplane of the RadICS platform provides interfaces to power supplies, process I/Os, communication links, as well as inputs and indicators. The internal backplane provides interfaces to the various modules installed within each chassis by means of a dedicated, isolated, point-to-point low-voltage differential signaling (LVDS) interface.

For application development, Radiy provides a tool called Radiy Product Configuration Toolset (RPCT). This tool can be used to configure logic for various applications using the functional block library (FBL).

Figure 5. RadICS Platform

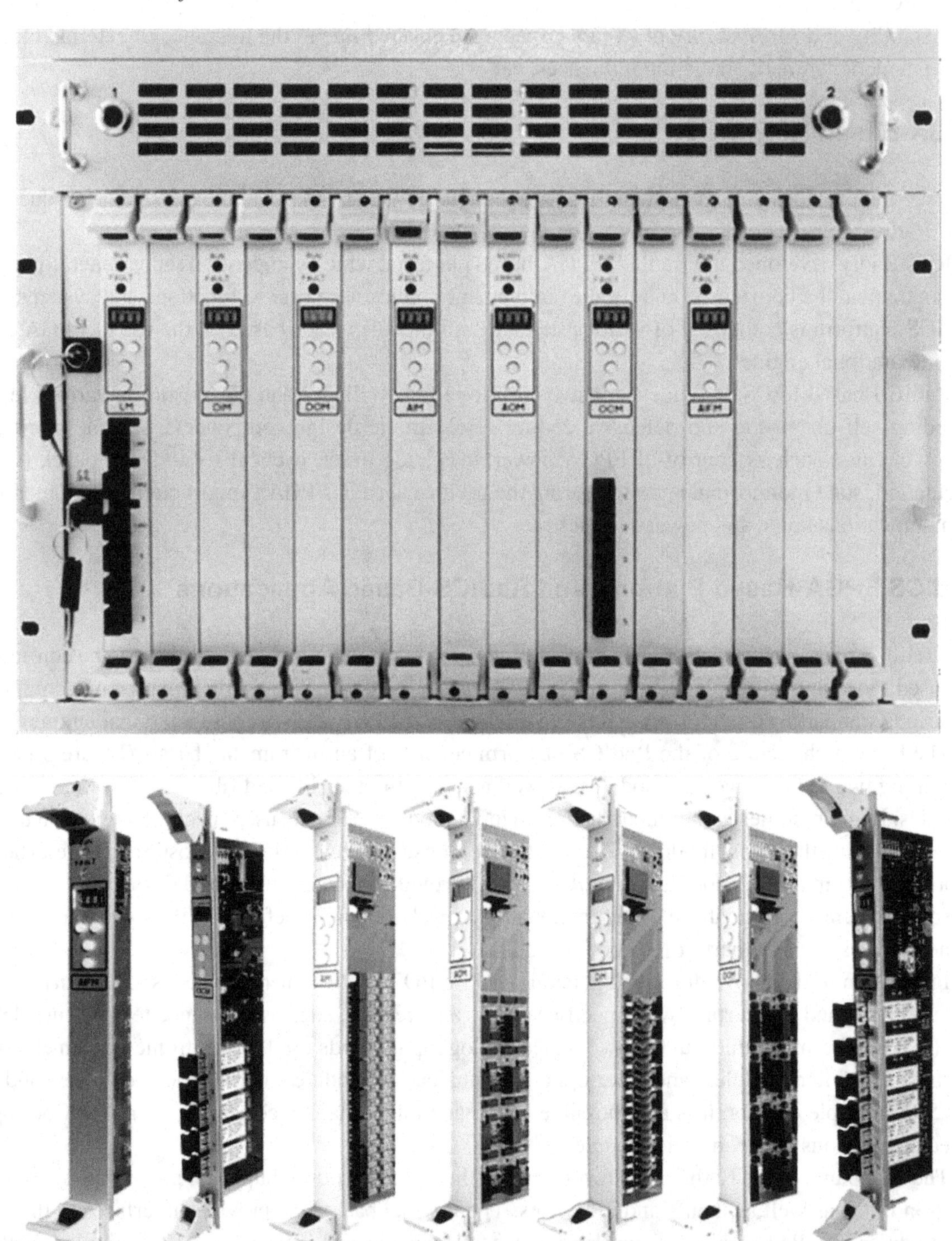

In addition, the RadICS platform includes extensive on-line self-surveillance and diagnostics at various levels, including control of FPGA power, watchdog timer, CRC calculations and monitoring of the performance of FPGA support circuits, I/O modules, communications units, and power supplies.

The RadICS platform can also be represented as a hierarchy with several levels, which could be arranged into two main groups: software and hardware blocks. The RadICS platform high level representation is shown in Figure 6 below.

Figure 6. RadICS platform high level representation

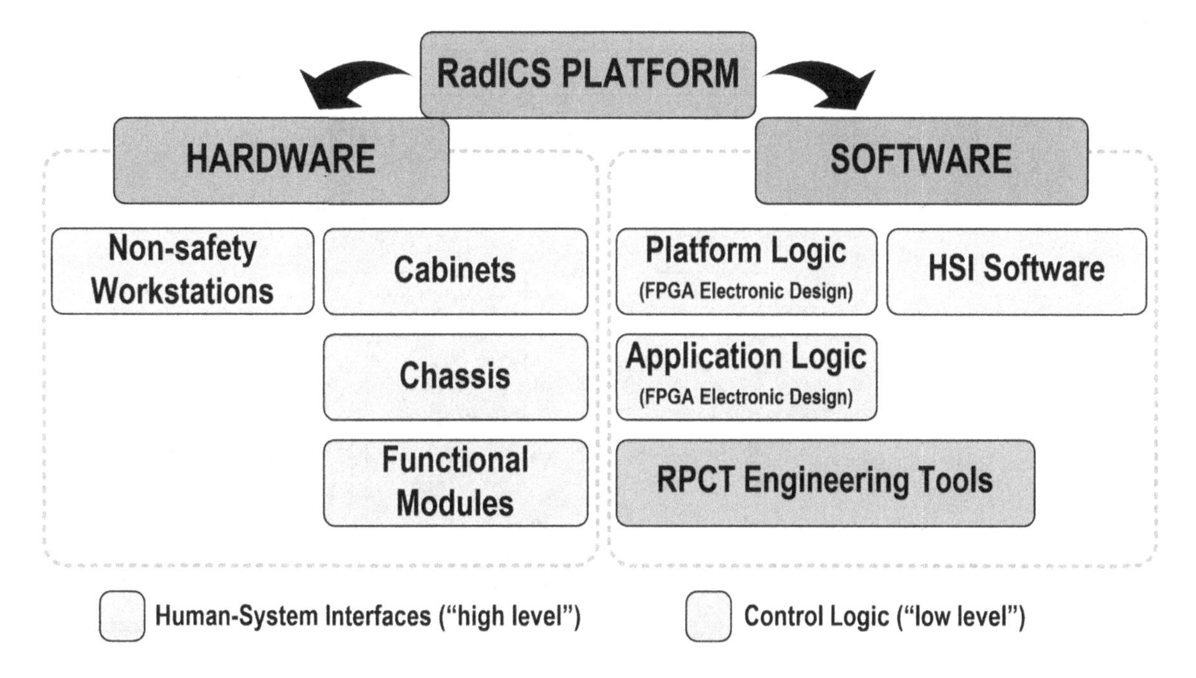

The diagnostic functions are separated from the logic functions and both are executed concurrently. In case of fault detection, the system is placed in a safe state as predefined for each application during configuration.

I&CSs based on the FPGA-based platforms produced by RPC Radiy include the most critical and high-reliability applications in NPPs, such as Reactor Trip, Reactor Power Control and Limitation, Engineered Safety Features Actuation, and Rod Control. Other examples include Nuclear Island Control Systems, Turbine Island Control Systems and Automatic Regulation, Control, Operation and Protection (ARCOP) of Research Reactor.

Reactor Trip System (RTS). The RTS continuously monitors the actual values of neutron flux and other process variables, and it conditions shutdown signals in case these variables reach their setpoints. RTS transmits all vital information necessary for surveillance and monitoring to the control room and other safety and non-safety systems (e.g., initiation status, plant and diagnostic data). RTS can have 3 or 4 redundant channels depending on the design basis of the nuclear reactor, and it can implement a voting logic of two-out-of-three (2oo3) or two-out-of-four (2oo4). Example of 2oo3 configuration is shown in Figure 7. The external interfaces of the RTS provide interfaces to power supplies, process I/Os, communication links, local inputs, and indicators.

Figure 7. Reactor Trip System Configuration (2oo3 voting logic version)

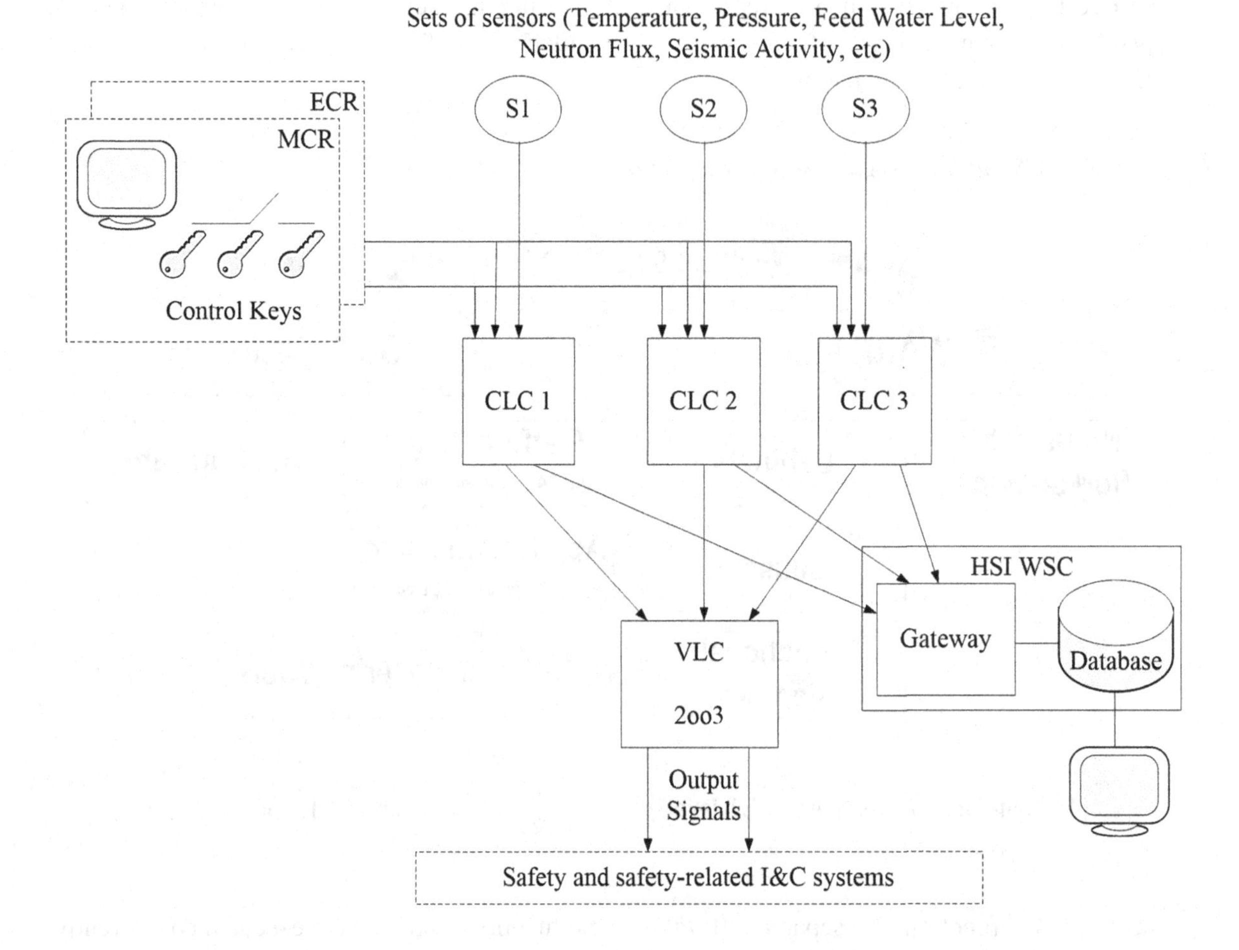

A typical RTS (see Figure 8) has on-line monitoring and maintenance capabilities. It can correct its voting logic in case faults are detected, so that system availability is optimized without compromising safety. RTS has a self-diagnostic subsystem, which includes troubleshooting assistance functions for easy localization of faults. In case of failure, RTS puts itself in the safe state, signalling actuation for shutdown. RTS also supports manual actuation of shutdown logic from the Main Control Room (MCR) or Emergency Control Room (ECR). The FPGA-based RTS architecture can be adapted to various reactor types (e.g., PWR, BWR, PHWR).

There are 30 RTSs produced by RPC Radiy in operation at Zaporozhe NPP, Rovno NPP, Khmelnitsky NPP and South-Ukrainian NPP.

Figure 8. Reactor Trip System

Reactor Power Control and Limitation Systems (RPCLS): RPCLSs (Figure 9) perform the following main functions:

- Automatic and continuous regulation of reactor neutron power and/or pressure in the main steam line of NPP power unit turbine.
- Control of reactor power at levels corresponding to the range of NPP power unit main licensing limitations, from start up through full-power operation.
- Fast-responding preventative protection of the reactor (runback at 40-50% of full power within 3 to 4 seconds).

To increase the reliability of the protection functions, output signals implement in a 2oo3 voting logic. If licensing schemes require, the system can be designed in a 2oo4 configuration with full reliability and quality compliance.

During the design phase of any specific RPCLS, divisional principles are implemented within the control and protection functions. In order to achieve high reliability and independences, different groups of protection functions are realized in separate galvanically isolated subunits.

From 2004 to 2012, nine Reactor Power Control and Limitation Systems were put in operation in Ukrainian NPPs.

Figure 9. Reactor Power Control and Limitation System

Figure 10. Rod Control System

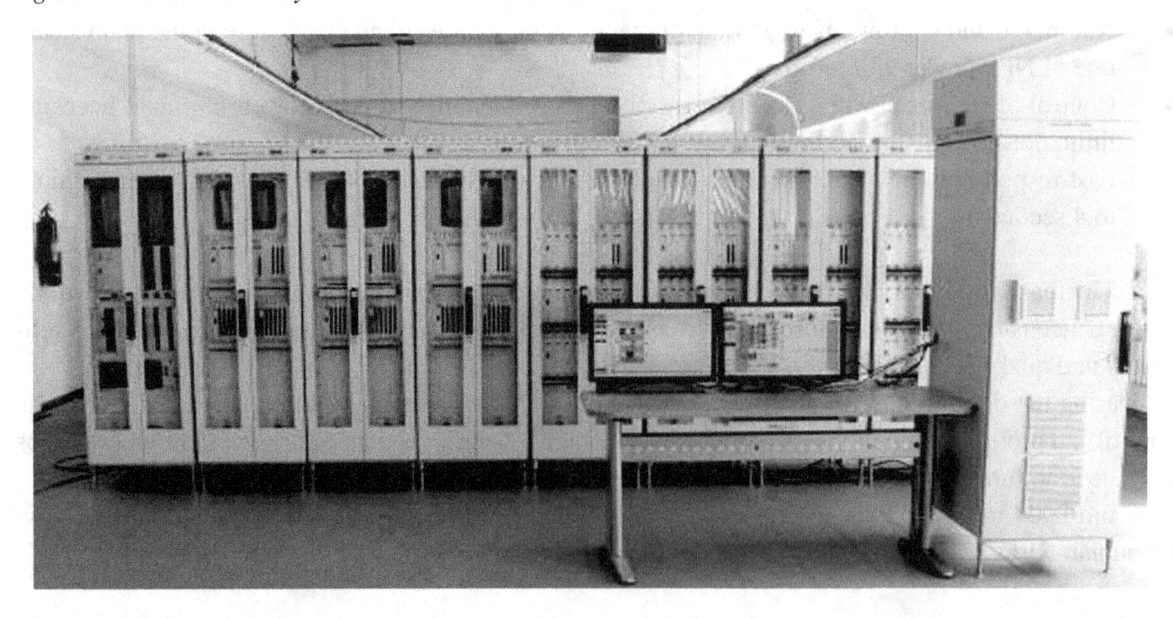

Rod Control System (RCS): The RCS (Figure 10), in general, consists of Rods Position Indication System / Subsystem (RPIS) and Rods Drives Control System / Subsystem (RDCS) with control logic processing equipment power supply subsystem and also can include its own Rod Drives Electric Power Supply Subsystem (RDEPSS) made by RPC Radiy (or any other type of RDEPSS).

RPIS can indicate all reactor control and safety rods position operation parameters and real rods position. RDCS performs all Rod Drives control functions and include Trip Portion (set of the Rod Drives power supply breakers).

RDEPSS provides the following functions:

- Uninterrupted electric power supply of Rod Drives in normal operation mode.
- Switching off the Rod Drives electric power supply by Emergency Protection (EP) signals in case of normal operation failure which requires placing the reactor into a subcritical state.

RCS can have 2 or 3 redundant channels depending on the design basis of the nuclear reactor, and it can implement a voting logic of 1oo2 or 2oo3. Generic architecture of RCS configuration in 1oo2 voting logic version for PWR Unit is shown in Figure 11.

Figure 11. Generic architecture of RCS configuration in 1oo2 voting logic version for PWR Unit

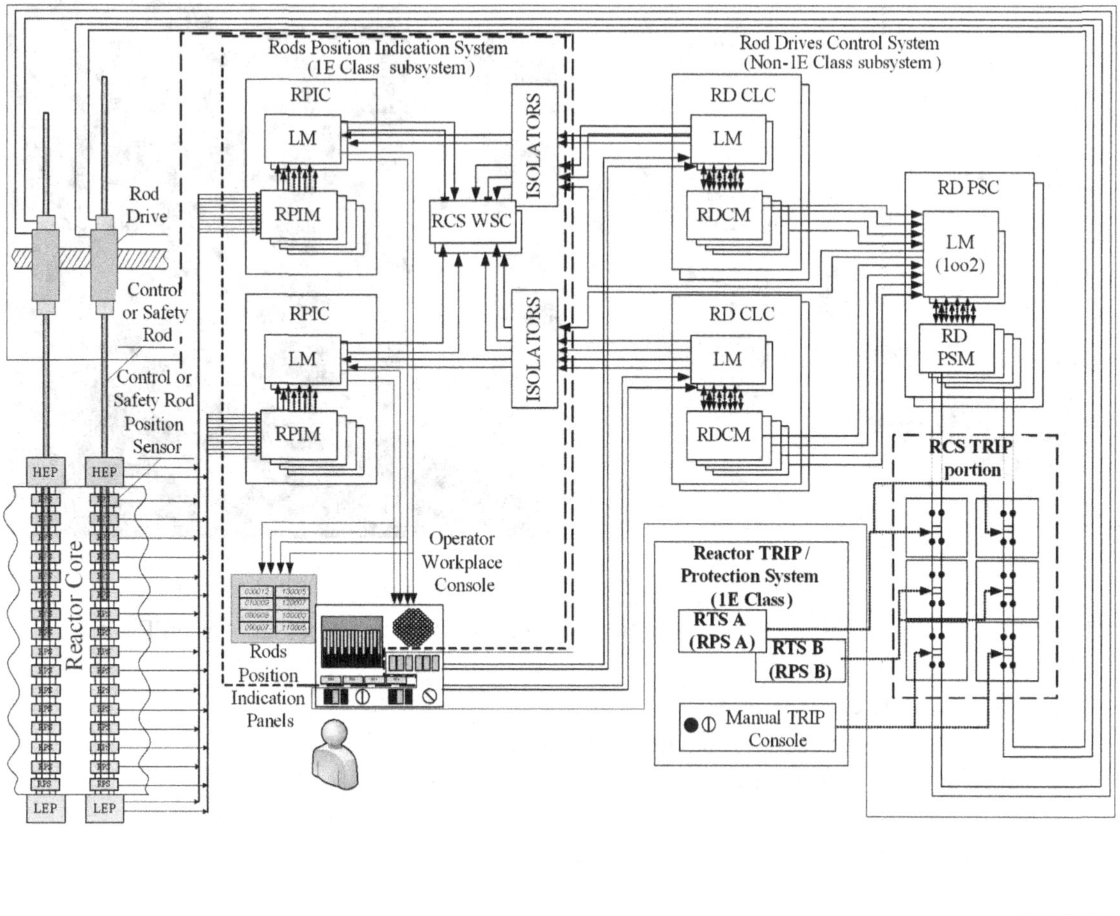

In 2012, the first full set of RCS has been successfully put in operation in the Unit 1 of the South-Ukrainian NPP with WWER-1000 PWR-type reactor.

Engineered Safety Features Actuation System (ESFAS): The ESFAS (see Figure 12) produced by RPC Radiy executes the following main functions:

- Protection, interlocking and monitoring of the automated operation of actuators.
- Automatic process control.
- Manual remote control of actuators.

Figure 12. Engineered Safety Features Actuation System

The ESFAS also provides the implementation of functions that are necessary for NPP safety:

- Information and data acquisition.
- Signal conditioning and control of safety signals, detectors, and sensor.
- Full-scope systems diagnostics.

The following design principles are applied in the ESFAS:

- Diversity of input signals (e.g., current, voltage, resistance, "dry contact").
- System size scalability accommodating needs for increased number of inputs and outputs.
- Simple and controlled ways of code modification of protection, interlocking and control algorithms.
- Adaptability of interfacing capabilities for communication and integration with other control, monitoring and regulating systems.

Figure 13. Nuclear Island System

The ESFAS can be supplied in single-, two-, three-, or four-channel installations. The ESFAS conforms to safety class 2, can be designed and built in accordance with applicable national standards in the EU countries and the USA.

Twenty one ESFASs are in operation now at Rovno NPP, South-Ukrainian NPP, Khemlnitsky NPP and Kozloduy NPP (Bulgaria).

Nuclear Island and Conventional (Turbine) Island Systems: Nuclear Island and Conventional (Turbine) Island Systems (see Figure 13) have the following main functions:

- Conditioning and initiation of protection, interlocks and alarm commands.
- Conditioning and initiation of automatic regulation commands when process values deviate from setpoints.
- Initiation of remote control commands based on operators' instructions.
- Indicate current states, positions and operating modes of actuators in control rooms.

There are ten Nuclear Island and Conventional (Turbine) Island Systems in operation now at Ukrainian NPPs.

Automatic Regulation, Control, Operation and Protection for Research Reactors (ARCOP): ARCOP system is designed to implement safe operation of research reactors. ARCOP performs the following functions:

- Measurement and monitoring of neutron physical reactor parameters.
- Measurement and monitoring of thermal physical parameters.
- Generating the emergency protection and preventative signaling.
- Automatic reactor power regulation.
- Remote and automatic control of actuators.
- Diagnostics and information display support.

In 2006, an ARCOP system was installed at the WWR-M type research reactor in the Institute of Nuclear Research at the National Science Academy of Ukraine, Kiev.

ANALYSIS ASPECTS OF FPGA-BASED NPP I&CS

Verification and Validation of FPGA-based NPP I&CSs

FPGAs were first introduced in non-safety systems in NPPs, where no specific process over general FPGA development process is required. However, to use FPGAs for safety systems, more strict processes are imposed by nuclear regulators to ensure the reliability and safety of the systems.

Since the development process of FPGA is similar to that of software for microprocessor-based systems, the conventional safety software development process including V&V methods can be applied. I&CSs supplied by the RPC Radiy were subjected to V&V processes to ensure their reliability and safety.

For example, for US commercial NPPs, the US NRC endorses IEEE Standard 7-4.3.2-2003 as the methods for high functional reliability and design requirements for computers, whereas IEEE Standard 1012-1998 as the methods of V&V.

IEEE Standard 1012-1998 postulates a phased software life cycle, and defines a number of V&V activities to be performed throughout the software lifecycle. The V&V activities include the following types of activities:

- Software requirements evaluation.
- Design evaluation.
- Interface analysis.
- Requirements traceability analysis.
- Source code and source code documentation evaluation.
- Validation testing.
- Hazard analysis.

Figure 14. Combined usage of dependability analysis techniques

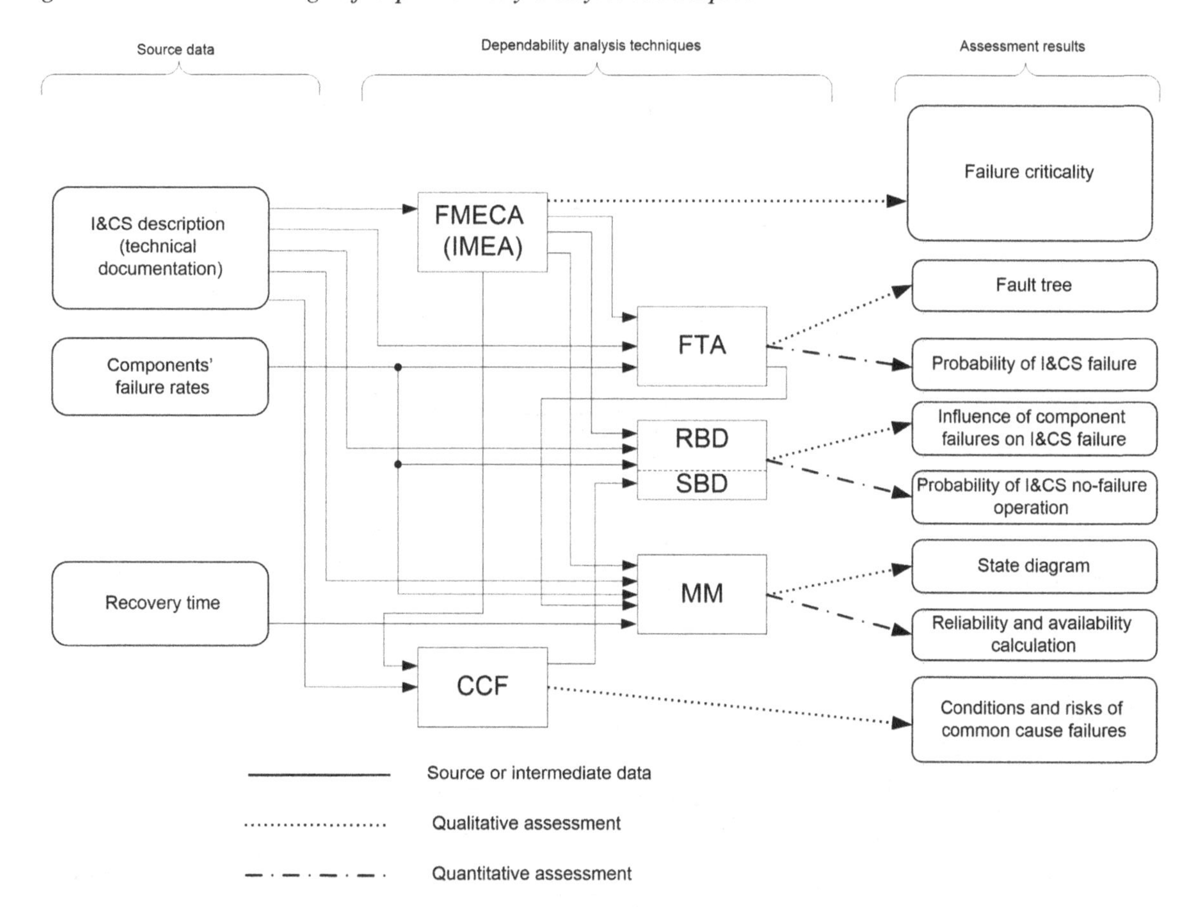

Combined Usage of Analysis Techniques

There are a lot of well-known techniques that can be used for NPP I&CS dependability analysis and assessment of its attributes. Using these techniques it is possible to perform quantitative and/or qualitative assessments. Qualitative assessments though lacking the ability to account, are very effective in identifying potential failures within the I&CS. We have performed some work to identify possible combination of techniques, results are shown in Figure 14. To carry out dependability analysis it is necessary to have I&CS technical documentation (this information is obtained from I&CS project) and reliability data of I&CS components (is obtained from component vendors).

The first stage of NPP I&CS dependability analysis is FMECA (Failure modes, effects and criticality analysis). During this stage all possible failure mechanisms and failure rates for all components involved and quantify failure contribution to overall NPP reliability and safety are analyzed.

In FMECA qualitative and quantitative results (see Figure 15) are obtained. Failure mode in FMECA refers to the way a failure might occur. Failure effect is the consequence of failure from the system's point of view. Failure criticality is assigned to each failure mode to get quantitative parameters.

FMECA is carried out early in the NPP I&CS development life cycle to find ways of mitigating failures and thereby enhancing reliability through design.

Figure 15. Reliability and safety block diagrams: principles of development

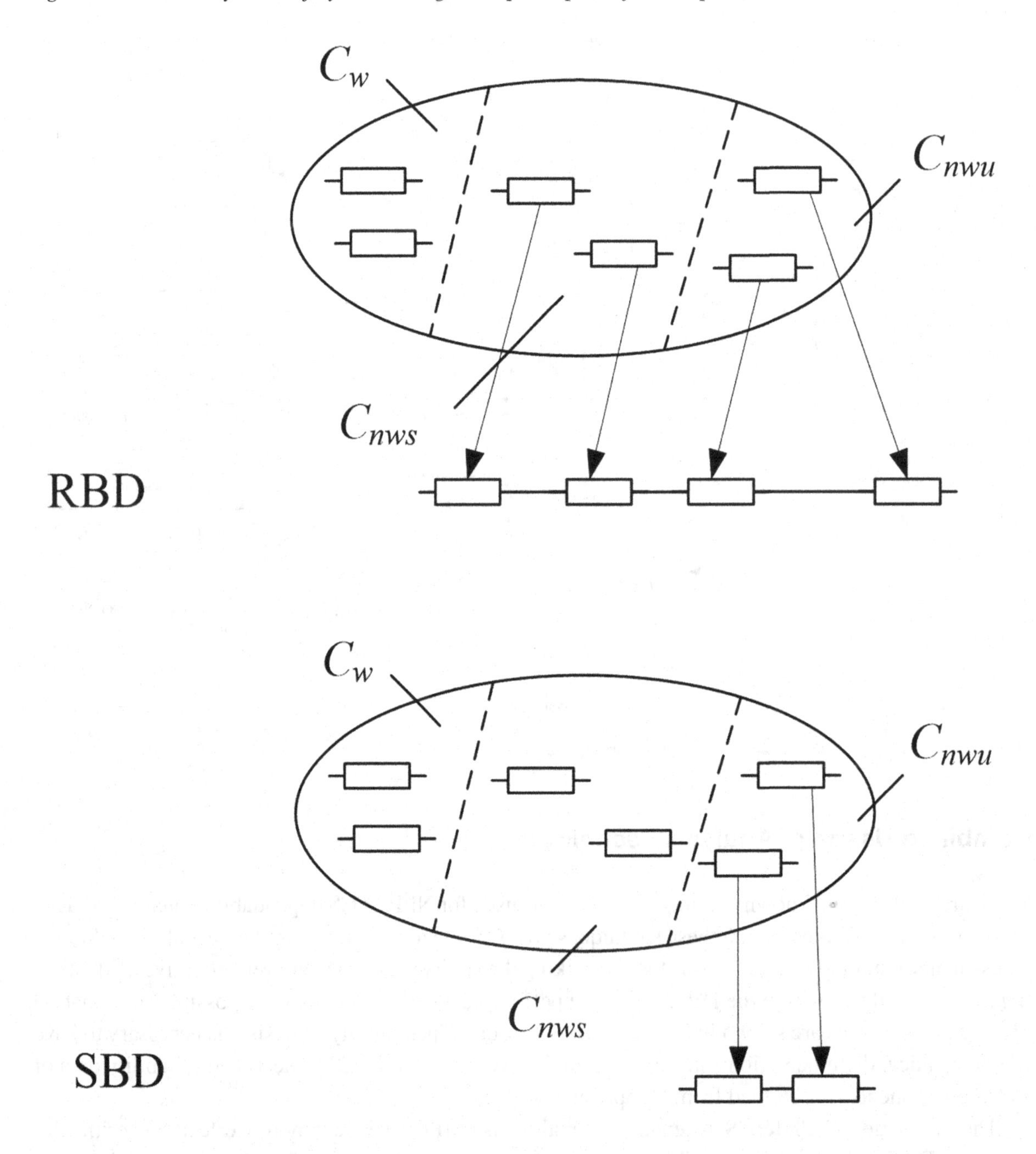

A traditional FMECA uses potential component failures as the basis of analysis. Component failures are analyzed one by one, and therefore important combinations of component failures might be overlooked. Environmental conditions, external impacts and other such factors are analyzed in FMECA only if they produce component failures; external influences that do not produce component failures (but may still produce I&CS failure) are often overlooked.

That's why it is not sufficient to use only FMECA during NPP I&CS analysis.

To take into account external impacts it is possible to use IMEA (Intrusion Modes and Effects Analysis). IMEA is a modification of FMECA that takes into account possible intrusions to the system, examples of this analysis are shown in (Babeshko, E. et al., 2010; Babeshko, E. et al., 2011).

Results of FMECA and IMEA are used during further FTA (Fault Tree Analysis), RBD / SBD (Reliability / Safety Block Diagram), CCF (Common Cause Failure Analysis), and also during Markov modeling.

Reliability block diagram (RBD) is a graphical analysis technique, which expresses the concerned system as connections of a number of components in accordance with their logical relation of reliability. Safety block diagram (SBD) is a similar technique that treats safety aspects.

Figure 15 shows RBD and SBD principles. Set of NPP I&CS components is split into the following groups:

- Components that can't lead to NPP I&CS failure C_w.
- Components that can lead to I&CS failure, but system state would be safe C_{nws}.
- Components that can lead to I&CS failure, but system state would be unsafe C_{unws}.

While RBD treats all possible failures (both C_{nws} and C_{nwu} are included into RBD), SBD treats only components that can lead to unsafe situation (only C_{nwu} are included). That gives possibility to concentrate on safety aspect and to simplify all following calculations.

During RBD (SBD) it is possible to use list of all components that can cause I&CS failure which has been obtained during FMECA. Then we take into account I&CS architecture (number of components, software and hardware versions, type of diversity, check and reconfiguration means) and sets of different faults and calculate reliability and safety indicators.

FMECA results are used in FTA to get list of all possible failures.

To perform Markov modeling it is required to know component's failure rates and recovery time so as to get state-to-state transitions. In most cases the NPP I&CS operation may be analyzed using a Markov model.

APPROACH TO THE MODERNIZATION OF NPP I&CS

Modernization Goals

NPPs modernization goals and objectives are defined by the utilities based on their operating experience, equipment obsolescence and commitments to their regulators, within the constraints imposed by their long and short term business plans. Examples of such goals and objectives are to minimize production costs by reducing the number of unplanned outages and decreasing the length of planned outages. Examples of external commitments are those made to licensing authorities, local governments, and international organizations, as well as with the public and other stakeholders.

In general, I&CS modernization goals fall under one or more of the following types:

- Increase the reliability of NPP I&CSs.
- Improving safety in the operation of NPP units.
- Life Extension of operating units.

- Environmentally safe and efficient production of electricity at minimum cost.
- Compliance with new, more stringent safety standards and design requirements than those to which the plant was originally designed, e.g., more demanding seismic, environmental, fire protection, reliability and human factors related requirements.
- Catering to equipment obsolescence.

The following are examples of factors that are typically to be considered when supporting utilities in the implementation of their I&CSs within planned or dedicated refurbishment outages:

- Time required for equipment replacement (demolition and installation times) and commissioning.
- Suitability of the new equipment to work with other related (unchanged) systems in the existing environment.
- Compliance of the replacement equipment with new requirements.

Approaches to Modernization

Entire System Modernization. This approach involves a complete replacement of the existing I&CS. In this case, all components, from the sensor to the I/O and logic processing equipment are replaced with modern equipment.

During the design process of the new system, its operation experience is taken into account in order to ensure that its strengths remain and its weaknesses are eliminated, thus improving the reliability and safety of the refurbished system.

FPGA-based platforms lend themselves very well to complete system modernization strategies due to the following reasons:

- No costs for the design or redesign of complex computer operating systems.
- Straightforward conformance to modern requirements (e.g., cyber security, diversity).
- Easier implementation of functions that require fast response time.

Partial Modernization: This involves the replacement of only the most critical or obsolete components via a Form Fit and Function (FFF) equivalent.

The new component may include additional useful functionality, such as self-diagnostics and more comprehensive displays. The new component must meet the physical and functional constraints of the old component, such as size, power consumption and pin-out.

When properly implemented, an FFF solution should allow the user to simply replace the old component with the new one without having to change operating procedures and interfaces with the rest of the equipment. Maintainers will have to be trained to troubleshoot the new equipment. Factors to be considered when pursuing partial modernization are:

- Portability of the old software into the new component.
- Impact of the eventual obsolescence of the new equipment.
- Reliability of the new equipment and its impact on safety analysis.

Adoption of FPGA technology could help resolve or mitigate potential problems associated with the above factors, because of its capability to replicate the functionality of electronic components.

In general, FPGA technology is very effective for module to module replacements. This approach is much more cost-effective compared to the cost of entire system modernization.

Modernization Activities

Typically, detailed work plans shall be developed for all phases of projects, i.e. design, procurement, manufacturing, certification, testing, installation and commissioning as required. The plans are discussed and agreed with customer, contractors and suppliers during contract negotiations and they are later used during the project execution phase. Deviations from the original schedule and reasons for these are documented and archived in a readily retrievable manner.

Project execution activities include, but are not limited to the following:

- Establishment of quality procedures, organization and work processes commensurate with project requirements.
- Control, administration and support of design work.
- Coordination, updating and monitoring of work execution to ensure compliance with project deadlines and requirements.
- Establishment of effective interfaces, work processes and means of communication (through meetings and correspondence) between project participants, in order to ensure achievement of the needed degree of involvement of all necessary parties and organizations.
- Controlled transfer of documentation to the Customer.
- Issue of documentation in language required by Customer.
- Implementation of a Lessons Learned process in order to avoid repeating mistakes in future projects.
- Periodic reporting to the customer on project progress, quality, cost (where applicable), schedule and safety issues.
- Provision of technical and other support in accordance with the warranty terms stated in the contract.

Reverse Engineering

Whether introducing new equipment or systems as part of NPPs refurbishment or life extension activities, in most cases there is a need to ensure that new components are compatible and interface correctly with existing plant components. However, it is frequently the case that equipment documentation may be either inexistent or lack all necessary details.

In order to avoid additional replacement costs and associated licensing effort, utilities and other clients resort to reverse engineer the existing equipment.

Reverse engineering activities include detailed analysis of existing system behavior in order to reproduce its functionality. FPGA-based I&CS platforms are very well suited for activities of this nature because they can be easily programmed to emulate and implement most of the functionality exhibited by existing equipment.

Reverse engineering could also be necessary to preserve the ''feel and look'' of existing human-machine interfaces, while replacing the rest of the system with a new one, including additional functionality or the same functionality as the old one.

Another benefit of reverse engineering is a possibility to determine reasons why the performance and / or the behaviour characteristics of the old module are the way they are. Sometimes reverse engineering confirms that they are just consequences of the initial design and implementation, and not due to functional requirements.

Experience of FPGA Usage in Nuclear Applications

Nuclear industry is moving toward wide application of FPGAs in control systems, and now FPGAs are being treated as a natural part of almost any NPP I&CS. For example, RPC Radiy (and its subsidiary Radics LLC) has put in operation more than 15000 of different FPGA chips in RadICS platform modules since early 2000s, and there were no module failures due to FPGA failure since than.

RadICS is SIL 3 (in a single channel configuration) digital safety system platform certified according to IEC 61508 requirements, as well as certified US NRC. This certifications confirm that FPGA based safety platform like RadICS is suitable for installation in any nuclear power generating facility to replace existing digital or analog safety systems. Such system is both hardware and software diverse making it fully deterministic solution available today. Also, FPGA-based platform is designed to easily fulfill all cyber requirements. By virtue of its diverse hardware and software and its "fail-to-safe-state" design, it is immune to digital system Common Cause Failures. With over 100 systems installed to-date and successful SIL 3 certification for the RadICS platform, Radiy's demonstrated technological expertise is paving the way for increased safety and efficiency in the nuclear field. The RadICS design is both hardware, utilizing FPGA technology, and software diverse eliminating common cause failure licensing concerns.

SOLUTION AND RECOMMENDATIONS

Nowadays FPGAs are widely used in different fields, including critical ones, and there is a trend that situation will remain like this in the nearest future. Therefore, it is necessary to work at reliability, safety and cyber security analysis of FPGA-based systems so as to enhance existing techniques by considering FPGA peculiarities.

Experience shows that combined usage of analysis methods provides better results, therefore such approach should be developed further.

FUTURE RESEARCH DIRECTIONS

More detailed analysis of advantages and risks of FPGA technology application should be fulfilled taking into account an experience of modern companies producing and implementing FPGA-based NPP I&CSs. Proposed approaches can be strengthened by development of support tools that will allow to automate analysis process. In such a way, the tools can support joint application of the different techniques, as well as calculate the metrics in order to choosing the optimal set of applicable methods to ensure reliability, safety and cyber security of FPGA-based I&CSs.

CONCLUSION

FPGA technology is an alternative to microprocessor based technologies and other types of programmable devices. FPGAs are semiconductor-based programmable devices which can be configured to perform custom-designed functions. One of the important advantages why FPGA technology is gaining acceptance for nuclear applications is short response time.

The common approach for the development of an FPGA-based NPP I&CS is to use a pre-developed and verified FPGA-based platform, whose modules can be used to implement specific applications just by configuring them. Nowadays, FPGA-based platforms are used in I&CS modernization projects at various NPPs for a wide range of safety and control functions and systems, such as reactor trip system, reactor power control and limitation system, engineered safety features actuation system, rod control system, nuclear island control system, and turbine island control system.

The above applications represented large-scale modernization projects, however, the technology can provide solutions for an even larger variety of applications, such as 'pin-to-pin' or like-for-like type replacement of obsolete circuit board components, reverse engineering, emulation of functions performed by obsolete computers, replacement of components and sub-systems, and building full I&CSs or diverse back-up systems in new NPP designs. FPGA technology allows implementing any safety and control functions that are typical in existing NPPs or in any new designs, therefore providing a technology-neutral implementation tool. FPGAs are very convenient for modernization of obsolete systems, partial or complete.

One of the example of FPGA-based platorms is RadICS Platform. It consists of a set of general-purpose building blocks that can be configured and used to implement application-specific functions and systems. For application development, there is a special tool (RPCT) that can be used to configure logic for various applications using the functional block library. In addition, the RadICS platform includes extensive on-line self-surveillance and diagnostics at various levels.

REFERENCES

Andrashov, A., Bakhmach, I., Kharchenko, V., & Kovalenko, A. (2019). Equipment qualification of FPGA-based platform RadICS to meet US NRC requirements. *Proceeding of the 11th International Topical Meeting on Nuclear Plant Instrumentation, Control, and Human-Machine Interface Technologies*, 327-335.

Andrashov, A., Bakhmach, I., Leontiiev, K., Babeshko, E., Kovalenko, A., & Kharchenko, V. (2019). Diversity in FPGA-based platform and platform based I&CS applications: Strategy and implementation. *Proceeding of the 11th International Topical Meeting on Nuclear Plant Instrumentation, Control, and Human-Machine Interface Technologies*, 174-182.

Barkalov, A. (2006). *Design of Control Units with Programmable Logic*. University of Zelena Gura.

Illiashenko, O., Kharchenko, V., & Kovalenko, A. (2012). Cyber security lifecycle and assessment technique for FPGA-based I&C systems. *Proceedings of IEEE East-West Design & Test Symposium (EWDTS) 2012*, 432-436.

Intel Corporation. (2018). *Intel® Arria® 10 Device Overview*. Author.

Kharchenko, V., Kovalenko, A., Leontiiev, K., Panarin, A., & Duzhy, V. (2018). Multi-diversity for FPGA platform based NPP I&C systems: new possibilities and assessment technique. *Proceedings of 26th International Conference on Nuclear Engineering*. 10.1115/ICONE26-82377

Kharchenko, V., Siora, A., Andrashov, A., & Kovalenko, A. (2012). Cyber Security of FPGA-Based NPP I&C Systems: Challenges and Solutions. *Proceeding of the 8th International Conference on Nuclear Plant Instrumentation, Control, and Human-Machine Interface Technologies (NPIC & HMIT 2012)*, 1338-1349.

Kharchenko, V. S., Illiashenko, O. A., Kovalenko, A. A., Sklyar, V. V., & Boyarchuk, A. V. (2014). Security informed safety assessment of NPP I&C systems: Gap-IMECA technique. *Proceedings of the 2014 22nd International Conference on Nuclear Engineering ICONE22*, 1-9. 10.1115/ICONE22-31175

Kovalenko, A., Andrashov, A., Bakhmach, E., & Sklyar, V. (2015). FPGA-based I&C Applications in NPP's Modernization Projects: Case Study. *Proceeding of the 9th International Conference on Nuclear Plant Instrumentation, Control, and Human-Machine Interface Technologies*, 1, 113-120.

Kovalenko, A., Kuchuk, G., Kharchenko, V., & Shamraev, A. (2017). Resource-Oriented Approaches to Implementation of Traffic Control Technologies in Safety-Critical I&C Systems. In *Green IT Engineering: Components, Networks and Systems Implementation. Studies in Systems, Decision and Control series. Springer.*

Lyu, M. R. (1996). *Handbook of Software Reliability Engineering*. McGraw-Hill Company.

Maerani, R. (2018). V&V Plan for FPGA-based ESF-CCS Using System Engineering Approach. *Journal of Physics: Conference Series*, ▪▪▪, 962.

McCabe, T. A. (1976). Complexity Measure. *IEEE Transactions on Software Engineering*, *4*(4No SE-2), 308–320. doi:10.1109/TSE.1976.233837

Melnyk, A. (2007). Automatic generation of ASICS. *Proceedings of NASA/ESA Conference on Adaptive Hardware and Systems*, 311–317. 10.1109/AHS.2007.36

Menon, C., & Guerra, S. (2015). *Field programmable gate arrays in safety related instrumentation and control applications*. Energiforsk, AB: Adelard LLP.

Nuclear Energy Series No, I. A. E. A. NP-T-3.17. (2016). Application of Field Programmable Gate Arrays in Instrumentation and Control Systems of Nuclear Power Plants. International Atomic Energy Agency.

Scott, J., & Lawrence, J. (1994). *Testing Existing Software for Safety Related Applications*. Lawrence Livermore National Laboratory.

Tam, S. (2003). Error Detection and Correction in Virtex-II Pro Devices. Application Note: Virtex-II Pro Family. XAPP645 (v1.1).

KEY TERMS AND DEFINITIONS

FPGA (Field Programmable Gate Array): Programmable complex electronic component which includes two entities: FPGA chip and FPGA electronic design.

FPGA Electronic Design: A set of statesments in HDL which is appropriate for implementation in FPGA chip.

HDL (Hardware Description Language): Specialized computer language used to describe the structure, design and operation of digital logic circuits.

IP Core (Intellectual Property Core): Reusable unit of logic, cell, or chip layout design that can be used as building blocks within ASIC chip designs or FPGA logic designs.

JTAG: Integrated method for testing interconnects on printed circuit boards that are implemented at the integrated circuit level.

Logic Synthesis: Process by which an abstract form of desired circuit behavior is turned into a design implementation in terms of logic gates.

LUT: The key component of modern FPGAs that is used to encode any n-input Boolean function by modeling such functions as truth tables.

SRS (System Requirements Specification): A structured collection of information that embodies the requirements of the system.

Chapter 17
NPP Monitoring Missions via a Multi-Fleet of Drones:
Reliability Issues

Herman Fesenko
https://orcid.org/0000-0002-4084-2101
National Aerospace University KhAI, Ukraine

Ihor Kliushnikov
https://orcid.org/0000-0001-9419-7825
Kharkiv National University of Air Force, Ukraine

ABSTRACT

A drone-based system of monitoring of severe NPP accidents is described. A structure of a multi-fleet of drones, consisting of main drone fleets and a reserve drone fleet, is considered. A matrix of drone fleet reliability assessment attributes is presented. Various structures for systems of control stations for the multi-fleet of drones are suggested. Reliability models for the multi-fleet of drones with centralized (irredundant), centralized (redundant), decentralized, and partially decentralized systems of control stations are developed and analyzed.

INTRODUCTION

Over the last few years, drones have become a popular tool for a variety of applications at nuclear facilities, including both indoor and outdoor inspections and mapping (Connora et al., 2016, Han et al.,2013, Martin et al., 2016, MacFarlane et al., 2014). The possibility of using drones when creating new NPP accident monitoring system is generating a lot of interest. A wireless drone-based radiation monitoring system that capable of detecting beta radiation (electrons), gamma radiation (photons) and X-rays from a safe distance is presented by "UAV Drone Radiation Monitoring" (n.d.). The general structure and main principles of creating a multi-version post-severe NPP accident monitoring system consisting of one wired and three drone-based wireless network subsystems are proposed by Fesenko et al. (2018).

DOI: 10.4018/978-1-7998-3277-5.ch017

In this case a multi-fleet of drones is applied to provide additional communication subsystems, Internet of Drones and private cloud based data processing to support a crisis center decision making system. It is vital to note, that a drone fleet involved in performing NPP monitoring missions should maintain a high reliability level considering the presence of its drone failures. One of the widely used techniques for implementing dependable systems with high reliability and fault tolerance is standby redundancy, in which one or several elements are online and working with some redundant elements serving as standby spares. When an online element experiences a failure, a standby element is activated to replace it and take over the work [Levitin et al., 2014]. The k-out-of-n system structure is a very popular type of redundancy in fault-tolerant systems [Ram & Dohi, 2019]. In a fault-tolerant drone fleet, one part of drones should be under redundancy and other one should be used as redundant drones. A switching process allowing the drone fleet to activate a redundant drone can be carried out via a control system (control station). Hence, there are a few challenges related to application of drone fleets to assure reliable monitoring of severe accidents in the aggressive environment and to optimize a schedule of fleet application and control stations structure.

BACKGROUND

Aghaei et al. (2017) investigate the redundancy allocation problem for a k-out-of-n system with a choice of redundancy strategies, which can be applied for drone fleets and multi-fleets of drones. In contrast to the existing approaches that often consider a predetermined strategy for each subsystem, the authors consider both active and standby strategies and developed a model to select the best strategy for each subsystem. Eryilmaz et al. (2017) present the reliability analysis of a weighted-k-out-of-n:G system consisting of two types of components. The system is assumed to have n components which are classified into two groups with respect to their weight and reliability, and it is assumed to operate if the total weight of all working components exceeds a prespecified threshold. This model can be used in case of heterogeneous fleet reliability assessment. Byun et al. (2017) propose the matrix-based system reliability method extended to k-out-of-n systems by modifying the formulations of event and probability vectors. The proposed methods can incorporate statistical dependence between component failures for both homogeneous and non-homogeneous k-out-of-n systems, and can compute measures related to parameter sensitivity and relative importance of components. The described method can be used in case of drones failures and control stations of different fleets failures dependencies. Pascual-Ortigosa et al. (2018) review the different definitions that multistate k-out-of-n systems have received and show how the algebraic method is used to study their reliability in a general way with a single approach. Considering a system with different requirements on the number of working components (i.e., value of k) for different system state levels, Mo at al. (2015) propose a new analytical method based on multi-valued decision diagrams for the reliability analysis of such multi-state k-out-of-n systems. These results could be used in case of application of multi-functional drones with assumed degradation of the functions. Based on the results of this work and taking into account the control station reliability, the aim of the paper is to develop models for a multi-fleet of drones with two-level hot standby redundancy considering a control system structure features.

The objectives of the paper are:

- To present a matrix of drone fleet reliability assessment attributes.

- To analyze structures of control systems for the multi-fleet of drones.
- To develop reliability models for the MFD with one- and two level system of control stations.
- To formulate recommendation for choice of a structure of systems of control stations.

The main contribution of this study comprises a matrix of drone fleet reliability assessment, reliability block diagrams, and formulae for calculation of probability of failure-free operation for the multi-fleet of drones with both one- and two-level systems of control stations.

DRONE FLEET BASED SYSTEM OF MONITORING OF SEVERE NPP ACCIDENTS

Existing NPP post-accident monitoring systems (PAMSs) are based on wired networks that connect sensor areas directly with the crisis center. Reliability and survivability of such systems are assured by redundancy of equipment, cable communications, and other components. However, in the case of severe accidents, wired network-based PAMS can experience damaged sensors or broken cable connections. Under such conditions, the NPP PAMS is partially or totally rendered useless. In the normal operational mode, data and command exchanges run through the wired network. If this process is damaged during an accident, a fleet of communication drones acting as an auxiliary wireless network is created to support these activities. Drones are launched in the event of a wired network failure or in the detection of a possible severe accident. The drones are designed to autonomously form a stable flight formation (drone fleet), which is configured in a master/slaves' arrangement. From this vantage point, the drone fleets will cooperate to maintain the following functions: 1) to monitor and collect all data from sensor modules that are equipped with wireless connections; 2) to form a reliable mesh network for optimal data streaming between point-to-point transmissions; 3) to provide surveillance imaging for damage control, and search and rescue; 4) to summarize areas of contamination; and 5) to provide an unmanned observation platform for exploratory surveillance. Thus, we have a drone fleet based system of monitoring of severe NPP accidents. The structure of such a system is developed by Fesenko et al. (2018) and shown in Fig. 17.1. This system consists of the following components: 1) the NPP; 2) sensors composed of drones, Wi-Fi sensors (Wi-FiS), light fidelity sensors (Li-FiS) based on a bidirectional wireless technology similar to Wi-Fi, and Internet-of-Things sensors (IoTS); 3) several drone fleets (DF1 and DF2); and 4) a communication interface to a crisis center decision-making system (DMS), an autonomous decision-making support system (DMSS; groups of experts) to assure crisis center functionality, and an Internet cloud portal. Finally, the Internet cloud portal is made up of one IoT subsystem (IoT S) and three drones-of-things subsystems labeled DoT S1, DoT S2, and DoT S3.

Figure 1. Structure of the drone fleet based system of monitoring of severe NPP accidents

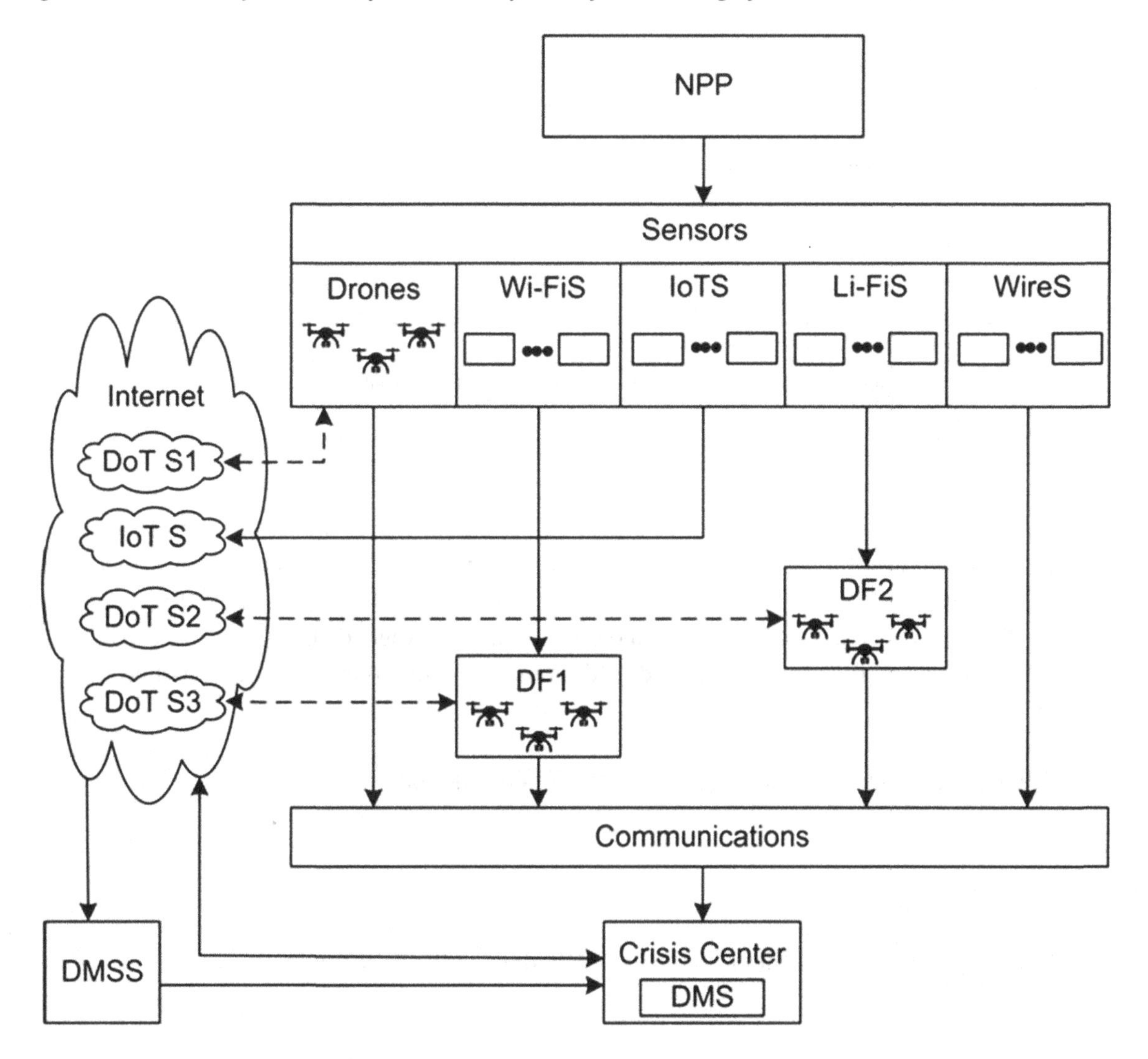

MATRIX OF DRONE FLEET RELIABILITY ASSESSMENT ATTRIBUTES

Abbreviations and Notations

To describe a matrix of drone fleet reliability assessment attributes, let us introduce abbreviations and notations presented in Table 1.

As a drone fleet/multi-fleet can use both various options for its formation and various standby redundancy techniques, let us present a set of reliability drone fleets assessment attributes as a matrix shown by means of Table 2.

The following attributes of drones and drone fleets are considered:

- **Type:** Single drone, fleet of drones, multi-fleet of drones.

Table 1. Abbreviations and notations needed to describe the matrix of drone fleet reliability assessment attributes

Abbreviations	Notations
DDL	The different distribution law for drones
DFT	Redundant drones have different flight time (FT) from a starting position to a failed drone/drone fleet. FT = Switching-on time of redundant drone(s) + Time of flight to change failed drone(s)
MG	The main group of drones (main drone fleet)
N	No
RG	The redundant group of drones (reserve drone fleet)
SDL	The same distribution law for drones
SFT	Redundant drones have the same flight time from a starting position to a failed drone/drone fleet
Y	Yes

- Parameters describing the number of performed functions (one or more than one function; for example, functions of video, measurement, control), state levels ("Y" is used to show that a fleet/multi-fleet is an MSS, "N" is used to show that a fleet/multi-fleet is not an MSS, i.e. a fleet/multi-fleet has two states only: up state and down state), heterogeneousness ("Y" is used to show that heterogeneous fleets are considered, "N" is used to show that heterogeneous fleets are not considered).

The following attributes of drone fleet reliability are considered:

- Irredundant.
- Redundant.

The following redundancy techniques (attribute "Redundant") are considered:

- Hot standby.
- Cold standby.
- Mixed standby.

Note that in order to assess the reliability of drone fleets, the following recovery policies can be analyzed:

- No recovery, no delays for changing failed drones by redundant ones (FT = 0).
- No recovery of failed drone(s). Unavailability is caused by delays of changing failed drones by redundant ones (if FT > 0).

Recovery of failed drones is carrying out at the starting position. In this case possibility/probability of returning failed/partially failed/maintained drones can be considered. For subcases when FT = 0 or FT > 0, assessment of availability can be done.

Table 2. Matrix of drone fleet reliability assessment attributes

Drones/Drone Fleets					Reliability														
Types	Parameters				Irredundant		Redundant												
					SFT	DFT	Hot standby				Cold standby				Mixed				
							SDL		DDL		SDL		DDL		SDL		DDL		
	Functions	State Levels	Heterogeneousness				SFT	DFT	SFT	DFT	SFT	DFT	SFT	DFT	SFT	DFT	SFT		
Single	1	-	-																
	>1	Y	-																
		N																	
Fleet	1	Y	Y																
			N																
		N	Y																
			N																
	>1	Y	Y																
			N																
		N	Y																
			N																
Multi-fleet	>1	Y	Y	MG															
				RG															
			N	MG															
				RG															
		N	Y	MG															
				RG															
			N	MG															
				RG															

MULTI-FLEET OF DRONES RELIABILITY MODELS

Abbreviations and Notations

To describe multi-fleet of drones reliability models, let us introduce abbreviations and notations presented in Table 3.

Table 3. Abbreviations and notations needed to describe the multi-fleet of drones reliability models

Abbreviations	Notations
MFD	multi-fleet of drones
NPP	nuclear power plant
RBD	reliability block diagram
RDF	reserve drone fleet
SCS	system of control stations
PFFO	probability of failure-free operation
DF1, DF2, …, DFn	main drone fleets
CS1	control station of the multi-fleet of drones
CS2	control station of a main drone fleet
$d_1, d_2, \ldots, d_k$	drones under redundancy of a main drone fleet
$d_{k+1}, d_{k+2}, \ldots, d_\omega$	redundant drones of a main drone fleet
$d_1, d_2, \ldots, d_r$	drones of the reserve drone fleet
$F(k,\omega)$	probability of failure-free operation of a main drone fleet
$P(n,k,\omega,0)$	probability of failure-free operation of the multi-fleet of drones with no drones in the reserve drone fleet
$P(n,k,\omega,1)$	probability of failure-free operation of the multi-fleet of drones with 1 drone in the reserve drone fleet
$P(n,k,\omega,2)$	probability of failure-free operation of the multi-fleet of drones with 2 drones in the reserve drone fleet
$P(n,k,\omega,r)$	probability of failure-free operation of the multi-fleet of drones with *r* drones in the reserve drone fleet
$R(n,k,\omega)$	probability of the state when 1 drones of the reserve drone fleet is used to change a failed drone of a main drone fleet
P_d	probability of failure-free operation of a drone of the multi-fleet of drones
p_{CS1}	probability of failure-free operation of a control station of the multi-fleet of drones
p_{CS2}	probability of failure-free operation of a main fleet control station

Description of Multi-Fleet of Drones

Consider a multi-fleet that is used for NPP monitoring missions. The multi-fleet comprises n main drone fleets and one reserve drone fleet. Each main drone fleet, consisting ω drones, has k drones under redundancy and ω-k redundant drones. The RDF consists of r drones. A drone of the RDF can be used to replace a failed drone of a main drone fleet. Using the matrix from Table 17.4, we can describe the MFD in the following way:

Figure 2. RBD for the MFD with the centralized (irredundant) SCS

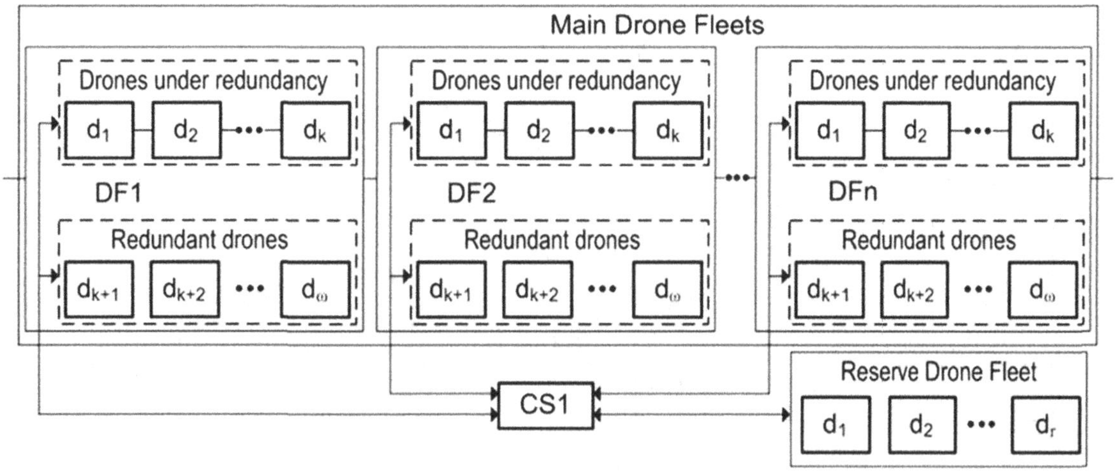

- Type of Drone Fleets: Multi-fleet.
- Functions: >1.
- State Levels: No.
- Heterogeneousness: No.
- Reliability: Redundant.
- Redundant: Hot standby.
- Hot standby: the same distribution law for drones.

Assumptions

The construction of the RBD is based on the following assumptions:

- Drone fleets are unrecoverable.
- Each of the drones has two states: up state and down state.
- Drones' failures are independent.
- Both a main drone fleet and the MFD have a structure of type "k-out-of-n".
- Replacement of failed drones of a main drone fleet is fulfilled by drones of the RDF after failure of a main drone fleet only.
- No delays for changing failed drones by redundant ones.
- All drones of the MFD are identical.

Structures of Systems of Control Stations for the Multi-Fleet of Drones

Depending on the number of control levels, SCS can have one of the following structures:

- Centralized (irredundant) SCS.
- Centralized (redundant) SCS.
- Decentralized SCS.

- Partially decentralized SCS (with z control stations for n groups of fleets, $z < n$, $n = az$, a – integer).

Figure 3. RBD for the MFD with the centralized (redundant) SCS

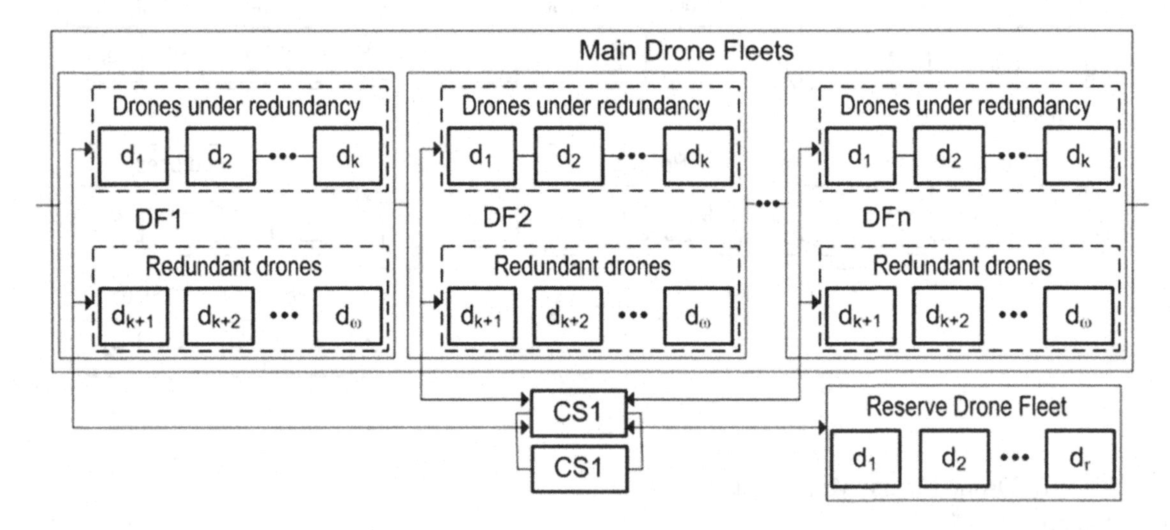

Reliability Block Diagrams

Reliability block diagrams for the described SCS structures are shown in Figure 2-5.

Figure 4. RBD for the MFD with the decentralized SCS

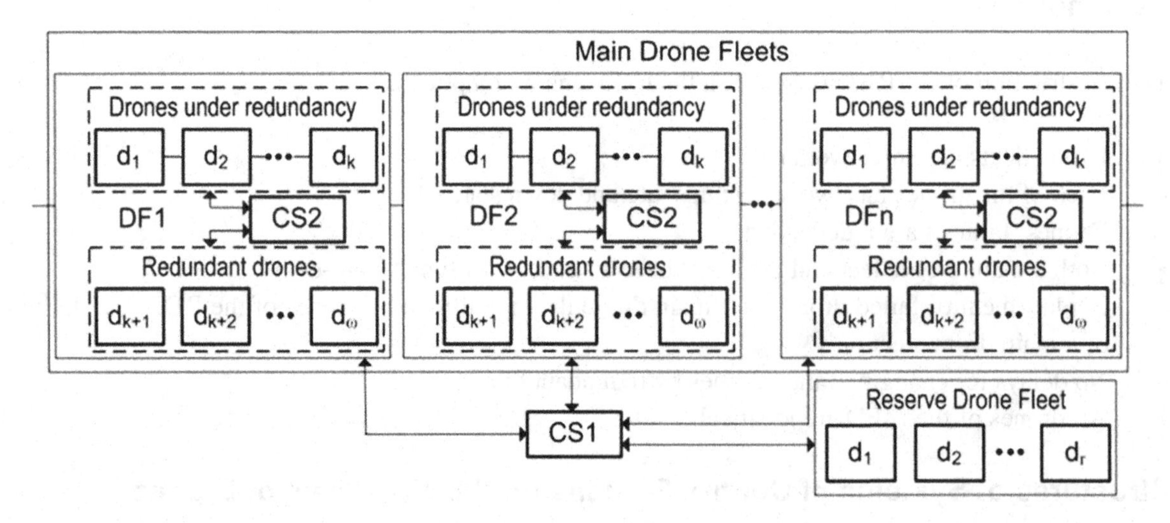

Consider an MFD performing monitoring missions near Zaporizhzhia NPP. The MFD consists of 2 main drone fleets (DF1 and DF2) and 1 RDF. Each MFD comprises 3 drones under redundancy and 1 redundant drone. RDF comprises 2 drones. Following Figure 2-5, various variants of the MFD are shown in Figure 6.

Figure 5. RBD for the MFD with the partially decentralized SCS (1 main fleet (DF1) is managed via CS1 and n-1 main fleets (DF2, ..., DFn) is managed via CS2)

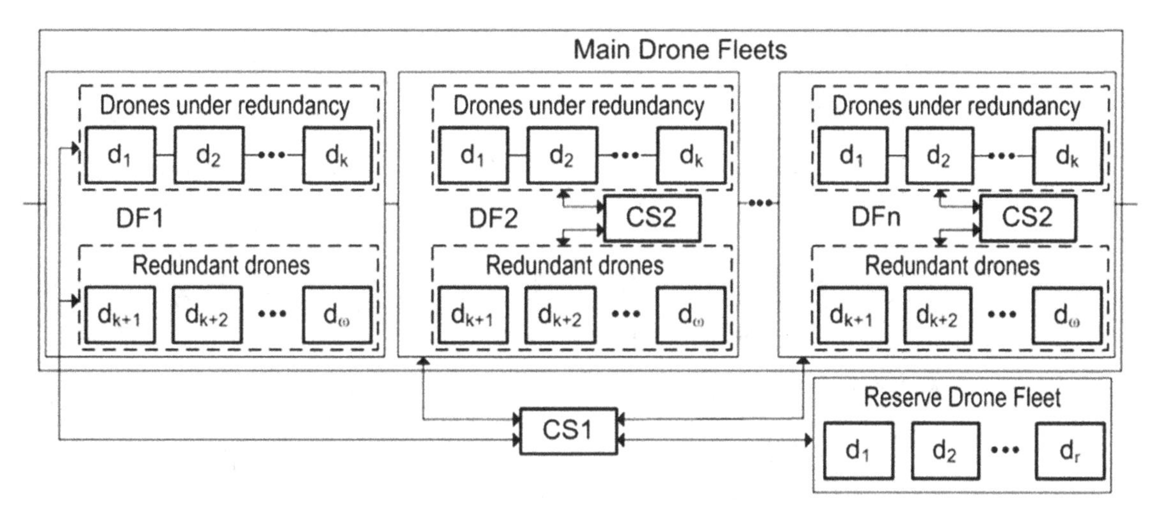

Formulae for the Probability of Failure-Free Operation of the Multi-Fleet of Drones

Based on the presented matrix and developed RBDs, formulae for the PFFO for all of the considered MFD, except for the MFD with the partially decentralized SCS (Figure 5), have been obtained (Table 4). These formulae allow making decisions and choosing one of the structures in accordance with the reliability criteria and cost.

RESULTS OBTAINED

Reliability Models Research Results for the Multi-Fleet of Drones with Decentralized SCS

Let's analyze reliability of the MFD with the decentralized SCS (Figure 4) using formulae from Table 4 regarding to this MFD. The following dependencies have been obtained:

A per cent increase in the PFFO for the multi-fleet by using 3 drones of the reserve drone fleet on the number of the main drone fleets ($p_{cs1} = 0.997$, $p_{cs2} = 0.999$, $k = 6$, $\omega = 9$) (Figure 7).

A per cent decrease in probability of the multi-fleet failure by using 3 drones of the reserve drone fleet instead of 1 drone/2 drones on the number of the main drone fleets ($p_d = 0.95$, $p_{cs1} = 0.997$, $p_{cs2} = 0.999$, $k = 6$, $\omega = 9$) (Fig. 8).

Figure 6. MFD performing monitoring missions near Zaporizhzhia NPP with a) centralized (irredundant) SCS, b) centralized (redundant) SCS), c) decentralized SCS, d) partially decentralized SCS

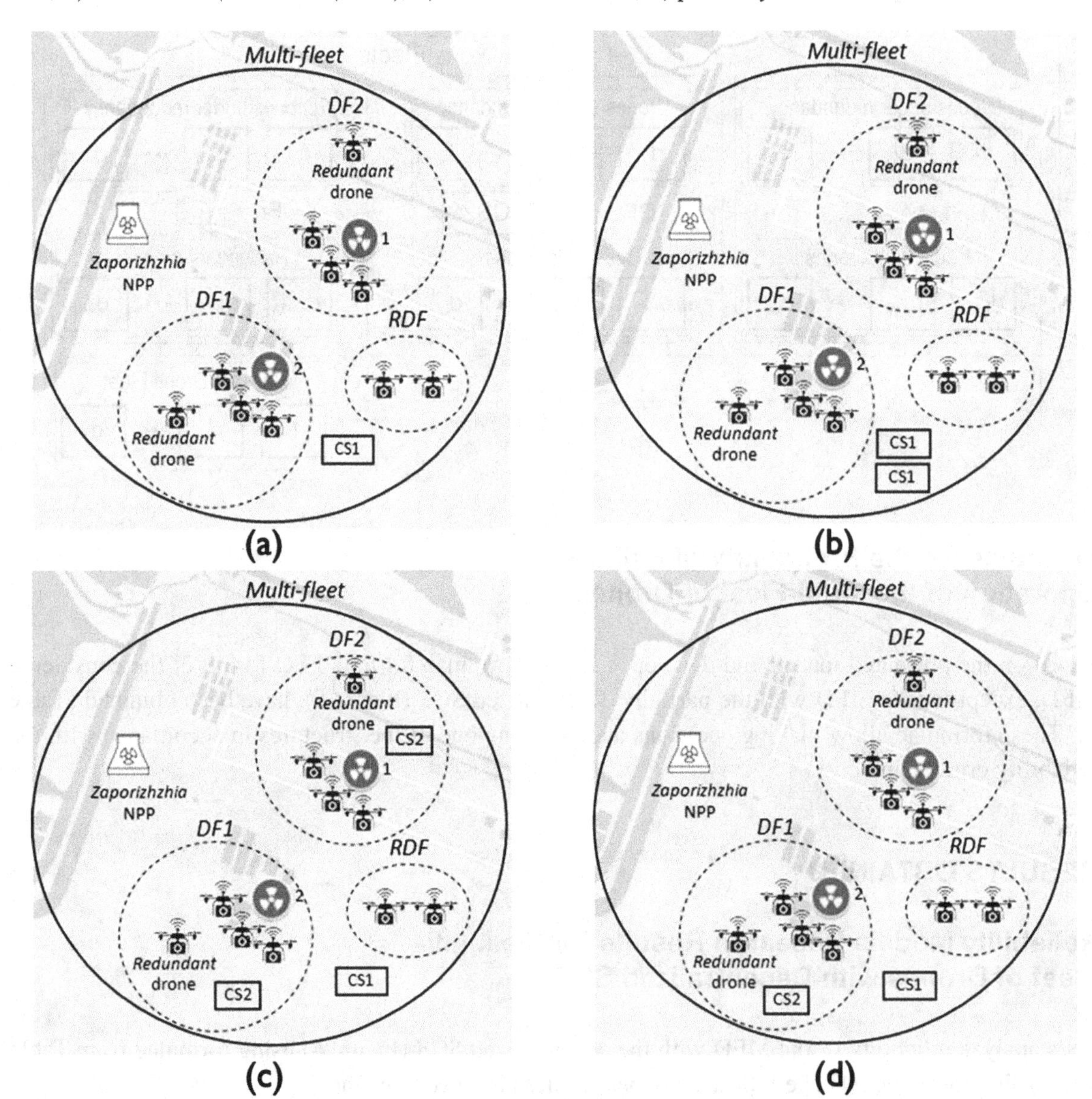

We can make the following conclusions based on the analysis of the dependencies shown in Figure 7 and 8.

- The maximum per cent increase in the PFFO for the multi-fleet by using 3 drones (7.4 per cent) is achieved when PFFO for a drone $p_d = 0.92$ and the number of the main drone fleets $n = 15$ (Figure 7), whereas the minimum per cent increase in the PFFO for the multi-fleet by using 3 drones (1.5 per cent) is achieved when PFFO for a drone $p_d = 0.99$ and the number of the main drone fleets $n = 2$.

- When 3 drones of the reserve drone fleet are used as redundant drones for the main drone fleets, growth in the number of the main drone fleets from 5 to 15 (Figure 8) leads to a per cent increase in the PFFO for the multi-fleet by 2.8, 3.4, and 4.7 per cent for $p_d = 0.99$, $p_d = 0.95$, and $p_d = 0.92$, respectively.
- A rise of 2 drones (3 drones instead of 1 drone) in the reserve drone fleet drones number (Figure 8) causes a decrease of between 3.7 per cent (the number of the main drone fleets $n = 5$) and 11,2 per cent (the number of the main drone fleets $n = 15$) in probability of the multi-fleet failure.

Table 4. Abbreviations and notations needed to describe the multi-fleet of drones reliability models

Parameter	Formula		
	MFD with the centralized (irredundant) SCS	MFD with the centralized (redundant) SCS	MFD with decentralized SCS
$F(k,\omega)$	$\sum_{i=0}^{\omega-k} C_\omega^i p_d^{\omega-i}(1-p_d)^i$	$\sum_{i=0}^{\omega-k} C_\omega^i p_d^{\omega-i}(1-p_d)^i$	$\sum_{i=0}^{\omega-k} C_\omega^i p_d^{\omega-i}(1-p_d)^i p_{CS2}$
$P(n,k,\omega,0)$	$F^n(k,\omega)p_{CS1}$	$F^n(k,\omega)\left[1-(1-p_{CS1})^2\right]$	$F^n(k,\omega)$
$R(n,k,\omega)$	$n\left[1-F(k,\omega)\right]p_d F^{n-1}(k,\omega)p_{CS1}$	$n\left[1-F(k,\omega)\right]p_d F^{n-1}(k,\omega)\times\left[1-(1-p_{CS1})^2\right]$	$n\left[1-F(k,\omega)\right]p_d F^{n-1}(k,\omega)p_{CS1}$

Figure 7. Per cent increase in the PFFO for the multi-fleet by using 3 drones of the reserve drone fleet on the number of the main drone fleets

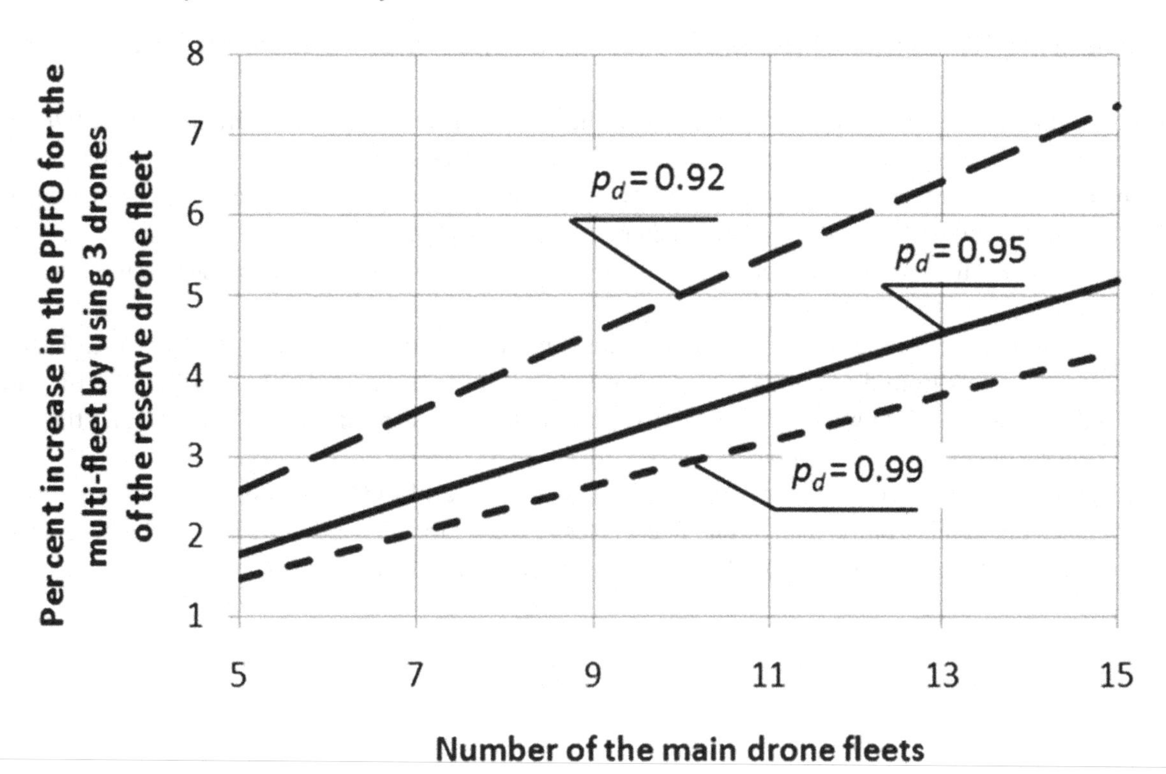

Figure 8. Per cent decrease in probability of the multi-fleet failure by using 3 drones of the reserve drone fleet instead of 1 drone/2 drones on the number of the main drone fleets

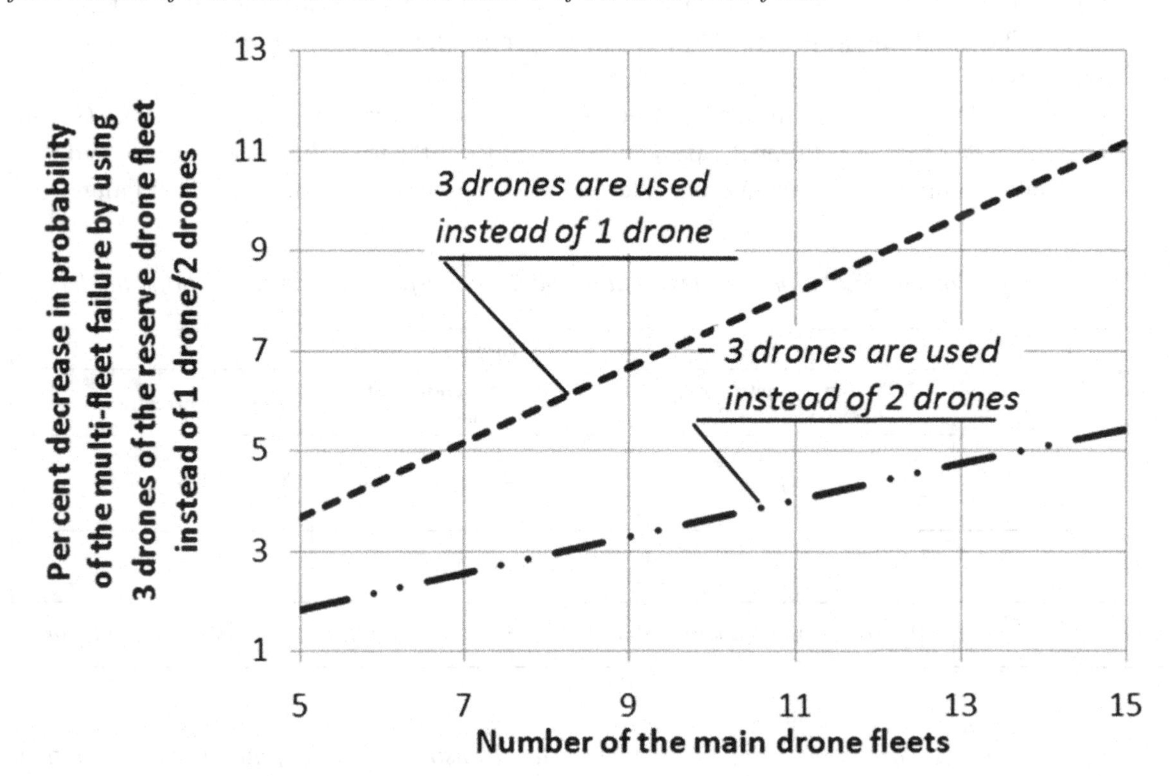

Comparing Reliability of the Multi-Fleet of Drones with Various SCS Structures

Using formulae from Table 4, the dependency of the PFFO for the MFD with 3 drones in the RDF on the number of the main drone fleets have been obtained (Figure 9) ($p_{cs1} = p_{cs2} = 0.99$, $p_d = 0.92$, $k = 6$, $\omega = 9$). The centralized (irredundant), centralized (redundant), and decentralized systems of control stations for the multi-fleet of drones have been taken into consideration.

We can make the following conclusions based on the analysis of the dependency shown in Figure 9.

- When 3 drones of the RDF are used as redundant ones for the main drone fleets, growth in the number fleets from 5 to 15 leads to decrease in the PFFO for the MFD by 0.0008, 0.0014, and 0.0012 for the MFD with the centralized (irredundant), the centralized (redundant), and the decentralized SCSs, respectively.
- The MFD with the centralized (redundant) SCS has the best PFFO value among other structures.
- When the number of the main drone fleets is less than 12, the use of the MFD with the decentralized SCS is preferable to the MFD with the centralized (irredundant) SCS.

Figure 9. Dependency of the PFFO for the MFD with 3 drones in the RDF on the number of the main drone fleets

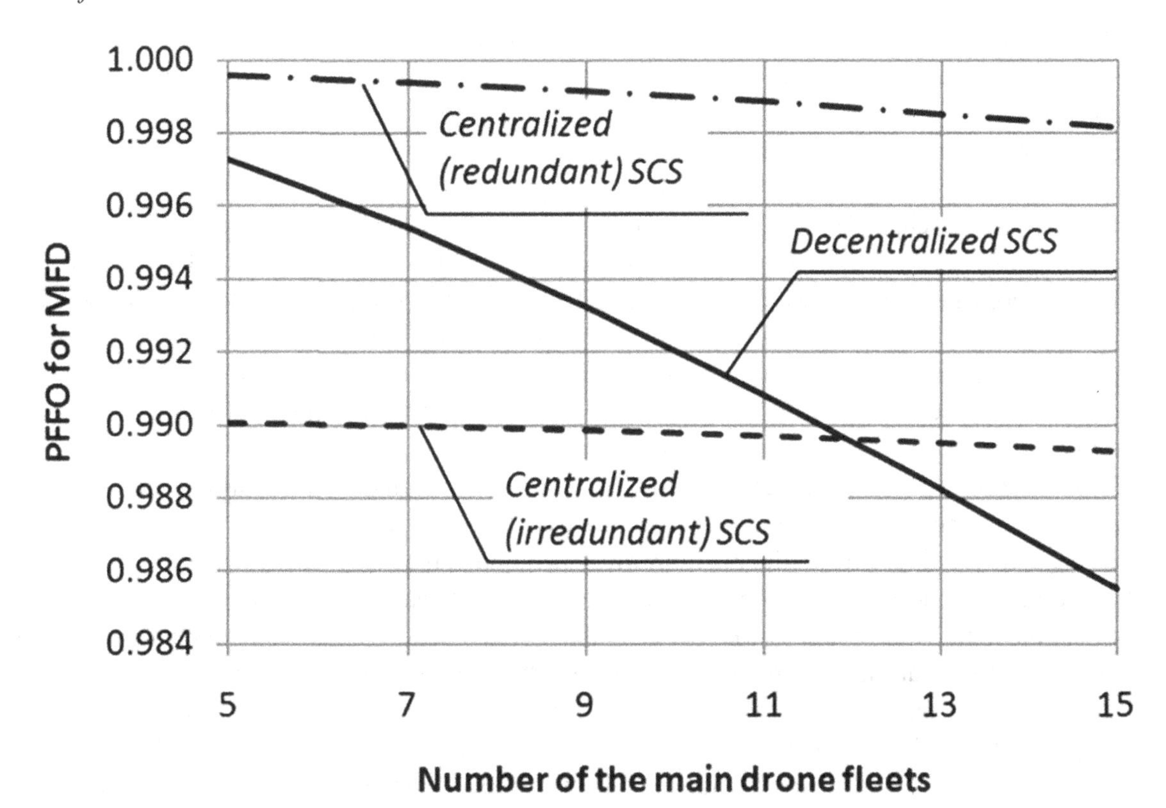

SOLUTIONS AND RECOMMENDATIONS

Based on the presented results the following recommendations on choice of the SCS structure can be formulated:

- To obtain the maximum complexity for the centralized (irredundant) MFD SCS allowing the MFD to be as reliable as the MFD with the decentralized SCS.
- To obtain the maximum complexity for the centralized (redundant) MFD SCS allowing the MFD to be as reliable as the MFD with the decentralized SCS.

FUTURE RESEARCH DIRECTIONS

Future research directions can be devoted to developing reliability models for a multi-fleet with different drones or for a recoverable multi-fleet.

CONCLUSION

A matrix of drone fleet reliability assessment attributes has been presented. The matrix covers the following attributes: standby redundancy techniques, types of drones/drone fleets, functions, state level, heterogeneousness.

A structure of a multi-fleet of drones, consisting of main drone fleets and a reserve drone fleet, has been considered.

The following dependencies are obtained for the multi-fleet of drones with the decentralized system of control stations: 1) a per cent increase in the probability of failure-free operation for the multi-fleet by using 3 drones of the reserve drone fleet on the number of the main drone fleets; 2) a per cent decrease in probability of the multi-fleet failure by using 3 drones of the reserve drone fleet instead of 1 drone/2 drones on the number of the main drone fleets.

It is shown that reliability of the multi-fleet can be improved both by adding drones to the reserve drone fleet and using more reliable drones for the multi-fleet.

The centralized (irredundant), centralized (redundant), decentralized, and partially decentralized systems of control stations for the multi-fleet of drones are suggested.

The dependency of the probability of failure-free operation of the multi-fleet of drones with 3 drones in the reserve drone fleet on the number of the main drone fleets have been obtained. The dependency shows that the multi-fleet of drones with the centralized (redundant) system of control stations has the best probability of failure-free operation value among other structures.

Based on the obtained results, tasks on choice of the systems of control stations structures are formulated and can be solved.

The new research will aim at developing the reliability models for a multi-state multi-fleet of drones with non-identical drones.

REFERENCES

Aghaei, M., Hamadani, A. Z., & Ardakan, M. A. (2017). Redundancy allocation problem for k-out-of-n systems with a choice of redundancy strategies. *J. Ind. Eng. Int.*, *1*(13), 81–92. doi:10.100740092-016-0169-3

Byun, J.-E., Noh, H.-M., & Song, J. (2017). Reliability growth analysis of k-out-of-n systems using matrix-based system reliability method. *Reliability Engineering & System Safety*, *165*, 410–421. doi:10.1016/j.ress.2017.05.001

Connora, D., Martin, P. G., & Scott, T. B. (2016). Airborne radiation mapping: Overview and application of current and future aerial systems. *International Journal of Remote Sensing*, *24*(37), 5953–5987. doi:10.1080/01431161.2016.1252474

Eryilmaz̧, S., & Sarikaya, K. (2017). Modeling and analysis of weighted-kout-of-n: G system consisting of two different types of components. *J. Risk and Reliability*, *3*(228), 265–271.

Fesenko, H., Kharchenko, V., Sachenko, A., Hiromoto, R., & Kochan, V. (2018). An Internet of Drone-based multi-version post-severe accident monitoring system: structures and reliability. In *Dependable IoT for Human and Industry Modeling, Architecting, Implementation* (pp. 197–217). Denmark, The Netherlands: River Publishers.

Han, J., Xu, Y., Di, L., & Chen, Y. (2013). Low-cost multi-UAV technologies for contour mapping of nuclear radiation field. *Int. J. Intelligent & Robotic Syst.*, *1–4*(70), 401–410. doi:10.100710846-012-9722-5

Levitin, G., Xing, L., & Dai, Y. (2014). Mission cost and reliability of 1-out-of-N warm standby systems with imperfect switching mechanisms. *IEEE Transactions on Systems, Man, and Cybernetics. Systems*, *9*(44), 1262–1271. doi:10.1109/TSMC.2013.2294328

MacFarlane, J. W., Payton, O. D., Keatley, A. C., Scott, G. P. T., Pullin, H., Crane, R. A., ... Scott, T. B. (2014). Lightweight aerial vehicles for monitoring, assessment and mapping of radiation anomalies. *Journal of Environmental Radioactivity*, *136*, 127–130. doi:10.1016/j.jenvrad.2014.05.008 PMID:24949582

Martin, P. G., Payton, O. D., Fardoulis, J. S., Richards, D. A., Yamashiki, Y., & Scott, T. B. (2016). Low altitude unmanned aerial vehicle for characterising remediation effectiveness following the FD-NPP accident. *Journal of Environmental Radioactivity*, *151*, 58–63. doi:10.1016/j.jenvrad.2015.09.007 PMID:26410790

Mo, Y., Xing, L., Amari, S. V., & Dugan, J. B. (2015). Efficient analysis of multi-state k-out-of-n system. *Reliability Engineering & System Safety*, *133*, 95–105. doi:10.1016/j.ress.2014.09.006

Pascual-Ortigosa, P., Sáenz-De-Cabezón, E., & Wynn, H. P. (2018). Algebraic analysis of multistate k-out-of-n systems. *Monografías de la Real Academia de Ciencias*, *43*, 131–134.

Ram, M., & Dohi, T. (2019). *Reliability Analysis Using k-out-of-n Structures*. Boca Raton, FL: CRC Press. doi:10.1201/9781351056465

UAV Drone Radiation Monitoring. (n.d.). Retrieved November 1, 2019, from http://www.aretasaerial.com/products/uav-drone-radiation-monitoring

Compilation of References

10 . CFR Part 50, Appendix B, Quality Assurance Criteria for Nuclear Power Plants and Fuel Reprocessing Plants.

Abrial, J.-R. (2010). *Modeling in Event-B*. Cambridge University Press. doi:10.1017/CBO9781139195881

Adziev, A. V. (1998). Myths about software safety: Lessons of famous disasters. *Open Systems, 6*. Retrieved December 16, 2012, from http://www.osp.ru/os/1998/06/179592/

Aghaei, M., Hamadani, A. Z., & Ardakan, M. A. (2017). Redundancy allocation problem for k-out-of-n systems with a choice of redundancy strategies. *J. Ind. Eng. Int.*, *1*(13), 81–92. doi:10.100740092-016-0169-3

Aizenberg, A., & Yastrebenetsky, M. (2002). Comparison of safety management principles for control systems of carrier rockets and nuclear power plants. *Space Science and Technology*, *1*, 55–60.

Aloul, F., Al-Ali, A. R., Al-Dalky, R., Al-Mardini, M., & El-Hajj, W. (2012). Smart Grid Security: Threats, Vulnerabilities and Solutions. *International Journal of Smart Grid and Clean Energy*, *1*(1), 1–6. doi:10.12720gce.1.1.1-6

Anderson, R. B. (1979). *Proving Programs Correct*. New York: Wiley.

Andrashov, A., Bakhmach, I., Kharchenko, V., & Kovalenko, A. (2019). Equipment qualification of FPGA-based platform RadICS to meet US NRC requirements. *Proceeding of the 11th International Topical Meeting on Nuclear Plant Instrumentation, Control, and Human-Machine Interface Technologies*, 327-335.

Andrashov, A., Bakhmach, I., Leontiiev, K., Babeshko, E., Kovalenko, A., & Kharchenko, V. (2019). Diversity in FPGA-based platform and platform based I&CS applications: Strategy and implementation. *Proceeding of the 11th International Topical Meeting on Nuclear Plant Instrumentation, Control, and Human-Machine Interface Technologies*, 174-182.

ANSI/ISA-99.00.01-2007. (2007). *Security for Industrial Automation and Control Systems: Terminology, Concepts, and Models*.

ANSI/ISA-99.00.02-2007. (2007). *Establishing an Industrial Automation and Control Systems Security Program*.

ANSI/ISA-99.00.03-2007. (2007). *Operating an Industrial Automation and Control Systems Security Program*.

ANSI/ISA-99.00.04-2007. (2007). *Specific Security Requirements for Industrial Automation and Control Systems*.

ANSI/ISA-99.02.01-2009. (2009). *Security for Industrial Automation and Control Systems: Establishing an Industrial Automation and Control Systems Security Program*.

ANSI/ISA-TR99.00.01-2007. (2007). *Security Technologies for Industrial Automation and Control Systems*.

Arushanyan, O. (1990). *Numerical Solution of Ordinary Differential Equations using FORTRAN*. Moscow: Moscow State University.

Avizienis, A., Laprie, J.-C., Randell, B., & Landwehr, C. (2004). Basic Concepts and Taxonomy of Dependable and Secure Computing. *IEEE Transactions on Dependable and Secure Computing*, *1*(1), 11–33. doi:10.1109/TDSC.2004.2

Babeshko, E., Kharchenko, V., Sklyar, V., Siora, A., Tokarev, V. (2011). Combined Implementation of Dependability Analysis Techniques for NPP I&C Systems Assessment. *Journal of Energy and Power Engineering*, *5*(42), 411-418.

Babeshko, E. (2008). Applying F(I)MEA-technique for SCADA-based Industrial Control Systems Dependability Assessment and Ensuring. *Proceedings of International Conference on Dependability of Computer Systems DepCoS–RELCOMEX 2008*, 309-315. 10.1109/DepCoS-RELCOMEX.2008.23

Babeshko, E., Bakhmach, I., Kharchenko, V., Ruchkov, E., & Siora, O. (2017). Operating Reliability Assessment of FPGA-Based NPP I&C Systems: Approach, Technique and Implementation. *Icon (London, England)*, E25–E66862. doi:10.1115/ICONE25-66862

Babeshko, E., Kharchenko, V., Odarushchenko, O., & Sklyar, V. (2015) Toward automated FMEDA for complex electronic product. *Proceedings of the International Conference on Information and Digital Technologies (IDT 2015)*, 22-27. 10.1109/DT.2015.7222945

Bach, J., Duthou, A., Monteil, P., & Burzynski, M. (2017). Computer security approach for ROLLS-ROYCE SPINLINE safety platform Rolls-Royce Civil Nuclear. NPIC&HMIT 2017, San Francisco, CA.

Badrignans, B. (2011). *Security Trends for FPGAS. From Secured to Secure Reconfigurable Systems*. Springer. doi:10.1007/978-94-007-1338-3

Bakhmach, E., Kharchenko, V., Siora, A., Sklyar, V., & Tokarev, V. (2009). Advanced I&C Systems for NPPS Based on FPGA Technology: European Experience. *Proceedings of 17th International Conference on Nuclear Engineering (ICONE 17).*

Barge, W. S. (2002). Autonomous Solution Methods for Large Markov Chains. *Pennsylvania State University CiteSeerX Archives*, 17.

Barkalov, A. (2006). *Design of Control Units with Programmable Logic*. University of Zelena Gura.

Basit, A., Sidhu, G. A. S., Mahmood, A., & Gao, F. (2017). Efficient and autonomous energy management techniques for the future smart homes. *IEEE Transactions on Smart Grid, Volume*, *8*(2), 917–926.

Ben-Ari, M. (2000). *Understanding programming languages*. Wiley.

Bhuvaneswari, T. (2013). A Survey on Software Development Life Cycle Models. *Monthly Journal of Computer Science and Information Technology, 2*(5), 262–267.

Biscoglio, I., & Fusani, M. (2010). Analyzing quality aspects in safety-related standards. In *Seventh American Nuclear Society International Topical Meeting on Nuclear Plant Instrumentation, Control and Human-Machine Interface Technologies NPIC&HMIT 2010*. Las Vegas, NV: American Nuclear Society.

Bobbio, A., & Trivedi. (1986). An Aggregation Technique for Transient Analysis of Stiff Markov Chains. *IEEE Transactions on Computers*, *C-35*(9), 803–814. doi:10.1109/TC.1986.1676840

Bobrek, M., Bouldin, D., & Holkomb, D. (2009). *Review Guidelines for FPGAs in Nuclera Power Plants Safety Systems*. NUREG/CR-7006 ORNL/TM-2009/020.

Brezhnev, E. (2010). Risk-analysis in critical information control system based on computing with words' model. *Proceeding of 7th International Workshop on Digital Technologies, Circuit Systems and Signal Processing*, 67-72.

Bryant, R. E. (1986). Graph-based Algorithms for Boolean Function Manipulation. *IEEE Transactions on Computers, 35*(8), 677–691. doi:10.1109/TC.1986.1676819

Buchholz, P. (1996). Structured analysis approach for Large Markov Chains. A Tutorial. Proc. of Performane'96.

Bukowsky, J., & Goble, W. (1994). An Extended Beta Model to Quantize the Effects of Common Cause Stressors. *Proceedings of ISAFECOMP*.

Butenko, V. (2014). Modeling of a Reactor Trip System Using Markov Chains: Case Study. *Proc. of 2014 22nd ICONE*, 5. 10.1115/ICONE22-31156

Byun, J.-E., Noh, H.-M., & Song, J. (2017). Reliability growth analysis of k-out-of-n systems using matrix-based system reliability method. *Reliability Engineering & System Safety, 165*, 410–421. doi:10.1016/j.ress.2017.05.001

CFR. (2015). *Title 10, Code of Federal Regulations*. Washington, DC: U.S. NRC.

Chen, Y. (2017). *The Application of FPGA-based FitRel Platform in Nuclear Power Plant Diverse Actuation System.* China Techenergy Co., Ltd (CTEC).

Christiansen, B. (2006). *Active FPGA Security through decoy circuits* (MS Thesis). Air Force Institute of Technology.

Clark, R. (2010). *Cyber War*. The Perfect Weapon.

Connora, D., Martin, P. G., & Scott, T. B. (2016). Airborne radiation mapping: Overview and application of current and future aerial systems. *International Journal of Remote Sensing, 24*(37), 5953–5987. doi:10.1080/01431161.2016.1252474

Cormen, T. (2001). *Introduction to Algorithms*. MIT Press.

Courtois, P. J. (1977). *Decomposability: Queueing and Computer Applications*. New York: Academic Press.

Cullen, W. (1990). *The Public Enquiry into the Piper Alpha Disaster*. Department of Energy.

Didenko, K., & Rozen, Yu. (1985). MicroDAT. Principles of construction, the main parameters and characteristics. *Instrumentation & Control Systems*, 11.

Drimer, S. (2009). *Security for volatile FPGAs*. Technical Report N 763, University of Cambridge Computer Laboratory.

DSTU-25010. (2016). *Systems and software engineering – Systems and software Quality Requirements and Evaluation - System and software quality models*. Author.

Dubois, D., & Prade, H. (1980). *Fuzzy Sets and Systems: Theory and Application*. New York: Academic.

Duzhyi, V., Kharchenko, V., Starov, O., & Rusin, D. (2010). Research Sports Programming Services as Multi-version Projects. *Radioelectronic and Computer Systems, 47*, 29–35.

ECSS (2008). ECSS-Q-ST-30-09. *European Cooperation for Space Standardization (ECSS): Availability analysis.*

Ehrlich, W., Lee, S. K., & Molisani, R. H. (1990). Applying Reliability Measurement: A Case Study. *IEEE Software, 1990*(2), 56–64. doi:10.1109/52.50774

EN. (2010a). *EN 61226. Nuclear power plants. Instrumentation and control important to safety. Classification of instrumentation and control functions.* EN.

EN. (2010b). *EN 55022. Information Technology Equipment - Radio Disturbance Characteristics - Limits and Methods of measurement.* EN.

EN. (2013). *EN 61513. Nuclear power plants - Instrumentation and control important to safety - General requirements for systems.* EN.

EN. (2015). *EN 60987. Nuclear power plants. Instrumentation and control important to safety. Hardware design requirements for computer-based systems.* EN.

EPRI TR1019181, *Guidelines on the Use of Field Programmable Gate Arrays (FPGAs) in Nuclear Power Plant I&C Systems*, Electric Power Research Institute, (2009).

EPRI TR1022983, *Recommended Approaches and Design Criteria for Application of Field Programmable Gate Arrays in Nuclear Power Plant I&C Systems*, Electric Power Research Institute, (2011).

EPRI. (1996). *TR 107330, Generic Requirements Specification for Qualifying a Commercially Available PLC for Safety Related Applications in Nuclear Power Plants.* Palo Alto, CA: Electric Power Research Institute.

Eryilmaz, S., & Sarikaya, K. (2017). Modeling and analysis of weighted-kout-of-n: G system consisting of two different types of components. *J. Risk and Reliability*, *3*(228), 265–271.

Everett, W., Keene, S., & Nikora, A. (1998). Applying Software Reliability Engineering in the 1990s. *IEEE Transactions on Reliability*, *47*(3-SP), 372-378.

Evidence: Using safety cases in industry and healthcare. (2012). Health Foundation.

EXIDA.com LLC. (n.d.). *Functional Safety - An IEC 61508 SIL 3 Compliant Development Process* (3rd ed.). Author.

Fesenko, H., Kharchenko, V., Sachenko, A., Hiromoto, R., & Kochan, V. (2018). An Internet of Drone-based multi-version post-severe accident monitoring system: structures and reliability. In *Dependable IoT for Human and Industry Modeling, Architecting, Implementation* (pp. 197–217). Denmark, The Netherlands: River Publishers.

Fink, R., Killian, C., & Nguyen, T. (Eds.). (2011). Recommended Approaches and Design Criteria for Application of Field Programmable Gate Arrays in Nuclear Power Plant Instrumentation and Control Systems. EPRI.

Firesmith, D., Capell, P., Elm, J., Gagliardi, M., Morrow, T., Roush, L., & Sha, L. (2006). *QUASAR: A Method for the Quality Assessment of Software-Intensive System Architectures.* CMU/SEI-2006-HB-002.

Gail, H. R. (1989). Calculating Availability and Performability Measures of Repairable Computer Systems. *Journal of the Association for Computing Machinery*, *36*(1), 171–193. doi:10.1145/58562.59307

GAO-04-321, *Cybersecurity for Critical Infrastructure Protection*, U.S. General Accounting Office, (2004).

Goddard, P. L. (2000) Software FMEA techniques. *Proceedings of 2000 Reliability and Maintainability Symposium.*

González, C. J. (2019). Reducing Soft Error Rate of SoCs Analog-to-Digital Interfaces with Design Diversity Redundancy. *IEEE Transactions on Nuclear Science.*

Goodenough, J., Weinstock, C., & Klein, A. (2015). *Eliminative Argumentation: A Basis for Arguing Confidence in System Properties, Technical Report, CMU/SEI-2015-TR-005.* Pittsburgh, PA: CMU/SEI.

Gorbenko, A., Kharchenko, V., & Romanovsky, A. (2009). Using Inherent Service Redundancy and Diversity to Ensure Web Services Dependability. In M. Butler, C. Jones, A. Romanovsky, & E. Troubitsyna (Eds.), *Methods, Models and Tools for Fault Tolerance* (pp. 324–341). Springer. doi:10.1007/978-3-642-00867-2_15

Gordieiev Oleksandr & Kharchenko Vyacheslav. (2018). IT-oriented software quality models and evolution of the prevailing characteristics. In *Proceedings of the 9 International Conference Dependable Systems, Services and Technologies - DESSERT'2018*, (pp. 390-395). Kyiv, Ukraine: Academic Press.

Grabenko, A., Bakhmach, E., Siora, A., & Tokarev, V. (2005). Cabinets modernized in a unified complex of technical equipment UCTM -MD, RT-MD. New developments of control and protection systems for PWR reactors. *Proceedings of the 4th international scientific and technical conference "Ensuring the Safety of NPPs with PWR".*

Grand, J. (2004, March). Practical Secure Hardware Design for Embedded Systems. *Proc. of the 2004 Embedded Systems Conference.*

Graydon, P., & Knight, J. (2009). *Assurance Based Development.* Technical Report CS-2009-10. University of Virginia.

GSN Community Standard. (2011). Version 1. Origin Consulting (York) Limited.

Hairer, E. (2010). Solving Ordinary Differential Equations II: Stiff and Differential-Algebraic Problems. Springer.

Han, J., Xu, Y., Di, L., & Chen, Y. (2013). Low-cost multi-UAV technologies for contour mapping of nuclear radiation field. *Int. J. Intelligent & Robotic Syst., 1–4*(70), 401–410. doi:10.100710846-012-9722-5

Hawkins, R., Habli, I., Kelly, T., & McDermid, J. (2013). Assurance Cases and Prescriptive Software Safety Certification: A Comparative Study. *Safety Science, 59*, 55–71. doi:10.1016/j.ssci.2013.04.007

Hayrer, E. (1999). *Solution of Ordinary Differential Equations. Stiff and Differential-Algebraic Poblems.* Mir.

HAZCADS. (2018). *Hazards and Consequences Analysis for Digital Systems.* Palo Alto, CA: EPRI.

Huang, H.-W., Wang, L.-H., Liao, B.-C., Chung, H.-H., & Jiin-Ming, L. (2011). Software safety analysis application of safety-related I&C systems in installation phase. *Progress in Nuclear Energy, 6*(53), 736–741. doi:.2011.04.002 doi:10.1016/j.pnucene

Huffmire, T. (2010). *Handbook of FPGA Design Security.* Springer. doi:10.1007/978-90-481-9157-4

Hurley, N., & Rickard, S. (2009). Comparing measures of sparsity. *IEEE Transactions on Information Theory, 55*(10), 4723–4741. doi:10.1109/TIT.2009.2027527

IAEA (2000). IAEA NS-G-1.1. Software for computer based systems important to safety in nuclear power plants.

IAEA. (1980). *50-SG-D3. Protection systems and related features in nuclear power plants.* Vienna, Austria: IAEA.

IAEA. (1984). *50-SG-D8. Safety related instrumentation and control systems for nuclear power plants.* Vienna, Austria: IAEA.

IAEA. (1988). *50-C-D. Code on the safety on nuclear power plants: Design.* Vienna, Austria: IAEA.

IAEA. (1998). *TECDOC-1016. Modernization of Instrumentation and control in nuclear power plants.* Vienna, Austria: IAEA.

IAEA. (1999). Modern instrumentation and control for nuclear power plants / Guidebook. Technical reports series, N°387. Vienna, Austria: IAEA.

IAEA. (1999a). INSAG-12. Basic safety principles for nuclear power plants. 75-INSAG-3, Rev. 1, Vienna, Austria: IAEA.

IAEA. (1999b). *Modern instrumentation and control for nuclear power plants: A Guidebook.* Technical report series N°387. Vienna, Austria: IAEA.

IAEA. (2000). *NS-G-1.1. Software for computer based systems important to safety in nuclear power plants: Safety guide.* Vienna, Austria: IAEA.

IAEA. (2001). The international nuclear event scale (INES). User's manual. Vienna, Austria: IAEA.

IAEA. (2002). *NS-G-1.3. Instrumentation and control systems important to safety in nuclear power plants: Safety guide*. Vienna, Austria: IAEA.

IAEA. (2006). *Safety Standards Series No. SF-1. Fundamental safety principles: safety fundamentals*. Vienna, Austria: IAEA.

IAEA. (2006). *SF-1. Fundamental safety principles*. Vienna, Austria: IAEA.

IAEA. (2007). *Terminology used in Nuclear Safety and Radiation Protection: IAEA Safety Glossary*. Vienna, Austria: IAEA.

IAEA. (2009). Safety assessment for facilities and activities: General safety requirements. IAEA safety standards series No. GSR Part 4. Vienna, Austria: IAEA.

IAEA. (2010). Governmental, legal and regulatory framework for safety: General safety requirements. IAEA safety standards series No. GSR Part 1, Vienna, Austria; IAEA.

IAEA. (2010). *SSG-90. Seismic Hazards in Site Evaluation for Nuclear Installations*. IAEA.

IAEA. (2011). *NP-T-3.12. Core knowledge on instrumentation and control systems in nuclear power plants*. Vienna, Austria: IAEA.

IAEA. (2011). *Nuclear Security Series No. 17, Computer security at nuclear facilities: reference manual: technical guidance*. Vienna, Austria: IAEA.

IAEA. (2011). *Nuclear Security Series No. 17. Computer security at nuclear facilities: reference manual: technical guidance*. Vienna, Austria: IAEA.

IAEA. (2011, a). SSR-2/2. Safety of nuclear power plants: commissioning and operation. Vienna, Austria: IAEA.

IAEA. (2012). *SSR-2/1. Safety of nuclear power plants: design*. Vienna, Austria: IAEA.

IAEA. (2013b). *The Statute of the IAEA*. www.iaea.org/About/statute.html

IAEA. (2014). IEC (1980). IEC 60780. Nuclear power plants – Electrical equipment of the safety system – Qualification. Ed. 2. IAEA.

IAEA. (2014). *SSG-30. Safety classification of structures, systems and components in nuclear power plants*. Vienna, Austria: IAEA.

IAEA. (2015). *Nuclear Security Series No. 23-G. Security of nuclear information*. Vienna, Austria: IAEA.

IAEA. (2016). *Conducting Computer Security Assessment at Nuclear Facilities*. Vienna, Austria: IAEA.

IAEA. (2016). *Safety Standards Series No. SSG-39. Design of Instrumentation and Control Systems for Nuclear Power Plants*. Vienna, Austria: IAEA.

IAEA. (2016). *SSG-39. Design of instrumentation and control systems for nuclear power plants*. Vienna, Austria: IAEA.

IAEA. (2016a). *IAEA SSG-39. Design of Instrumentation and Control Systems for Nuclear Power Plants*. IAEA.

IAEA. (2016b). *IAEA SSR-2/1. Safety of Nuclear Power Plants: Design*. IAEA.

IAEA. (2016c). *IAEA SSR-2/2. Safety of Nuclear Power Plants: Commissioning and Operation*. IAEA.

IAEA. (2017). *NST047. Computer security techniques for nuclear facilities*. Draft.

IAEA. (2017). *Nuclear Security Series No. 28-T, Self-assessment of nuclear security culture in facilities and activities: technical guidance*. Vienna, Austria: IAEA.

IAEA. (2018). *Nuclear Security Series No. 33-T, Computer security of instrumentation and control systems at nuclear facilities*. Vienna, Austria: IAEA.

IAEA. (2018). *Nuclear Security Series No. 33-T. Computer security of instrumentation and control systems at nuclear facilities*. Vienna, Austria: IAEA.

IEC (2004) IEC 62138, Nuclear power plants – Instrumentation and control important for safety – Software aspects for computer-based systems performing category B or C functions.

IEC (2006) IEC 60880, Nuclear power plants – Instrumentation and control systems important to safety – Software aspects for computer-based systems performing category A functions.

IEC (2006, a). IEC 60800. Nuclear power plants – Instrumentation and control system important for safety – Software aspects for computer-based systems performing category A functions.

IEC (2006, b). IEC 61165. *Application of Markov technqiues*.

IEC (2006,a). IEC 60812. *Analysis technique for system reliability – Procedure for Failure Mode and Effects Analysis (FMEA)*.

IEC (2006,b). IEC 60880. *Nuclear power plants – Instrumentation and control systems important to safety – Software aspects for computer-based systems performing category A functions*.

IEC (2006,c). IEC 61025. *Fault tree analysis*.

IEC (2008). IEC 60300-3-3. Application guide – Analysis techniques for dependability – Guide on methodology.

IEC (2008). IEC 61508. *Functional Safety of Electrical/Electronic/Programmable Electronic Safetyrelated Systems*.

IEC (2009) IEC 61226, Nuclear power plants - Instrumentation and control important to safety - Classification of instrumentation and control functions.

IEC (2010) IEC 61508, Electric / Electronic / Programmable Electronic safety-related systems, parts 1-7.

IEC (2010). IEC 61508. *Functional Safety of Electrical/Electronic/ Programmable Electronic Safety-Related Systems*.

IEC 60880. (2006). *Nuclear power plants – Instrumentation and control systems important to safety – Software aspects for computer-based systems performing category A functions*.

IEC 61508: Edition 2. (2010). *Functional Safety of Electrical/Electronic/Programmable Electronic Safety-related Systems*.

IEC 61513. (2011). *Nuclear power plants – instrumentation and control important to safety – General requirements for systems*. Ed. 2.

IEC 62138. (2004). *Nuclear power plants – Instrumentation and control important for safety – Software aspects for computer-based systems performing category B or C functions*.

IEC 62566. (2010). *Nuclear Power Plants – Instrumentation and control important to safety – Hardware language aspects for systems performing category A functions*. Ed.1.

IEC 62645. (2011). *Nuclear power plants - Instrumentation and control systems - Requirements for security programmes for computer-based systems*. Ed. 1.

IEC. (1991). *60821, VME bus – Microprocessor system bus for 1 byte to 4-byte data.*

IEC. (2004a). IEC 60709. Nuclear power plants – Instrumentation and control systems important to safety – Separation. Ed. 2. IEC.

IEC. (2004b). *IEC 62138. Nuclear power plants – Instrumentation and control important for safety – Software aspects for computer-based systems performing category B or C functions.* IEC.

IEC. (2005). *IEC 62138. Nuclear power plants- Instrumentation and control important for safety. Software aspects for computer-based systems performing category B or C function.* IEC.

IEC. (2006). *60880, Nuclear power plants – Instrumentation and Control Systems important to Safety – Software aspects for computer-based systems performing category A functions.*

IEC. (2006). *IEC 60880. Nuclear power plants - Instrumentation and control systems important to safety - Software aspects for computer-based systems performing category A functions.* IEC.

IEC. (2006). *IEC 60880. Nuclear power plants – Instrumentation and control systems important to safety – Software aspects for computer-based systems performing category A functions.* IEC.

IEC. (2007). *IEC 60987. Nuclear power plants - Instrumentation and control important for safety - Programmed digital computers important to safety for nuclear power stations.* IEC.

IEC. (2007a). *IEC 62342. Nuclear power plants – Instrumentation and control systems important to safety – Management of aging.* IEC.

IEC. (2007b). *IEC 62340. Nuclear power plants – Instrumentation and control systems important to safety – Requirements for coping with common cause failure (CCF).* IEC.

IEC. (2007c). *IEC 60987. Nuclear power plants – Instrumentation and control important to safety – Hardware design requirements for computer-based systems.* IEC.

IEC. (2008). 60297-3-100, Mechanical structures for electronic equipment – Dimensions of mechanical structures – Basic dimensions of front panels, subracks, chassis, racks and cabinets.

IEC. (2008). IEC 61508. Ed.2. Functional safety of electrical/electronic/programmable electronic safety related systems. IEC.

IEC. (2008). *IEC 61508. Functional safety of Electrical/Electronic/Programmable Electronic safety-related systems.* IEC.

IEC. (2009). *61226. Nuclear power plants – Instrumentation and control important to safety — Classification of instrumentation and control functions.* Geneva, Switzerland: IEC.

IEC. (2009). *IEC 61226. Nuclear power plants - Instrumentation and control systems important to safety – Classification.* IEC.

IEC. (2009a). *61226, Nuclear power plants – I&C systems important to safety – Classification of I&C functions.*

IEC. (2009a). *IEC 60964. Nuclear power plants – Control rooms – Design.* IEC.

IEC. (2009b). *62003, Nuclear power plants – Instrumentation and control important to safety – Requirements for electromagnetic compatibility testing.*

IEC. (2009b). IEC 61226 Nuclear Power Plants – Instrumentation and control important to safety–Classification of instrumentation and control functions. Ed. 3. IEC.

IEC. (2009c). IEC 61500. Nuclear power plants – Instrumentation and control important to safety – Data communication in systems performing category A functions. Ed. 2. IEC.

IEC. (2011). *61513, Nuclear power plants – Instrumentation and control for systems important to safety – General requirements for systems.*

IEC. (2011). IEC 61513. Ed.2. Nuclear power plants – Instrumentation and control important to safety – General requirements for systems. IEC.

IEC. (2011). *IEC 61513. Nuclear power plants - Instrumentation and control important to safety - General requirements for systems.* IEC.

IEC. (2011a). IEC 61513. Nuclear power plants – instrumentation and control important to safety – General requirements for systems. Ed. 2. IEC.

IEC. (2012a). *60987, Nuclear power plants – Instrumentation and control important to safety – Hardware design requirements for computer-based systems.*

IEC. (2015). *62808, Nuclear power plants – Instrumentation and control systems important to safety – Design and qualification of isolation devices.*

IEC. (2016). *68859 + AMD1:2019 CSV. Nuclear power plants — Instrumentation and control systems — Requirements for coordinating safety and cybersecurity.* Geneva, Switzerland: IEC.

IEC. (2019). *62645, Nuclear power plants — Instrumentation, control and electrical power systems — Cybersecurity requirements.* Geneva, Switzerland: IEC.

IEC. (2019). *62645. Nuclear power plants — Instrumentation, control and electrical power systems — Cybersecurity requirements.* Geneva, Switzerland: IEC.

IEC. (2019). *IEC 60050–395. International electrotechnical vocabulary - Part 395: Nuclear instrumentation: Physical phenomena, basic concepts, instruments, systems, equipment and detectors.* IEC.

IEC. (2019). *TR 63084, Nuclear power plants – Instrumentation and control systems important to safety-Qualification of platforms for systems important to safety.*

IEC/IEEE. (2011). *IEC/IEEE 62582 Nuclear power plants – Instrumentation and control important to safety – Electrical equipment condition monitoring methods – Part 1: General. Part 2: Indenter Modulus. Part 4.* IEC/IEEE.

IEC/IEEE. (2016). *IEC/IEEE 60780-323 - Nuclear facilities - Electrical equipment important to safety-Qualification.* IEC/IEEE.

IEC/IEEE. (2017). IEC 63147 Edition 1.0 2017-12 IEEE Std 497: IEEE/IEC International Standard - Criteria for accident instrumentation for nuclear power generation stations. IEC/IEEE.

IEEE (1988,a). IEEE 982.1. *Standard Dictionary of Measures to Produce Reliable Software.*

IEEE (1988,b). IEEE. 982.1. *Standard Guide of Measures to Produce Reliable Software.*

IEEE (1990). IEEE 610.12. *Standard Glossary of Software Engineering Terminology.*

IEEE. (2003). *Std. 323, Qualifying Class 1E equipment for nuclear power generating stations.* Piscataway, NJ: IEEE.

IEEE. (2004a). *Std. 344, IEEE Standard for Seismic Qualification of Equipment for Nuclear Power Generating Stations.* Piscataway, NJ: IEEE.

IEEE. (2004b). *Std. 1012, Software Verification and Validation*. Piscataway, NJ: IEEE.

IEEE. (2010). *Std. 7-4.3.2, Standard criteria for digital computers in safety systems of nuclear power generating stations*. Piscataway, NJ: IEEE.

Illiashenko, O., Kharchenko, V., & Kovalenko, A. (2012). Cyber security lifecycle and assessment technique for FPGA-based I&C systems. *Proceedings of IEEE East-West Design & Test Symposium (EWDTS) 2012*, 432-436.

Illiashenko, O., & Babeshko, E. (2011). Choice and Complexation of Techniques and Tools for Assessment of NPP I&C Systems Safety. *Icon (London, England)*, *2011*(19), E19–E43484. doi:10.1299/jsmeicone.2011.19._ICONE1943_194

Intel Corporation. (2018). *Intel® Arria® 10 Device Overview*. Author.

Isaias, P. (2015). High Level Models and Methodologies for Information Systems. Springer Science+Business Media. doi:10.1007/978-1-4614-9254-2

ISO/IEC (2010). ISO/IEC 25010. *Systems and software engineering – Systems and software Quality Requirements and Evaluation (SQuaRE) – System and software quality models*.

ISO/IEC (2011). ISO/IEC 25023. *Systems and software engineering – Systems and software Quality Requirements and Evaluation (SQuaRE) – Measurement of system and software product quality*.

ISO/IEC 15408. (2009). *Information technology – Security techniques – Evaluation criteria for IT security*. Ed. 3.

ISO/IEC 17799. (2005). *Information technology – Security techniques – Code of practice for information security management*.

ISO/IEC 27000. (2009). *Information technology – Security techniques – Information security management systems – Overview and vocabulary*.

ISO/IEC 27001. (2005). *Information technology – Security techniques – Information security management systems – Requirements*.

ISO/IEC 27002. (2005). *Information technology – Security techniques – Code of practice for information security management*.

ISO/IEC 27003. (2010). *Information technology – Security techniques – Information security management system implementation guidance*.

ISO/IEC 27004. (2009). *Information technology – Security techniques – Information security management – Measurement*.

ISO/IEC 27005. (2011). *Information technology – Security techniques – Information security risk management*. Ed. 2.

ISO/IEC 27006. (2011). *Information technology – Security techniques – Requirements for bodies providing audit and certification of information security management systems*.

ISO/IEC 27007. (2011). *Information technology – Security techniques – Guidelines for information security management systems auditing*.

ISO/IEC 27008. (2011). *Information technology – Security techniques – Guidelines for auditors on information security management systems controls*.

ISO/IEC. (2004). *13335-1, Information technology – Security techniques – Management of information and communications technology security – Part 1: Concepts and models for information and communications technology security management*. Geneva, Switzerland: ISO/IEC.

ISO/IEC. (2018). *27005, Information technology — Security techniques — Information security risk management.* Geneva, Switzerland: ISO/IEC.

Johnson, G. (2002). Comparison of IEC and IEEE standards for computer-based control systems important to safety. In *CNRA/CSNI workshop on licensing and operating experience of computer-based I&C systems. Workshop Proceedings.* AEN/NEA.

Johnson, G., & Duchac, A. (2017). The Development of the New Idea Safety Guide for Design of Instrumentation and Control Systems for Nuclear Power Plants. *Reliability, Theory & Applications*, *1*, 2017.

Jonson, G. (2010). The INSAG Defense in Depth Concept and D-in-D&D in I&C. *Proceedings of 7th ANS Topical Meeting on NPIC-HMIT.*

Karry, R. (2010, October). Trustworthy Hardware: Identifying and Classifying Hardware Trojans. *Computing. The Magazine*, *43*(10), 39–46. doi:10.1109/MC.2010.299

Kelly, T. (1998). *Arguing Safety: A Systematic Approach to Managing Safety Cases* (PhD thesis). University of York.

Kemikem, D. (2018) Quantitative and Qualitative Reliability Assessment of Reparable Electrical Power Supply Systems using Fault Tree Method and Importance Factors. *2018 13th Annual Conference on System of Systems Engineering (SoSE).*

Kersken, M. (2001). *Qualification of pre-developed software for safety-critical I&C application in NPP's.* Paper presented at CNRA/CSNI Workshop on Licensing and Operating Experience of Computer-Based I&C Systems, Hluboka-nad-Vltavou, Czech Republic.

Kharchenko, V. (2012c). Gap-and-IMECA-based Assessment of I&C Systems Cyber Security. In *Complex Systems and Dependability*. Springer-Verlag.

Kharchenko, V. (2014). Markov's Model and Tool-Based Assessment of Safety-Critical I&C Systems: Gaps of the IEC 61508. *Proc. 12th Int. Conf. PSAM*, 16.

Kharchenko, V. (2017). Green IT Engineering: Concepts, Models, Complex Systems Architectures. In Studies in Systems, Decision and Control, (vol. 74). Berlin: Springer International Publishing. doi:10.1007/978-3-319-44162-7

Kharchenko, V. S. (Ed.). (2012). CASE-assessment of critical software systems. Quality. Reliability. Safety. Kharkiv, Ukraine: National Aerospace University KhAI.

Kharchenko, V. S., & Vilkomir, S. A. (2000). *The Formalized Models of Software Verification Assessment.* Paper presented at 5th International Conference Probabilistic Safety Assessment and Management, Osaka, Japan.

Kharchenko, V., Bakhmach, E., & Siora, A. (2009). Diversity-scalable decisions for FPGA-based safety-critical I&C systems: From theory to implementation. *Proceedings of the 6th Conference NPIC&HMIT.*

Kharchenko, V., Siora, A., & Sklyar, V. (2011). Multi-Version FPGA-Based NPP I&Cs: Evolution of Safety. In NPP – Control, Reliability and Human Factors. InTech.

Kharchenko, V., Siora, A., Andrashov, A., & Kovalenko, A. (2012). Cyber Security of FPGA-Based NPP I&C Systems: Challenges and Solutions. *Proceeding of the 8th International Conference on Nuclear Plant Instrumentation, Control, and Human-Machine Interface Technologies (NPIC & HMIT 2012)*, 1338-1349.

Kharchenko, V. (1999). Multi-version Systems: Models, Reliability, Design Technologies. *Proceeding of 10th ESREL Conference*, 73-77.

Kharchenko, V. (2012). GAP- and HTT-based analysis of safety-critical systems. *Radioelectronic and Computer Systems*, *7*(59), 198–204.

Kharchenko, V. (2012d). Cyber Security Lifecycle and Assessment Technique for FPGA-based I&C Systems. *Proceeding of IEEE East-West Design & Test Symposium (EWDTS'2012)*, 432-436.

Kharchenko, V. (2015). Security Assessment of FPGA-based Safety-Critical Systems: US NRC Requirements Context. *Proceedings of the International Conference on Information and Digital Technologies (IDT 2015)*, 117-123. 10.1109/DT.2015.7222963

Kharchenko, V. (2018). Multi-Diversity for FPGA Platform Based NPP I&C Systems: New Possibilities and Assessment Technique. *Proceedings of the 2018 26th International Conference on Nuclear Engineering*. 10.1115/ICONE26-82377

Kharchenko, V. S., Illiashenko, O. A., Kovalenko, A. A., Sklyar, V. V., & Boyarchuk, A. V. (2014). Security informed safety assessment of NPP I&C systems: Gap-IMECA technique. *Proceedings of the 2014 22nd International Conference on Nuclear Engineering ICONE22*. 10.1115/ICONE22-31175

Kharchenko, V., Duzhyi, V., Sklyar, V., & Volkoviy, A. (2012). Safety Assessment of Multi-version FPGA-based NPP I&C Systems: Theoretical and Practical Issues. *Proceedings of the 5th International Workshop on the Applications of FPGA in Nuclear Power Plants.*

Kharchenko, V., & Illiashenko, O. (2016). Diversity for security: case assessment for FPGA-based safety-critical systems. *MATEC Web Conf.*, *76*. 10.1051/matecconf/20167602051

Kharchenko, V., Odarushchenko, O., & Odarushchenko, V. (2011). Multi-fragmental Availability Models of Critical Infrastructures with Variable Parameters of System Depend*ability. International Journal of Information Security*, *28*, 248–265. doi:10.11610/isij.2820

Kharchenko, V., Odarushchenko, O., Odarushchenko, V., & Popov, P. (2013). Availability Assessment of Computer Systems Described by Stiff Markov Chains: Case Study. *Springer, CCIS*, *412*, 112–135. doi:10.1007/978-3-319-03998-5_7

Kharchenko, V., Siora, A., & Bakhmach, E. (2008). Diversity-scalable decisions for FPGA-based safety-critical I&C systems: from Theory to Implementation. *Proceedings of the 6th Conference NPIC&HMIT.*

Kharchenko, V., Siora, A., Sklyar, V., & Volkoviy, A. (2012). Defence-in-Depth and Diversity Analysis of FPGA-based NPP I&C Systems: Conception, Technique and Tool. *Proceedings of the ICONE20*. 10.1115/ICONE20-POWER2012-54349

Kharchenko, V., Siora, A., Sklyar, V., Volkoviy, V., & Bezsaliy, V. (2010). Multi-Diversity Versus Common Cause Failures: FPGA-Based Multi-Version NPP I&C Systems. *Proceedings of the 7th Conference NPIC&HMIT, Las-Vegas, USA, November, 2010.*

Kharchenko, V., & Sklyar, V. (Eds.). (2008). *FPGA-based NPP Instrumentation and Control Systems: Development and Safety Assessment. RPC Radiy, National Aerospace University "KhAI"*. State Scientific and Technical Center for Nuclear and Radiation Safety.

Kharchenko, V., Sklyar, V., Siora, A., & Tokarev, V. (2008). Scalable Diversity-oriented Decisions and Technologies for Dependable SoPC-based Safety-Critical Computer Systems and Infrastructures, *Proceeding of IEEE International Conference on Dependability of Computer Systems*, 339-346. 10.1109/DepCoS-RELCOMEX.2008.21

Kharchenko, V., Sklyar, V., & Volkoviy, A. (2007). Multi-Version Information Technologies and Development of Dependable Systems out of Undependable Components. *Proceedings of International Conference on Dependability of Computer Systems*, 43-50. 10.1109/DEPCOS-RELCOMEX.2007.34

Kharchenko, V., Yastrebenetsky, M., & Sklyar, V. (2004). Diversity Assessment of Nuclear Power Plants Instrumentation and Control Systems, *Proceeding of 7th International Conference on PSAM and ESREL Conference*, 1351-1356. 10.1007/978-0-85729-410-4_218

Kobayashi, N., Nakamoto, A., Kawase, N., Sussan, F., & Shirasaka, S. (2018). What Model(s) of Assurance Cases Will Increase the Feasibility of Accomplishing Both Vision and Strategy? *Review of Integrative Business and Economics Research*, *7*(3), 1–17.

Kovalenko, A., Andrashov, A., Bakhmach, E., & Sklyar, V. (2015). FPGA-based I&C Applications in NPP's Modernization Projects: Case Study. *Proceeding of the 9th International Conference on Nuclear Plant Instrumentation, Control, and Human-Machine Interface Technologies*, 1, 113-120.

Kovalenko, A., Kuchuk, G., Kharchenko, V., & Shamraev, A. (2017). Resource-Oriented Approaches to Implementation of Traffic Control Technologies in Safety-Critical I&C Systems. In *Green IT Engineering: Components, Networks and Systems Implementation. Studies in Systems, Decision and Control series. Springer.*

Kovalenko, A., Kuchuk, G., Kharchenko, V., & Shamraev, A. (2017). *Resource-Oriented Approaches to Implementation of Traffic Control Technologies in Safety-Critical I&C Systems. Springer.*

Kyun, L. M., Whan, S. S., & Yun, D. H. (2009). *Development and Application of POSAFE Q PLC Platform.* Retrieved from: https://inis.iaea.org/collection/NCLCollectionStore/_Public/43/130/43130436.pdf?r=1&r=1

Lahtinen, J., Valkonen, J., Bjorkman, K., Frits, J., & Niemela, I. (2010). *Model checking methodology for supporting safety critical software development and verification.* Paper presented at ESREL 2010 Annual Conference, Rhodes, Greece.

Langner, R. (2013). *To Kill a Centrifuge.* Retrieved from https://www.langner.com/wp-content/uploads/2017/03/

Lawrence, S., Hatton, L., & Howell, C. (2002). *Solid Software.* Prentice Hall.

Lee, S. J., Choi, J. G., Kang, H. G., & Jang, S.-C. (2010). Reliability assessment method for NPP digital I&C systems considering the effect of automatic periodic tests. *Annals of Nuclear Energy*, *37*(11), 1527–1533. doi:10.1016/j.anucene.2010.06.009

Levitin, G., Xing, L., & Dai, Y. (2014). Mission cost and reliability of 1-out-of-N warm standby systems with imperfect switching mechanisms. *IEEE Transactions on Systems, Man, and Cybernetics. Systems*, *9*(44), 1262–1271. doi:10.1109/TSMC.2013.2294328

Lopez, C., Sargolzaei, A., Santana, H., & Huerta, C. (2015). Smart Grid Cyber Security: An Overview of Threats and Countermeasures. *Journal of Energy and Power Engineering*, *9*, 632–647.

Lyu, M. R. (1996). *Handbook of software reliability engineering*. McGraw-Hill Company.

Lyu, M. R. (1996). *Handbook of Software Reliability Engineering*. McGraw-Hill Company.

MacFarlane, J. W., Payton, O. D., Keatley, A. C., Scott, G. P. T., Pullin, H., Crane, R. A., ... Scott, T. B. (2014). Lightweight aerial vehicles for monitoring, assessment and mapping of radiation anomalies. *Journal of Environmental Radioactivity*, *136*, 127–130. doi:10.1016/j.jenvrad.2014.05.008 PMID:24949582

Madonna, M., Martella, G., Monica, L., Pichini Maini, E., & Tomassini, L. (2009). The human factor in risk assessment: Methodological comparison between human reliability analysis techniques. *Prevention Today*, *5*(1/2), 67–83.

Maerani, R. (2018). V&V Plan for FPGA-based ESF-CCS Using System Engineering Approach. *Journal of Physics: Conference Series*, ▪▪▪, 962.

Maksimov, M., Fung, N., Kokaly, S., & Chechik, M. (2018). Two decades of assurance case tools: A survey. In B. Gallina, A. Skavhaug, E. Schoitsch, & F. Bitsch (Eds.), *Computer Safety, Reliability, and Security*. Springer.

Malhotra, M., Muppala, J. K., & Trivedi, K. S. (1994). Stiffness-Tolerant Methods for Transient Analysis of Stiff Markov Chains. *Microelectronics and Reliability*, *34*(11), 1825–1841. doi:10.1016/0026-2714(94)90137-6

Marszal, E., & McGlone, J. (2019). *Security PHA Review for Consequence-Based Cybersecurity*. International Society of Automation.

Martin, P. G., Payton, O. D., Fardoulis, J. S., Richards, D. A., Yamashiki, Y., & Scott, T. B. (2016). Low altitude unmanned aerial vehicle for characterising remediation effectiveness following the FDNPP accident. *Journal of Environmental Radioactivity*, *151*, 58–63. doi:10.1016/j.jenvrad.2015.09.007 PMID:26410790

Masood, R. (2016). *Report GW-CSPRI-2016-03. Assessment of Cyber Security Challenges in Nuclear Power Plants.* Washington, DC: The George Washington University.

McCabe, T. A. (1976). Complexity Measure. *IEEE Transactions on Software Engineering*, *4*(4No SE-2), 308–320. doi:10.1109/TSE.1976.233837

Medoff, M., & Faller, R. (2010). *Functional Safety – An IEC 61508 SIL 3 Compatible Development Process. Exida L.L.C.*

Melnyk, A. (2007). Automatic generation of ASICS. *Proceedings of NASA/ESA Conference on Adaptive Hardware and Systems*, 311–317. 10.1109/AHS.2007.36

Mendel, J. M. (2002). An architecture of making judgment using computing with words. *International Journal of Applied Mathematics and Computer Science*, *12*(3), 325–335.

Menon, C., & Guerra, S. (2015). *Field programmable gate arrays in safety related instrumentation and control applications*. Energiforsk, AB: Adelard LLP.

MIL-HDBK-217F, Military Handbook: Reliability Prediction of Electronic Equipment

Misztal, A. (n.d.). Connecting and applying the FTA and FMEA methods together. *Some problems and methods of ergonomics and quality management,* 153-163.

Mo, Y., Xing, L., Amari, S. V., & Dugan, J. B. (2015). Efficient analysis of multi-state k-out-of-n system. *Reliability Engineering & System Safety*, *133*, 95–105. doi:10.1016/j.ress.2014.09.006

NAPB. (2002). *NAPB 03.005. Fire protection. Development of firefighting standards for nuclear power plants with pressurized water reactors*. Kyiv, Ukraine: Ministry of Fuel and Energy of Ukraine.

Naser. (Ed.). (2009). *Guidelines on the Use of Field Programmable Gate Arrays (FPGAs) in Nuclear Power Plant I&C Systems*. EPRI.

NEI 08-09, *Cyber Security Plan for Nuclear Power Reactors*, Rev. 6, Nuclear Energy Institute, (2010).

Nicola, V. F. (1982). *Markovian Models of Transactional System Supported by Check Pointing and Recovery Strategies, Part 1: a Model with State-Dependent Parameters*. Eindhoven Univ. Technol., Eindhoven, The Netherlands, EUT Rep. 82-E-128.

NIST SP 800-30, *Risk Management Guide for Information Technology Systems*, National Institute of Standards and Technology, (2002).

NIST SP 800-53, *Recommended Security Controls for Federal Information Systems and Organizations*, Rev. 3, National Institute of Standards and Technology, (2009).

NIST. (2002). *Special Publication 800-30, Risk Management Guide for Information Technology Systems.* Gaithersburg, MD: NIST.

NIST. (2008). *Special Publication 800-115, Technical Guide to Information Security Testing and Assessment.* Gaithersburg, MD: NIST.

NP 306.2.202-2015. Nuclear and Radiation Safety Requirements for Instrumentation and Control Systems Important to NPP Safety (Ukraine).

NP. (2000). *NP 306.5.02/3.035. Nuclear and radiation safety requirements to instrumentation and control systems important to nuclear power plants safety*. Kiev, Ukraine: State Nuclear Regulatory Committee.

NP. (2000). NP 306.5.02/3.035. Requirements for nuclear and radiation safety to information and control systems important to safety of nuclear power plants. Kyiv, Ukraine: NP.

NP. (2004). *NP 306.2.100. Provisions on procedure of investigation and calculation of the disturbances in nuclear plants operation*. Kiev, Ukraine: State Nuclear Regulatory Committee.

NP. (2008). *NP 306.2.141. General principle safety of nuclear power plants*. Kiev, Ukraine: State Nuclear Regulatory Inspectorate of Ukraine.

NP. (2008). *NP 306.2.141. General provisions on the safety of nuclear power plants*. Kiev, Ukraine: State Nuclear Regulatory Inspectorate.

NP. (2015). *306.2.202. Requirements on nuclear and radiation safety to instrumentation and control systems, important for safety of nuclear power plants*. Kyiv, Ukraine: SNRIU.

NP. (2015). NP 306.2.202 Requirements for nuclear and radiation safety to information and control systems important to safety of nuclear power plants. Kyiv, Ukraine: NP.

NP. (2015). *NP 306.2.202. Nuclear and radiation safety requirements to information and control systems important to of nuclear power plants safety*. Kiev, Ukraine: State Nuclear Regulatory Inspectorate.

NP. (2015). *NP 306.2.202. Requirements for nuclear and radiation safety instrumentation and control systems important to safety of nuclear power plants*. Kiev, Ukraine: State Nuclear Regulatory Inspectorate of Ukraine.

NP. (2016a). *NP 306.2.205. Requirements for power systems important for the safety of nuclear power plants*. Kiev, Ukraine: State Nuclear Regulatory Inspectorate of Ukraine.

NP. (2016b). *NP 306.2.208. Requirements for seismic design and assessment of seismic safety of nuclear power plant units*. Kiev, Ukraine: State Nuclear Regulatory Inspectorate of Ukraine.

Nuclear Energy Series No, I. A. E. A. NP-T-3.17. (2016). Application of Field Programmable Gate Arrays in Instrumentation and Control Systems of Nuclear Power Plants. International Atomic Energy Agency.

Nuclear Security Series No, I. A. E. A. (2011). *17, Computer security at nuclear facilities: reference manual: technical guidance*. Vienna: IAEA.

NUREG. (2004). *CR-6847, Cyber Security Self-Assessment Method for U.S. Nuclear Power Plants*. Richland: PNNL.

NUREG. (2016). *NUREG-0800. U.S. Nuclear Regulatory Commission Regulations: Standard Review Plan for the Review of Safety. Analysis Reports for Nuclear Power Plants: LWR Edition*. NUREG.

NUREG/CR-6003. Method for Performing Diversity and Defense-in-Depth Analyses of Reactor Protection Systems, United States Nuclear Regulatory Commission, 1994.

NUREG/CR-7006, *Review Guidelines for Field-Programmable Gate Arrays in Nuclear Power Plant Safety Systems*, U.S. Nuclear Regulatory Commission, (2010).

NUREG/CR-7007. Diversity Strategies for NPP I&Cs, United States Nuclear Regulatory Commission, 2009.

NUREG/CR-7141. Cyber Security Regulatory Framework for Nuclear Power Reactors, United States Nuclear Regulatory Commission, 2014.

Oleksandr, G., Vyacheslav, K., & Kate, V. (2017). Usable Security Versus Secure Usability: an Assessment of Attributes Interaction. In *Proceedings of the 13th International Conference on ICT in Education, Research and Industrial Applications. Integration, Harmonization and Knowledge Transfer*, (pp.727-740). Kyiv, Ukraine: Academic Press.

Oleksandr, G., Vyacheslav, K., & Mario, F. (2015). Evolution of software quality models: usability, security and greenness issues. In *Proceedings of the 19-th International Conference on Computers (part of CSCC 15),* (pp. 519-523). Zakynthos Island, Greece: Academic Press.

Oleksandr, G., Vyacheslav, K., Nataliia, F., & Vladimir, S. (2014). Evolution of software Quality Models in Context of the Standard ISO 25010. In Proceedings of Dependability on Complex Systems DepCoS – RELCOMEX (DepCOS) (pp. 223-233). Brunow, Poland: Academic Press.

Panfilo, F. (2011). *Westinghouse Nuclear Automation. Digital Feed Water Upgrades Experience, Validation & Lessons Learned.* Retrieved from: http://fsrug.org/Presentations%202011/Westinghouse%20-%20Ovation%20FW%20Presentation.pdf

Pascual-Ortigosa, P., Sáenz-De-Cabezón, E., & Wynn, H. P. (2018). Algebraic analysis of multistate k-out-of-n systems. *Monografías de la Real Academia de Ciencias, 43*, 131–134.

Pekka, P. (2000). Human reliability analysis methods for probabilistic safety assessment. Technical Research Centre of Finland.

Pressman, R. S. (1997). *Software Engineering: A Practioner's Approach.* McGraw-Hill Company.

Press, W. H. (2007). *Numerical Recipes. The Art of Scientific Computing.* Cambridge University Press.

Prokhorova, Y., Kharchenko, V., Ostroumov, B., Ostroumov, S., & Sidorenko, N. (2008). Dependable SoPC-Based Onboard Ice Protection System: from Research Project to Implementation. *Proceeding of IEEE International Conference on Dependability of Computer Systems*, 312-317. 10.1109/DepCoS-RELCOMEX.2008.43

Pullum, L. (2001). *Software Fault Tolerance Techniques and Implementation.* Artech House Computing Library.

Ram, M., & Dohi, T. (2019). *Reliability Analysis Using k-out-of-n Structures.* Boca Raton, FL: CRC Press. doi:10.1201/9781351056465

Ravi, S., Raghunathan, A., Kocher, P., & Hattangady, S. (2004). Security in Embedded Systems: Design Challenges. *ACM Transactions on Embedded Computing Systems, 3*(3), 461–491. doi:10.1145/1015047.1015049

Regulatory Guide (RG) 1.152-2011, Revision 3, Criteria for Use of Computers in Safety Systems of Nuclear Power Plants.

Reibman, A. (1989). Analysis of Stiff Markov Chains. ORSA Journal on Computing, 1(2), 126-133. doi:10.1287/ijoc.1.2.126

Reibman, A., & Trivedi, K. (1988). Numerical Transient Analysis of Markov models. *Comput. Opns. Res., 15*(1), 19–36. doi:10.1016/0305-0548(88)90026-3

RG 5.71, *Cyber security programs for nuclear facilities,* U.S. Nuclear Regulatory Commission, (2010).

RG. (1988). *1.100. Seismic Qualification of Electric and Mechanical Equipment for Nuclear Power Plants.* Washington, DC: U.S. NRC.

RG. (2010). *5.71. Cyber security programs for nuclear facilities.* Washington, USA: U.S. NRC.

RG. (2011). *1.152, Revision 3. Criteria for use of computers in safety systems of nuclear power plants.* Washington, USA: U.S. NRC.

RG. (2015). *5.83. Cyber security event notifications.* Washington, DC: U.S. NRC.

Rolls-Royce. (2018). *Technical sheet civil nuclear – systems – instrumentation & control. Hardline. Non-programmed safety I&C platform dedicated to nuclear safety.* Retrieved from: https://www.rolls-royce.com/~/media/Files/R/Rolls-Royce/documents/customers/nuclear/UK_Hardline.pdf

Rolls-Royce. (2019a). *Technical sheet civil nuclear – systems – instrumentation & control. Spinline. Modular I&C digital platform dedicated to nuclear safety.* Retrieved from: https://www.rolls-royce.com/~/media/Files/R/Rolls-Royce/documents/customers/nuclear/UK_Spinline.pdf

Rolls-Royce. (2019b). *Technical sheet civil nuclear – systems – instrumentation & control. Rodline. The most-used digital Rod Control System in the world.* Retrieved from: https://www.rolls-royce.com/~/media/Files/R/Rolls-Royce/documents/customers/nuclear/UK_Rodline.pdf

Rosenberg, M., & Bobryakov, S. (2003). *Elsevier's Dictionary on Nuclear Engineering.* London: Elsevier Science.

Rozen, Y. (2007). Electromagnetic compatibility of instrumentation and control systems components (1). Rules for regulations and estimation. *Nuclear and Radiation Safety, 2.*

Rozen, Y. (2008). Electromagnetic compatibility of instrumentation and control systems components (2). Electromagnetic interference resistance. *Nuclear and Radiation Safety, 4.*

Rushby, J. (2015). *The Interpretation and Evaluation of Assurance Cases.* Technical Report SRI-CSL-15-01. SRI International.

Sadeghi, A.-R. (2011). *Towards Hardware-Intrinsic Security: Foundations and Practice.* Springer.

Safety. (n.d.). In *Wikipedia.* Retrieved from https://en.wikipedia.org/wiki/safety/

Saleem, Y., Crespi, N., Rehmani, M., & Copeland, R. (2019). Internet of Things-aided Smart Grid: Technologies, Architectures, Applications, Prototypes, and Future Research Directions. *IEEE Access: Practical Innovations, Open Solutions, 7,* 62962–63003. doi:10.1109/ACCESS.2019.2913984

Sanders, W. H. (1998). An Efficient Disk-based Tool for Solving Large Markov Models. *Performance Evaluation, 33*(1), 67–84. doi:10.1016/S0166-5316(98)00010-8

Sanders, W. H. (2003). Optimal State-space Lumping in Markov Chains. *Information Processing Letters, 87*(6), 309–315. doi:10.1016/S0020-0190(03)00343-0

Sanger, D. (2018). *Confront and Conceal.* Gardner Books.

Scott, J., & Lawrence, J. (1994). *Testing Existing Software for Safety Related Applications.* Lawrence Livermore National Laboratory.

Shamraev, A. (2017). Green Microcontrollers in Control Systems for Magnetic Elements of Linear Electron Accelerators. In Green IT Engineering: Concepts, Models, Complex Systems Architectures. doi:10.1007/978-3-319-44162-7_15

Shu-wen, W. (2011). Research on the Key Technologies of IOT Applied on Smart Grid. *International Conference on Electronics, Communications and Control (ICECC),* 2809–2812. 10.1109/ICECC.2011.6066418

Siora, A., Krasnobaev, V., & Kharchenko, V. (2009). Fault-Tolerance Systems with Version-Information Redundancy. Ministry of Education and Science of Ukraine, National Aerospace University KhAI.

Siora, A., Sklyar, V., Rozen, Yu., Vinogradskaya, S., & Yastrebenetsky, M. (2009). Licensing Principles of FPGA-Based NPP I&C Systems. *Proceedings of 17th International Conference on Nuclear Engineering (ICONE 17)*. 10.1115/ICONE17-75270

Sklyar, V. (2016). Safety-critical Certification of FPGA-based Platform against Requirements of U.S. Nuclear Regulatory Commission (NRC): Industrial Case Study. *Proceedings of the 12th International Conference on ICT in Education, Research and Industrial Applications*, 129-136.

Sklyar, V., & Kharchenko, V. (2017). Assurance Case Driven Design based on the Harmonized Framework of Safety and Security Requirements. *Proceedings of the 13th International Conference on ICT in Education, Research and Industrial Applications*, 670-685.

Smith, D., & Simpson, K. (2004). *Functional Safety. A Straightforward Guide to applying IEC 61508 and Related Standards*. Oxford, UK: Elsevier Butterworth–Heinemann.

Smith, J., & Simpson, K. (2001). *Functional safety: A Straight forward Guide to IEC 61508 and related standards*. Oxford, UK: Butterworth Heinemann.

Sommerville, J. (2011). Software Engineering (9th ed.). Addison-Wesley.

SOU. (2016). SOU NAEK 100. Engineering, scientific and technical support. Instrumentation and control systems important for the safety of nuclear power plants. General technical requirements. Kiev, Ukraine: National Nuclear Power Company "Energoatom".

Srinivasan, A. (1990). Algorithms for Discrete Functions Manipulation. *Proc. Int'l Conf. on CAD (ICCAD'90)*, 92-95.

StJohn-Green, M., Piggin, R., McDermid, J. A., & Oates, R. (2015) Combined security and safety risk assessment - What needs to be done for ICS and the IoT. *Proceedings of 10th IET System Safety and Cyber-Security Conference*. 10.1049/cp.2015.0284

Structured Assurance Case Metamodel. (2016). v2.0. Object Management Group.

Swain, A. (1964). *THERP*. Sandia National Laboratories, SC-R-64-1338.

Tam, S. (2003). Error Detection and Correction in Virtex-II Pro Devices. Application Note: Virtex-II Pro Family. XAPP645 (v1.1).

Tarasyuk, O., Gorbenko, A., Kharchenko, V., Ruban, V., & Zasukha, S. (2011). Safety of Rocket-Space Engineering and Reliability of Computer Systems: 2000-2009 Years. Radio-Electronic and Computer Systems, 11, 23-45.

Tehranipoor, M. (2010). A Survey of Hardware Trojan Taxonomy and Detection. *Proc. of IEEE Design & Test of Computers*, 10-25. 10.1109/MDT.2010.7

Toulmin, S. (1958). *The Uses of Argument*. Cambridge University Press.

TR. (2013). *Advanced Logic System Topical Report*. Rockville, MD: U.S. NRC.

Trivedi, K. S. (2000). Availability Models in Practice. *Proc. Int. Workshop on Fault-Tolerant Control and Computing (FTCC-1)*.

UAV Drone Radiation Monitoring. (n.d.). Retrieved November 1, 2019, from http://www.aretasaerial.com/products/uav-drone-radiation-monitoring

Vilkomir, S. A., & Kharchenko, V. S. (1999). Methodology of the review of software for safety important systems. In G. I. Schueller, P. Kafka (Eds.), *Safety and Reliability. Proceedings of ESREL'99 - The Tenth European Conference on Safety and Reliability* (pp. 593-596). Munich-Garching, Germany: Academic Press.

Vilkomir, S., & Kharchenko, V. (2000). *An "asymmetric" approach to the assessment of safety-critical software during certification and licensing*. Paper presented at ESCOM-SCOPE 2000 Conference, Munich, Germany.

Vilkomir, S. (2009). Statistical testing for NPP I&C system reliability evaluation. *Proceedings of the 6th American Nuclear Society International Topical Meeting on Nuclear Plant Instrumentation, Controls, and Human Machine Interface Technology (ICHMI 2009).*

Vilkomir, S., Swain, T., & Poore, J. (2009). Software Input Space Modeling with Constraints among Parameters. *Proceedings of the 33rd Annual IEEE International Computer Software and Applications Conference COMPSAC*, 136-141. 10.1109/COMPSAC.2009.27

Volkoviy, A., Lysenko, I., Kharchenko, V., & Shurygin, O. (2008). Multi-Version Systems and Technologies for Critical Applications. National Aerospace University KhAI.

Wang, Y., Lin, W., Zhang, T., & Ma, Y. (2012). Research on Application and Security Protection of Internet of Things in Smart Grid. In *International Conference on Information Science and Control Engineering (ICISCE).* TLP: White Analysis of the Cyber Attack on the Ukrainian Power Grid. https://www.nerc.com/pa/CI/ESISAC/Documents/E-ISAC_SANS_Ukraine_DUC_18Mar2016.pdf

Wang, W., & Lum, Z. (2013). Cyber security in the Smart Grid: Survey and challenges. *Computer Networks, 57*(5), 1344–1371. doi:10.1016/j.comnet.2012.12.017

Webster's New World Dictionary. (1991). (3rd college ed.). Prentice Hall.

WENRA. (2012). *Research of WENRA Reactor Harmonization Working Group. Design safety of new nuclear power plants.* WENRA.

WENRA. (2014a). *WENRA Safety Reference Levels for Existing Reactors. Update in relation to lessons earned from Tepco Fukushima Dai-Ichi Accident.* WENRA.

WENRA. (2014b). *WENRA Safety Reference Levels for Existing Reactors.* WENRA.

Westinghouse. (2018). *Westinghouse COMMON Q Platform.* Retrieved from: http://www.westinghousenuclear.com/Portals/0/operating%20plant%20services/automation/safety%20related%20platforms/NA-0113%20Common%20Q%20Platform.pdf

WESTRON. (2017). *The VULCANVULCAN/VULCANVULCAN-M hardware platform for the control systems of nuclear power plants and thermal power plants.* Retrieved from: http://www.westron.kharkov.ua/files/Vulcan-Vulcan-M_Presentation_Rus.pdf

Wood, R., Belles, R., & Cetiner, M. (2009). *Diversity Strategies for NPP I&C Systems.* NUREG/CR-7007 ORNL/TM-2009/302.

Xu, Z., & Jiang, D. (2016). A New ALS Based PMS Design and Its Evaluations. *World Journal of Nuclear Science and Technology, 6*(1).

Yastrebenetsky, M. (2014). Nuclear Power Plant Instrumentation and Control Systems for Safety and Security. IGI Global. doi:10.4018/978-1-4666-5133-3

Yastrebenetsky, M. (Ed.). (2004). Safety of Nuclear Power Plants: Instrumentation and Control Systems. Technika.

Yastrebenetsky, M., Butova, O., Inyishev, V., & Spector, L. (2005). Factors of functional safety of control systems for power units of NPP. *Nuclear Measurement and Information Technologies, 3*(15).

Yastrebenetsky, M. (Ed.). (2004). *Nuclear power plants safety. Instrumentation and control systems.* Kyiv, Ukraine: Technica.

Yastrebenetsky, M. (Ed.). (2011). *Nuclear power plants safety. Nuclear reactors control and protection systems.* Kyiv, Ukraine: Osnova-Print.

Yastrebenetsky, M., Rozen, Yu., Vinogradska, S., & Johnson, G. (2011). *Nuclear power plants safety: nuclear reactor control and protection systems.* Kiev: Osnova - Print.

Yastrebenetsky, M., Vasilchenko, V., & Vinogradska, S. (2004). *Nuclear power plants safety: instrumentation and control systems.* Kiev: Technika.

Ye, F., & Cleland, G. (2012). *Weapons Operating Centre Approved Code of Practice for Electronic Safety Cases.* Adelard LLP.

Zadeh, L. (2009). From computing with numbers to computing with words-from manipulation of measurements to manipulation of perceptions. IEEE Trans. Circ. Syst, Fund. Theory Applic., 4(1), 105 –119.

Zadeh, L., & Kacprzyk, J. (1999). *Computing with Words in Information/Intelligent Systems – Part 1:Foundation; Part 2: Applications. Physica-Verlag.*

About the Contributors

Michael Yastrebenetsky PhD, Doctor of Technical Sciences, Professor, Honored Scientist of Ukraine, Honored Laborer of Nuclear Energetics of Ukraine. Until 1992 – Head of Department "Reliability of Automated Control Systems" of the Central Scientific Research Institute of Complex Automation. Head of the Department "NPP Control and Information Systems Safety Analysis" of State Scientific and Technical Centre of Nuclear and Radiation Safety (1993-2015), Leading Researcher of this Department (2015- at the present time). Professor of the Department "Systems Analysis and Control" of the Kharkov National Technical University (1986- 2018). Author of 14 boors, 300 articles, 45 international and national standards and regulations on control systems. Organizer and Chair of 1-5 International Scientific Technical Conferences "NPP I&C Systems: Safety Aspects". President of International Association of Reliability specialists "Gnedenko-Forum" (2015- at the present time). Expert of International Electrotechnical Commission -Technical Committee "Instrumentation, control and electrical power systems of nuclear facilities" (2000- at the present time). Member of IAEA International Working Group in Control and Instrumentation (1999-2011).

Vyacheslav Kharchenko is Professor (1991), Doctor of Science (1995). Honor Inventor of Ukraine (1991). Leading researcher of the Department "NPP Control and Information Systems Safety Analysis" (1997-2008), Head of Computer Systems, Networks and Cyber security, National Aerospace University "KhAI", Kharkiv, Ukraine (2001- at the present time). Head of Centre for Safety Infrastructure-Oriented Research and Analysis, RPC Radiy (2007- at the present time). President of Ukrainian Scientific and Educational IT-Society (2018- at the present time). Supervisor of 50 PhD and DrS dissertations (defended). Invited researcher and professor (UK, 2004, 2016, 2017; Germany, 2006; Poland, 2020; Slovakia, 2010, 2019; USA, 2011). National coordinator of 10 EU projects on dependable computing, infrastructure safety and green IT-engineering. General Chair of Dependable Systems, Services and Technologies (DESSERT) Conferences (2006-2020). Author of 30 books and chapters (Springer, IGI-Global, River Publishers), 268 papers in Journals and Proceedings indexed by Web of Science and Scopus, more 700 patents,14 national and branch standards on NPP I&C and aerospace systems reliability and safety.

* * *

Anton Andrashov is head of international projects division at RPC Radiy. He received his B.S. (2005) and MSc (2007) with honor degrees in Computer Engineering from National Aerospace University KhAI. PhD (National Aerospace University KhAI, 2019). Lecturer assistant at the Department of computer systems and networks, KhAI (2007- 2011). Senior researcher at the Centre for safety infrastructure-oriented research and analysis (2008-2011). Currently involved in implementation of several international projects including SIL3 and NRC certification of FPGA-based safety platform for NPP I&C systems. The author of 5 monographs and textbooks, more than 60 articles, including articles published in the Japan, Germany, USA, and other countries.

Ievgen Babeshko, MSc (National Aerospace University KhAI, 2007), PhD (National Aerospace University KhAI, 2019). Centre for Safety Infrastructure-Oriented Research and Analysis, senior researcher (2009-now). National Aerospace University KhAI, senior lecturer of the Computer Systems, Networks and Cyber Security Department (2007-2019), associated professor of the Computer Systems, Networks and Cyber Security Department (2019-at the present). Author of 3 monographs, 25 articles. Expert of Ukrainian Technical Committee 185 "Industrial Automation" (2017- at the present). Research and development activities include reliability and safety modelling and assessment, software development and verification for critical industries.

Ievgenii Bakhmach, MSc (Kirovohrad Agricultural Engineering Institute, 1973). General Director of industrial association "Kirovohradnerudprom" (1986-1992); Chairman of the Board of Public Company "Kirovohradgranit" (1992-2001); Chairman of the Supervisory Council of RPC Radiy (since 2001); Chairman of the Board of regional department of Ukrainian Union of Industrialists and Entrepreneurs (since 2005). Awardee of Certificate of Honour from the Cabinet of Ministers, Certificate of Honour from the Verkhovna Rada of Ukraine and honorary degree of the "Honoured worker of industry in Ukraine". Author of 10 monographs and more than 50 articles.

Eugene Brezhnev, MSc (Kharkiv Military University, 1995) PhD (Kharkiv Military University, 2000), lecturer, senior researcher (Kharkiv Military University, 2000-2010), Doctoral of technical science student (National Aerospace University KhAI, 2010-2013), associate professor and professor at the Department of computer system and network, National Aerospace University KhAI (2012-present). Senior researcher at the Center for safety infrastructure-oriented research and analysis, RPC Radiy (2011-2017). Head of department of quality management, PRC Radics (2017-at the present time). The author of 5 monographies, 90 articles.

Valentina Butenko, MSc in Applied Mathematics (Poltava National Technical University, 2012) PhD in Information Technology (National Aerospace University "KhAI", 2015), Assistant Professor at Department of Information Technologies and Systems (Poltava National Technical University named after Yuriy Kondratyuk, 2013), Assistant Professor at Department of Computer Systems, Networks and Cybersecurity (National Aerospace University "KhAI", 2013-2020). Author of 6 monographs, 16 articles.

Mikhail Chernyshov, after graduation from Moscow Institute of Physics and Technology in 1977, began to work in Scientific and Production Association Electropribor (Kharkov) and was involved in development of guidance systems for missiles and satellites. Since 1994 he is the General Director of LLC "Westron", managing development of platforms and I&C systems for nuclear power plants and other industrial objects. His main sphere of interest is information and control system design.

Vyacheslav Duzhyi, MSc on Computing (Kharkiv Aviation Institute (KhAI) named after N.E. Zhukovsky, 1985), PhD on Information Technology (National Aerospace University KhAI, 2015). Engineer at the Department of Computer Engineering (KhAI, 1985-1986), Assistant (1986-1988), Senior Lecturer (1988-2015), Associate Professor at the Department of Computer Systems, Networks and Cybersecurity (2015-present time). Author of 7 monographs and textbooks, 17 papers and 20 conference reports, a lot of technical industry reports.

Herman Fesenko is Associate Professor of the Department of Computer Systems, Networks and Cybersecurity at the National Aerospace University "Kharkiv Aviation Institute". He has received M.S. degree in Automation Control Systems in 1995 and Ph.D. degree in Technical Sciences in 2002 from the Kharkiv Military University. His main expertise is Industrial Internet of Things, including Internet of Flying Things, development of multi-version Internet of Drone based NPP monitoring systems, Big Data methodologies. He is an author of more than 160 papers in refereed conferences and journals. 12 papers are devoted to developing drone fleet based NPP monitoring systems. Dr. Fesenko has been involved in a number of national and international research and academic projects and has supervised more than 25 M.S. students. He is currently working under a thesis to get a National Ukrainian academic degree of Doctor of Technical Sciences.

Kostyantyn Gerasimenko, PhD (2013). Since 2000 has worked in Severodonetsk Research and Production Association "Impulse" (Ukraine) as a lead developer of safety and normal operation I&C systems for nuclear power plants, then as a deputy director on nuclear automation. He is a renowned expert in the field of NPP I&C systems, especially for NPPs with the WWER-type reactor, author of many articles for automation journals and several scientific works.

Oleg Ivanchenko, PhD (Sevastopol Navy institute, 2004). Senior lecturer, Associate Professor of the Department of Radioelectronic systems, Sevastopol Navy institute, 2004-2010), Associate Professor of the Department of Cybernetics and Computer Engineering, Sevastopol National Technical University (2010-2014). Associate Professor of Department of Information Systems and Technologies (2014-2018); Associate Professor of Department of Computer Sciences and Software Engineering (2018-2019), Associate Professor of Department of Transport Systems and Technologies, University of Customs and Finance (2019-at the present). Author of 4 monographs and more than 60 articles.

Gary Johnson, BS Electronics Engineering (California Polytechnic State University, San Luis Obispo, 1974). US Department of Energy. Customer engineer responsible for Instrumentation, Control, and Electrical Systems of the Fast Flux Test Facility (1974-1980). Portland General Electric Company. Group leader responsible for responding to new US Nuclear Regulatory Commission requirements involving electrical and electronic systems (1980-1984). Lawrence Livermore National Laboratory. Task Area Leader supporting research and inspections for the US Nuclear Regulatory Commission and support to the US Department of Energy oversight of operating facilities (1984-2007). University of California at Berkeley, Nuclear Engineering Department. Visiting Scholar managing graduate students performing research in digital instrumentation and control systems (1998-2000). Computer Dependability Associates. Principle, supporting commercial and government customers activities involving digital Instrumentation and control systems. Chairman, International Electrotechnical Commission - Instrumentation, Control, and Electrical Power Systems of Nuclear Facilities (2010-2017). International Atomic Energy Agency. Senior Safety Officer (2018-2013). Independent consultant (2013 to present).

Oleksandr Klevtsov, PhD (2010). Since 1999 has worked in Kharkiv subsidiary of the State Scientific and Technical Center for Nuclear and Radiation Safety (SSTC NRS). Since 2014 – Head of "Department of safety analysis of information and control systems" of SSTC NRS. Carry out the expert analysis of nuclear and radiation safety and reliability assessment of more than 100 NPP instrumentation and control systems. Author of 5 Ukrainian national standards, more than 35 articles in scientific-technical journals, 1 chapter of book about NPP instrumentation and control systems, and 5 books about programming languages. Is involved in research activity in field of cyber security of NPP instrumentation and control systems.

Ihor Kliushnikov is a Leading Researcher of Scientific Center of Air Force for Kharkiv National University of Air Force. He has received MSc degree in Automation Control Systems in 1995 and PhD degree in Technical Sciences in 2003 from the Kharkiv Military University. His main expertise is UAV-enabled wireless networks, including Internet of Flying Things, development of models on using automated battery replacement stations for the persistent operation of UAV-enabled wireless networks during NPP post-accident monitoring. He is an author of more than 76 papers in refereed conferences and journals. Dr. Kliushnikov has been involved in a number of national research projects and standards on civil and military UAV applications.

Andriy Kovalenko is Head of Department of Electronic Computers, Kharkiv National University of Radio Electronics. PhD (Kharkiv National University of Radio Electronics, 2008), Doctor of Technical Sciences (topic is "Models and methods for synthesis and reconfiguration of computer systems and networks architectures of critical application objects", National Technical University "Kharkiv Polytechnical Institute", 2018), Professor (2019). Senior Researcher at Center for safety infrastructure-oriented research and analysis, RPC Radiy (2009-at the present). Author of more than 190 papers (including monographs and conference proceedings). Active member of Editorial Board of the following journals: "Innovative technologies and scientific solutions for industries", "Advanced Information Systems" and "Control, Navigation and Communication Systems". Active participant of workshops and conference devoted to safety and security problems of critical infrastructures, as well as computer networking. Research and development activities: computer systems and networks, safety and security of critical instrumentation and control systems, FPGA technology, Embedded systems.

Kostiantyn Leontiiev, Master's degree in control and automatic systems (Kirovohrad National Technical University "KNTU", 2002), system engineer (PC "RPC Radiy", 2004-2016), technical director, RPC Radiy (2016-at the present). Part time PhD student (National Aerospace University, 2018-at the present). Author of 12 articles.

Elena Odarushchenko, PhD (National Aerospace University "KhAI", 2008), Assistant Professor (Kharkov Military University, 1994-1999), Assistant Professor (Poltava Military Institute of Communications, 1999-2004), Head of the Department of Mathematics (Poltava Military Institute of Communications, 2004-2007), Assistant Professor (National Technical University of Ukraine "Kyiv Polytechnic Institute", 2007-2018), Assistant Professor (Poltava State Agrarian Academy, 2018-2020). Author of 5monographs, (2 published in English), 28 articles (5 published in English).

Oleg Odarushchenko, PhD (Kharkov Military University. 1998), Assistant Professor (Poltava Military Institute of Communications, 2005), Head of the Engineering Faculty (Poltava Military Institute of Communications, 2006-2007), Dean of the Faculty of Information Systems and Technologies (Poltava National Technical University named after Yuriy Kondratyuk, 2008-2012), Head of the Verification and Validation Department (Research and Production Corporation Radics, 2012-2020). Author of 7 monographs, (2 published in English), 61 articles.

Artem Panarin, Master's degree in system programming (Kirovohrad National Technical University "KNTU", 2006), software engineer (PC "RPC Radiy", 2005-at the present). Part time PhD student (National Aerospace University, 2018-at the present). Author of 8 articles.

Yuri Rozen was, until 1995, head of a department, deputy director for scientific work, chief designer of the Scientific and Manufacturing Enterprise on Automated Control Systems. Since 2001 – Head of "Laboratory of control systems safety analysis" of "Department of safety analysis of information and control systems" of the State Scientific and Technical Center for Nuclear and Radiation Safety. Twice was a winner of a prize of the USSR Council of Ministers. Held the rank "best inventor of instrument construction". Author of 3 books, more than 80 articles, 26 inventions, 48 intergovernmental and national standards and normative documents on nuclear safety.

Yevhen Ruchkov, Master's degree in electronic systems (Zaporizhia State Engineering Academy, 2002), lead engineer (ZNPP, 1997-2002), head of transaction services department (PC "RPC Radiy", 2005-2020). Part time PhD student (National Aerospace University, 2018-at the present). Author of 4 articles.

Oleksandr Siora, MSc (Kirovohrad Agricultural Engineering Institute, 1978), PhD (National Aerospace University "KhAI", 2005). General Director of RPC Radiy (since 1998); Member of Coordinating Expert Council of SE NNEGC Energoatom (since 2000); Corresponding Member of Engineering Academy of Ukraine (2005). Awardee of Certificate of Honour from the Cabinet of Ministers, Certificate of Honour from the Ministry of Fuel and Energy and Honours of National Academy of Sciences of Ukraine. Author of 10 monographs and more than 70 articles.

Vladimir Sklyar, PhD (Kharkiv Military University, 2001), Doctor of Technical Sciences (National Aerospace University "KhAI", 2012), Professor (Department of Computer Systems, Networks and Cybersecurity, 2014). State Scientific and Technical Centre of Nuclear and Radiation Safety; Head of Laboratoty (until 2011); Technical Director of Company Radiy (2011-2015), National Aerospace University "KhAI" of Computer Systems, Networks and Cybersecurity; Professor of the Department of Computer Systems, Networks and Cybersecurity (since 2013). Author of 10 monographs and more than 100 articles. Member of IAEA Technical Working Group on Nuclear Power Plants Instrumentation and Control (2011-2015); Expert of International Electrotechnical Commission Technical Committee 45A "Instrumentation, control and electrical power systems of nuclear facilities" (since 2009).

Artem Symonov, Master's Degree (2012). Since 2016 has worked in Kharkiv subsidiary of the State Scientific and Technical Center for Nuclear and Radiation Safety (SSTC NRS). Since 2018 – Head of "Laboratory of information systems safety analysis and cybersecurity" of "Department of safety analysis of information and control systems" of SSTC NRS. Carry out the expert analysis of nuclear and radiation safety a lot of different NPP instrumentation and control systems. Author of five articles in scientific-technical journals. Is involved in research activity in field of cyber security of NPP instrumentation and control systems.

Viktor Tokarev, PhD (National Aerospace University "KhAI", 2005), the general designer of I&C system projects (PC "RPC Radiy", 2002-at the present time). Author of 20 articles.

Serhii Trubchaninov, since 2008 has worked in Kharkiv subsidiary of the State Scientific and Technical Center for Nuclear and Radiation Safety (SSTC NRS). Since 2014 – Director of Kharkiv Subsidiary of SSTC NRS. Carry out the expert analysis of nuclear and radiation safety and reliability assessment of more than 100 NPP instrumentation and control systems. Author of 3 Ukrainian national standards, 15 articles in scientific-technical journals.

Andriy Volkoviy, PhD (National Aerospace University named after N. Ye. Zhukovskiy "Kharkiv Aviation Institute", 2006), Associate Professor (Computer systems and Networks, 2012). Author of 50 articles and proceedings (16 in English). National Aerospace University "Kharkiv Aviation Institute" – assistant lecturer (2000-2004), senior lecturer (2004-2009) and associate professor (2009-2012); Research and Production Corporation "Radiy" – Senior Research Fellow (2008-2012); Samsung R&D Institute Ukraine – Senior Software Engineer (2012-2017); Mellanox Technologies – Senior Firmware Engineer (2017-at the present).

Index

J

L

M

N

O

P

R

S

T

V

ates
isher Services